AF213351

THEORETICAL AND APPLIED MECHANICS

PROCEEDINGS OF
THE THIRTEENTH INTERNATIONAL CONGRESS OF
THEORETICAL AND APPLIED MECHANICS

MOSCOW UNIVERSITY
AUGUST 21-26, 1972

EDITORS

E. BECKER · G. K. MIKHAILOV

SPRINGER-VERLAG BERLIN · HEIDELBERG · NEW YORK 1973

Dr. rer. nat. ERNST BECKER

Professor an der
Technischen Hochschule
Darmstadt

Professor GLEB K. MIKHAILOV

USSR National Committee on
Theoretical and Applied Mechanics
Moscow

With 203 Figures

ISBN-13:978-3-642-65592-0 e-ISBN-13:978-3-642-65590-6
DOI: 10.1007/978-3-642-65590-6

This work is subject to copyright. All rights are reserved, whether the whole or part of the material is concerned, specifically those of translation, reprinting, re-use of illustrations, broadcasting, reproduction by photocopying machine or similar means, and storage in data banks.
Under § 54 of the German Copyright Law where copies are made for other than private use, a fee is payable to the publisher, the amount of the fee to be determined by agreement with the publisher.
© by Springer-Verlag, Berlin · Heidelberg 1973.
Softcover reprint of the hardcover 1st edition 1973

Library of Congress Catalog Card Number 58-2749

The use of registered names, trademarks, etc. in this publication does not imply, even in the absence of a specific statement, that such names are exempt from the relevant protective laws and regulations and therefore free for general use.

Preface

The Thirteenth International Congress of Theoretical and Applied Mechanics was held in Moscow from Monday, 21 August, to Saturday, 26 August 1972. About 2500 participants from 37 countries all over the world attended the congress that was convened by the Congress Committee of the International Union of Theoretical and Applied Mechanics. The local organization lay in the hands of the Organizing Committee, established by the USSR National Committee on Theoretical and Applied Mechanics. The USSR Academy of Sciences rendered partial financial help to the organization of the congress. The Organizing Committee was assisted by the Institute of Problems of Mechanics of the USSR Academy of Sciences, by the Research Institute for Mechanics of Moscow University, and by the Computing Center and the Institute of Applied Mathematics of the USSR Academy of Sciences. The Bureau of IUTAM had allocated a considerable sum for partial financial support of young scientists attending the congress.

The Thirteenth Congress was officially opened on Monday morning at the Kremlin Palace of Congresses by Academician N. I. Muskhelishvili, President of the Congress, and Professor W. T. Koiter, President of IUTAM. Greeting addresses were offered by:

Mr. K. N. Rudnev, Minister, member of the Council of Ministers of the USSR,

Academician M. V. Keldysh, President of the USSR Academy of Sciences,

Mr. L. N. Efremov, First Deputy Chairman of the State Committee for Science and Technology under the Council of Ministers of the USSR,

Mr. V. P. Isaev, First Deputy Chairman of the Executive Committee of the Moscow City Council of Deputies.

At the opening ceremony, Academician L. I. Sedov presented to Academician N. I. Muskhelishvili the book "Continuum Mechanics and Related Problems of Analysis—To the 80th Birthday of N. I. Muskhelishvili." The book contains 67 papers that were dedicated to Academician Muskhelishvili by scientists from sixteen countries.—Professor Koiter informed the participants of the opening ceremony that the Congress Committee had decided to send a telegram with greetings to Professor C. B. Biezeno, Delft, who was President of the First Congress, held at Delft 48 years ago. He also greeted the Secretary of that First Congress, Professor J. M. Burgers, who was among the participants of the Thirteenth Congress.

Immediately after the opening ceremony, the first general lecture of the congress was presented by Academician V. V. Novozhilov. The following scientific events all took place at the main building of Moscow University on the Lenin Hills. The Thirteenth Congress was officially closed on Saturday afternoon at the aula of Moscow University. On that occasion members of the Congress Committee expressed their gratitude and praise to the Organizing Committee for the competent arrangement of the Thirteenth Congress, which was, by number of attendants, the biggest ever held. The audience was informed that the General Assembly of IUTAM had elected, in its meeting on 24 August 1972, Professor H. Görtler President of IUTAM. According to the statutes of IUTAM, Professor W. T. Koiter thereby became Vice-President, and Professor M. Roy retired from the vice-presidency. Professor D. C. Drucker was elected Treasurer of IUTAM. The Academy of Sciences of the German Democratic Republic was admitted, by vote of the General Assembly, as an adhering organization of IUTAM. Furthermore, Professors G. K. Batchelor, N. J. Hoff, and W. T. Koiter ended their term as members of the Executive Committee of the Congress Committee. Professors B. A. Boley, W. Fiszdon, and N. I. Muskhelishvili became new members of the Executive Committee; Professor H. Görtler will be ex officio President of the Committee. Finally, it was disclosed that the Congress Committee had gratefully accepted the invitation by the Section for Mechanics of the Division of Applied Sciences of the Royal Institute of Engineers (The Netherlands) to hold the Fourteenth Congress at Delft during the week commencing on 30 August 1976.

Five general lectures on topics of broad interest and twenty sectional lectures on more specialized topics, each of one hour duration, were presented at the Thirteenth Congress upon invitation by the Congress Committee. In addition, 225 contributed papers were presented in six parallel sessions. The authors' names and the titles of the contributed papers are listed at the end of this volume. The number of contributed papers had to be strictly limited in order to allow not less than thirty minutes for presentation and discussion of each paper without unduly increasing the number of parallel sessions. Therefore an International Papers Committee, established by the Congress Committee, selected the contributed papers on the basis of extended summaries submitted by the authors. The decisions of the Committee were strictly guided by considerations of the scientific merit of a paper. On the average, the number of papers submitted was about six times as high as the number that could be accepted. This made the task of the International Papers Committee rather difficult, and a large number of valuable papers had to be rejected. The work of the International Papers Committee was alleviated by National Papers Committees of USSR, USA, UK, France, and the Federal Republic of Germany. These national committees preselected the papers submitted by authors from their countries. The authors of contributed papers are free to publish their manuscripts in journals of their own choice.

This volume contains the five general lectures and seventeen, out of twenty, sectional lectures. The following sectional lectures are, upon the authors' request, not included in this volume:

G. F. Carrier (Cambridge, Mass.): The dynamics of atmospheric vortices;
S. S. Grigoryan (Moscow): Some problems of mechanics in oil industry;
L. E. Scriven (Minneapolis, Minn.): Molecules, particles, interfaces and continua.

Professor Carrier has informed us that his manuscript should not be included because he has not yet completed the quantitative aspects of those parts of the study which are not already published.

In order to make all contributions to this volume accessible to as many readers as possible, all articles are in English. Six manuscripts that were originally submitted in Russian had therefore to be translated.

As for two previous congresses, Springer-Verlag has again taken up the task of publishing the Proceedings of the Thirteenth Congress. We appreciate the competence and helpfulness of Springer-Verlag in their cooperation with us.

Darmstadt and Moscow, June 1973

Ernst Becker **Gleb K. Mikhailov**

Opening ceremony at the Kremlin Palace of Congresses; Academician M. V. Keldysh at the speakers rostrum.

Academician N. I. Muskhelishvili delivering opening address.

International Union of Theoretical and Applied Mechanics

Congress Committee

J. Ackeret
**G. K. Batchelor*
**E. Becker* — Secretary
B. A. Boley
J. M. Burgers
A. Caquot
S. Goldstein
H. Görtler
**N. J. Hoff*
**W. T. Koiter* — President
N. I. Muskhelishvili
F. K. G. Odqvist
H. Parkus
M. Roy
W. Rubinowicz
L. I. Sedov
H. Solberg
I. Tani
G. I. Taylor
G. Temple
W. Weibull

(* members of the Executive Committee).

USSR National Committee on Theoretical and Applied Mechanics

Organizing Committee

N. I. Muskhelishvili — Chairman
L. I. Sedov — Deputy Chairman
G. G. Chakhmakhchev — Deputy Chairman
G. K. Mikhailov — Secretary-General
N. Kh. Arutyunyan
G. G. Chernyi
A. A. Dorodnitsyn
A. Yu. Ishlinsky
M. V. Keldysh
S. G. Korneev
M. A. Lavrent'ev
L. G. Loitsyanskii
M. M. Maslov
M. D. Millionshchikov
D. E. Okhotsimskii
B. N. Petrov
G. S. Pisarenko
G. N. Savin
V. V. Struminskii
V. V. Sychev

List of Authors of Papers in this Volume

General Lectures

Prof. *Arthur E. Bryson*, Chairman, Department of Applied Mechanics, School of Engineering, Stanford, Cal. 94305, USA.
Prof. *B. M. Fraeijs de Veubeke*, Laboratoire des Techniques Aéronautiques et Spatiales, Université de Liège, 75, Rue du Val-Benoit Liège, Belgium.
Prof. Sir *M. J. Lighthill*, University of Cambridge, Department of Applied Mathematics and Theoretical Physics, Silver Street, Cambridge CB 3 9 EW, UK.
Prof. *V. V. Novozhilov*, Faculty of Mathematics and Mechanics, University of Leningrad, B. O. 10 line 33, Leningrad B 178, USSR.
Prof. Dr. *K. Wieghardt*, Institut für Schiffbau der Universität Hamburg, 2 Hamburg 33, Lämmersieht 90.

Sectional Lectures

Prof. *A. A. Deribas*, Novosibirsk, USSR.
Dr. *H. L. Goldsmith*, University Medical Clinic, The Montreal General Hospital, 1650 Cedar Avenue, Montreal 109, Quebec, Canada.
Prof. *J.-P. Guiraud*, Université de Paris, Faculté des Science Mécanique Théorique, Tour 66-9. Quai Saint-Bernard Paris 5, France.
Prof. *N. J. Hoff*, Department of Aeronautics and Astronautics, Stanford University, Stanford, Cal. 94305, USA.
Prof. *A. Yu. Ishlinsky*, Director, Institute for Problems in Mechanics, Academy of Sciences of the USSR Moscow A-40, Leningrader Avenue 7, USSR.
Prof. *W. T. Koiter*, Technische Hogeschool, Mekelweg 2, Delft, Netherlands.
Prof. *M. Landahl*, Department of Aeronautics and Astronautics, School of Engineering, Massachussetts Institute of Technology Cambridge, Mass. 02139, USA, and Division of Mechanics, The Royal Institute of Technology, Stockholm 70, Sweden.
Prof. *H. W. Liepmann*, Graduate Aeronautical Laboratories, California Institute of Technology, Pasadena, Cal. 91109, USA.
Prof. *M. S. Longuet-Higgins*, Department of Applied Mathematics and Theoretical Physics, University of Cambridge, Silver Street, Cambridge CB 3 9 EW, UK, and Institute of Oceanographic Sciences, Wormley, Godalming, U.K.
Prof. *E. H. Mansfield*, Structures Department, Procurement Executive, Ministry of Defence, Royal Aircraft Establishment, Farnborough, Hants, England.
Prof. *J. R. Melcher*, University of Cambridge, Department of Applied Mathematics and Theoretical Physics, Silver Street, Cambridge CB 3 9 EW, UK.
Prof. *F. I. Niordson*, The Technical University of Denmark, Department of Solid Mechanics, Building 404, 2800 Lyngby, Denmark.
Dr. *P. Pedersen*, The Technical University of Denmark, Department of Solid Mechanics, Building 404, 2800 Lyngby, Denmark.
Dr. *J. R. Philip*, Chief of Division CSIRO, Division of Environmental Mechanics, P. O. Box 821, Canberra City, A.C.T., 2601, Australia.
Dr.-Ing. E. h. *J. C. Rotta*, Deutsche Forschungs- und Versuchsanstalt für Luft- und Raumfahrt E.V., Aerodynamische Versuchsanstalt, Bunsenstraße 10, D-34 Göttingen, FRG.
Prof. *L. I. Sedov*, Moscow, USSR.
Prof. *J. G. Simmonds*, University of Virginia, Charlottesville, Virginia 22901, USA.
Prof. *I. N. Vekua*, Rector, Tbilisi State University, l, Chavchavadze Avenue, Tbilisi, Georgian SSR, USSR.
Prof. *I. I. Vorovich*, Rostov-on-Don, USSR.

Contents

General Lectures

Sectional Lectures

Control theory for random systems

Arthur E. Bryson, Jr.

Stanford, CA, USA

I. Introduction

Automatic control is used in connection with *goal-oriented, man-made systems*. There are several, usually overlapping, reasons for using automatic control: (a) to relieve human operators from tedious, repetitive tasks (e.g., dial telephones, traffic signals, automatic elevators, autopilots), (b) to speed up and/or lower costs of production processes (e.g., chemical process control, automatic cutting and drilling machines), and (c) to control rapidly-changing or complicated systems accurately and safely (e.g., guidance and control of spacecraft, automatic landing of aircraft).

There are two distinctly different types of automatic control: *programmed* (or open-loop) *control* and *feedback* (or closed-loop) *control* (see Fig. 1). The difference is that the output affects the signals to the actuator when using feedback control but does not affect them when using programmed control. Typical programmed control systems are those used with traffic signals or automatic washing machines where a sequence of events is scheduled by a clock-timer. Examples of simple feedback control systems are the temperature controllers used in buildings, ovens, and hot water heaters, where the temperature error (the difference between sensed and desired temperature) is used with some controller logic, to switch on and off the heating element.

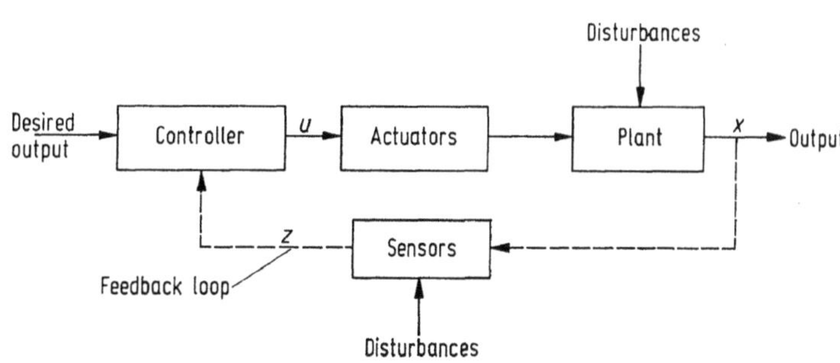

Fig. 1. Block diagram; open-loop (programmed) control has no feedback loop; closed-loop control has feedback loop.

The purpose of feedback control is to reduce to acceptable levels the effects of unpredictable disturbances and uncertain parameters, using acceptable levels of control action. Thus, it is important for the designer of a feedback control system to have reasonably accurate models of (a) the system to be controlled and (b) the expected disturbances. Until recently, engineers had better tools for modeling the system than they had for modeling the expected disturbances; they used only rather simple descriptions such as constant values and impulse functions (or integrals of the latter—step functions, ramp functions, etc.). While such descriptions are often adequate to yield satisfactory designs when only one disturbance source acts on the system, more realistic descriptions are needed when several disturbance sources act on the system simultaneously.

Many systems of practical interest can be modeled by linear ordinary differential equations. A convenient form for such a model is a set of coupled first-order ordinary equations:

$$\dot{x} = F(t)\,x + G(t)\,u + \Gamma(t)\,w; \quad x(t_0) = x_0, \tag{1.1}$$

$$y = M(t)\,x, \quad \xi = Nx(t_f), \tag{1.2}$$

$$z = H(t)\,x + v, \tag{1.3}$$

where

$x(t)$ = n-vector of state variables,
$u(t)$ = m-vector of control variables,
$w(t)$ = r-vector of process disturbance variables,
ξ = s-vector of terminal state parameters ($s \leq n$),
$y(t)$ = p-vector of output variables ($p \leq n$),
$z(t)$ = q-vector of measurement variables,
$v(t)$ = q-vector of measurement disturbance variables,
$F(t)$ = $n \times n$ open-loop dynamics matrix,
$G(t)$ = $n \times m$ control distribution matrix,
$\Gamma(t)$ = $n \times r$ disturbance distribution matrix,
$M(t)$ = $p \times n$ output distribution matrix,
N = $s \times n$ terminal state distribution matrix, and
$H(t)$ = $q \times n$ measurement distribution matrix.

In most problems, the vectors are deviations from nominal values and the matrix elements are partial derivatives evaluated on the nominal "path".

The control problem is to design a dynamic system (a "compensator") with measurements, z, as input and controls, u, as output, so that $y(t)$ and ξ of the controlled system are kept acceptably close to zero in the presence of anticipated disturbances w, v, and x_0, while using acceptable amounts of control, u.

During the last ten years it has become clear that excellent compensators can be designed in two completely separate steps: (a) *a statefeedback controller*, $u = -Cx$, is designed to minimize an appropriate quadratic error criterion, assuming the availability of perfect measurements of the system state variables and no disturbances, i.e., $z = x$, $v = 0$, $w = 0$; (b) a *measurement-feedback estimator* (or *filter*) is designed to yield maximum likelihood estimates of the state variables of the system, $\hat{x}$, from the available sensor measurements, z, taking into account w, v, x_0, and assuming perfect knowledge of the control variables, $u(t)$. The state-feedback controller is then used with the state estimates from the filter, i.e., $u = -C(t)\,\hat{x}$. The resulting control system forms a "dynamic compensator" (see Fig. 2) which minimizes the quadratic error criterion on the average (*Joseph* and *Tou* (1961, *Gunckel* and *Franklin* (1963), and *Potter* (1964))[1].

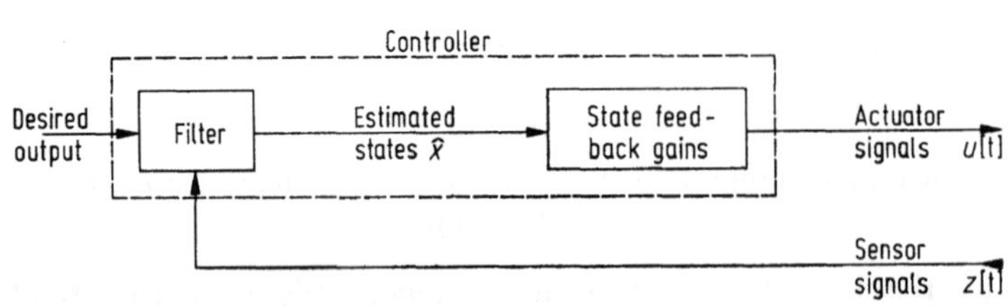

Fig. 2. Block diagram; dynamic compensator formed from filter and state feedback gains.

In Section II we discuss state feedback controller design, and in Section IV we discuss the design of state estimators (measurement-feedback filters). Section III is a review of the concept of a Gauss-Markov random process which forms the basis of the filter design. Section V discusses the behavior of systems controlled by such filter/state-feedback compensators.

[1] *Joseph* and *Tou*, as well as *Gunckel* and *Franklin*, showed this for Gauss-Markov sequences (discrete-step processes); *Potter* extended the proof to continuous Gauss-Markov processes.

II. Design of state-feedback controllers

For $w = v = 0$, and $z \equiv x$, a rather general design problem for the system (1.1)—(1.2) is to try to find an acceptable $u(t)$ program in some time interval, $t_0 \leq t \leq t_f$, so that the output variables, $y(t)$ and ξ stay acceptably close to zero for a given x_0.

A useful way to judge one design relative to another is to specify a *quadratic criterion* of the form:

$$J = \frac{1}{2} \xi^T \Omega \xi + \frac{1}{2} \int_{t_0}^{t_f} (y^T A y + u^T B u)\, dt \tag{2.1}$$

where the positive definite weighting matrices Ω, $A(t)$, and $B(t)$ are chosen to express relative preferences of the designer, and $[\ ]^T$ indicates matrix transpose of $[\ ]$, i.e., interchange of rows and columns.

Using the calculus of variations (or dynamic programming), it is straightforward to show that the state-feedback controller that minimizes J is given by[2]:

$$u = -C(t)\, x \tag{2.2}$$

where the $m \times n$ feedback gain matrix, $C(t)$, is given by

$$C = B^{-1} G^T S \tag{2.3}$$

and

$$\dot{S} = -SF - F^T S + C^T B C - M^T A M; \quad S(t_f) = N^T \Omega N. \tag{2.4}$$

The nonlinear equation, (2.4), for the symmetric $n \times n$ matrix, $S(t)$, is called a *backward matrix Riccati equation* because of the term $C^T B C = S G B^{-1} G^T S$ which is quadratic in the elements of S, and because it must be integrated backward from the final time, t_f, to the initial time, t_0, to generate $C(t)$. The significance of S is that

$$J_{\min} = \frac{1}{2} x^T(t_0)\, S(t_0)\, x(t_0). \tag{2.5}$$

Examples of such time-varying *terminal control problems* and their solutions are given elsewhere (e.g., *Bryson* and *Ho* (1969), Section 14.6).

Obviously it is not always possible to control all of the outputs with a given set of controls. A "controllability" criterion was given by *Kalman, Ho,* and *Narendra* (1963).

Of special interest is the class of problems with *stationary solutions*. For such problems, the matrices F, G, M, A, B must be constant, and $t_f - t$ must be large compared to characteristic times of the controlled system. S may then have a constant asymptotic value, which means that the feedback gain matrix tends to a constant value:

$$\left.\begin{aligned} S(t) &\to S_\infty \quad &(2.6)\\ C_\infty &= B^{-1} G^T S_\infty \quad &(2.7) \end{aligned}\right\} \text{ as } t_f - t \to \infty.$$

A controller of the form (2.2) with constant gains, C_∞, is called a *regulator*. In principle, the regulator gains C_∞ could be found by solving sets of quadratic equations (Eqs. (2.4) with $\dot{S} = 0$) to find the elements of the steady-state matrix, S_∞. However, this turns out to be an impractical procedure for $n \geq 3$. Until recently, the procedure most often used was to integrate the matrix Riccati equation, (2.4), backward on a computer until $S(t)$ reached a steady-state [see, e.g., *Kalman* and *Englar* (1966)]. This turns out to be expensive and involves several numerical difficulties and hazards. *MacFarlane* (1963) and *Potter* (1966) showed another ingenious way to find S_∞ using *eigenvector decomposition*. It does not seem to be widely known or used, perhaps because it requires an accurate computer program to find the eigenvalues and eigenvectors of a large $(2n \times 2n)$ matrix. However, *Francis* (1962) developed a remarkably fast and accurate algorithm for eigenvalue-eigenvector determination (the "QR algorithm"), and *Hall* (1971) applied it to the MacFarlane-Potter method for finding S_∞.

---

[1] See, e.g., *Bryson* and *Ho* (1969), Section 5.2. If $x(t_0)$ is known ahead of time, the optimum $u(t)$ can be pre-calculated and programmed as an open-loop input; this is a peculiar feature of deterministic control design—it predicts no difference between open-loop and closed-loop control.

The eigenvector decomposition method is based on finding the eigenvectors of the Euler-Lagrange equations for the problem of minimizing (2.1):

$$\begin{bmatrix} \dot{x} \\ \dot{\lambda} \end{bmatrix} = \begin{bmatrix} F, & -GB^{-1}G^T \\ -M^TAM, & -F^T \end{bmatrix} \begin{bmatrix} x \\ \lambda \end{bmatrix}. \tag{2.8}$$

The eigenvalues of this system are symmetric about the imaginary axis in the complex plane. (If s_i is an eigenvalue, $-s_i$ is also an eigenvalue). Hence, the characteristic equation of (2.8) is called the *symmetric-root characteristic equation* (see *Letov* (1960), *Chang* (1961), and *Rynaski* and *Whitbeck* (1966)). If we let T_- be the $2n \times n$ matrix consisting of the n eigenvectors of (2.8) corresponding to the eigenvalues with negative real parts, then *Potter* (1966) has shown that:

$$S_\infty = \Lambda_-(X_-)^{-1} \quad \text{where} \quad T_- = \begin{bmatrix} X_- \\ \Lambda_- \end{bmatrix}. \tag{2.9}$$

Furthermore, the eigenvalues of the controlled system (i.e., the eigenvalues of $F - GC$) are the eigenvalues of (2.8) that have negative real parts, and the eigenvectors of the controlled system are the columns of X_-.

A revealing alternative form for (2.8), in terms of the Laplace transform of the control variables, $u(s)$, was suggested by *Rynaski* and *Whitbeck* (1966):

$$[B + Y^T(-s)\, A\, Y(s)]\, u(s) = 0 \tag{2.10}$$

where $Y(s) = M[sI - F]^{-1}G$ is the matrix of transfer functions from the control variables to the outputs, i.e.,

$$y(s) = Y(s)\, u(s) \tag{2.11}$$

and s is the Laplace transform variable. The corresponding equation for the system using feedback control gains, C, may be written

$$[I + C(sI - F)^{-1}G]\, u(s) = 0. \tag{2.12}$$

It follows from (2.10) and (2.12) that

$$[B + Y^T(-s)\, A\, Y(s)] = [I + C(-sI - F)^{-1}G]^T\, B[I + C(sI - F)^{-1}\, G] \tag{2.13}$$

which is an $m \times m$ matrix equation (where u is an m-vector). For $s = i\omega$, where ω = frequency, (2.13) is the well-known "spectral factorization" problem encountered in solving certain Wiener-Hopf integral equations. The problem here is to find the $m \times n$ matrix C that satisfies (2.13), i.e., to "factor" the left side of (2.13) into two $m \times n$ matrices, one with eigenvalues in the left half of the complex s plane and the other with exactly the negative of those eigenvalues which are then in the right half plane.

This factorization can be done very quickly and accurately by eigenvector decomposition, using (2.8) and a digital computer [see, e.g., *Bryson* and *Hall* (1971)]. It can also be done as follows: (a) find the n eigenvalues of (2.8)[1] that have negative real parts and the corresponding n eigenvectors of (2.10) (each with only m components); then (b), require that the eigenvalues and eigenvectors of (2.12) be the same as those in (a). If $m = 1$, all that is needed is to equate coefficients of like powers of s of the characteristic equation of (2.12) and the closed-loop characteristic equation formed from the n eigenvalues of (2.8) that have negative real parts; this gives n linear equations for the n components of C. If $m > 1$, the requirement that the eigenvectors be parallel adds $n(m - 1)$ equations, giving a total of nm linear equations for determining the nm components of C.

As an *example*, we design a *regulator* for a fourth-order model of a *helicopter in hover*. Fig. 3 shows the nomenclature and reference axes. The four state variables are θ = pitch angle of the fuselage, q = pitch rate ($q = \dot{\theta}$), x = horizontal distance of mass center from the hover point, and u = horizontal velocity ($u = \dot{x}$). The control variable is δ = tilt angle of the rotor-thrust vector with respect to the fuselage. The primary disturbance is horizontal wind, w,

[1] For $m > 1$, the characteristic equation of (2.10) is of order $2mn$ instead of $2n$, and hence contains extraneous roots (pointed out to author by *J. V. Breakwell*).

which we shall consider in a later section. The vertical motions (controlled by increasing or decreasing rotor thrust), the lateral motions (controlled by tilting the rotor thrust laterally), and the yawing motions (controlled by the tail rotor) are all nearly uncoupled from the longitudinal motions considered here and may be treated as separate regulator design problems. The longitudinal equations of motion in state variable form may be written as:

$$\begin{bmatrix} \dot\theta \\ \dot q \\ \dot u \\ \dot x \end{bmatrix} = \begin{bmatrix} 0 & 1 & 0 & 0 \\ 0 & -\sigma_1 & -\alpha_1 & 0 \\ g & -\alpha_2 & -\sigma_2 & 0 \\ 0 & 0 & 1 & 0 \end{bmatrix} \begin{bmatrix} \theta \\ q \\ u \\ x \end{bmatrix} + \begin{bmatrix} 0 \\ n \\ g \\ 0 \end{bmatrix} \delta + \begin{bmatrix} 0 \\ -\alpha_1 \\ -\sigma_2 \\ 0 \end{bmatrix} w. \tag{2.14}$$

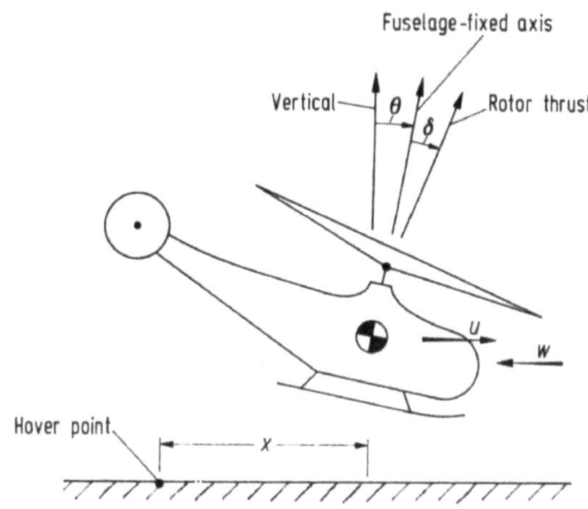

Fig. 3. Nomenclature and reference axes for helicopter hover example.

A reasonable quadratic performance index is:

$$J = \frac{1}{2} \int_0^\infty (A_x x^2 + B\,\delta^2)\,dt \tag{2.15}$$

which reflects a desire to keep x and δ near zero.

The root square characteristic equation in the form (2.10) is given by:

$$B + A_x Y_x(s)\, Y_x(-s) = 0, \tag{2.16}$$

where

$$Y_x(s) = \frac{x(s)}{\delta(s)} = \frac{g\left[s^2 + \left(\sigma_1 - \frac{\alpha_2 n}{g}\right) s + n\right]}{\Delta_0(s)}$$

$$\Delta_0(s) = s[s^3 + (\sigma_1 + \sigma_2)\, s^2 + (\sigma_1\sigma_2 - \alpha_1\alpha_2)\, s + \alpha_1 g]$$

and $\Delta_0(s) = 0$ is the open-loop characteristic equation. Thus (2.16) becomes:

$$\Delta_0(s)\,\Delta_0(-s) + \frac{A_x}{B} g^2 \left[s^2 + \left(\sigma_1 - \frac{\alpha_2 n}{g}\right) s + n\right]\left[s^2 - \left(\sigma_1 - \frac{\alpha_2 n}{g}\right) + n\right] = 0. \tag{2.17}$$

Coefficients for a typical transport helicopter are given in Table 1, along with the open-loop eigenvalues and eigenvectors. Note the uncontrolled helicopter is *slightly unstable* in hover, which is fairly typical. The weighting factor A_x/B may be selected as follows: rewrite (2.15) as

$$J = \frac{1}{2} \int_0^\infty \left[\left(\frac{x}{x_0}\right)^2 + \left(\frac{\delta}{\delta_0}\right)^2\right] dt. \tag{2.15a}$$

Suppose $\delta_0 = 5°$ is the maximum amount of rotor tilt we would like to commit to position control, and we would be willing to use this amount when the position error is 2 meters. It follows then that we should choose $\delta = 5.0/57.3$ rad and $x_0 = 2.0$ meters. Hence

$$\frac{A_x}{B} = \frac{\delta_0^2}{x_0^2} = \left(\frac{5.0}{57.3}\right)^2 \frac{1}{(2)^2} \cong .0020 \text{ m}^{-2}. \tag{2.18}$$

It is often helpful to the designer to see how the eigenvalues change as the weighting factors change. The concept of a *root locus plot* was introduced by *Evans* (1950); *Chang* (1961) applied it to the root square characteristic equation (2.10) and *Rynaski* and *Whitbeck* (1965) extended the application to multi-input, multi-output systems. Fig. 4 shows a root locus of (2.17) with A_x/B as the parameter. The open-loop poles and their reflections across the imaginary axis are the "poles" (shown as x's) of the root locus; the "zeros" (shown as 0's) are at $\pm(.25 \pm 2.50j)\,\text{sec}^{-1}$.

For the weighting factor given in (2.18) the closed-loop eigenvalues and eigenvectors, as well as the corresponding feedback gains, are given in Table 1.

Table 1. Regulator design for a typical transport helicopter

Coefficients: $\sigma_1 = .415\ \text{sec}^{-1}$, $\sigma_2 = .0198\ \text{sec}^{-1}$, $n = 6.27\ \text{sec}^{-2}$,
$\alpha_1 = .0111\,\text{m}^{-1}\,\text{sec}^{-1}$, $\alpha_2 = 1.43\,\text{m}\,\text{sec}^{-1}$, $g = 9.80\,\text{m}\,\text{sec}^{-2}$

Open-loop eigenvalues and eigenvectors[1]

s, sec^{-1}	0	$-.68$	$+.12 \pm .38j$
θ, deg.	0	2.4	$.79 \pm .56j$
x, meters	1.0	1.0	1.0

Closed-loop eigenvalues and eigenvectors[1] *for* $(A_x/B) = .002\ \text{m}^{-2}$

s, sec^{-1}	$-1.24 \pm .56j$	$-.42 \pm 1.14j$
θ, deg.	$8.0 \mp 4.9j$	$-5.4 \mp 8.1j$
x, meters	1.0	1.0

Feedback gains for $(A_x/B) = .002\ \text{m}^{-2}$

$C_\theta = .864$ $C_q = .325$ sec. $C_u = .0869\ \text{m}^{-1}$ sec. $C_x = .0447\ \text{m}^{-1}$

where $\delta = -C_\theta\theta - C_q q - C_u u - C_x x$

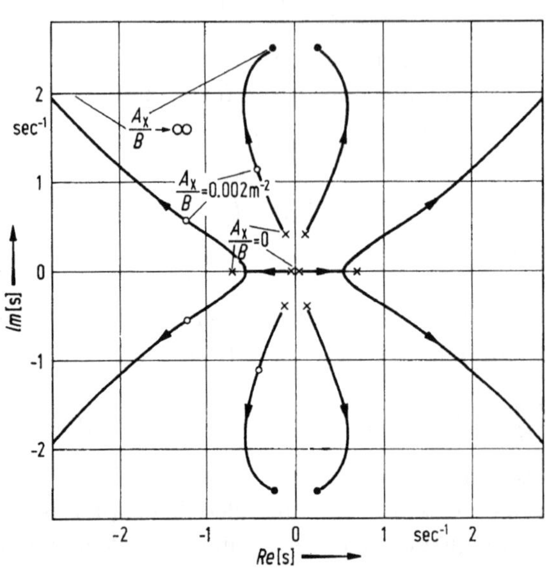

Fig. 4. Root locus for position regulator, helicopter in hover, with A_x/B as parameter.

Even for $A_x = 0$, $(B \neq 0)$ the θ, q, u states of the helicopter are stabilized by this technique; the two unstable complex roots of the open-loop system are simply "reflected" across the imaginary axis, while the stable real root and the zero root remain unchanged. The corresponding gains are:

$$C_\theta = .0535, \quad C_q = .0785\ \text{sec.}, \quad C_u = C_x = 0. \tag{2.19}$$

[1] The q and u components of the eigenvectors are given by $q = s\theta$, $u = sx$.

III. Modeling random disturbances as stochastic processes

How can "unpredictable disturbances" be modeled realistically? One very useful way is to describe them as *stochastic processes*, a concept which was largely developed in this century by *Einstein* ("Brownian motion", 1905), *Langevin* (1908), *Markov* (1912), *Planck* (1917), *Wiener* (1930), *Uhlenbeck* and *Ornstein* (1930), *Kolmogorov* (1931), *Khintchine* (1938), *Wold* (1938), *Chandrasekhar* (1943), *Rice* (1944), *Wang* and *Uhlenbeck* (1945), *Phillips* (1947), *Laning* and *Battin* (1958), *Newton*, *Gould*, and *Kaiser* (1957), and *Kalman* and *Bucy* (1961), to mention only a few.

A special class of stochastic processes, the *Gauss-Markov process* (GMP), has emerged as a particularly simple and useful one in desribing random disturbances. Many natural and man-made dynamic processes can be modeled quite well by GMP's. A GMP may be described by giving its *mean value vector*, $\bar{x}(t)$, and its correlation matrix $C(t, \tau)$, where

$$\bar{x}(t) \stackrel{\Delta}{=} E[x(t)] \tag{3.1}$$

$$C(t, \tau) \stackrel{\Delta}{=} E\{[x(t) - \bar{x}(t)]\,[x(\tau) - \bar{x}(\tau)]^T\}, \tag{3.2}$$

$E[\ ]$ means "expected value of", and x is a column vector with n components. The operation on the right side of (3.2) is an "outer product" or "dyadic product"; in Cartesian tensor notation,

$$C_{ij}(t, \tau) \stackrel{\Delta}{=} E\{[x_i(t) - \bar{x}_i(t)]\,[x_j(\tau) - \bar{x}_j(\tau)]\}.$$

The "expected value of [ ]" may be thought of as an "ensemble average", i.e., the average of [ ] over a large number of samples of the random process $x(t)$. If $x^{(i)}(t)$ is the ith sample history, then

$$\bar{x}(t) = \lim_{N\to\infty} \frac{1}{N} \sum_{i=1}^{N} x^{(i)}(t) \tag{3.3}$$

$$C(t, \tau) = \lim_{N\to\infty} \frac{1}{N} \sum_{i=1}^{N} [x^{(i)}(t) - \bar{x}(t)]\,[x^{(i)}(\tau) - \bar{x}(\tau)]^T. \tag{3.4}$$

The *covariance matrix*, a measure, of the mean-square deviation of the random process from its mean value, is given by

$$X(t) \stackrel{\Delta}{=} E\{[x(t) - \bar{x}(t)]\,[x(t) - \bar{x}(t)]^T\}. \tag{3.5}$$

Obviously $X(t) \equiv C(t, t)$. The probability density of $x(t)$ at any time t is gaussian; i.e.

$$p[x(t)] = \frac{1}{(\pi)^{n/2}\,|X(t)|^{1/2}} \exp\left\{-\frac{1}{2}\,[x(t) - \bar{x}(t)]^T X^{-1}(t)\,[x(t) - \bar{x}(t)]\right\} \tag{3.6}$$

where $p(\xi)\,d(\xi)$ is the probability that the value of x lies between ξ and $\xi + d\xi$.

The basic GMP is the *purely-random process*; it is an idealized, very jittery process that is completely uncorrelated from one time to the next. Its correlation matrix has the form

$$Q(t)\,\delta(t - \tau) \tag{3.7}$$

where $\delta(\cdot)$ is the Dirac delta function. The purely random process may be thought of as the limit of a sequence of impulses, random in both magnitude and time of occurrence, where the average magnitude of the impulses is zero (as many positive as negative impulses), but the *mean* square magnitude is $[\sigma(t)]^2$. The power spectral density is then

$$Q(t) \cong 2[\sigma(t)]^2\,\beta(t) \tag{3.7a}$$

where $\beta(t)$ is the average number of impulses per unit time that occur at time t.

A GMP can always be represented by the state vector, $x(t)$, of a linear dynamic system forced by a gaussian purely random process where the initial state vector is gaussian:

$$\dot{x} = F(t)\,x + \Gamma(t)\,w$$

where

$$\begin{aligned} &E[w(t)] = \bar{w}, \quad E\{[w(t) - \bar{w}]\,[w(\tau) - \bar{w}]^T\} = Q(t)\,\delta(t - \tau) \\ &E[x(t_0)] = \bar{x}_0, \quad E\{[x(t_0) - \bar{x}_0]\,[x(t_0 - \bar{x}_0]^T\} = X_0, \\ &E\{[w(t) - \bar{w}]\,[x(t_0) - \bar{x}_0]^T\} = 0. \end{aligned} \tag{3.8}$$

This *markov representation* shows clearly that the state property for *deterministic* processes and the markov property for *random* processes are precisely analogous: (a) if knowledge of a finite-dimensional vector at any time, t_0, makes the future of a deterministic process independent of the past, then it has the *state property*. For example, if $w(t)$ in (3.8) is known, then knowledge of $x(t_0)$, the initial state vector, is sufficient to predict $x(t)$ for $t > t_0$; (b) if knowledge of the probability density function for a finite-dimensional vector at any time, t_0, makes the future of a random process independent of the past, then it has the *markov property*. For example, if $Q(t)$ in (3.8) is known, then knowledge of $p[x(t_0)]$, the initial probability density function, is sufficient to predict $p[x(t)]$ for $t > t_0$.

Since the gaussian probability density of a vector, x, is completely specified by giving its mean value $\bar{x}$ and its covariance matrix, X (see 3.6), a GMP is specified for $t \geq t_0$ by giving $\bar{x}(t_0)$, $X(t_0)$, $F(t)$, $\Gamma(t)$, $\bar{w}(t)$, and $Q(t)$. In fact, it can be shown that

$$d\dot{\bar{x}}/dt = F\bar{x} + \Gamma\bar{w};\ \bar{x}(t_0) = \bar{x}_0, \tag{3.9}$$

$$\dot{X} = FX + XF^T + \Gamma Q \Gamma^T, \quad X(t_0) = X_0{}^{((1))}. \tag{3.10}$$

Note that $\bar{x}(t)$ and $X(t)$ are completely independent of each other. It can also be shown (e.g., *Bryson* and *Ho*, p. 333) that the correlation matrix is given by:

$$C(t,\tau) = \begin{matrix} \Phi(t,\tau)\, X(\tau); & t \geq \tau, \\ X(t)\, \Phi^T(\tau, t); & t \leq \tau, \end{matrix} \tag{3.11}$$

where $\Phi(t,\tau)$ is the $n \times n$ *transition matrix* of $\dot{x} = Fx$, i.e.,

$$\frac{d}{dt}\Phi(t,\tau) = F(t)\,\Phi(t,\tau); \quad \Phi(\tau,\tau) = I. \tag{3.12}$$

One of the simplest nonstationary GMP's is the so-called *Wiener process*, which is the integral of stationary white noise:

$$\begin{aligned} &\dot{x} = w \\ &E[x(0)] = 0, \quad E[x(0)]^2 = 0, \\ &E[w(t)] = 0, \quad E[w(t)\, w(\tau)] = Q\,\delta(t-\tau), \quad Q = \text{constant}. \end{aligned} \tag{3.13}$$

From (3.9) and (3.10)

$$\begin{matrix} \dot{\bar{x}} = 0, & \bar{x}(0) = 0 \\ \dot{X}(t) = Q, & X(0) = 0 \end{matrix} \Rightarrow \begin{cases} \bar{x}(t) = 0 \\ X(t) = Qt. \end{cases} \tag{3.14}$$

The root-mean-square (RMS) value $x(t)$ is $\sqrt{X(t)} = \sqrt{Qt}$, which shows the diffusive or random walk character of GMP's. The Wiener process may be thought of as a series of random step functions of average magnitude zero occurring at random times but with a certain average number of steps per unit time, β. It follows that $Q = 2\overline{c^2}\beta$ where $\overline{c^2}$ is the mean square magnitude of the step functions.

Another well-known nonstationary GMP is the *underdamped oscillator forced by white noise* (*Wang* and *Uhlenbeck* (1945)). This process could be interpreted, for example, as a pendulum in a random wind (see sketch below). The equations of motion in state variable form may be written as

$$\begin{aligned} \dot{\theta} &= q \\ \dot{q} &= -\omega^2\theta - 2\zeta\omega q + w \end{aligned} \tag{3.15}$$

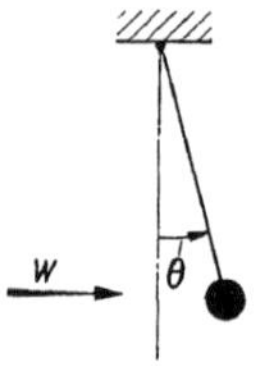

((1)) If F = constant matrix with eigenvalues s_i, $i = 1, \ldots, n$, then it can be shown that the $n(n+1)/2$ eigenvalues of (3.10) are all possible combinations $s_i + s_j$, $i, j = 1, \ldots, n$.

where $w(t)$ is proportional to wind velocity, ω = undamped natural frequency of the pendulum, and ζ = damping ratio. Suppose the motion starts with $\theta(0) = \theta_0$, $q(0) = 0$ and $\overline{w}(t) = 0$, $E[w(t)\, w(\tau)] = Q\, \delta(t - \tau)$, i.e., we know the initial state exactly $[X_{\theta\theta}(0) = X_{qq}(0) = X_{q\theta}(0) = 0]$ and the wind has zero mean velocity but has a constant spectral density proportional to Q. The mean value history is given by the familiar equations:

$$\begin{aligned} \dot{\overline{\theta}} &= \overline{q}; & \overline{\theta}(0) &= \theta_0, \\ \dot{\overline{q}} &= \omega^2\overline{\theta} - 2\zeta\omega\overline{q}; & \overline{q}(0) &= 0. \end{aligned} \tag{3.16}$$

The elements of the covariance matrix are given by

$$\begin{aligned} \dot{X}_{\theta\theta} &= 2X_{\theta q}; & X_{\theta\theta}(0) &= 0, \\ \dot{X}_{\theta q} &= -\omega^2 X_{\theta\theta} - 2\zeta\omega X_{\theta q} + X_{qq}; & X_{\theta q}(0) &= 0, \\ \dot{X}_{qq} &= -2\omega^2 X_{\theta q} - 4\zeta\omega X_{qq} + Q; & X_{qq}(0) &= 0. \end{aligned} \tag{3.17}$$

For $\zeta = 0.2$, the time histories are shown in Fig. 5. Note the two variances and the covariance change with twice the frequency that $\overline{q}$ and $\overline{\theta}$ change. A more revealing display of the time histories is given in Fig. 6, where $\overline{q}/\omega$ is plotted against $\overline{\theta}$ and probability ellipses are shown at selected points on the path. The axes of the ellipses are in the directions of the eigenvectors of X and the major and minor axes are proportional to the eigenvalues of X. Note the angular velocity

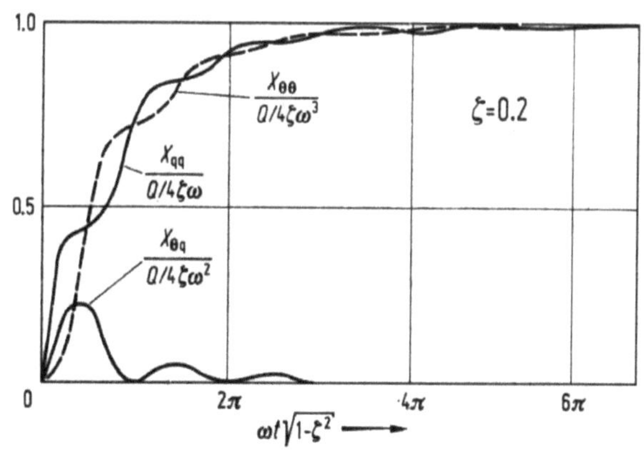

Fig. 5. Pendulum in random wind of spectral density Q; variance and covariance histories, starting with zero variance.

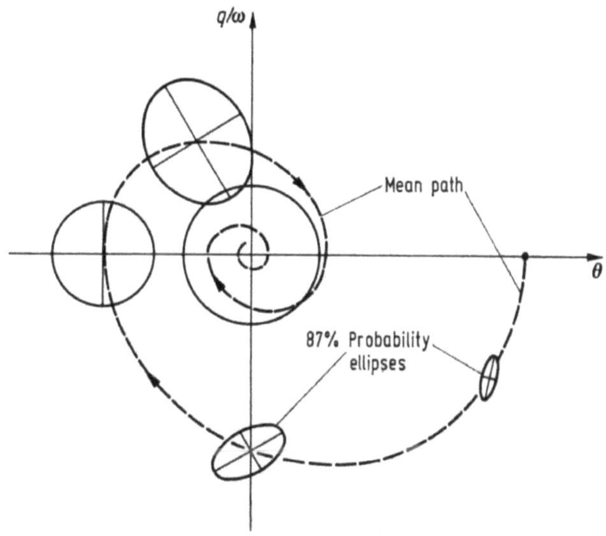

Fig. 6. Pendulum in random wind; mean angular velocity (q) vs. angular deflection (θ), with 87% probability ellipses at selected points on path.

q becomes uncertain first (elongated ellipse parallel to q axis) but the uncertainty quickly extends to θ. For $\sqrt{1-\zeta^2}\omega t \gg 2\pi$,

$$\bar{\theta} \to 0, \quad \bar{q} \to 0, \quad X_{\theta q} \to 0, \quad X_{\theta\theta} \to \frac{Q}{4\zeta\omega^3}, \quad X_{qq} \to \frac{Q}{4\zeta\omega}. \tag{3.18}$$

Thus, it is the presence of damping ($\zeta > 0$) that produces a *statistical steady-state*. The uncertainty created by the random wind is just balanced by the effects of damping. Note that X does not always increase with time; if $X(0) > X(\infty)$, then X would decrease with time toward $X(\infty)$.

For a *stationary Gauss-Markov process*,

$$C(t, \tau) = C(t - \tau), \tag{3.19}$$

that is, C is a function of $(t - \tau)$ only. The *Fourier transform* of the correlation matrix of a stationary GMP is called its *spectral density matrix*, $Q(\omega)$, i.e.,

$$Q(\omega) \stackrel{\Delta}{=} \int_{-\infty}^{\infty} C(t)\, e^{-j\omega t}\, dt, \tag{3.20}$$

which implies that

$$C(t) = \frac{1}{2\pi} \int_{-\infty}^{\infty} Q(\omega)\, e^{j\omega t}\, d\omega. \tag{3.21}$$

The scalar, *stationary, purely random GMP* has a constant spectral density and for this reason is often called *white noise*, since it is made up of all frequencies from 0 to ∞. Its correlation is therefore

$$Q\, \delta(t - \tau), \tag{3.22}$$

where Q = spectral density (constant).

At the opposite extreme from white noise is a *random bias*; it is simply an unpredictable constant. It has a constant correlation and hence its spectral density is given by

$$C\, \delta(\omega) \tag{3.23}$$

where $C/2\pi$ = correlation (constant). Its spectrum is thus a "line" at $\omega = 0$ (zero frequency). In between white noise (zero correlation time), and a random bias (infinite correlation time),

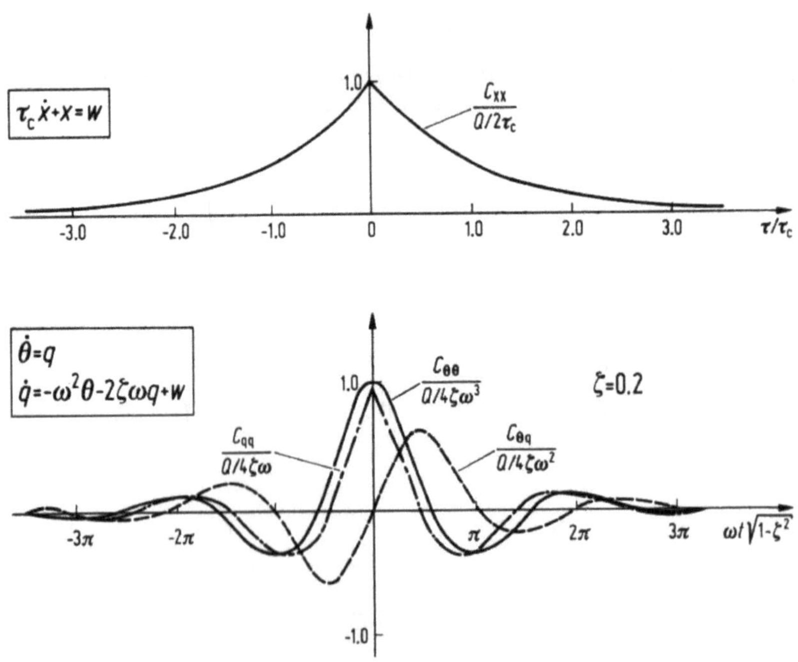

Fig. 7. (a) Stationary time correlation; first order system forced by gaussian white noise, w, with spectral density Q, (b) stationary time correlations and cross-correlation; second order underdamped system forced by gaussian white noise, w, with spectral density Q (e.g., pendulum in random wind).

stationary GMP's have finite correlation times, i.e., $C(t-\tau) \to 0$ as $|t-\tau| \to \infty$. For example, white noise forcing a first-order system produces a GMP with an exponential correlation (see Fig. 7). As another example, the components of the correlation matrix for the pendulum in a random wind (when it has reached a statistical steady state) are also shown in Fig. 7 for $\zeta = 0.2$.

IV. Design of state estimators

In the early 1960's a significant advance was made in control theory for random systems. *Kalman* and *Bucy* (1961), using the calculus of variations, developed the *maximum likelihood filter* for estimating the state of a nonstationary GMP in real time from measurements of the process that contain additive white noise. This extended and clarified the pioneering work of *Wiener* (1942) and *Kolmogorov* (1941), who had developed the minimum mean square error filter for stationary stochastic processes (not necessarily markovian or gaussian) using generalized harmonic analysis.

Kalman and *Bucy* assumed that a mathematical model of the GMP to be estimated was known, (1.1), and assumed that the process noise, $w(t)$, and the measurement noise, $v(t)$, were purely random processes[1]:

$$\begin{aligned} E[w(t)] &= 0, \quad E\{[w(t)]\,[w(\tau)]^T\} = Q(t)\,\delta(t-\tau) \\ E[v(t)] &= 0, \quad E\{[v(t)]\,[v(\tau)]^T\} = R(t)\,\delta(t-\tau)^{((1))} \end{aligned} \tag{4.1}$$

where $Q(t)$ is a known, possibly time-varying, $r \times r$ spectral density matrix for the process noise and $R(t)$ is a known, possibly time-varying, $q \times q$ spectral density matrix for the measurement noise.

The Kalman-Bucy (KB) filter is then given by:

$$d\hat{x}/dt = F\hat{x} + Gu + K(z - H\hat{x}); \quad \hat{x}(t_0) = \bar{x}(t_0), \tag{4.2}$$

where

$$\begin{aligned} K &= PH^TR^{-1} \\ \dot{P} &= FP + PF^T + \Gamma Q\Gamma^T - KRK^T; \quad P(t_0) = X(t_0). \end{aligned} \tag{4.3}$$

Note F, Γ, K, H, P, R, Q can all be time-varying quantities, and the process can start at any time, t_0. Gu is assumed to be a *known* forcing function[2]. The filter (4.2) is similar to the mean value model of (3.9), with the addition of some known forcing functions, Gu, and some *feedback* terms proportional to $(z - H\hat{x})$, the difference between the actual measurements and the expected value of the measurements based on the current estimate of the state. (The expected value of the measurement noise, v, is zero since a purely random process cannot be predicted.) The rate of change of the estimated state, $d\hat{x}/dt$, is thus proportional to the measurement prediction error.

The *errors* in the estimates given by (4.2) are defined as

$$\tilde{x} \overset{\Delta}{=} \hat{x} - x. \tag{4.4}$$

By subtracting (1.1) from (4.2), a set of differential equations for the errors in the estimates can be found:

$$d\tilde{x}/dt = (F - KH)\,\tilde{x} + Kv - \Gamma w; \quad \tilde{x}(t_0) = \bar{x}(t_0) - x(t_0). \tag{4.5}$$

From (4.5) it is clear that errors in the estimates are caused by (a) random errors in the measurement signals, $v(t)$; (b) random forcing functions, $w(t)$, and random initial conditions, $x(t_0)$.

The $n \times n$ symmetric matrix $P(t)$ in (4.3) is the *covariance matrix of the error in the estimates*, i.e.,

$$P(t) = E\{[\hat{x}(t) - x(t)]\,[\hat{x}(t) - x(t)]^T\}. \tag{4.6}$$

---

[1] Note that if $q < n$, $z(t)$ is, in general, *not* a markov process since $z(t_0)$ does not enable one to predict $z(t)$ for $t > t_0$.

((1)) Obviously, there is no lack of generality in assuming $\overline{w}(t) = \overline{v}(t) = 0$.

[2] If there is uncertainty in implementing $u(t)$, it can be included in $w(t)$.

It may be computed from (4.3), which is a *forward matrix Riccati equation*, since $KRK^T \equiv PH^TR^{-1}HP$ is *quadratic* in the elements of P. Comparing (4.3) with (3.10) we see that they are the same except for the *information rate term* ($-KRK^T$). Obviously P is less than or equal to X, i.e., with measurement information the uncertainty in the state is usually reduced.

The $n \times q$ gain matrix, $K(t)$, is proportional to the uncertainty in the estimate of the state (the covariance matrix, P) and inversely proportional to the uncertainty in the measurements (the spectral density matrix, R). Thus, if the uncertainty in state is large compared to uncertainty in measurements, the filter weights the measurement prediction error heavily and vice-versa.

Obviously it is not always possible to estimate all the state variables from a given set of measurements (4.1). An "observability" criterion was given by *Kalman, Ho,* and *Narendra* (1963).

As an *example*, suppose we had a measurement of the angular velocity of the pendulum in a random wind, i.e.,

$$z = q + v, \tag{4.7}$$

so that $H = [0, 1]$, and assume $R =$ constant. The KB filter is then

$$\begin{aligned} d\hat{\theta}/dt &= \hat{q} + K_\theta(z - \hat{q}); & \hat{\theta}(0) &= \theta_0, \\ d\hat{q}/dt &= -\omega^2\hat{\theta} - 2\zeta\omega\hat{q} + K_q(z - \hat{q}); & \hat{q}(0) &= 0, \end{aligned} \tag{4.8}$$

where

$$K_\theta = \frac{1}{R} P_{\theta q}, \quad K_q = \frac{1}{R} P_{qq}, \tag{4.9}$$

and

$$\begin{aligned} \dot{P}_{\theta\theta} &= 2P_{\theta q} - \frac{1}{R} P_{\theta q}^2; & P_{\theta\theta}(0) &= 0 \\ \dot{P}_{\theta q} &= -\omega^2 P_{\theta\theta} - 2\zeta\omega P_{\theta q} + P_{qq} - \frac{1}{R} P_{\theta q} P_{qq}; & P_{\theta q}(0) &= 0 \\ \dot{P}_{qq} &= -2\omega^2 P_{\theta q} - 4\zeta\omega P_{qq} + Q - \frac{1}{R} P_{qq}^2; & P_{qq}(0) &= 0. \end{aligned} \tag{4.10}$$

The estimates $\hat{\theta}$ and $\hat{q}$ from (4.8) depend on the measurement history, $z(t)$, of the particular run. However, we can predict $P(t)$, the mean square error of the estimates, ahead of time. For $\zeta = 0.2$ and $Q/\omega^2 R = 2.0$, the time histories of $P_{\theta\theta}$, $P_{\theta q}$, P_{qq} are shown in Fig. 8. They are similar to the time histories of $X_{\theta\theta}$, $X_{\theta q}$, X_{qq} in Fig. 5 but are *less*. The uncertainty created by the random wind is balanced not only by damping but also by the information coming from the measurement of q.

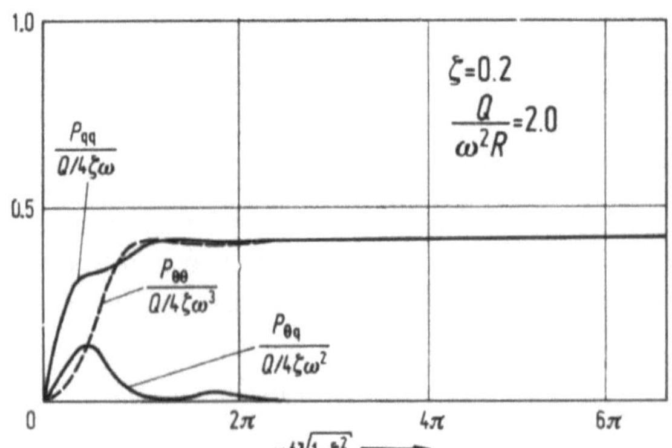

Fig. 8. Pendulum in random wind—estimate error variance and covariance histories using measurement of q with additive white noise having spectral density R.

Again a statistical steady-state is reached for $\sqrt{1-\zeta^2}\omega t \gg 2\pi$:

$$
\begin{aligned}
&P_{\theta q} \to 0 \\
&P_{\theta\theta} \to \frac{Q}{4\omega^3}\frac{2}{\zeta+\zeta'} \\
&P_{qq} \to \frac{Q}{4\omega}\frac{2}{\zeta+\zeta'}
\end{aligned}
\qquad \text{where} \quad \zeta' = \sqrt{\zeta^2 + \frac{Q}{4\omega^2 R}}. \tag{4.11}
$$

For $Q/\omega^2 R \to 0$ (noise in the measurement $\gg$ wind noise), Eqs. (4.11) indicate no change from the unmeasured situation (3.18). For $Q/\omega^2 R \to \infty$ (noise in the measurement $\ll$ wind noise), $P_{\theta\theta} \to 0$, $P_{qq} \to 0$, i.e., the errors in the estimates become negligibly small.

Stationary filters. The Wiener-Kolmogorov (WK) filtering technique for stationary gaussian processes[1], while remarkably creative, was difficult to understand for most engineers. It involved solving an integral equation (of the Wiener-Hopf type) for the Fourier transform of a kernel function, $W(t-\tau)$, such that a scalar "signal" is estimated with minimum mean-square error, by

$$
\int_{-\infty}^{t} W(t-\tau)\, z(\tau)\, dt \tag{4.12}
$$

where $z(t)$ is the signal plus noise; the time correlations (hence spectral densities) of the signal and the noise are assumed known. Even after finding $W(t-\tau)$, the problem remained of implementing (4.12), which, being a convolution integral, is not useful for real-time computation of the estimated signal. A dynamic system had to be found whose impulse-response closely matched $W(t-\tau)$.

In engineering circles it gradually became known that almost the only practical way to design WK filters was to assume that both the signal and the noise had "rational spectra" (see, e.g., *Phillips*, Section 7.4, or *Laning* and *Battin*, Section 7.5), which is equivalent to assuming that they are components of vector GMP's. The Kalman-Bucy development *starts* with this assumption and yields the dynamic system for the filter *directly* ((4.2) with all coefficient matrices constant); finding the steady-state solution of (4.3) is thus equivalent to solving a Wiener-Hopf integral equation for rational spectra. Unlike the WK approach, the KB approach handles the multiple-output, multiple-measurement case with ease. However, the WK filter handles measurement noise with rational spectra in a relatively straightforward manner, whereas the KB filter, as originally presented (1961), *required* that the measurement noise be white (constant spectra). Fortunately, a simple extension of the KB technique enables one to handle "colored" measurement noise even for nonstationary GMP's [cf. *Bryson* and *Johansen* (1965)]. In this case the measurement appears directly as part of the estimated state ("feedforward") and, if the measurement noise is sufficiently smooth, the optimum filter may even include differentiators.

Stationary KB filter design may be done using *eigenvector decomposition*. The Euler-Lagrange equations for the filter are (see *Bryson* and *Ho*, p. 396):

$$
\begin{bmatrix} \dot{x} \\ \dot{\lambda} \end{bmatrix} = \begin{bmatrix} F, & -\Gamma Q \Gamma^T \\ -H^T R^{-1} H, & -F^T \end{bmatrix} \begin{bmatrix} x \\ \lambda \end{bmatrix} + \begin{bmatrix} 0 \\ H^T R^{-1} z \end{bmatrix}. \tag{4.13}
$$

If we let T_+ be the $2n \times n$ matrix consisting of the n eigenvectors of (4.13) corresponding to the eigenvalues with positive real parts, then

$$
P_\infty = -X_+(\Lambda_+)^{-1} \quad \text{where} \quad T_+ = \begin{bmatrix} X_+ \\ \Lambda_+ \end{bmatrix}. \tag{4.14}
$$

The eigenvalues of the error estimates, $\tilde{x} \overset{\Delta}{=} \hat{x} - x$, are the eigenvalues of (4.13) with negative real parts and the eigenvectors are the columns of $[\Lambda_+]^{-1}$ (see, e.g., *Bryson* and *Hall* (1971)).

[1] Their treatment did *not* assume that the process was Markovian.

Again, a more revealing form for (4.13) may be given in terms of the Laplace transform of the measurements, $z(s)$, and the measurement estimates, $H\hat{x}(s)$:

$$[R + Z(s)\,QZ^T(-s)]\,R^{-1}[z(s) - H\hat{x}(s)] = z(s) \tag{4.15}$$

where $Z(s) = H[sI - F]^{-1}\Gamma$ is the matrix of transfer functions from the disturbance variables to the measurement variables, i.e.,

$$z(s) = Z(s)\,w(s). \tag{4.16}$$

The corresponding equation for the filter using measurement feedback gains, K, may be written:

$$[I + H(sI - F)^{-1}K]\,[z - H\hat{x}(s)] = z(s). \tag{4.17}$$

It follows that the "spectral factorization" equation is:

$$[R + Z(s)\,QZ^T(-s)] = [I + H(sI - F)^{-1}K]\,R[I + H(-sI - F)^{-1}K]^T \tag{4.18}$$

which is a $q\times q$ matrix equation (where z is a q-vector). The problem here is to find the $n\times q$ filter gain matrix K. Again, by requiring the eigenvalues with negative real parts of (4.13) to be the same as the eigenvalues of (4.17), the single measurement case is solved ($q = 1$). For $q > 1$, we require, in addition, that the eigenvectors of (4.17) be parallel to the eigenvectors of (4.15) that correspond to the eigenvalues of (4.13) with positive real parts (see the discussion following (2.13)).

As an *example*, we design a *steady-state filter* for the *helicopter in hover* (see the last part of Section II). We assume a position measurement and a pitch angle measurement, both with jittery noise having correlation times short compared to decay times of the closed-loop eigenvalues; thus they can be modeled as white noise processes. For an RMS position measurement error of 1 meter with a correlation time of .136 sec,. the spectral density is

$$R_x = 2(1)^2\,(.136) = .272\ \text{m}^2\text{sec}. \tag{4.19}$$

For an RMS pitch angle measurement error of 0.5 deg. with a correlation time of 0.10 sec., the spectral density is

$$R_\theta = 2\left(\frac{.5}{57.3}\right)^2(0.1) = 1.53\times 10^{-5}\ (\text{rad})^2\ \text{sec}. \tag{4.20}$$

On the other hand, the correlation time of wind fluctuations is comparable to, or larger than, the decay times of the closed-loop eigenvalues so we model the wind with a first-order shaping filter as

$$\tau_c\dot{w} + w = q_w, \tag{4.21}$$

where q_w is white noise. We have taken $\tau_c = 5$ sec. and RMS wind magnitude as 7 m sec^{-1}; this implies that the spectral density of q_w should be

$$Q = 2(7)^2\,(5) = 490\ \text{m}^2\ \text{sec}^{-1}. \tag{4.22}$$

The wind velocity, w, is now an additional (fifth) state variable. The state-feedback gains on θ, q, u, and x are not changed by adding w as a state variable (in the root locus, Fig. 4, a zero is added that just cancels the additional pole at $s = -(1/\tau_c)$), but an additional gain is added for w, i.e., $\delta = -C_\theta\theta - C_q\dot{\theta} - C_u\dot{x} - C_x x - C_w w$. It is straight-forward to show that

$$C_w = B^{-1}G^T S_w \tag{4.23}$$

where

$$\left[\frac{1}{\tau_c}I - (F - GB^{-1}G^T S_\infty)^T\right]S_w = S_\infty\Gamma.$$

For the numbers used here, this gives

$$C_w = -.00195\ \text{m}^{-1}\ \text{sec}. \tag{4.24}$$

The filter design, on the other hand, is significantly affected by adding the shaping filter. The root square characteristic equation in the form (4.15) is given by

$$|R + Z(s)\,QZ^T(-s)| = 0 \tag{4.25}$$

where

$$R = \begin{bmatrix} R_x & 0 \\ 0 & R_\theta \end{bmatrix}, \quad Z(s) = \begin{bmatrix} Z_x(s) \\ Z_\theta(s) \end{bmatrix},$$

$$Z_x(s) \stackrel{\Delta}{=} \frac{x(s)}{q_w(s)} = \frac{-\sigma_2\left[s^2 + \left(\sigma_1 - \dfrac{a_1 a_2}{\sigma_2}\right)s + \dfrac{a_1 g}{\sigma_2}\right]}{(\tau_c s + 1)\,\Delta_0(s)},$$

$$Z_\theta(s) = \frac{\theta(s)}{q_w(s)} = \frac{-a_1 s^2}{(\tau_c s + 1)\,\Delta_0(s)},$$

and $\Delta_0(s)$ is given below (2.16). Thus (4.23) may be written as:

$$\frac{R_\theta}{Q} + Z_\theta(s)\,Z_\theta(-s) + \frac{R_\theta}{R_x} Z_x(s)\,Z_x(-s) = 0\,. \tag{4.25a}$$

A root locus of (4.25a) with Q/R_θ as the parameter is shown in Fig. 9 for the numerical values of Table 1,

$$\frac{R_\theta}{R_x} = \frac{1.525\times 10^{-5}}{.272}\,\mathrm{m}^{-2},$$

and $\tau_c = 5$ sec. The "zeros" of the root locus are at $.190 \pm .193j$ sec^{-1}, and the "poles" are the open-loop poles of Table 1, the "wind pole" at $s = -0.2$ sec^{-1}, and their reflections in the imaginary axis. For $Q = 490$ m^2 sec^{-1}, the eigenvalues of the filter estimate errors are given in Table 2, as well as at the corresponding filter gains and predicted RMS values for the filter estimate errors. Table 3 shows the dynamic compensator, comprised of the filter and the *estimated* state feedback to rotor tilt, δ.

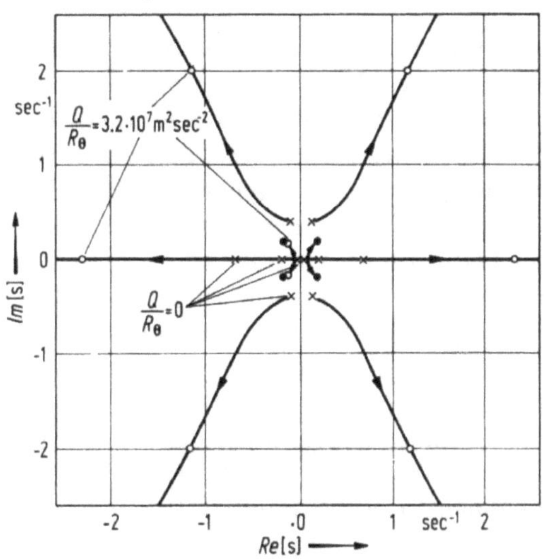

Fig. 9. Root locus for (x, θ) measurement filter, helicopter in hover, with Q/R_θ as parameter, $(R_x/R_\theta) = 5.6\times 10^{-5}$ m^{-2}.

Table 2. Filter design for a typical transport helicopter
[RMS wind 7 m sec^{-1} with correl. time 5 sec, $R_x = .272$ m^2 sec, $R_\theta = 1.52\times 10^{-5}$ sec]

Eigenvalues of the filter estimate errors (sec^{-1})

-2.34 $\quad -0.19 \pm 0.19j$ $\quad -1.17 \pm 2.00j$

Filter gains (units in meters, sec, radians)

n	θ	q	u	x	w
K_{nx}	.000327	.000251	.0735	.381	$-.128$
$K_{n\theta}$	4.05	8.21	21.4	5.94	-954

RMS Estimate errors

$\tilde\theta = 0.45$ deg $\quad \tilde q = 1.15$ deg sec^{-1} $\quad \tilde u = .092$ m sec^{-1} $\quad \tilde x = 0.32$ m $\quad \tilde w = 3.78$ m sec^{-1}

Table 3. Dynamic compensator for a typical transport helicopter

$$\begin{bmatrix} \dot{\hat{\theta}} \\ \dot{\hat{q}} \\ \dot{\hat{u}} \\ \dot{\hat{x}} \\ \dot{\hat{w}} \end{bmatrix} = \begin{bmatrix} 0 & 1 & 0 & 0 & 0 \\ 0 & -\sigma_1 & -\alpha_1 & 0 & -\alpha_1 \\ g & -\alpha_2 & -\sigma_2 & 0 & -\sigma_2 \\ 0 & 0 & 0 & 1 & 0 \\ 0 & 0 & 0 & 0 & -\dfrac{1}{\tau_c} \end{bmatrix} \begin{bmatrix} \hat{\theta} \\ \hat{q} \\ \hat{u} \\ \hat{x} \\ \hat{w} \end{bmatrix} + \begin{bmatrix} 0 \\ n \\ g \\ 0 \\ 0 \end{bmatrix} \delta + \begin{bmatrix} K_{\theta x} & K_{\theta\theta} \\ K_{qx} & K_{q\theta} \\ K_{ux} & K_{u\theta} \\ K_{xx} & K_{x\theta} \\ K_{wx} & K_{w\theta} \end{bmatrix} \begin{bmatrix} z_x - \hat{x} \\ z_\theta - \hat{\theta} \end{bmatrix}$$

$\delta = -C_\theta\hat{\theta} - C_q\hat{q} - C_u\hat{u} - C_x\hat{x} - C_w\hat{w} =$ commanded rotor tilt,

$z_x =$ measurement of x,

$z_\theta =$ measurement of θ, and

C_n, K_{nx}. $K_{n\theta}$ given in Tables 1 and 2; $\tau_c = 5$ sec, $C_w = -.00195\ \text{m}^{-1}$ sec

Even for $Q = 0$, a stable filter is designed by this technique. The two unstable complex roots of the open-loop system are "reflected" across the imaginary axis, while the stable real roots and the zero root remain unchanged. The corresponding filter gains and RMS estimate errors are:

$$K^T = \begin{bmatrix} -.00126, & -.000635, & .0197, & .104 \\ .388, & .0893, & .261, & -22.5 \end{bmatrix},$$

$$\tilde{\theta} = .083 \text{ deg}, \ \tilde{q} = .029 \text{ deg sec}^{-1}, \ \tilde{u} = .026 \text{ msec}^{-1}, \ \tilde{x} = .072 \text{ m}. \qquad (4.26)$$

It is not always possible to find a satisfactory steady-state filter by these techniques, even when the system is completely observable. If there are any stable (or neutrally stable) modes of the open-loop system that are "uncontrollable" by the disturbances[1], then the steady-state variance and covariances of these modes will be zero, i.e., the optimum time-varying filter will have estimated them exactly by the time it reaches the statistical steady-state. As a result, the steady-state gains for this mode will be zero; this may produce an unsatisfactory filter, particularly if the undisturbable mode is neutrally stable or only marginally stable (this will show up as a zero or very small eigenvalue for the steady-state filter). A straightforward remedy for this situation is to divide the system into disturbable and undisturbable modes, and then (a) design a KB filter for the disturbable modes neglecting the coupling to the undisturbable modes (i.e., treat them as known forcing functions); then (b) design an "observer" for the undisturbable modes, neglecting the coupling to the disturbable modes, with eigenvalues very much less than the eigenvalues of the filter in (a), but large enough so that errors in these modes will die out rapidly enough to be satisfactory [see, e.g., *Bryson* and *Luenberger* (1970) for observer design].

The following example illustrates a *neutrally stable, undisturbable mode*. An *inverted pendulum* is to be *stabilized by torquing a wheel* mounted at the tip of the pendulum as shown in the sketch below. A random wind blows on the pendulum rod but produces no torque on the wheel, so the system model is reasonably well described by

$$\begin{aligned} \dot{\theta} &= q \\ \dot{q} &= n^2\theta + u + w, \\ \dot{r} &= -au \end{aligned}$$

where $r =$ angular velocity of wheel, u is proportional to the torque applied to the wheel, and w is proportional to the wind velocity. Clearly, the wheel is undisturbable by the wind but r is observable if r or $r - q$ is measured; in principle, then, a nonstationary filter could bring $\hat{r} - r$ to zero exactly after a long time even with a very noisy measurement. The steady-state

[1] "Undisturbable" would be a better word than "uncontrollable" in this context.

of such a filter would have zero gain on the measurements in the $d\hat{r}/dt$ channel, i.e., no attention would be paid to new measurement data.

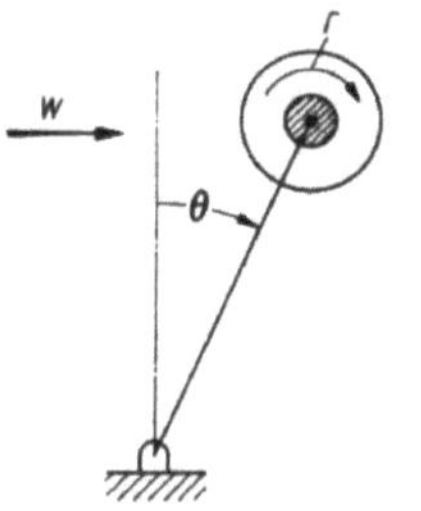

V. Mean square behavior of systems with filter/state-feedback controllers

As discussed in the introduction, the control system that minimizes the quadratic criterion (2.1) *on the average*, in the presence of random disturbances, is constructed by using the state-feedback control gains of (2.5) on the *estimated state* from the filter (4.2), i.e.,

$$u = -C\hat{x}. \tag{5.1}$$

This combination of filter and state-feedback gains (see Fig. 2) constitutes a "*compensator*" in the language of classical controller design.

The system, controlled by this compensator, with random forcing functions, $w(t)$, random errors in the sensor signals, $v(t)$, and random initial conditions, $x(t_0)$, is itself a gauss-markov process. Hence, its mean and mean-square behavior can be predicted using (3.9) and (3.10), by considering the state vector to be $\begin{bmatrix} x \\ \hat{x} \end{bmatrix}$ and the disturbance vector to be $\begin{bmatrix} w \\ v \end{bmatrix}$. $\hat{x}$ is thus treated as the "state" of the compensator. However, this prediction is actually simpler than just indicated if we use $\begin{bmatrix} \tilde{x} \\ \hat{x} \end{bmatrix}$ as the state vector, where $\tilde{x} = \hat{x} - x =$ filter estimate error, because $\tilde{x}$ does *not* depend on $\hat{x}$:

$$\begin{aligned} d\tilde{x}/dt &= (F - KH)\,\tilde{x} + Kv - \Gamma w;\;\; \tilde{x}(t_0) = -x(t_0) \\ d\hat{x}/dt &= (F - GC)\,\hat{x} - KH\tilde{x} + Kv;\; \hat{x}(t_0) = 0 \end{aligned} \tag{5.2}$$

where we have used

$$\begin{aligned} z &= Hx + v, \\ u &= -C\hat{x}. \end{aligned} \tag{5.3}$$

Since $E\{[\tilde{x}(t_0)]\,[\hat{x}(t_0)]^T\} = 0$, it follows from (5.2) that

$$E\{[\tilde{x}(t)]\,[\hat{x}(t)]^T\} = 0 \quad \text{for} \quad t \geq t_0. \tag{5.4}$$

Therefore

$$E\{[x(t)]\,[x(t)]^T\} \equiv E\{(\hat{x} - x)\,(\hat{x} - x)^T\} = E(\hat{x}\hat{x}^T) + E(\tilde{x}\tilde{x}^T),$$

or

$$X = \hat{X} + P, \tag{5.5}$$

where

$$X = E(xx^T), \quad \hat{X} = E(\hat{x}\hat{x}^T), \quad P = E(\tilde{x}\tilde{x}^T).$$

Now P is given by (4.3) and it follows directly from (3.8), (5.2), and (5.4) that

$$d\hat{X}/dt = (F - GC)\,\hat{X} + \hat{X}(F - GC)^T + PH^TR^{-1}HP;\;\; \hat{X}(t)_0 = 0, \tag{5.6}$$

which is a *linear* matrix equation, since $PH^TR^{-1}HP$ is a known forcing function (i.e., it can be pre-computed from (4.3)). From (5.3) and (5.5) it follows that

$$E[u(t)\,u^T(t)] = C\hat{X}C^T. \tag{5.7}$$

For the statistically stationary case, $\dot{P} = 0$, $d\hat{X}/dt = 0$, and (5.6) becomes a set of $n(n+1)/2$ simultaneous linear equations. For even moderate values of n, solution of that many linear equations is not a trivial task.

For the *helicopter* example used in Sections II and IV, with the data of Tables 1 and 2, the predicted RMS state and control deviations in the gusty wind (RMS 7 m $\sec^{-1}$, correlation time 5 sec.) using the state-feedback filter compensator are given in Table 4.

Table 4. Predicted RMS state and control deviations of typical transport helicopter in a gusty wind using controller of tables 1 and 2

$\theta = 1.95$ deg	$q = 2.92$ deg $\sec^{-1}$	$u = 0.25$ m $\sec^{-1}$	$x = 0.51$ m	$\delta = 1.33$ deg

To show the importance of *both* sensors in the example, we designed a controller using only the position sensor (it is not possible, of course, to use only the pitch angle sensor, since then the position error can not be estimated). For the same position sensor accuracy and the same gusty wind, the predicted RMS deviations were $\theta = 24.4$ deg., $x = 4.76$ m, $\delta = 10.4$ deg. which would be quite unacceptable. This shows that the pitch angle sensor is essential.

The overall *compensator transfer function*, from the measurements, $z(s)$, to the controls, $u(s)$, for the stationary case is given by

$$u(s) = -C[sI - F + GC + KH]^{-1}Kz(s). \tag{5.8}$$

Some of the poles and zeroes of this transfer function may have positive real parts, i.e., the compensator does not necessarily have "minimum phase". (Most classical compensator designs have minimum phase.) Thus, the compensator (5.8) will, in general, require active logic components, i.e., it cannot be built just from passive components like resistors and capacitors. However, the eigenvalues (poles) of the controlled system are simply the eigenvalues of the state feedback controller, $F - GC$, plus the eigenvalues of the filter, $F - KH$, all of which have negative real parts[1].

VI. Summary

The current theory available for designing controllers for dynamic systems subjected to random disturbances may be summarized as follows:

(a) Model the system by a set of coupled first-order ordinary differential equations (i.e., put them in state vector form).

(b) Linearize the system about a nominal path or operating condition.

(c) Design a linear state-feedback controller using weighting factors in a quadratic performance index as described in Section II.

(d) Model the random disturbances as Gauss-Markov processes, using shaping filters if necessary.

(e) Design a Kalman-Bucy filter to estimate the state vector from the available measurements as described in Section IV.

(f) Cascade the filter and the state-feedback controller, i.e., use state-feedback on the estimated state vector. This is the optimal "compensator" for the system.

(g) Predict the RMS (root mean square) response of the controlled system in the presence of the random disturbances using the methods of Section V. If the response is unsatisfactory, modify the compensator by changing weighting factors in (c) and/or use additional or more accurate measurements.

Acknowledgements

The author wishes to thank his colleagues, *J. V. Breakwell, D. B. DeBra, J. D. Powell, W. E. Hall,* and his students for many valuable suggestions and criticisms of this paper.

[1] Thus, (5.8) may obscure the stability of the controlled system.

References

Bryson, A. E., Ho, Y. C.: Applied Optimal Control, Xerox-Blaisdell, Lexington, Mass., 1969.

Bryson, A. E., Hall, W. E.: Stanford University, SUDAAR Rep. No. 446, November 1971.

Bryson, A. E., Johansen, D. E.: IEEE Trans. Auto. Control 10 (1965) 4—10.

Bryson, A. E., Luenberger, D. G.: Proc. IEEE 58 (1970) 1803—1811.

Chandrasekar, S.: Reviews of Modern Physics 15 (1943) 1—89.

Chang, S. S. L.: Synthesis of Optimum Control Systems, New York: McGraw-Hill 1961.

Einstein, A.: Ann. d. Physik 17 (1905) 549 and 19 (1906) 371.

Evans, W. R.: AIEE Trans. Part II 69 (1950) 66—69.

Francis, J. G. F.: Computer J. 4 (1961) 265—271; 4 (1962) 332—345.

Gunckel, T. L., Franklin, G. F.: Trans. ASME J. Basic Engr., 85-D (1963) 197—201.

Wiener, N.: The Extrapolation, Interpolation, and Smoothing of Stationary Time Series with Engineering Applications, MIT Radiation Lab. Rep., Feb. 1942 (published as a book: New York: Wiley 1949.

Wold, H.: A Study in the Analysis of Stationary Time Series, Uppsala: Almquist and Wiskell 1938.

Bibliography

Aström, K. J.: Introduction to Stochastic Control Theory.

Boguslavski, I. A.: Methods of Navigation and Control with Incomplete Statistical Information, Moscow, 1970 (In Russian). New York: Academic Press 1970.

Leondes, C. T. (Ed.): AGARDograph No. 139, Nat. Techn. Inform. Serv., Springfield, VA, Feb. 1970.

Letov, A. M.: Dynamics and Control of Flight, Moscow, 1969 (in Russian).

Meditch, J. S.: Stochastic Optimal Linear Estimation and Control, New York: McGraw-Hill 1969.

Wilkerson, J. H., Martin, R. S., Peters, G.: Numer. Math. 14 (1970) 219—231.

Hall, W. E.: Ph. D. Diss., Stanford Univ., Stanford, Calif., 1971.

Joseph, P. D., Tou, J. T.: Trans. AIEE (Applications and Industry) 80 (1961) 193—196.

Kalman, R. E., Bertram, J. E.: Trans. AIEE 77 (1958).

Kalman, R. E., Bucy, R.: Trans. ASME 83D (1961) 95—108.

Kalman, R. E., Englar, T. S.: NASA CR-475, June 1966.

Kalman, R. E., Ho, Y. C., Narendra, K. S.: Contrib. to Differential Eqns. 1 (1963) 189—213.

Khintchine, A. Ya.: Math. Ann. 109 (1934) 604—615. Also Progr. of Math. Sci., USSR, No. 5, 42—51 (1938).

Kolmogorov, A. N.: Foundations of the Theory of Probability. New York: Chelsea 1931.

Kolmogorov, A. N.: Bull. Moscow Univ., USSR, Ser. Math. 5 (1941) 13—14.

Langevin, P.: Comptes Rendus 146 (1908) 530.

Laning, J. H., Battin, R. H.: Random Processes in Automatic Control, New York: McGraw-Hill 1958.

Letov, A. M.: (3 parts), Automation and Remote Control 21, 1960.

Markov, A. A.: Wahrscheinlichkeitsrechnung, Leipzig 1912.

McFarlane, A. G. J.: J. Electron. and Control 14 (1963) 643—654.

Newton, G. C., Gould, L. A., Kaiser, J. F.: Analytical Design of Linear Feedback Control, New York: Wiley 1957.

Planck, M.: Sitz. der Preuss., Akad. 324, 1917.

Phillips, R. S.: Chapters 6 and 7 of Theory of Servomechanisms, ed. by James, Nichols, and Phillips: New York: McGraw-Hill 1947.

Potter, J. E.: MIT Expt. Astronomy Lab. Rep. RE-11, Aug. 1964.

Potter, J. E.: SIAM J. Appl. Math. 14 (1966) 496—501.

Rice, S. O.: Bell System Techn. J. 23 and 24 (1944—1945) 1—162.

Rynaski, E. G., Whitbeck, R. F.: Cornell Aero. Lab. Rep. IH-1943-F-1, 193, Oct. 1965. (Also AFFDL-TR-65-28).

Uhlenbeck, G. E., Ornstein, L. S.: Phys. Rev. 36 (1930) 823—841.

Wang, M. C., Uhlenbeck, G. E.: Reviews of Modern Physics 17 (1945) 323—342.

Wiener, N.: Acta Mathematica 55 (1930) 117—258.

The numerical analysis of structures

B. Fraeijs de Veubeke

Liège, Belgium

Introduction

Elasticity theory, even in its simplest guise, linear with respect to geometrical deformations and with respect to material properties, has few so-called "closed form" solutions. The need for solving numerically elasticity problems and the advent of fast and powerful computing machines seem even to have partly robbed such types of solutions of their interest. It is now often easier and faster to produce numerical results of a given problem by using modern discretization methods than to have to compute the special functions involved in a closed form presentation. In the end the value of such presentations lies in the ease with which they can be manipulated to exhibit the influence of the design parameters, whereas a similar computer study of the influence of design is still a long and costly process. Structural optimization, which is the goal of such studies, receives nowadays a good deal of attention; its costeffective implementation will largely depend on the speed and accuracy with which the more classical problems can be solved, namely the determination, for a given design, of its statical and dynamical response to external loads or displacements of supports.

Matrix structural analysis

Not so long ago the concept of a "structure" was essentially that of an assemblage of simple types of elements, mostly bars and beams capable of transmitting axial loads, transverse shear loads, bending and possibly torsion moments. For such elements the classical theory of elasticity provided accurate foundations for good engineering approximations, allowing their behavior as "transmitters" to be described by a finite number of displacement and conjugate force coordinates. Thus a structure could be naturally conceived as a system with a finite, although possibly large, number of degrees of freedom, and design in force transmission problems, like bridge engineering, consisted mainly in adopting a more or less complicated topology of interconnexions between the constitutive elements.

Matrix structural analysis is essentially concerned with the mathematical formulation of the interconnexions and the resultant properties of the structure in terms of the kinematics of straining and external and internal equilibrium. Its extensive development was stimulated by its natural compatibility with electronic computer software. Its achievements remain of utmost importance in the more general context of the finite element method.

Indeed it may be suspected that the finite element idea emerged from a desire to adapt the available arsenal of matrix methods to structures of less particular nature. Many structures are determined primarily by constraints due to the functions they perform; the shape of aircraft wings is largely dominated by aerodynamic considerations: torsional stiffness requirements are generally more efficiently met by using stressed-skin construction; containers and roofs lead naturally to shell-type construction. Thus, to save the concept of a structure as a finite-dimensional system made of interconnected elements, it becomes necessary to conceive the element as a discretized model of a piece of elastic continuum and to generalize the concept of connexion to extensive, and no more punctual, interfaces.

There are of course other advantages to this scheme that explain its growing success. Modern aerospace structures have very complex geometries; this does not inhibit the finite element method, whose pieces can be taylored to fit all the geometrical accidents. Furthermore the simpler the individual pieces, the easier a discrete representation of their elastic behaviour in terms of shaping functions. Local changes in the nature of the material or modifications in its behaviour are also easier to take into account by the subdivision process. Finally, in the presence of unsatisfactory results, it is often possible to localize the defects in idealization or discretization and to correct them while saving a large part of the previous calculations.

Force and displacement method

One of the distinctions that matrix structural analysis contributed to underline is that in the choice of either displacements or redundant forces as basic unknowns to solve for. Denoting by

q the column vector of generalized displacements
p the column vector of conjugate external forces,
e the column vector of generalized strains,
s a column vector of conjugate generalized stresses,

we may write the basic equations of matrix structural analysis as:

$$\text{kinematics} \qquad S^T q = e, \tag{1}$$

$$\text{statics} \qquad Ss = p, \tag{2}$$

$$\text{stress-strain} \qquad s = Je, \quad e = Fs. \tag{3}$$

Equation (1) is a discretized form of the linearized strain measures of a continuum

$$\begin{bmatrix} \varepsilon_{11} \\ \varepsilon_{22} \\ \varepsilon_{33} \\ \gamma_{23} \\ \gamma_{31} \\ \gamma_{12} \end{bmatrix} = \begin{bmatrix} D_1 & 0 & 0 \\ 0 & D_2 & 0 \\ 0 & 0 & D_3 \\ 0 & D_3 & D_2 \\ D_3 & 0 & D_1 \\ D_2 & D_1 & 0 \end{bmatrix} \begin{bmatrix} u_1 \\ u_2 \\ u_3 \end{bmatrix} \quad \text{or} \quad \varepsilon = Du \tag{4}$$

presented here in cartesian coordinates (x_1, x_2, x_3) and introducing the matrix differential operator D whose elements are partial differential operators $D_i = \partial/\partial x_i$.

Equation (2) is a discretized form of the equilibrium equations of a continuum: volume equilibrium

$$D^T\sigma + \phi = 0 \tag{5}$$

with σ the stress vector conjugate to ε and ϕ the vector of body loads, and surface equilibrium

$$N^T\sigma = t \tag{6}$$

with t the vector of surface tractions. N is derived from the operator D by substituting the direction cosines n_i of the outward normal in place of D_i. Equation (3) is a discretized form of the stress-strain relations

$$\sigma = H\varepsilon, \tag{7}$$

where H is a Hookean matrix of elasticity moduli. The duality between kinematics and statics in a continuum is prefigurated by the appearance of the same differential operator D in (4) and (5), and this is confirmed by the equality between internal and external virtual work:

$$\int_R \sigma^T\varepsilon \, dV = \int_R \sigma^T Du \, dV = \int_S (N^T\sigma)^T u \, dS - \int_R (D^T\sigma)^T u \, dV = \int_S t^T u \, dS + \int_R \phi^T u \, dV \tag{8}$$

by, in succession, using (4), integrating by parts and then using (5) and (6). If the region R over which the integration is performed is subdivided into finite elements, the result implies that there is no contribution to virtual work from the interfaces. This will be the case if we adhere to the exact transition conditions between face + and face − of an interface:

$$u^+ = u^-, \tag{9}$$

$$t^+ + t^- = 0 \tag{10}$$

expressing respectively continuity of displacements and reciprocity of surface tractions. This explains why the discretized forms (1) and (2) show the same duality by the presence of a single matrix operator S^T, the *structural kinematical matrix*. Passing from internal to external virtual work in discretized form

$$e^T s = (S^T q)^T s = q^T S s = q^T p \tag{11}$$

we can see that their equality implies the truth of (2) if (1) is known and conversely.

The study of kinematical and statical properties of a structure is fundamentally related to the problems of inverting Eqs. (1) and (2). The solutions to the homogeneous problems introduce important concepts

$$S^T q = 0 \to q = Rr \qquad r \text{ arbitrary}, \tag{12}$$

$$Ss = 0 \to s = Xx \qquad x \text{ arbitrary}. \tag{13}$$

In (12) R is a base matrix of kinematical modes of the structure (all modes involving no strain such as rigid body modes and possible mechanisms). In (13) X is a base matrix of self-stressing modes of the structure (all modes involving no external forces). The columns of R and X are maximal sets of linearly independent solutions. Between the number of elements of the different vectors

$$\nu = n(q), \quad \sigma = n(s), \quad \varrho = n(r) \text{ and } \xi = n(x)$$

the following relation holds

$$\nu - \varrho = \sigma - \xi. \tag{14}$$

From (12) and (13) follow, by definition

$$S^T R = 0, \tag{15}$$

$$SX = 0. \tag{16}$$

Since problem (1) for q is singular, there are existence conditions for a solution

$$X^T S^T q = 0 \to X^T e = 0. \tag{17}$$

They are, in discretized form the compatibility equations for strains. Similarly the existence conditions for a solution of problem (2) are

$$R^T S s = 0 \to R^T p = 0. \tag{18}$$

They are the global equilibrium conditions for the applied external forces. Assuming the existence conditions to be satisfied, the problems have the following general solutions

$$q = S^{T*} e + Rr, \tag{19}$$

$$S = S^* p + Xx, \tag{20}$$

where the first terms are particular solutions. Computing virtual works

$$s^T e = p^T S^{*T} e + x^T X^T e = p^T S^{*T} e$$

by virtue of (17), and

$$p^T q = p^T S^{T*} e + p^T R r = p^T S^{T*} e$$

by virtue of (18).

As strains are compatible and external forces in equilibrium, both expressions must be equal. Consequently if a pseudo-inverse matrix S^* has been found for problem (2), its transpose S^{*T} can serve as pseudo-inverse S^{T*} for the solution of problem (1) and conversely. The automatic computation of the matrices involved in the solution (19) and (20) will be discussed in the next section. *The displacement method* consists in presenting the elastic problem in the form

$$p = Ss = SJe = (SJS^T)\, q = Kq \tag{21}$$

resulting from a successive application of (1), (3) and (2).

$$K = SJS^T$$

is the symmetrical *structural stiffness matrix*. Its *core* J is normally positive definite so that the solutions of the homogeneous problem $Kq = 0$ are the same as those of (12), and the existence conditions are the same as (18): global equilibrium of the applied loads. In most applications $R = 0$, that is the structure is at least isostatically restrained, in which case K is non singular and invertible. In practice the solution is carried out on computer by any form of Gaussian elimination procedure or by conjugate gradient methods. For very large problems recourse has to be made to substructure techniques. If the structure has kinematical freedoms, the pseudo-inversion of the now singular problem (21), leading to a solution of type

$$q = K^*p + Rr \tag{22}$$

akin to (19), can be carried out by the same techniques as will be discussed in the next section. Once q is determined, the strains and stresses follow from (1) and (3).

The force method is most easily presented as an application of the principle of minimum complementary energy. The most general state of stress of the structure being given by (20), the stress energy is

$$\frac{1}{2} s^T F s = \frac{1}{2} p^T S^{*T} F S^* p + x^T X^T F S^* p + \frac{1}{2} x^T X^T F X x$$

the first part of which may be ignored as independent of the unknown redundancy coordinates x.

Minimization of the energy with respect to x gives

$$X^T F X x = -X^T F S^* p \tag{23}$$

and this can be inverted because $X^T F X$ is positive definite (F is positive definite and X has linearly independent columns). Backsubstitution of the solution into (20) solves the problem of stress determination. If displacements are needed, they can be found through (3) and (19).

The combined method consists in minimizing the total complementary energy

$$\frac{1}{2} s^T F s - p^T q$$

thus assuming the displacements to be imposed and incorporating the equilibrium constraints (2) by means of a Lagrangian vector multiplier λ:

$$\frac{1}{2} s^T F s + \lambda^T (p - Ss) - p^T q = \min!$$

Minimization with respect to p identifies $\lambda - q$. Minimization with respect to s yields

$$Fs = S^T \lambda = S^T q$$

a relation which is directly obvious from (1) and (3). This and the equilibrium Eq. (2) are then solved together

$$\begin{bmatrix} F & S^T \\ S & 0 \end{bmatrix} \begin{bmatrix} s \\ -q \end{bmatrix} = \begin{bmatrix} 0 \\ p \end{bmatrix}. \tag{24}$$

The partioned matrix remains symmetrical, and is positive semi-definite.

The Gauss-Jordan algorithm

For structures made of assembled beams the determination of redundancies was traditionally done by *releases*, that is by cutting members or at least some of their load transmission capabilities, until the internal forces in the released structure could be determined by statics. This procedure which largely relies for its success on physical intuition becomes too slow, tedious and prone to error for the large arrays of more complicated finite elements. The automatic determination of all the self-stressing states by the computer itself was a considerable step towards salvation of the force method, it is due independently to *Denke* [12] and *Robinson* [18]. As will be shown the method used is still richer in the answers it provides, being capable of determining simultaneously the kinematical degrees of freedom that can play an important role in dynamic analysis, and the pseudo-inverse matrices of singular problems. The Jordan version of the Gauss elimination algorithm can be presented as follows when applied to the transpose of the structural kinematical matrix: a $\nu \times \nu$ non singular premultiplier P is determined step by step by selection of pivots to transform S into a matrix

$$PS = \begin{bmatrix} I_{\sigma-\xi} & Q \\ 0 & 0 \end{bmatrix} \tag{25}$$

having ϱ last rows of zeros (whereby ϱ is actually determined by the algorithm and consequently $\xi = \sigma - \nu + \varrho$) and where $I_{\sigma-\xi}$ is the identity matrix in $\sigma - \xi = \nu - \varrho$ dimensions, Q being a $(\sigma - \xi)\times\xi$ matrix. P itself is structured as

$$P = \begin{bmatrix} N & 0 \\ T & L \end{bmatrix}, \tag{26}$$

where N is a non singular $(\nu - \varrho)\times(\nu - \varrho)$ matrix, L is non singular $\varrho\times\varrho$ and lower triangular, T is $\varrho\times(\nu - \varrho)$. Problem (2) becomes equivalent to

$$s_{(1)} + Qs_{(2)} = Np_{(1)}, \tag{27}$$

$$0 = Tp_{(1)} + Lp_{(2)}, \tag{28}$$

where s and p have been split to suit the partitioning of P and PS. The homogeneous problem (13) has clearly, from (27), the solution $s_{(2)} = x$ arbitrary $s_{(1)} = -Qx$ and we can take

$$X = \begin{bmatrix} -Q \\ I_\xi \end{bmatrix}. \tag{29}$$

Equation (28) is also clearly the existence conditions (18) and we can take

$$R^T = (T\ L). \tag{30}$$

Finally the complete Eq. (27) compared to (20) yields

$$S^* = \begin{bmatrix} N & 0 \\ 0 & 0 \end{bmatrix} \tag{31}$$

and it is easily verified that by using in (19), (30) and

$$S^{T*} = S^{*T} = \begin{bmatrix} N^T & 0 \\ 0 & 0 \end{bmatrix} \tag{32}$$

Eq. (1) is correctly satisfied.

The Gauss-Jordan algorithm gives at least one computationally automatic answer to the inversion problems of (1) and (2) and consequently to the use of a Force program. It does not however provide the answer to the search of *compact* self-stressing states [6]. Compact self-stressing states are those which concern a minimum number of elements within the structure, their use minimizes coupling between self-stressing states and provides optimum conditioning of Eq. (23). Again, if the structure is dismantled and then reassembled by progressive addition of elements, the compact self-stressings can be discovered, the problem is to replace tedious manual processes of search by some improved version of the Gauss-Jordan algorithms. Research has proceeded in the direction of weighting of the matrix elements and pivot choices [25], proper sequencing of the elements is certainly one of the important factors.

The option between displacements methods and force methods depends at first glance on the relative sizes of the systems of Eqs. (21) or (23) and this, in turn, depends primarily on the topology of interelement connexions. There are other important factors to consider. The displacement method can bypass the construction of the structural kinematical matrix by setting up directly the structural stiffness matrix and this may be viewed as economical. On the other hand the structural kinematical matrix is independent of the elasticity characteristics of the elements, hence of those dimensions, thicknesses ... areas, that do not affect the geometry of interconnexions. A single Gauss-Jordan analysis remains valid in optimization studies with respect to such design parameters. This may give the force method a regain of favour.

Displacement and equilibrium models

Except for the early introduction of the concept of shear-panel to deal in a simple manner with stressed-skin construction, a scientific elaboration of finite element models did not start before 1960, about 10 years after matrix methods were already well established.

The variational principles of elasticity theory played an important role in the mathematical theory of finite element models.

Discretization of the displacement field by a Rayleigh-Ritz process was of course directly applicable to each element, the new problem was to establish the correct transitional conditions at interfaces. The simplest concept in theory, not always so easy to realize in practice, is to adopt the continuity condition (9) suggested by the continuum approach. This was certainly intuitive in the first real two-dimensional finite element of the literature [10].

The property of continuity of displacements at interfaces is called *conformity*. Accordingly, *conforming displacement models* have their displacement patterns defined by local (nodal) displacements on their boundaries in such a way that identification of nodal displacements on an interface produces complete interface continuity.

The continuous, piecewise differentiable approximation to the displacement field, provided by conforming displacement models, has the interesting property of underestimating the strain energy of a structure submitted to prescribed loads under zero displacements of supports. It overestimates it when the loads are removed but the supports are given prescribed displacements.

Another, rather intuitive, property that is usually incorporated in displacement models is the existence of rigid body modes amongst the displacement degrees of freedom. Although experience has shown that this requirement is not necessary for convergence of the displacement field with reduction in element size, its satisfaction has invariably produced better results.

The idea that another mathematical model of finite element could produce reverse bounds to the strain energy was investigated later [14, 17] on the basis of the variational principle of stresses, or minimum of the complementary energy. Indeed the work of *Friedrichs* [2], showing how the minimum problem of one functional could be transformed into the maximum of another, then that of *Prager* and *Synge* on approximations in function space [5, 11] generalizing early considerations by *Trefftz* [1], led immediately to this possibility. Its application to finite elements required again the correct type of transitional conditions at interfaces. The one that was directly suggested by continuum theory was the reciprocity condition (10). When it is enforced, finite elements are called *diffusive*.

Accordingly *diffusive equilibrium models* have their discretized stress field, in equilibrium with body loads, defined entirely by stress resultants (the generalized forces) of boundary surface tractions.

Reciprocity of the stress resultants entails complete reciprocity of the boundary surface tractions.

Building up a library of elements of both types, which conform to displacement and diffusive equilibrium, enables the same structure to be analyzed twice with complementary effects; it is the principle of dual analysis [27].

The exact and unknown value of the strain energy can be bracketed between the numerical estimates provided by each finite element analysis, thus one has available a quantitative estimate of *energy convergence*. Numerical brackets for local displacements and even for local stress values are obtainable in principle by introducing corresponding singularities in the problem [9]. This procedure has however up to now shown itself to be too expensive for current use.

Because the displacement method and conforming displacement models are both intimately related to the variational principle of displacements, or minimum total potential, and similarly with respect to the variational principle of stresses for the force method and diffusive equilibrium models, there has been an unfortunate tendency to obliterate the distinction between models and methods. Let it be stressed that diffusive equilibrium models are easily provided with a stiffness matrix so that their assemblages can be solved, and have been solved, using a displacement method. Conversely, self-stressing states of arrays of conforming displacement models can be found by Gauss-Jordan techniques and a force program used for their numerical evaluation. This allows dual analysis to be performed with only one type of computer software.

There is however the interesting observation that the topology of element connexions has a tendency to favour the force method for equilibrium models and the displacement method for displacement models, although the differences in problem sizes diminish with increasing sophistication of the models.

Strong and weak nodal connexions

Another important distinction brought forward by the nature of the models is the so-called strong or weak character of the generalized variables. In displacement models the nodal displacements are called strong because, once their numerical values are known, any local value of the displacement field is determined. On the contrary the equivalent nodal forces are weak because they are determined from virtual work considerations. If q denotes the boundary nodal displacements of an isolated element, the cartesian components of the displacement field of that element is representable in the form

$$\begin{bmatrix} u_1(x) \\ u_2(x) \\ u_3(x) \end{bmatrix} = u(x) = Q(x)\, q + B(x)\, b\,, \tag{33}$$

where $Q(x)$ is a matrix of assumed functions (generally low order polynomials) called *shaping functions* or weighting functions and $B(x)$ a matrix of *bubble functions*, so-called because they vanish on the boundary and do not consequently influence the motion of the element boundaries. The bubble coordinates b can obviously be determined by energy minimization at the element level, since they are unconnected to the neighboring elements. The generalized boundary forces g, conjugate to q, and the generalized body loads f, conjugate to the bubble coordinates, are defined by virtual work computation over the domain E of the elements and its boundary ∂E:

$$\int_E u^T \phi\, dV + \int_{\partial E} u^T t\, dS = q^T g + b^T f \tag{34}$$

with, as a consequence of (33),

$$g = \int_E Q^T \phi\, dV + \int_{\partial E} Q^T t\, dS\,, \tag{35}$$

$$f = \int_E B^T \phi\, dV\,. \tag{36}$$

Thus the generalized loads are weighted averages of the real body load and surface traction distributions. The knowledge of their numerical values determines the distributions only within equivalence classes.

Recognition of this weak character of generalized loads suggests at once that the stress information might be of the same nature. As a matter of fact the strong stress information deducible by differentiation of the displacement field through (4) and (7) is sometimes difficult to interpretate.

In correspondance with (35) and (36) it is easily demonstrated that the effect of discretizing displacements is to replace the local equilibrium Eqs. (5) and (6) by their weighted averages using the same weighting functions. Thus, as is well known, the Rayleigh-Ritz process becomes identical with a Galerkin process of weighted residuals of the equilibrium equations. This has led to the obvious extension of using the more general Galerkin approach with finite elements in field problems where no variational principle is known to exist.

In equilibrium models the situation is reversed. The local stresses are completely fixed by the generalized forces and consequently the body loads and surface tractions with which they are in equilibrium. The virtual work process now yields weighted averages for the conjugate displacements which become the weak variables. Perhaps the idea that from the kinematical viewpoint the elements are only weakly interconnected does not appeal to engineers, which might explain the relative dearth of equilibrium elements in present day literature. However it would seem that in the majority of applications the engineer is more concerned about stresses and that a certain fuzziness in the displacement information is no repealing disadvantage. While equilibrium equations are here exactly satisfied, the Galerkin type of weighted residual approach applies to the compatibility equations for strains.

Hybrid models

The Friedrichs transformations, when applied to elasticity theory furnish a host of more general types of variational principles. From the finite element viewpoint the liberation from transitional constraints like (8) and (9) by Lagrangian multipliers is of equal importance as the liberation from the strain displacement constraints (4) inside of the element. The different possible combinations lead to a bewildering variety of hybrid elements, only few of which have been experimented. In this connexion one should mention the contributions of *Pian* [16], *Jones* [15] and *Prager* [19].

Elements of hybrid character may possess interesting convergence properties, unfortunately they do no more provide bound estimates for the strain energy.

Plate bending models. Static-geometric analogies

The acceptance of the Kirchhoff-Love assumption in the bending theory of thin plates or shells has posed real difficulties in attempts to enforce conformity. Not only the transverse deflection of the plate but also its normal slope must be continuous at an interface. To obtain this it is necessary to construct a superelement, that is a small aggregate of elements whose outer boundary motion can be determined in terms of local deflections and slopes ensuring conformity [22].

It is characteristic of such elements that, at their corners, the curvature tensor is not single-valued. Several authors have produced conforming plate or shell bending elements of apparently simpler type. However they use higher order connexions, adding partly or in toto local values of the curvature tensor. Not only does this preclude loading modes associated with curvature discontinuities but it makes it impossible to connect rationally elements which are not coplanar. This limits severely their use in complex geometrical situations. By contrast, the plate bending diffusive type of equilibrium element is quite simple [20, 23].

Kirchhoff plate bending and plate stretching are known to be related by static geometric analogies. Exchanging the roles of stresses and strains, the equilibrium equations of bending go over into the compatibility equations of stretching and conversely. As a result the transverse deflection distribution of bending can serve as Airy function to give an equilibrium state in stretching. The cartesian component of displacement in stretching can serve as stress-functions to satisfy the bending equilibrium equations. The recognition that conformity and diffusitivity for finite elements also exchange their roles in the analogies [21] led directly to the concept of setting up diffusive equilibrium models from their analogues, that is discretizing directly the stress-functions. As a matter of fact the discretization of stress functions brings us back to the "Gegenstück-Verfahren" of *Trefftz*, where it was applied to the Prandtl torsion function. Because the volume equilibrium equations were automatically satisfied by the very existence and differentiability of the torsion function, *Trefftz* was able to show that an underestimate of the torsional rigidity occured when the torsion function satisfied simply the boundary conditions, that in fact satisfies surface equilibrium. His method for obtaining an overestimate is however too complicated. Instead of requiring satisfaction of the differential equation governing the torsion function, which is tantamount to requiring compatibility of strains, discretization of the conjugate warping function in the conjugate variational principle of displacements yields the same overestimates with much less effort.

Conclusions

In this brief survey, there are many topics that could not be touched upon by lack of time and space. The isoparametric elements developped mainly by the finite element group at Swansea [30]; they furnish conforming displacement models with curved boundaries. Elastodynamics, where the extension to a continuum of a variational principle of *Toupin* [8], the complementary energy analogue to Hamilton's principle, allows diffusive equilibrium elements to prove their efficiency as eigenvalue economizers [26]. The recent penetration of finite element procedures into geometrically non-linear elasticity problems and in the much wider field of material non-linearities. Finally the controversial subjects of convergence and finite differences versus finite element methods. Those are important aspects, where conjugate efforts by mathematicians and engineers are needed to bring in a far more mature state of the art some fundamental guidelines, that satisfy both mathematical rigour and physical understanding.

Lack of space is also responsible for the limitations in the list of references. A much wider cross-section of the enormous body of papers available can be found by consulting the references [18, 24, 28 and 29].

References

1. Trefftz, E.: Proc. 2nd Int. Cong. Appl. Mech. Zürich, 1927, 131.
2. Friedrichs, K.: Nachr. der Ges. d. Wiss. zu Göttingen, 1929, 13—20.
3. Courant, R., Hilbert, D.: Methoden der Mathematischen Physik IV. Die Grundtatsachen der Variationsrechnung, 139—233, Berlin: Springer 1931.
4. Ebner, H., Koeller, H.: Luftfahrtforschung 14 (1937) 12.
5. Prager, W., Synge, J. L.: Quart. Appl. Math. 5 (1947) 241—269.
6. de Veubeke, B. F.: Bull. 24 du Service Technique de l'Aéronautique, Bruxelles 1951.
7. Langefors, B.: J. Aer. Sci. 19 (1952) 451—458.
8. Toupin, R. A.: J. Appl. Mech., June (1952) 151—152.
9. Washizu, K.: J. Math. Phys. 32 (1953) 117—128.
10. Turner, M. J., Clough, R. W., Martin, M. C., Topp, L. J.: J. Aero. Sci. 23 (1956) 805—823.
11. Synge, J. L.: The hypercircle in mathematical physics, Cambridge: Cambridge University Press 1957.
12. Denke, P. H.: NASA T.N. D-1666, 1959.
13. Argyris, J. H., Kelsey, S.: Energy theorems and structural analysis, London: Butterworth 1960.
14. de Veubeke, B. F.: AGARDograph 72, Oxford/London: Pergamon Press, 1963, 165—201.
15. Jones, R. E.: AIAA J. 2 (1964) 821—826.
16. Pian, T. H. H.: AIAA J. 2 (1964) 1232—1336.
17. de Veubeke, B. F.: Displacement and equilibrium models in the finite element method, Ch. 9, Stress Analysis, ed. O. C. Zienkiewicz and G. Holister, New York: Wiley 1965.
18. Robinson, J.: Structural matrix analysis for the engineer, New York: Wiley 1966.
19. Prager, W.: Recent Progr. in Appl. Mech. The Folke Odqvist Volume, 1967, 463—474.
20. Morley, L. S. D.: J. Roy. Aero. Soc. 71 (1967) 715—719.
21. de Veubeke, B. F., Zienkiewicz, O. C.: J. Strain Analysis 2 (1967) 265—271.
22. de Veubeke, B. F.: Int. J. Solids and Str. 4 (1968) 95—108.
23. de Veubeke, B. F., Sander, G.: Int. J. Solids and Str. 4 (1968) 447—468.
24. Przemieniecki, J. S.: Theory of matrix structural analysis. New York: McGraw-Hill 1968.
25. Robinson, J., Haggenmacher, G. W.: Some new developments in matrix force analysis. Japan-US seminar on matrix methods of structural analysis and design, Tokyo 1969.
26. Geradin, M.: Proc. IUTAM Coll. on High Speed Computing of Elastic Structure Liège. Congrès et Colloques de l'Université de Liège, Place du XX Août 16 (1970) 589—623.
27. Sander, G.: Proc. IUTAM Coll. on High Speed Computing of Elastic Structures Liège, Congrès et Colloques de l'Université de Liège, Place du XX Août 16 (1970) 167— 207.
28. Zienkiewicz, O. C.: Appl. Mech. Rev. (1970) 249—256.
29. Zienkiewicz, O. C.: The finite element method in Engineering Science, New York: McGraw-Hill 1971.
30. Irons, B. M., Zienkiewicz, O. C.: Proc. Conf. Recent Advances in Stress Analysis, Roy. Aero. Soc. (1968).

Aquatic animal locomotion

Sir James Lighthill

Cambridge, U.K.

1. Introduction

To this Congress of Theoretical and Applied Mechanics I want to report about an aspect of Zoologically Applied Mechanics: namely, the Theoretical Mechanics of Interactions between an aqueous medium and the external surfaces of a totally immersed animal swimming through it.

Generally speaking, progress in Zoologically Applied Mechanics can be made only through close collaboration between a zoologist and a mechanician. The animal kingdom is vast, and the problem of discerning and describing what types of animal motion have in different parts of it been significant, above all in the evolutionary sense, is one for the zoologist alone. At the same time, study of the complicated interactions between those motions and a surrounding aqueous medium is one for the hydrodynamicist alone. It is therefore when a zoologist and a hydrodynamicist have got to know each other well enough to be able to talk together about the problems, and gradually to learn enough of each other's language so as to be able to communicate effectively, that collaborative progress involving hydrodynamically sound analysis of zoologically significant motions becomes possible.

Especially significant for aquatic animal locomotion was such a collaboration which started around 1950 between Sir *Geoffrey Taylor* with his enormous mechanics experience and Sir *James Gray* whose zoology school at Cambridge had for over 20 years been a prime source of knowledge, and of research workers, in almost every aspect of animal locomotion (*Gray* 1968). It was in fact on *aquatic* animal locomotion and its hydrodynamics that Taylor and Gray particularly talked, and this led to important papers by *Taylor* (1951, 1952a, 1952b) on cases of such motion both at very low Reynolds numbers and at high Reynolds numbers.

These, like all Sir Geoffrey Taylor's papers in hydrodynamics, were so interesting that various younger workers on reading them were impelled to initiate work in imitation or extension of them. I was one of those both in respect of my own work (*Lighthill* 1952) and that of my research student *Hancock* (1953). However, that work of ours at Manchester on motions at very low Reynolds numbers failed to be influential because we had not learned the art of communicating with zoologists: either to discover from them the problems of zoological significance or to indicate to them our mechanically significant conclusions. Fortunately, *G. J. Hancock* subsequently went to work with Sir *James Gray* and this collaboration produced some really significant results (*Gray* and *Hancock* 1955).

I personally did not meet *James Gray* until 1960 when I had the good fortune to spend a fascinating day with him discussing aquatic animal locomotion at high Reynolds number. This convinced me that it would be valuable to work out the application of hydrodynamical slender-body theory to fish swimming movements. I did this rather hastily and published it (*Lighthill* 1960) but I failed to maintain the collaboration with the Cambridge zoologists which would have been necessary to follow through the significance of the work, partly because duties as Director of the Royal Aircraft Establishment left little time for personal studies and partly because I still did not appreciate how essential it was to immerse myself in the zoological literature and the language of zoologists if I was to grasp the true nature of their problems or to indicate to them the meaning of my hydrodynamical investigations. Those were followed up by parallel studies in the application of two-dimensional aerofoil theory carried out with great hydrodynamical brilliance by Dr *T. Y. Wu* (1961) but again without clear identification of the areas of zoological relevance.

Much later, in 1968, I returned to the subject in response to a proposal by the editors of Annual Reviews of Fluid Mechanics to include a survey article on it by me in their first volume. By then I had at last found out what kind of closeness of communication was needed between biologists and hydrodynamicists, through several combined studies in the physiological fluid dynamics of circulation and respiration: studies which I have recently described in summary form (*Lighthill* 1972). This pertinent experience, together with a reading of the literature of aquatic animal locomotion which exposed the general failure of communication between the two bodies of expertise evident in most previous work, convinced me that if I was to help in clarifying the hydrodynamics of aquatic animal locomotion I must talk to zoologists and go on talking to them; read their works and go on reading them; study their collections (in museums and aquaria) and go on studying them!

Finally with a great deal of help not only from Sir *James Gray* but from his colleague Dr *R. Bainbridge* as well as from Dr *N. B. Marshall* of the British Museum of Natural History and Dr *Quentin Bone* of the Marine Biological Association at Plymouth, I produced (*Lighthill* 1969) a survey that included a paragraph as follows: "It is likely that most readers may possess specialised knowledge in either zoology or the mechanics of fluids, but not both. The survey is written as far as possible to be equally intelligible to either kind of reader." In fact I tried to produce, with the help of all that talking and reading and study of collections, a conducted tour through the animal kingdom in respect of the methods used for aquatic animal propulsion. It called upon voluminous work by zoologists on the data and their evolutionary interpretations, and attempted to relate to them such findings in the hydrodynamicists' papers as were of true zoological significance, identifying areas within the animal kingdom where each of those was specially relevant. That Review article tried to identify also the areas where further hydrodynamical researches would be particularly valuable.

Several of the researches there proposed have since been carried out and described in new papers of my own (*Lighthill* 1970, 1971), in new work by Dr *Wu* (1971a, 1971b) and in work by two of my new research colleagues Dr *J. R. Blake* and Dr *D. Weihs*. It is this new work which makes up the main subject matter of this lecture. It has been a pleasure to see how effective my colleagues Dr *Blake* and Dr *Weihs* have been in learning to collaborate with zoologists and to use their language; much quicker than I had been! I shall be quoting from their work, including some still unpublished material, with their much appreciated permission.

This has been work where a considerable amount of observational data by Sir *James Gray* and other British zoologists gathered over the past 40 years has for the first time been analysed by hydrodynamic theory. At the same time zoologists are beginning to publish new observational material which they are themselves analysing by the new hydrodynamical methods (see for example *Webb* 1971).

The plan of this lecture assumes then, as stated in my Abstract available in advance, that the material in my survey article in Annual Reviews of Fluid Mechanics (*Lighthill* 1969) may be taken as a starting point. Those of you who have not read this but would like to appreciate more clearly how what I tell you about the hydrodynamics today fits into the zoological context, may find out something about it in that Review article, which is readily accessible. In fact I must urge you to do so, since the material is given importance and significance only by its zoological context. I feel justified, however, in leaving it out in a lecture to this 13th Congress of Theoretical and Applied Mechanics, since including the zoology in detail would leave no time for any mechanics, and also since it is my duty to concentrate on work done since the 12th Congress; in other words, on the new work that has appeared since that Review article was written.

2. The subdivisions of hydrodynamic theory relevant to aquatic animal locomotion

I shall mention, however, before describing that work in detail, the main kinds of distinction that exist between the different types of hydrodynamic theory relevant to aquatic animal locomotion (Fig. 1). The distinction whose importance is most obvious to any hydrodynamicist is that already mentioned, based on the value of the Reynolds number. The locomotion of animals whose size is of the order of 1 mm or less is a phenomenon where inertial effects are negligible because Reynolds number is very low (of order 1 or less). Theories of them have to find distributions of pressure and viscous stress satisfying *equilibrium* equations, leading to equations for velocities in a biharmonic form. The hydrodynamicist's problems for larger animals are, however, harder since fluid inertia is important as well as fluid stresses, and the equations of motion are fundamentally nonlinear.

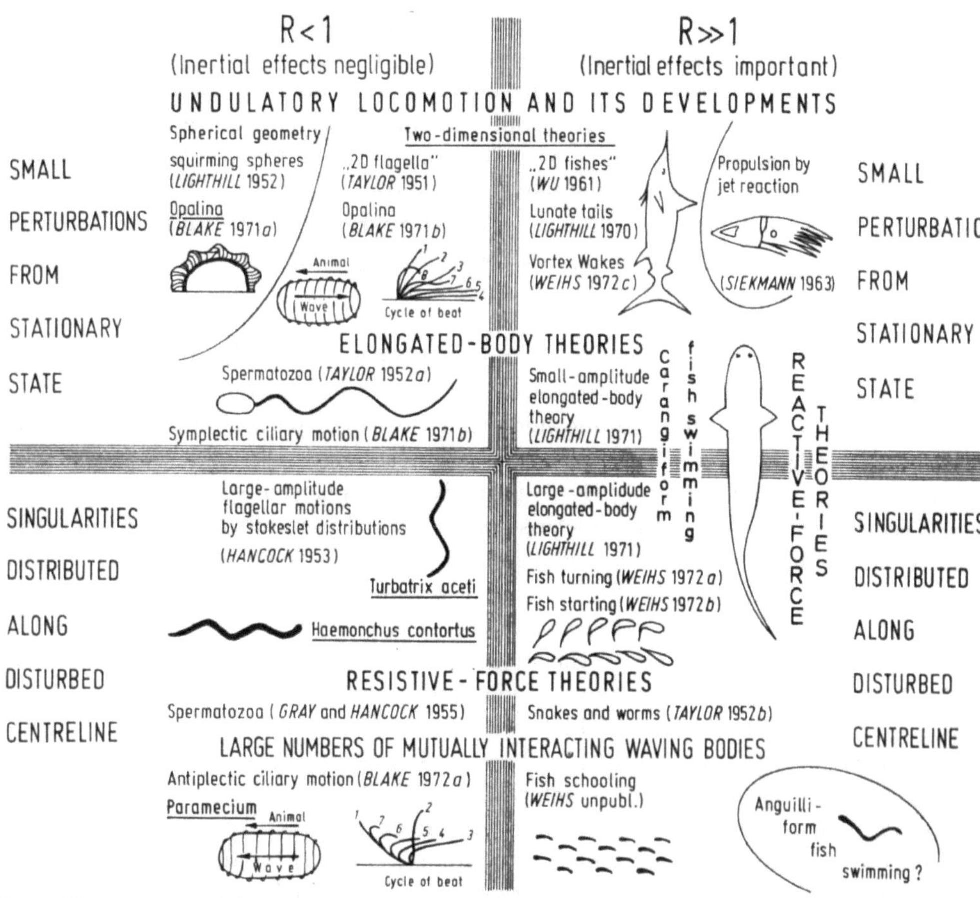

Fig. 1. Illustrating subdivisions of hydrodynamic theory relevant to aquatic animal locomotion.

One consequence of this distinction is a further subdivision that appears only at high Reynolds number: between on the one hand those types of aquatic animal propulsion on which I shall here concentrate, types that can be viewed as various kinds of developments from the *undulatory* methods of propulsion normal at very low Reynolds number, and on the other hand quite different types of propulsion using the inertial reaction of a jet. Little research work on these jet-propulsion mechanisms seems to have emerged since my 1969 Review. They are in many groups of invertebrates available as an escape reaction which depends on a sudden muscular contraction expelling a high-velocity jet of water so that the animal shoots off in the opposite direction. Much less widespread are the animals using such contraction in a regular periodic sequence to produce a "pulse-jet" type of propulsion, although I might remark that those include not only various squids and other animals noted by *Lighthill* (1969) but also certain bivalve molluscs that can make rather startling progress by repeated sharp closures of the gap between their two shells!

Within the topic of undulatory methods of propulsion and developments from them, there are two more distinctions, each important *both* at very low Reynolds number and at high Reynolds number: distinctions apparent already in the papers of the early nineteenfifties. They are both concerned with the kinds of method used for handling the complex geometry of the animals' motions, and the fact that those motions involve three space-dimensions and one time dimension.

As an instance of one of the distinctions, *Taylor* (1952a) when he analysed the propulsion of a microscopic single-celled organism by a waving tail or flagellum modelled it as an elongated body of revolution, thus improving on an earlier study (*Taylor* 1951) with a much cruder two-dimensional model in which the third dimension of the space in which the animal swims is completely suppressed! Most subsequent theories have been based either on two-dimensional analysis or on some sort of elongated-body theory: the latter is certainly to be preferred when the body really is elongated and moves in three dimensions. Some aquatic animals, however, do (as we shall see) possess propulsive surfaces of genuinely high aspect-ratio (like the marlin's tail in Fig. 1) to which an approximate two-dimensional analysis is quite appropriate.

Both these early papers possessed at the same time one *common* feature: they further approximated the effects of the animal's complex undulatory motions by regarding them as a small perturbation of the stationary stretched-straight state. That sort of small-perturbation approach has subsequently been much used also at high Reynolds number: especially by *Lighthill* (1960) and *Wu* (1961) for elongated-body theory and two-dimensional theory respectively. Obviously, however, it is not ideal since rather large perturbations are involved in most really interesting swimming motions.

For example the Taylor theory made the velocity of a swimming spermatozoon increase indefinitely as the square of the amplitude of its swimming motions. My research student *Hancock* (1953) used a quite different method from the small-perturbation approach: one in which the fluid velocity field was modelled as that of a line-distribution of singularities but the singularities were distibuted along the actual distorted position of the animal's centreline (line of cross-section centres). We gave at this time a special name to the most important kind of singularity that characterizes these solutions of the Stokes equations of low-Reynolds-number flow: we called this singularity (first used by *Stokes* 1851) the stokeslet.

Hancock's solutions involved distributions of stokeslets and ordinary source doublets along the instantaneous centre-line of the organism at each instant. For determining their strengths he derived an integral equation, which it was feasible to solve numerically although very arduous (with the aids then available), and whose solutions gave a reasonable levelling-off of spermatozoon speed as the amplitude of swimming movements became large. He applied his work also to nematode worms, of which two are illustrated in Fig. 1. However the whole analysis was mathematically far too subtle and intricate to have any chance of appealing to zoologists and the calculations inconveniently complex, perhaps because the level of accuracy attempted was too high in relation to the accuracy of the observations that can be made in such a field.

It was the collaboration between Dr *Hancock* and Sir *James Gray* that was finally successful in hammering out a method of adequate accuracy which is easily comprehensible to all concerned and gives quantitative results relatively easily (*Gray* and *Hancock* 1953). The approximation was relatively crude in that in a suitably chosen form of the integral equation the integral was suppressed altogether, but the results derived are in reasonable agreement with observations on spermatozoa. Physically, the approximation meant regarding the interaction between a short segment of the elongated animal and the surrounding aqueous medium as consisting of a localised resistive force depending only upon the relative velocity between them, but with components of that relative velocity tangential to the line of centres resisted less than components normal to the line of centres.

The general idea of that approximation had been suggested earlier by *Taylor* (1952b) in work in motions at high Reynolds number of animals like worms and snakes whose body shapes are not very well adapted to aquatic locomotion. He similarly proposed the use of simple resistive forces with tangential or normal component depending on the tangential or normal component of relative velocity respectively. The only difference was that the dependence was linear at very low Reynolds number but quadratic at high Reynolds number, while perhaps at intermediate Reynolds number it ought to be given an intermediate form. I propose to call all theories "resistive" if they make the local instantaneous force per unit length of animal depend, either linearly or nonlinearly, on the instantaneous velocity of that section of animal relative to the water.

Sir *James Gray* subsequently developed the resistive-force theories still further and showed in his book "Animal Locomotion" (*Gray* 1968) what a wide range of phenomena they will adequately account for. Here I shall take just a few minutes to expound them in by far their simplest form when the resistance relationships (whatever the Reynolds number) *are* approximated by linear laws and the cosine of the angle between the animal's line of centres and its direction of motion is approximated by 1.

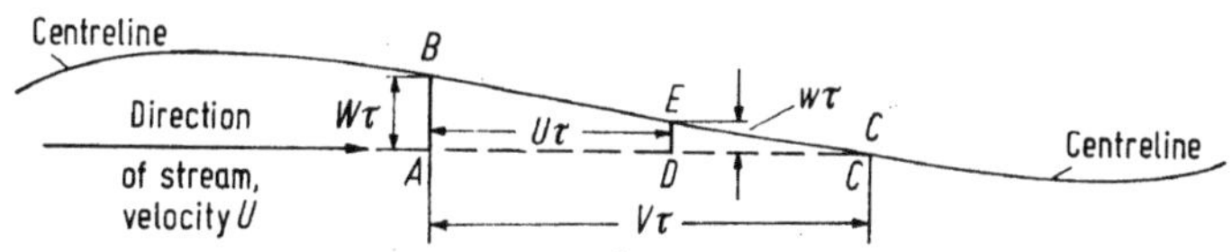

Fig. 2. The connection between lateral velocities W relative to animal and w relative to water in undulatory motion.

Fig. 2 shows a short segment BEC of the centreline of an elongated animal supposed to be performing undulatory motions from side to side in a stream of velocity U in which those swimming motions just stop it slipping backwards. (Thus U is its mean swimming speed relative to the stream). The undulation passes backwards at a different speed V. The lateral velocity is W so that in time τ the centre-line moves sideways a distance $W\tau$ from A to B, while the wave moves backwards a distance $V\tau$ from A to C and therefore C is at the lateral position occupied by A a time τ earlier. During the time interval τ the water moves a distance $U\tau$ so that relative to the water the centre-line moves a distance DE $= w\tau$, where w, its normal velocity relative to the water, is

$$w = W\left(1 - \frac{U}{V}\right). \qquad (1)$$

This conclusion is as in Eq. (2) and Fig. 5 of *Lighthill* (1969); note also an alternative, analytical way of appreciating it using $h(x, t)$ for lateral displacement as in *Lighthill* (1960) and the equations

$$W = \partial h/\partial t, \quad w = \partial h/\partial t + U\, \partial h/\partial x, \quad 0 = \partial h/\partial t + V\, \partial h/\partial x. \qquad (2)$$

We take force per unit length of centre-line resisting this normal velocity w as $K_N w$, where K_N is a normal-force coefficient, and multiply that by the slope W/V (see Fig. 2) to obtain the corresponding component of forward *thrust* per unit length $K_N w(W/V)$. The same slope means however that the lateral velocity W includes a tangential component $W(W/V)$ which with K_T as the tangential-force coefficient *reduces* the thrust per unit length by $K_T W(W/V)$ to the net amount

$$P = (K_N w - K_T W)\,(W/V). \qquad (3)$$

This thrust per unit length (3) averaged along the whole centreline must balance the drag per unit length $K_T U$.

Equation (3) is a formula of which two uses can be made. The useful thrusting work UP done per unit length can be compared with the rate $W(K_N w)$ of mechanical energy expenditure by the lateral motions with velocity W against the resisting normal force $K_N w$ per unit length. The ratio is an efficiency

$$\eta = \frac{UP}{W(K_N w)} = \left(1 - \frac{w}{W}\right)\left(1 - r\frac{W}{w}\right), \qquad (4)$$

where

$$r = K_T/K_N \qquad (5)$$

is the ratio of tangential to normal force coefficients. Maximum efficiency is achieved when

$$w/W = r^{1/2}, \quad \text{giving} \quad U/V = 1 - r^{1/2} \quad \text{and} \quad \eta = (1 - r^{1/2})^2. \tag{6}$$

This agrees with observations that at a Reynolds number of 1 or less when r is close to 0.5 (*Gray* and *Hancock* 1955) we have the ratio U/V of animal speed to wave speed around 0.3; note that maximum efficiencies are low (around 0.1) at these Reynolds numbers. A grass-snake (*Tropidonotus natrix*) with Reynolds number around 10^6 might however have r as low as 0.1, and thus U/V around 0.7 again as observed (*Gray* 1939); maximum efficiency might then be greater (around 0.5). Alternatively, as *Taylor* (1952b) pointed out, a polychaete worm with numerous parapodial attachments as in Fig. 3 might have tangential-force coefficient greater than normal-force coefficient, e.g., $r = 1.5$. This would give U/V around -0.2, again as observed; such polychaete worms adopt a vigorous forward-travelling undulation, although probably with maximum efficiency less than 0.05.

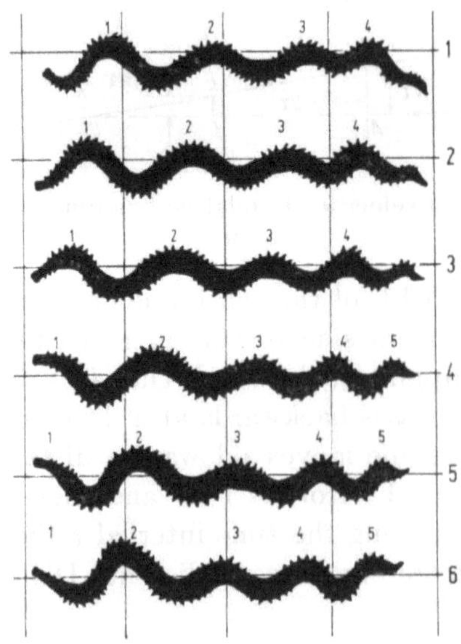

Fig. 3. Successive positions, at intervals of 0.05 s, of a polychaete worm (*Nereis diversicolor*) swimming to the left. Numbers attached to moving "crests" show that the wave travels in the same direction as the worm, although considerably faster (*Gray* 1939).

Furthermore, in all these cases the resulting swimming speed U can be deduced from a balance between the mean of the thrust per unit length P and the drag per unit length $K_T U$. This gives

$$U^2 = \overline{W^2}\left(1 - \frac{w}{W}\right)\left(\frac{w}{W} - r\right)\frac{1}{r}, \tag{7}$$

and maximum swimming speed is achieved when

$$w/W = 1/2(1 + r), \quad \text{giving} \quad U^2/\overline{W^2} = (1 - r)^2/4r. \tag{8}$$

This varies from 2 for $r = 0.1$ (grass-snake) through 0.12 for $r = 0.5$ (nematode) to 0.04 for $r = 1.5$ (polychaete worm) in general agreement with the trend of ratios of swimming speed squared to mean square lateral velocity. Actually in the maximum-efficiency condition (6) the swimming speed is given by

$$U^2/\overline{W^2} = (1 - r^{1/2})^2/r^{1/2} \tag{9}$$

which is not much less than the maximum given by (8) even when $r = 0.1$, perhaps explaining why the higher value of w/W is not observed in such a case.

These very simple inferences from the resistive theory in the most straightforward approximation of all can be improved upon, either by not replacing cosines by 1 or by not linearising resistance laws, and yet involve calculations that are still quite manageable. The method is, however, essentially a method for organisms that propel themselves by means relatively wasteful of energy, as a hydrodynamicist will appreciate from the relatively low values of "Froude efficiency" quoted.

Considerable improvements in the efficiency of the undulatory mode at high Reynolds number were achieved by the appearance in the fishes and other vertebrates of a deep, laterally compressed tail. Those advantages accrued partly because such a shape of cross-section was well adapted to interact with the water, *not* primarily through a resistive force depending on instantaneous relative velocity, but rather with a *reactive* force depending on rate of change of that velocity. At high Reynolds number such a reactive force arises from the rate of change of momentum of water due to the rate of change of either the velocity or the virtual mass of surfaces in contact with it.

Anguilliform swimming, that is the swimming of creatures like eels that maintain a markedly undulatory waveform all the way from head to tail so that considerably more than a whole wavelength is recognizably present, but where the mean swimming thrust is augmented by the mean rate of shedding of momentum into the wake at the laterally compressed tail, can be regarded as utilising both resistive and reactive forces. The development of a different mode of swimming, the carangiform swimming which appears in a wide variety of groups of fishes and other vertebrates, may however derive from the fact that it makes more *exclusive* use of the reactive forces which inherently can operate at a better Froude efficiency.

Carangiform swimming retains that feature of a backward-moving wave which requires the oscillations of the posterior end to lag behind those of an anterior section, but the amplitude is far more variable, being practically zero over the front half or even two-thirds of the length of the fish, and rising very steeply towards the posterior end. A fish using such swimming motions to move forwards produces very sudden accelerations in the water that it "flicks into motion", and the resulting thrust is derived principally from the reactive rate of change of momentum.

The reactive theory for swimming of animals with elongated bodies was given by *Lighthill* (1960, 1969, 1970) in a series of papers based on a small-perturbation expansion about the stationary state. This theory is a development of the slender-body theory in aerodynamics, but "elongated body" seems to be a phrase more readily understood in the world of zoology. The special feature that arises in the *zoological* application to carangiform motion is that any unbalanced sideforces in the posterior region of the fish may produce wasteful, undesired "recoil" motions of sideslip of the fish's mass centre, and yaw about it. *Lighthill* (1969, 1970) showed that two morphological features which are almost universally found together with carangiform motion greatly minimise these undesired recoil motions; first, the body *depth* is large for an extensive region around the mass centre so that any recoil effects involve a large virtual mass of water and are consequently reduced in amplitude, and secondly there is a major reduction in body depth and hence of virtual mass just before the caudal fin, which is where especially large sideforces would otherwise result from the large spatial gradient in oscillation amplitude.

The ultimate development in speed and efficiency of aquatic animal locomotion was found in the combination of carangiform movement with a tailfin of high aspect-ratio as in certain scombroid fishes, in certain sharks, and in cetaceans. This crescent-moon-shaped or "lunate" tail with its well rounded leading edge has many of the excellent qualities of aircraft wings (though it possesses their *horizontal* arrangement and oscillates up and down only in the cetaceans, that is, whales and dolphins, and of course is vertical and oscillates from side to side in the relevant sharks and scombroid fishes). The thrust and efficiency of these animals' posterior propulsive units can be approximately analysed on a two-dimensional theory, which here comes into its own (*Lighthill* 1969, section 8).

In the last two sections of this lecture I shall be reporting upon two developments in the theory of carangiform locomotion. First, for the slender or elongated animals, there is the extension (*Lighthill* 1971) of the "reactive" theory so that it no longer relies on a small-perturbation expansion but works like the developed forms of the resistive theory by placing the necessary singularities along the actual disturbed position of the animals' centreline. I developed this large-amplitude elongated-body theory to be able to improve the accuracy of thrust prediction in normal carangiform swimming, but Dr *Weihs* (1972a, 1972b) went on to add more accuracy to the method and then apply it to problems where no small-perturbation approach would have been possible: problems of turning and starting. (The sequence of 10 pictures in Fig. 1 shows a goldfish *Cyprinus auratus simultaneously* starting and turning!) Swimming fishes, indeed, have succeeded in developing refined methods of turning through (say) 90° in a distance of half their length with practically no loss of kinetic energy, and it is interesting to find out what movements they make to achieve this and why those movements are mechanically effective.

Finally I shall give some recent developments in two-dimensional theory that are relevant to lunate tails and their vortex wakes. They include studies of my own (*Lighthill* 1970) on why it should have been this lunate form of tail on which strains so different as whales, sharks and tuna should have "converged" in a common search for speed and propulsive efficency and also some recent studies on vortex wakes by Dr *Weihs* (1972c) which are relevant to wakes of individual fishes with lunate tails and also to the problems of assemblages of slender fishes that are grouped together in schooling patterns. Here we have an extension of animal locomotion theory at high Reynolds number to a study of large numbers of mutually interacting waving bodies.

It is this last-named feature which will be paralleled in the middle section of my lecture, where I describe the important recent extensions by Dr *J. R. Blake* of theories of motion of individual flagella at very low Reynolds number to the case of ciliary propulsion, where the movements of large numbers of attached cilia in something close to synchronism produce relative movement between an animal and the surrounding fluid. Two different types of theory were found appropriate for this study of large numbers of mutually interacting waving bodies at low Reynolds numbers: one derived from small-perturbation theory applied to symplectic ciliary motion as in the organism *Opalina,* and one derived from large-amplitude resistive-force theory applied to antiplectic ciliary motion as in *Paramecium.*

3. Hydrodynamics of ciliary propulsion

The Ciliophora are a class of single-celled creatures which propel themselves by the nearly synchronised motion of large numbers of hair-like organalles attached to the cell surface. These organelles, the cilia, wave not in exact synchronism but in so-called *metachronism* with the phase of one row of cilia lagging by a certain amount behind that of an adjacent row. The effect is that of a wave of movements passing across the bank of cilia. The motion of each individual cilium consists of a fast "effective stroke" in the direction opposite to the animal's motion, with the cilium almost rigid during this effective stroke, and a slower "recovery stroke" with the cilium relatively limp. The metachronism is called symplectic when the wave propagates approximately in the same direction as the effective stroke. This is reminiscent of the undulatory mode of propulsion in general since the wave is then moving in the direction opposite to the animal's motion.

A typical animal exhibiting symplectic metachronism is *Opalina,* a disk-shaped multinucleate protozoon about 200 μm wide and about 20 μm thick. Individual cilia are 10 μm long or a little more and are arranged in rows 3 μm apart in which the cilia may be as close together as 0.3 μm. The tips of the cilia never become far apart from one another. This is particularly because during the fast effective stroke which might separate a cilium from its neighbours it is moving towards a cilium at an earlier stage in the cycle and away from one at a later stage. The general appearance is of "a field of waving corn" with a clear undulating envelope embracing the tips of all the cilia. It is as if a disk-shaped body with a highly flexible surface were propelling itself by passing undulations in that surface backwards so as to exercise the necessary traction on the fluid.

Amusingly enough that was exactly the hydrodynamical problem I had tackled 20 years ago (*Lighthill* 1952) using spherical geometry to simplify the necessary mechanics. At that time in my naive way I had thought "Ah well! Taylor is working out how cells can propel themselves if they have tails; but let's see if they could get along at all without tails by some sort of squirming motion ..." whereas of course the only sensible thing to do was to talk to zoologists and find out how protozoans without a flagellum do propel themselves! After my 1969 Review it was obvious that it was to symplectic motion in Ciliophora that such a method of analysis should really be applied, and accordingly the first piece of work that Blake undertook after I suggested as a subject for his PhD thesis the hydrodynamics of ciliary propulsion was the improvement and extension of my old spherical-geometry theory and its application to motions like those of *Opalina* (*Blake* 1971a).

He quickly had good success with this, and an encouraging feature of his results was that, with amplitudes of motion corresponding to ciliary lengths just quoted, the organism was predicted as able to achieve a velocity of about 100 μm/s, close to those actually achieved. The rate of working needed could be expressed, interestingly enough, as an efficiency of under 0.03 when compared with the rate of working by a constant external force propelling the organism at the same speed. (This corresponds to 0.1 at low Reynolds number estimated in Section 2 for a waving elongated body.) Furthermore, not all waveforms were calculated to produce such large velocities, or even positive velocities of swimming, though it was encouraging that those with relatively pointed crests and shallow troughs like those observed in nature were predicted to be the effective ones.

Another model used by *Blake* (1971b) for *Opalina* was an infinite plane sheet carrying a moving envelope of ciliary tips, his reasoning being that the real finite disc might have properties intermediate between the finite sphere and the infinite plane sheet. Comparing the two models with the behaviour of the animal he came to the conclusion that "planeness is more important than finiteness", showing clearly that cilia attached to a plane surface parallel to the direction of desired motion were more effective than cilia arranged all round a spherical shape. In nature most organisms employing ciliary propulsion do seem to have the cilia attached to surfaces parallel to the direction of motion. Note that the predicted speed of swimming would not be expected to change as the disk area increased to infinity since the thrust and the drag would equally increase in proportion to area, and so the infinite-sheet case may really be useful.

The plane model has several other advantages. It enables a deeper study of the relative merits of different waveforms, showing that a given swimming speed is achieved in symplectic motion with lowest rate of working if the tips of the cilia trace out the same kind of pattern as do the surface particles in a water wave. The motion of the tips comes fairly close to this in *Opalina*. Another advantage of the plane model is that the small-perturbation expansion can be carried to higher orders, and *Blake* (1971b, 1972b) was able to show by those means that even with a wave of the maximum possible amplitude corresponding to cusped crests the ratio of swimming speed to wave velocity (what I called U/V in section 2) could not exceed about 0.5. This agrees with maximum values observed in symplectic metachronism.

I leave that subject now with a mere note of a third model used by *Blake* (1971b) based on a body of revolution, and pass to antiplectic metachronism. This is different in important ways: the swimming velocities produced are larger relative to either the wave velocity or the square root of the rate of working. Primarily this is because the individual cilium has greater freedom of movement during its effective stroke than in the symplectic case. Now in its effective stroke it is moving towards a cilium at a later stage in the cycle which is already stretched forward away from it, and its neighbour behind is similarly stretched backward away from it (being at an earlier stage). Thus, the critical act by each cilium is performed in splendid isolation, and a model which regards the tips of the cilia as if they were a flexible solid surface ceases to have any potential value. *Paramecium*, another multinucleate protozoan, is a typical animal exhibiting antiplectic metachronism, being an elongated prolate spheroid about 250 μm long with cilia 10 μm long and a metachronal wavelength of 20 μm, but its typical velocity is 1000 μm/s, an order of magnitude greater than for *Opalina* (and U/V for *Paramecium* even exceeds 1).

Blake (1972a) developed a completely new type of theory for antiplectic metachronism although building it up on the sound foundations laid by *Gray* and *Hancock* (1955). Thus he regarded each cilium as a distribution of stokelets along its centreline corresponding to the distribution of the force of interaction between the cilium and the fluid. He had however to make two important steps forward beyond the Gray and Hancock model.

First, he had to consider the effect of each stokeslet in the presence of the nearby solid wall of the organism. He could treat this problem as that of a stokeslet near an infinite plane wall and showed (*Blake* 1971c) that the solution to the Stokes flow equations in this case may elegantly be expressed in terms of an image system. The image system is different for stokeslets corresponding to force components parallel to and perpendicular to the boundary, as illustrated in Figs. 4 and 5 respectively. In the first case the image system consists of an equal and opposite stokeslet plus a stokes-doublet (consisting of two equal and opposite stokeslets oriented as

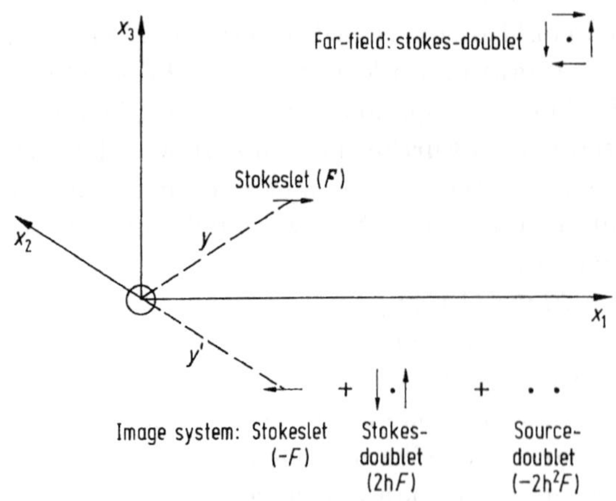

Fig. 4. Image system for a stokeslet of strength F directed parallel to, and at a distance h from, the solid plane wall $x_3 = 0$ (*Blake* 1971c).

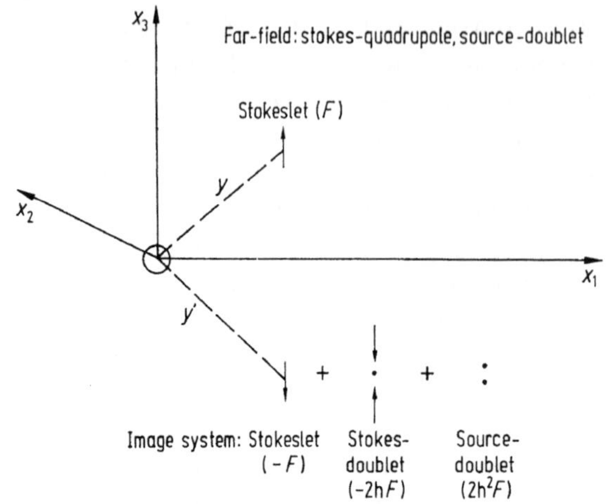

Fig. 5. Image system for a stokeslet of strength F directed perpendicular to, and at a distance h from, the solid plane wall $x_3 = 0$ (*Blake* 1971c).

shown) plus a source doublet. The far field is substantial, being of the combined stokes-doublet form shown with velocities falling off like r^{-2}. In the second case the type of stokes-doublet in the image system is different and cancels out the equivalent stokes-doublet provided by the original stokeslet and its equal and opposite image. The far-field in consequence is much weaker, falling off like r^{-3}. For ciliary propulsion this means that relative motion between the water and the wall is contributed to principally by ciliary forces *tangential* to the boundary rather than by forces perpendicular to it. Blake subsequently found out that the image system had previously been discovered by the great *Lorentz* (1896), but noone seems to have made any effective use of it till Blake came along!

Secondly, he had to take into account that each cilium was acting in the presence of a velocity field brought about by the actions of all of the other cilia. This statistical element of the effect of the large numbers of cilia involved was obviously the hardest aspect of the problem.

The method of solution chosen was similar to that used in atomic theory to study the structure of atoms of high atomic number; that is, containing large number of electrons. For this purpose the atomic theorists use a "self-consistent field" approximation where the radial distribution of electrostatic field resulting from the electronic distribution is regarded as the key unknown function, to be determined by the fact the Schrödinger's equation, with that radial field distribution, must possess such solutions for the distributions of electrons as are *consistent* with the electrostatic field postulated. Exactly such a self-consistent field model was found to give good results in Blake's problem.

There, the distribution of the velocity field parallel to the wall as a function of distance from it is regarded as the key unknown function $U(x_3)$; this represents velocities relative to the wall and so its limit as $x_3 \to \infty$ is the swimming velocity of the organism while its lower values as x_3 reduces to its wall value of zero describe the shearing that occurs in the ciliary sublayer. The unknown function $U(x_3)$ is now determined as a self-consistent field as follows. *Gray* and *Hancock* (1955) give the force with which each cilium acts on the fluid in terms of the velocity of the cilium *relative* to the fluid, here taken as moving with the velocity $(U(x_3), 0, 0)$. The effect of this force is represented by a stokeslet together with its image system discussed earlier and the velocity fields of all the cilia are thus written down and summed up, the sum being simplified by use of Poisson's summation formula. Then the self-consistency condition requires of course that the velocity field so obtained is the one that we first thought of! This leads to an integral equation for $U(x_3)$ which lends itself to numerical solution by iterative means.

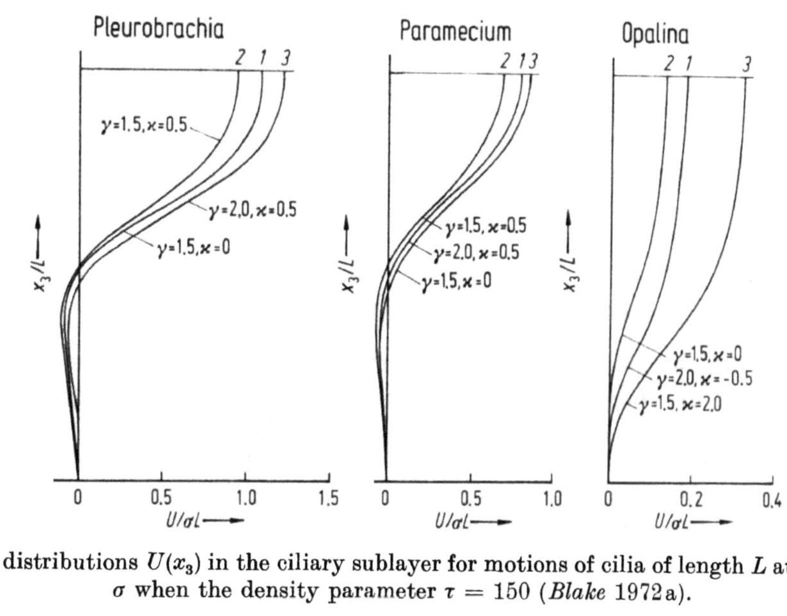

Fig. 6. Velocity distributions $U(x_3)$ in the ciliary sublayer for motions of cilia of length L at radian frequency σ when the density parameter $\tau = 150$ (*Blake* 1972a).

Figure 6 shows some results of such calculations with x_3 nondimensionalised in terms of the length L of a cilium and U nondimensionalised in terms of the product of L with the radian frequency σ. The method was carried out for various values of a frequency parameter $\varkappa$, which is the ratio of σL to the wave velocity c, and of γ which is the ratio of normal-force coefficient to tangential-force coefficient so that it is the reciprocal of the r used in section 2. Note the much higher values of velocity produced in *Paramecium* with its antiplectic, than in *Opalina* with its symplectic, metachronism. The third animal *Pleurobrachia* is one of the ctenophores, small organisms related to the jellyfishes which propel themselves by eight "combs" of cilia, waving in antiplectic metachronism.

Blake (1972a) uses one more parameter τ, proportional to the density of the cilia per unit area. The data in Fig. 6 are for $\tau = 150$, whereas Fig. 7 shows how U^∞, the velocity of fluid relative to the organism outside the ciliary sublayer, varies as a function of this density parameter τ. Time does not permit reference to any more results from this paper which include calculations of many of the quantities that the zoologists most need to know: particularly, how the forces and bending moments applied by the animal to a cilium vary during the cycle as well as of overall rate of working. I will remark, however, that elsewhere *Blake* (1972b) shows good agreement, *for symplectic metachronism only*, between the envelope method and this much more comprehensive method of analysis.

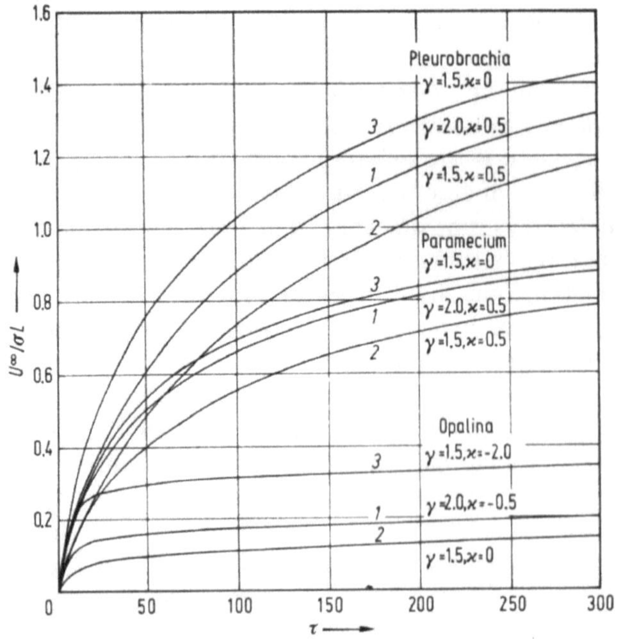

Fig. 7. Variation of $\lim_{x_3 \to \infty} U(x_3) = U^\infty$, the swimming speed of the organism relative to fluid at a large distance, as a function of the density parameter τ (*Blake* 1972a).

4. Large-amplitude elongated-body theory

Now I must return to motions at high Reynolds number and reactive force theories, and describe my recent work (*Lighthill* 1971) in developing these so that they can be applied to motions of elongated bodies with arbitrary amplitude. This new method is based, like the successful low-Reynolds-number methods, on distributing along the disturbed centreline those singularities that are relevant: here, the momentum carrying dipoles (source-doublets). The only difference is that since the present method is applied exclusively to fishes I shall call the centreline the spinal column or backbone!

After many searches for the right approach, I recognized finally that what I wanted could be obtained by use of the following three principles:

(i) Water momentum near a section of fish is in a direction perpendicular to the backbone and has magnitude equal to the virtual mass, m per unit length, times the component w of fish velocity in that direction.

(ii) Thrust can be obtained by considering rate of change of momentum within a volume V enclosing the fish whose boundary at each instant includes a flat surface Π perpendicular to the caudal fin through its posterior end.

(iii) In the momentum balance it is necessary to take into account transfer of momentum across Π not only by convection but also by the action of the resultant $mw^2/2$ of the pressures generated by the motions within the plane Π.

One advantage of considering only rate of change of momentum within the volume V which excludes the complicated vortex wake is that all the complex changes of momentum due to movements of vortices in that wake, each under the influence of the others, can be ignored. A simple hydrodynamical theorem equates the resultant pressure force on such a plane, due to motions in the plane with virtual mass m per unit length, to the expression $mw^2/2$.

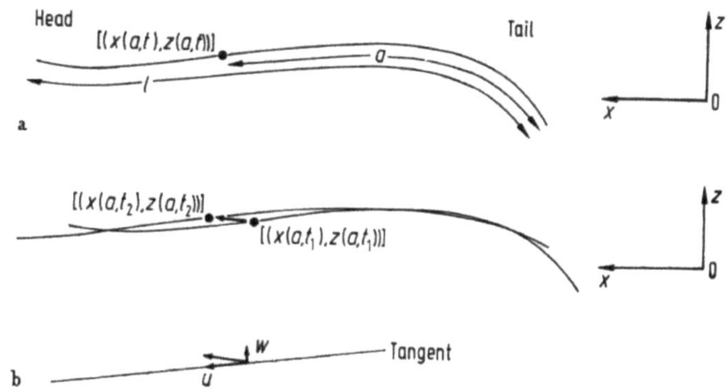

Fig. 8. Notation used by *Lighthill* (1971) to describe large-amplitude motions of fishes in a frame of reference (x, y, z) that is stationary relative to fluid at a large distance.

To turn these principles into a method for calculation, we use a coordinate system as defined in Fig. 8, where a point on the spinal column is defined by a Lagrangian coordinate a, measured along the spinal column from its tail. The y-axis is vertical and the x- and z-axes horizontal and otherwise arbitrary except that in cases of steady swimming we take the x-axis as the direction of the mean swimming velocity. The velocity component perpendicular to the spinal column is w as already stated while the tangential component is u.

The spinal column is effectively inextensible which in analytical terms means

$$\left(\frac{\partial x}{\partial a}\right)^2 + \left(\frac{\partial z}{\partial a}\right)^2 = 1. \tag{10}$$

Thus unit vectors tangential and perpendicular to the backbone in the horizontal plane $y = 0$ have components

$$\left(\frac{\partial x}{\partial a}, \frac{\partial z}{\partial a}\right) \quad \text{and} \quad \left(-\frac{\partial z}{\partial a}, \frac{\partial x}{\partial a}\right) \tag{11}$$

respectively, from which we deduce that

$$u = \frac{\partial x}{\partial t}\frac{\partial x}{\partial a} + \frac{\partial z}{\partial t}\frac{\partial z}{\partial a}, \qquad w = \frac{\partial z}{\partial t}\frac{\partial x}{\partial a} - \frac{\partial x}{\partial t}\frac{\partial z}{\partial a}. \tag{12}$$

Now we can write down the equation for the instantaneous reactive force (P, Q) due to the momentum mw per unit length brought instantaneously into being by the fish's perpendicular motions. This reactive force (P, Q) must on the average provide a thrust to balance the viscous drag resisting the fish's tangential motions. In other words the whole idea of this method is that tangential will motions be treated resistively but perpendicular motions reactively.

This reactive part (P, Q) of the total force, then, is determined by the equation

$$\frac{d}{dt}\int_0^l mw\left(-\frac{\partial z}{\partial a}, \frac{\partial x}{\partial a}\right) da = \left[-umw\left(-\frac{\partial z}{\partial a}, \frac{\partial x}{\partial a}\right) + \frac{1}{2}mw^2\left(\frac{\partial x}{\partial a}, \frac{\partial z}{\partial a}\right)\right]_{a=0} - (P, Q). \tag{13}$$

The left-hand side is the rate of change, in a volume V that includes the fish but not the wake, of a total water momentum whose value per unit length has magnitude mw and direction [see (11)] perpendicular to the backbone. The right-hand side adds up three elements contributing to this rate of change: (i) loss by convection across the plane Π with velocity u; (ii) gain by action of the pressure force $mw^2/2$ acting across Π; (iii) loss due to the water acting with a force (P, Q) on the fish.

A minor rearrangement using (12) enables us to write (13) as

$$(P, Q) = \left[mw \left(\frac{\partial z}{\partial t}, -\frac{\partial x}{\partial t} \right) - \frac{1}{2} mw^2 \left(\frac{\partial x}{\partial a}, \frac{\partial z}{\partial a} \right) \right]_{a=0} - \frac{d}{dt} \int_0^l mw \left(-\frac{\partial z}{\partial a}, \frac{\partial x}{\partial a} \right) da. \tag{14}$$

In the case of periodic swimming movements the mean thrust $\overline{P}$ can be calculated using the fact that the mean of the time-derivative must be zero as

$$\overline{P} = \overline{\left[mw \frac{\partial z}{\partial t} - \frac{1}{2} mw^2 \frac{\partial x}{\partial a} \right]_{a=0}}, \tag{15}$$

where the bar denotes a mean. The time-derivative term in (14) is important, however, because without it the thrust would fluctuate enormously, falling to zero whenever $w = 0$ at $a = 0$, whereas *Lighthill* (1971) shows that the time-derivative term acts to smooth out those fluctuations. The expression for Q is important also in leading to the conclusions regarding recoil mentioned in section 2.

The same method gives an expression for the mean rate of energy wastage

$$\overline{T} = \overline{\left[\frac{1}{2} mw^2 u \right]_{a=0}} \tag{16}$$

by shedding of kinetic energy $mw^2/2$ per unit length at velocity u across Π into the vortex wake. Thus all the terms in the Froude efficiency

$$\eta = \overline{U}\overline{P}/(\overline{U}\overline{P} + \overline{T}) \tag{17}$$

including the mean swimming speed

$$\overline{U} = \overline{\left[u \frac{\partial x}{\partial a} \right]_{a=0}} \tag{18}$$

can be written in terms of mean values of quantities evaluated at the posterior end $a = 0$. This makes it easy to infer how Froude efficiencies in the 80% region can be expected, in sharp contrast to the values around 50% (see Section 2) expected at high Reynolds number from a resistive interaction with perpendicular motions.

In balancing these reactively produced thrusts against the resistive reactions to tangential motions, as *Lighthill* (1971) does in detail for the dace *Leuciscus*, it must be remembered that boundary layers tend to remain very thin towards the back of a swimming fish due to the lateral movements bringing the fish in contact with continually new fluid, and accordingly the drags are greater once the fish is swimming than when it is gliding. The relatively large amplitudes of tail motion exhibited by most swimming fishes may thus be interpreted in terms of the need to produce reactive thrusts sufficient to balance the greatly enhanced viscous drag that appears once lateral movements commence. Steady swimming with *small* amplitudes of tail movement may be rather futile and indeed less efficient as a means of achieving a moderate rate of progress than production of thrust by those large amplitude movements, interspersed with periods of gliding, such as are commonly observed among fishes in tanks.

These theoretical approaches were found suitable also for the work of *Weihs* (1972a, 1972b) on fish turning and fish starting, of which I shall describe the turning studies here. Sir *James Gray* had taken some excellent films of fishes turning in 1933 but they were only analysed from a hydrodynamical point of view when Dr *Weihs* came to work with me in Cambridge in 1971. *Weihs* made an improvement to my method in the details of his treatment of fins: thus, he used the virtual-mass approach for forces associated with the fish body but used slightly improved aerodynamic coefficients for forces on fins which he felt were not *sufficiently* "slender" for the fish studied to be treated accurately enough by "slender-body theory". He was able to calculate the forces and moments at each of the three stages in the observed turning process and show how they relate to the observed motion of the fish's mass-centre and to the rotation of the fish about it.

Figure 9 shows a sequence of shots from Sir James Gray's film of a turning rudd *Cyprinus erythrophthalmus*. The fish's success in making the turn in an extremely short distance with little kinetic energy loss is most noticeable. It seems to be attributable according to Dr Weihs' analysis to the fact that the earliest big movement is a *nose turn* and that vigorous activity with the tail fin only follows after this to complete the job. The benefits of this *alternate* use of fore and aft control surfaces seem very clear.

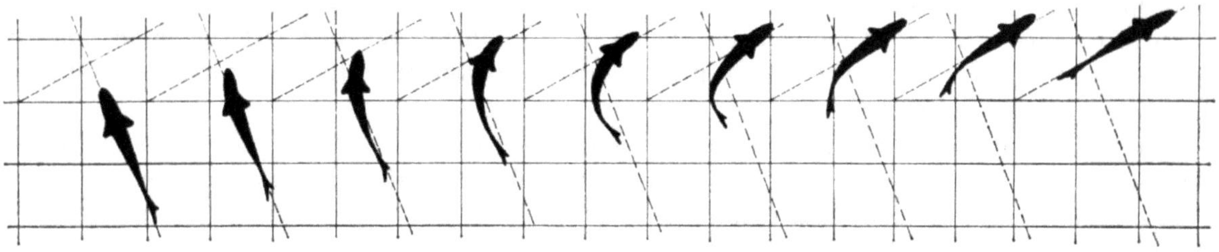

Fig. 9. Successive positions at intervals of 0.04 s (time sequence from left to right) of a rudd (*Cyprinus erythrophthalmus*) making a turn to the right (*Weihs* 1972a).

Figure 10 shows successive positions of the spinal column of the turning rudd measured up carefully from the photographs. The circles indicate successive positions of the mass centre. The first stage of the turning process (shots 1, 2 and 3) is one in which the mass centre continues in a straight line because a substantial turn of the nose to the right produces momentum-change forces (like the d/dt terms in Eq. (14)) that are balanced by the inherently more effective momentum-shedding forces from fins due to smaller movements of the posterior part of the body to the left. The total side force is zero and the net effect is a growing couple permitting rotation about the mass centre.

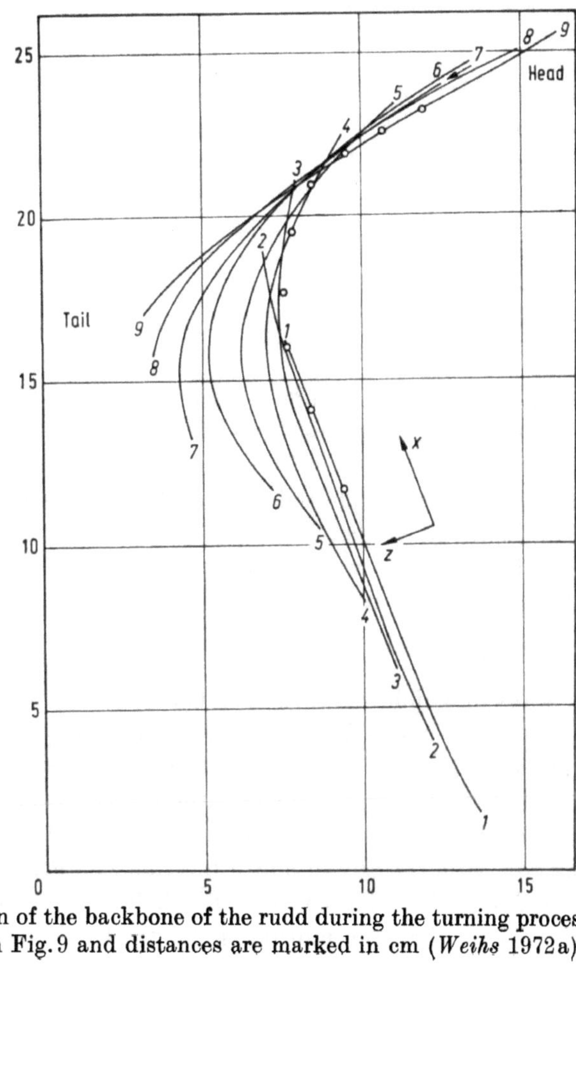

Fig. 10. Tracings of position of the backbone of the rudd during the turning process. Numbers refer to frames in Fig. 9 and distances are marked in cm (*Weihs* 1972a).

The second stage of the turning process (shots 4, 5 and 6) is one in which the mass centre itself follows a highly curved path, indicating a substantial force component towards the centre of turn. This is contributed by the increasing dominance of the forces provided by the substantial turning of the posterior parts, and the forces calculated from that observed turning agree with the observed centripetal acceleration.

The third stage (shots 7, 8 and 9) is one in which the tail is doing all the work and the hydrodynamic "lift" (really sideforce) on the caudal fin is accelerating the fish *along* its new path to balance any loss of momentum due to friction during the turn. At the same time the moment of that force about the mass centre is such as to stop the rotation of the fish and leave it moving in a straight line.

All the detailed calculations of forces and moments made by Dr *Weihs* bore out this analysis except that the calculated moment during stage 2 seemed enough to bring about a much *greater* retardation of rate of turn than seemed necessary. Then the noticed an important feature of the film record, namely that at the onset of stage 2 the fish extends and twists its right pectoral fin to exert a brake action on the right-hand side, reducing speed and increasing rate of turn. This explains the rather enhanced rate of turn at the beginning of stage 2 and the fact that larger moments than would otherwise be expected are required to damp it down. The minor speed reduction reduces the radius of turn generated by a given sideforce in proportion to the square of the speed reduction and then that final tail flick gets back most of the lost speed relatively easily! The airbrakelike action of that pectoral fin on the side towards which the fish is turning comes as a sudden movement between stage 1 (when nose turning dominates the scene) and stage 2 with its extensive tail deflection and nose straightening. Finally stage 3 with a sharp tail straightening completes the process. It is intriguing to speculate whether ships will ever do as well!

5. Vortex wakes

Time permits now only a very brief discussion of my final topic: vortex wakes, which appropriately come last because they are what aquatic animal locomotion leaves behind! At this series of Congresses it is a pleasure to begin this last discussion with a contribution by *Karman* and *Burgers* (1934): Figure 11 is adapted in fact from a diagram in their famous joint article in the series "Aerodynamic Theory".

Fig. 11. Illustrating vortices cast off by the caudal-fin trailing edge as the fish moves to the left, and also the jet-like streamline pattern induced by those vortices (*Lighthill* 1969, after *Karman* and *Burgers* 1934).

We see here Karman and Burgers emphasizing that the famous Karman vortex street can also work in reverse: it can, with the sense of rotation of all the vortices altered, be responsible for thrust instead of drag. Between the vortices with that altered sense of rotation a jet-like motion is induced, and the momentum it carries is related to the thrust on the fish ... So after all there is rather more relationship between the undulatory mode and jet propulsion than first appeared!

I shall mention a few ways in which this simple picture might be improved on and two ways in which the idea might have zoological application. The picture as it stands shows a purely two-dimensional motion so that it would at first sight be most nearly applicable to the wakes of deep high-aspect-ratio lunate tails (see section 2), although Dr *Weihs* suggests that also fishes swimming in formation one above another might produce such two-dimensional wakes and would be able in this way to assure the advantage of high aspect-ratio just when it suited them!

Be that as it may, *Weihs* (1971c) has studied the behaviour of these hitherto neglected thrust-generating vortex streets when the motion is strictly two-dimensional, and found that they have interesting tendencies to change their form with distance downstream, becoming wider by a factor around 1.5 and somewhat altering also their longitudinal spacing.

Somewhere, of course, this behaviour must be modified by interaction with the drag wake though for a fish with lunate tail that would be more concentrated and differently located (around the locus of the fin *centre*). The changes it may bring to the vorticity distribution must, however, at some downstream location, reduce to zero the momentum carried by the wake, since the average values of thrust and drag cancel.

I want to mention now three-dimensional effects and their possible relation to why the fastest marine animals from widely different lines of evolution should have converged on to the lunate tail as their method of propulsion. My detailed mathematical analysis of its merits by purely two-dimensional theory (*Lighthill* 1970) indicated merely that something like a *straight* trailing edge would be optimal: that is, a leading edge convex forwards accompanying a straight trailing edge. This suggests that any advantage of the trailing edge itself being somewhat convex forwards must stem from three-dimensional considerations.

Such three-dimensional considerations involve taking streamwise vorticity shed by a finite-span lunate tail into account alongside the cross-stream vorticity shown in Fig. 11. They then indicate that Fig. 11 must be viewed as a section by a horizontal plane of a sequence of vortex rings that are batted backwards diagonally alternately to the right and to the left by the animal. Actually the typical frequency parameters used by animals with lunate tails are such that those would be approximately *circular* rings, and evidently the curved trailing edge may assist in forming such circular rings, which in general are very good devices for carrying as much momentum as possible in relation to their energy.

Initially the vortex rings would move diagonally backwards but when the effect of the drag wake had mingled sufficiently with them they would be left with only a lateral motion. Dr *McCutcheon* mentioned to me that he had seen very recently such a pattern of laterally moving vortex rings forming the wake behind a scombroid fish. I continue to like the vortex-ring argument for lunate-tail optimality and am also continuing to seek ways of checking it more satisfactorily.

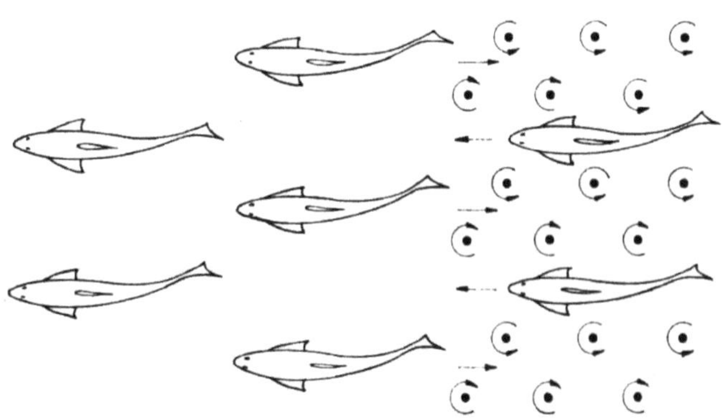

Fig. 12. Diamond-shaped fish schooling pattern.

Finally, I come to fish schooling. I mentioned already that the wake in Fig. 11 might be closely two-dimensional if several fishes were swimming in formation one above the other and that it would be to their advantage in respect of the power needed to swim at a given speed if they did so. It is also clear that fishes swimming diagonally behind the fish shown would particularly benefit from its vortices helping to drag them forward. This suggests a diamond pattern of fishes (Fig. 12) as having a particular advantage in respect of reduced power consumption to swim at a given speed, and Dr *Weihs* in unpublished work has been calculating the advantages which are in fact obtained.

There is a certain body of observational evidence that such diamond-shaped lattice patterns are in fact taken up by fish schools when moving fast, and the question arises whether *passive* forces bring the pattern into play or whether very elaborate control mechanisms act to maintain the lattice pattern. Dr *Weihs* can see certain mechanisms whereby passive forces might possibly be effective. I have jokingly suggested to him that the whole phenomenon may indeed be like the way in which the disordered molecules of a liquid form into a crystal lattice when the temperature is lowered: in other words, as the speed of the fish school increases, the *energy* advantage of the fishes dropping into this diamond-shaped lattice formation is suddenly able to outweigh the loss of entropy!!!

References

Blake, J. R.: J. Fluid. Mech. 46 (1971a) 199—208.
Blake, J. R.: J. Fluid Mech. 49 (1971b) 209—222.
Blake, J. R.: Proc. Camb. Phil. Soc. 70 (1971c) 303—310.
Blake, J. R.: J. Fluid Mech. (1972a) 1—23.
Blake, J. R.: Ph. D. thesis, University of Cambridge 1972b.
Gray, J.: J. Exptl. Biol. 16 (1939) 9—17.
Gray, J.: Animal Locomotion. London: Weidenfeld & Nicolson (1968).
Gray, J., Hancock, G. J.: J. Exptl. Biol. 32 (1955) 802—814.
Hancock, G. J.: Proc. Roy. Soc. A 217 (1953) 96—121.
Karman, T., Burgers, J. M.: General Aerodynamic Theory: Perfect Fluids, vol. 2 of Aerodynamic Theory (ed. Durand, W. F.). Leipzig: Springer 1934.
Lighthill, M. J.: Comm. Pure & Appl. Math. 5 (1952) 109—118.
Lighthill, M. J.: J. Fluid Mech. 9 (1960) 305—317.
Lighthill, M. J.: Ann. Rev. Fluid Mech. 1 (1969) 413—446.
Lighthill, M. J.: J. Fluid Mech. 44 (1970) 265—301.
Lighthill, M. J.: Proc. Roy. Soc. B 179 (1971) 125—138.
Lighthill, M. J.: J. Fluid Mech. 52 (1972) 475—497.
Lorentz, H. A.: Zittingsverlag Akad. v. Wet. 5 (1896) 168—187.
Siekmann, J.: J. Fluid Mech. 15 (1963) 399—418.
Stokes, G. G.: Trans. Camb. Phil. Soc. 9, pt. II (1851) 8—106.
Taylor, G. I.: Proc. Roy. Soc. A 209 (1951) 447—461.
Taylor, G. I.: Proc. Roy. Soc. A 211 (1952a) 225—239.
Taylor, G. I.: Proc. Roy. Soc. A 214 (1952b) 158—183.
Webb, P. W.: J. Exptl. Biol. 55 (1971) 489—540.
Weihs, D.: Proc. Roy. Soc. B 182 (1972a) 59—72.
Weihs, D.: submitted to 1st Internat. Congr. Biorheology (1972b).
Weihs, D.: J. Fluid Mech. 54 (1972c) 679—691.
Wu, T. Y.: J. Fluid Mech. 10 (1961) 321—344.
Wu, T. Y.: J. Fluid Mech. 46 (1971a) 337—355.
Wu, T. Y.: J. Fluid Mech. 46 (1971b) 521—568.

Perspective in phenomenological approach to the problem of fracture

V. V. Novozhilov

Leningrad, USSR

We come to know about fracture in our childhood when breaking toys, pottery and glass; already at that time we get a qualitative idea about cracks and their treacherous ability to rapidly extend. At that age we get also an idea about slow fractures due to gradual accumulation of damages and defects when wearing out shoes and clothes. As to the quantitative data on fracture, they have been collected every year and every hour for more than 100 years in a large number of scientific and industrial laboratories. The corresponding experimental information is being systematically published in the form of tables and graphs.

Thus, fracture is a well-known phenomenon, studied in detail by means of experimentation, but still, its theoretical description is far from being perfect. This is due to the fact that the fracture mechanism, though it seems to be simple, is very complex in reality, and it embraces an enormous range on the scale of dimensions. It begins on the scale of atoms and ends being observable visually.

The theory of fracture as well as the problems of plastic deformation (both being closely connected with each other) are the objects of the solid state physics and mechanics of continua. The first one tries to explain the facts taking into account the discreteness of solid body structures, that is their atomic structure.

The second one describes the facts using, instead of interatomic forces and distances, the averaged characteristics stresses and strains the relation between which, i.e., the equations of state of solid body, are to be determined from experiment.

The achievements of the solid state physics during the last several decades are very impressive. Various failures in atomic lattices have been revealed and classified and their fundamental role in the processes of plastic deformation and fracture ascertained. By means of theoretical considerations as well as experimentation the most important characteristics of failures have been studied, including their formation, elimination and interaction. But the panorama of phenomena found out during these studies was so complicated that, at the present time, we cannot even think of developing a statistical theory capable of quantitative determination of plastic strains and fracture in solid bodies. This is especially so because the information on structural defects and laws of their interactions is incomplete and not always reliable.

But the technical progress cannot stop to wait until all the buttons on the jacket of the theory of dislocations are sewed on. If it is impossible now to construct a theory of deformation and fracture of solid bodies starting, so to say, from the front door, then we have to start from the back door which is called the phenomenological method, and this means that we are going over to the concepts of the mechanics of continua.

The history of physics shows that the theories which are statistical in nature, at first usually pass through the phenomenological stage before they become really statistical. A classical example is the Boyle-Mariotte-Gay-Lussac law which had been established and had been existing for a long time as a phenomenological law; and it is only later that the parameters involved in it (pressure and temperature) were given a statistical interpretation. This example proves that phenomenological theories should not be opposed (as is, sadly, done now and again) against the theories that in a direct way take into account the atomic structure of matter.

At a certain level of knowledge this phenomenological approach is inevitable. The development of the solid state physics will, undoubtedly, not disprove the results obtained in the field of plasticity and strength using the approach of the mechanics of continua, but rather will help to explain and develop them just as the statistical gas theory did not cancel the Boyle-Mariotte theory but gave it a deeper interpretation and allowed it to be made more accurate.

Though at present solid state physics is not able to provide a quantitative description of plastic deformation and fracture phenomena, nevertheless, even now, it exerts a useful influence upon the development of the phenomenological theory of these processes contributing to the explanation of the formula structures and the interpretation of the parameters involved.

Although my paper deals with some problems related to the development of the phenomenological theory of fracture, at the beginning it is necessary to say some words about the theory of plasticity because the plastic deformation and the fracture processes are closely connected.

The modern theory of plasticity is a bulky construction, consisting of numerous variants. This is due to the above mentioned physical complexity of the plastic deformation mechanism. All the theoretical constructions that describe it are approximate and become more and more bulky as the requirements in accuracy increase. Because of this, an inductive way of development is caracteristic for the theory of plasticity, with the variants becoming more and more complicated and the possibilities thereof more and more extensive. This is why there are so many variants. If the aim of the theory of plasticity be the calculation of residual deformation in structures, then this aim may be considered as having been long since achieved because from the point of view of an engineer, the simplest variants of both the flow theory and deformational theory, intended for the case of proportional loading, are accurate enough. Nevertheless, the theory of plasticity continues to develop in the direction of complication of the stress-plastic deformation relationships by taking into account the second order terms. The purpose of doing this, though it seems to be useless from the practical point of view, is that by studying minor corrections in the equations of state one can get an indirect information about microstresses, the distribution of which is of crucial importance at the initial stage of the fracture process, i.e. at the stage of failure accumulation [1].

Microstresses exist in all the solid bodies, especially in polycrystals. The original cause for that lies in microheterogeneity and microanisotropy in the structure of solid bodies. Any deformation, and most of all the plastic one, greatly affects the field of microstresses. The plastic deformation is known to occur mostly along slip bands which take up a relatively small part of the body volume, whereas the major part of the body continues to be elastic even beyond the yield point. Due to the essential heterogeneity of the plastic deformation, it is followed by the appearence of a random field of microdeformations of turbulent nature.

The experimental comparison of the work expended on the plastic deformation with the heat energy thereby released [2] shows that the elastic energy of microdeformation is not only comparable, in value, to the elastic energy of the average deformation but that, in the case of well-expressed plasticity, it can also exceed it. Hence we may gain some idea of the significance of microdeformations and the associated microstresses. Where microstresses reach maximal values, micropores and microcracks are formed in the body. While the plastic deformation is going on, they increase both in number and size. This process, known as failure accumulation or, occasionally, as plastic dilatation, is the first stage of fracture. This stage terminates when, through merging of some microfailures, a macrocrack appears in the body. The second stage consists of the extension of the crack and terminates with the division of the body into parts.

In the case of brittle fracture, the first stage can be neglected, the strength of a body being defined by initial failures. But with fractures such as the fatigue fracture or the fracture caused by creep, the part played by the first stage is so important that in evaluating the limiting number of cycles, or fracture time, the fracture process, by and large, can be sometimes represented, to a first approximation, as failure accumulation.

In what follows, the ideas which form the basis of the phenomenological approach to the first stage are considered. Again, an analogy between the theory of failure accumulation and the theory of plasticity will be pointed out. The basis for the analogy stated is irreversibility of both the plastic deformation process and the fracture process, the two being thus governed not only by the running values of the parameters involved but also by the whole of the pre-history of changes in the parameters.

On the face of it, the latter statement is at variance with the fact that presently existing strength criteria (to be found in the textbooks on strength of materials) completely disregard the history of loading a body. All these criteria are formulated as particular algebraic equations relating the stress components at the moment of fracture. The reason for that is that the above criteria apply to a quite definite path of loading (this, however, is usually not stated) that is, to simple loading where all the components of the stress tensor increase proportionally to one and the same parameter.

With simple loading the deformational theory of plasticity which describes the residual deformation with functional relationships is known to be valid. In the general case of complex loading, however, where all the relations between the components of the stress tensor undergo changes with loading, the relationship between stress and plastic deformation is no longer just functional but assumes the form of differential equations where the argument is the length of the trajectory of plastic deformation. These equations are non-holonomic and can be integrated only when the path of loading is known. This general theory (usually referred to as the flow theory) comprises, as a special case, the deformational theory. It is natural to expect that, with complex loading (in the sense specified above), another irreversible process, namely that of failure accumulation, is also to be described with the non-holonomic differential equations rather than with functional relationships.

In the theory of cyclic loading, i.e., the kind of complex loading studied experimentally most thoroughly at the present time, such an approach, though in implicit form, has long been employed and won general recognition. It is the relationship between the parameters which control the process of failure accumulation being assumed differential that the familiar principle of linear summation of fatigue failure is based upon.

In looking for the earliest works employing this approach one should begin with the paper of *Arvid Palmgren* [3], published in 1924, where the theory of wear of ball bearings, taking into account the history of machine service, was proposed. In 1937 *B. F. Langer* [4] took over this approach into the theory of fatigue strength (see also [5]), and in 1952 it spread into the creep theory where it was used in estimating the fracture time when temperature varied [6]. In 1958 *L. M. Kachanov* applied an analogous idea to estimate the influence of stress level upon fracture time [7].

In essence, in all the works cited above the fracture or wear condition is taken in the following form:

$$\int_0^{\lambda_0} F\, d\lambda = 1, \tag{1}$$

where λ is a positive monotonically increasing parameter characterizing the extension of the irreversible process in question; λ_0 is the critical value of the parameter corresponding to the moment of completion of the process; $F > 0$ is a function of physical parameters which characterize the process.

If we now introduce a monotonically increasing auxiliary function

$$V = \int_0^{\lambda} F\, d\lambda \tag{2}$$

to be a measure of failure, then the integral (1) falls into two parts, the first being the equation describing the irreversible process according to the history of changes in the parameters involved:

$$dV/d\lambda = F \tag{3}$$

and the second being the condition of completion of the process:

$$V(\lambda) = 1 \tag{4}$$

which is the equation to define the critical value $\lambda = \lambda_0$.

In the problem dealing with the wear of bearings it is the total angle of rotation (or the number of rotations) reckoned from the beginning of the bearing's operation till the moment under consideration that is assumed to be λ. In the problem dealing with fracture caused by creep λ is the time reckoned from the onset of loading. In the problem dealing with fracture of an elastic-plastic body it is natural to take as λ the invariant introduced by *F. Odqvist*:

$$\lambda = \int_0^{\varepsilon_{ij}^p} \sqrt{d\varepsilon_{ij}^p \, d\varepsilon_{ij}^p} \tag{5}$$

which is usually referred to as the length of the trajectory of plastic deformation. The arguments for the function F are to be chosen on the basis of the available physical concepts of failure accumulation, the form of the function itself being defined from experiment.

We shall now outline the corresponding considerations as applied to the special case of deformation of a solid body. Where creep may be ignored, time being thus eliminated from the set of parameters characterizing the process, it is the length of the trajectory of plastic deformation (Eq. (5)) which should be taken as λ. Here changes in λ will result in the following simultaneous processes:

Loading: This is described by the stress tensor σ_{ij} whose course of changes is assumed to be given. This process defines all the rest.

Macroscopic deformation: This is defined by the elastic strain tensor ε_{ij}^e and the plastic strain tensor ε_{ij}^p. Both these tensors change with the stress tensor according to the equations of the theory of elasticity and the theory of plasticity.

Microscopic deformation: This is described by the fields of centered random microstress and microstrain tensors, their characteristics being dependent upon the macroscopic deformations.

Failure accumulation: This is described by the measure of failure, V, dependent upon the parameters of all the above given processes but mainly upon microstresses, their peak values causing formation and extension of micropores and microcracks.

The influence of microstresses upon failure accumulation is certain to be factor Number One; taking into account this factor, however, presents difficulties because of the lack of a statistical theory of plasticity of polycrystals at the present time. The situation, though, is not hopeless; for not only are microstresses affected by macroscopic plastic deformation but, as it was pointed out above, in turn they also affect the stress-strain relationship. This manifests itself in the form of corrections in the equation of state of the solid body. Of these corrections the most appreciable one relates to the so called Bauschinger effect, the gist of which consists in the fact that the material subjected to strengthening undergoes softening in the inverse direction of loading.

To the first approximation, this effect can be described by the translational theory of flow, the equations of which are of the form:

$$T_0 \, d\varepsilon_{ij}^p / d\lambda = T_{ij} = \sigma_{ij}' - s_{ij}, \tag{6}$$

where σ_{ij}' stands for the deviator of the stress tensor and s_{ij} stands for the tensor defining the position of the center of the boundary of the elastic region shifting with plastic deformation:

$$T_0 = \sqrt{T_{ij} \cdot T_{ij}} = \text{const}. \tag{7}$$

The tensor s_{ij}, with the dimension of stress, characterizes the influence of microstresses upon the macroscopic plastic deformation ε_{ij}^p [8]. Now, perhaps nothing more can be said for certain about this tensor for so far its relation to statistical characteristics of microstresses is still unclear. There can be no question, however, that it does express the macroscopic manifestation of these stresses.

Equation (6) has been studied experimentally fairly well and s_{ij} (to be termed below the microstress tensor) has been found to be a deviator depending, chiefly, upon plastic deformation.

Where e_p (the magnitude of the plastic deformation tensor) is less than a certain value e_p^*, called "the saturation threshold" of the Bauschinger effect [9], there, for many materials, the formula

$$s_{ij} \approx g \cdot \varepsilon_{ij}^p, \quad g = \text{const} \tag{8}$$

holds. Above the saturation threshold the character of the s_{ij} vs. ε_{ij}^p relationship sharply changes, the rise in microstresses with plastic deformation substantially slowing down. The experimentally obtained curve that is typical for the dependence of the magnitude of the microstress tensor, s, upon the magnitude of the plastic deformation tensor, e_p, is shown in Fig. 1. The bend seen on the curve physically denotes that beyond the saturation threshold there begins either a vigorous formation of new slip bands resulting in the diminuition of heterogeneity of plastic deformation, or intense microcracking leading to relaxation of elastic microdeformation, or both of these processes take place.

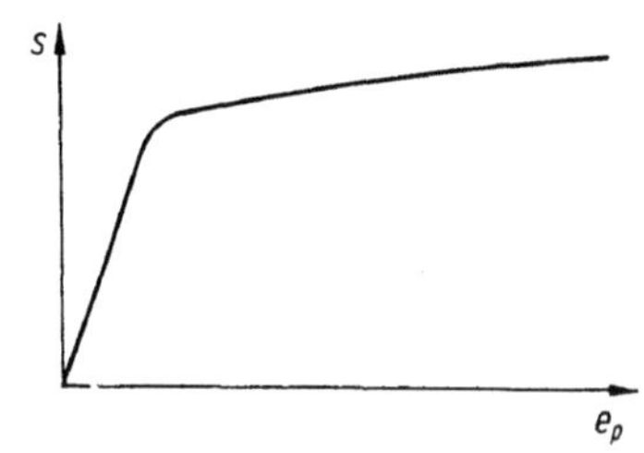

Fig. 1. Microstress s as function of plastic deformation e_p.

After this digression into the field of the theory of plasticity we are coming back to the equation of failure accumulation in order to define more concretely the right-hand side by taking into account the above considerations. In accordance with the idea of microstresses being the original cause for plastic dilatation, it is natural to regard the right-hand side as a function of the invariants of the tensor s_{ij} characterizing the Bauschinger effect and, thereby, microstresses.

O. G. Rybakina has pointed out [10] that in this case the good agreement with a wide range of experiments is provided by the hypothesis that states the rate of failure accumulation to be proportional to the magnitude of the above tensor. However, in the presentation below we shall use the assumption that the rate of failure accumulation is proportional to the magnitude of plastic deformation [11]; this assumption is ruder but it has the advantage to be more simple. It coincides with the preceding one only where plastic deformation is below the saturation threshold. A decrease in accuracy thus introduced is justified by the simplicity and comprehensibility of the formulas, qualities which are essential for a brief presentation of the point to a large audience. So, let us write the equation of failure accumulation in the form

$$dV/d\lambda = Ke_p, \quad K = \text{const}. \tag{9}$$

If we integrate this equation for a special case of symmetric cyclic deformation, with $V = 1$ being taken as fracture condition, the following formula is obtained:

$$N\varepsilon_p^2 = \text{const}, \tag{10}$$

where N stands for the limiting number of cycles and ε_p for the amplitude of plastic deformation. Expression (10) is nothing more than the Coffin criterion. The fact that this popular criterion, holding for many construction alloys over a wide range of numbers of cycles, turns out to be included, as a special case, in Eq. (9), allows to consider this form of equation as worthy of notice.

However, the theory of fracture in the form proposed here cannot be sufficiently general because of the grave disadvantage inherent in it, namely that neither the fracture condition $V = 1$ nor the equation of failure accumulation (9) respond to a change in sign of stress; hence they do not distinguish, in particular, between tension and compression. But it is obvious that both the rate of failure accumulation (i.e., the derivative $dV/d\lambda$) and the fracture condition must depend not only upon the stress magnitude (which is indirectly allowed for in (9) through e_p) but also upon the stress sign. It is a matter of common knowledge that to produce fracture of a body by means of compression is not only far more difficult than by means of extension, but that it is practically impossible to do this if the plasticity of the material is sufficiently high. It is in the fracture condition that due allowance for the stress sign is of particular importance. Let us attempt to make the corresponding allowance, if only approximately.

The basis for the theory under discussion is the concept that there exists a moment of deformation when drastic changes in the picture of failures in a solid body occur; this happens when many microfailures merge to develop a crack. This process is of short duration, and physically it represents a loss of equilibrium stability in the system of microfailures. As the crack appeared is taken to be a macroscopic object, conditions for its formation must depend more upon the averaged macroscopic stresses than upon the microstresses. At present, conditions of instability both for a single crack and a regular crack system are being intensively studied and a great body of theoretical and experimental information on this subject is now available. According to what we know now, the formulas to define critical dimensions of the cracks always have the structure as follows

$$l_{cr} = A/\sigma^2, \tag{11}$$

where σ stands for tensile stress in the field of which the crack occurs, and A is constant.

Consider now a crack of length l_0 which starts extending for some reason. Then from Formula (11) the following expression can be obtained:

$$(\Delta l/l_0)_{\mathrm{cr}} = A/(l_0\sigma^2) - 1 = S_0^2/\sigma^2 - 1 \tag{12}$$

to determine the critical value of crack length increment. The constant S_0 entering into this equation is, from the physical point of view, the stress causing instability in crack equilibrium even with the initial length. If in Formula (12) by l_0 is meant a measure of the length of initial microcracks preexisting in a solid body and by Δl is meant a measure of their increment caused by plastic deformation, then the ratio $\Delta l/l_0$ may be taken as dimensionless characteristic of failure accumulated in the body under deformation. Assuming this characteristic to be identical to V we obtain the fracture condition in the following form:

$$V = S_0^2/\sigma^2 - 1. \tag{13}$$

This treatment results in S_0 acquiring the meaning of brittle fracture stress, i.e., the normal tensile stress producing fracture of the material without preceding plastic deformation. Needless to say, that the considerations presented here should not be taken as a derivation of Formula (13). They are no more than likely observations that suggest a definite structure of the fracture condition. Solution of the algebraic Eq. (13) with respect to σ yields:

$$\sigma = S_0/\sqrt{1 + V} = S, \tag{14}$$

where S is the running value of brittle fracture stress which depends upon the accumulated failures; S is dependent upon the path of loading and decreases as the latter increases.

If we take a particular path of loading (to be more specific, let it be, say, uniaxial extension) and plot on a common plane two curves, one of which represents stresses and the other brittle fracture stresses, as a function of plastic deformation, then we shall get a picture shown in Fig. 2. *P. Ludwik* [12] was the first to give such a graph. He pictured the fracture condition as the cross-point of the deformation curve with the strength curve. Thus, Eq. (14) is the Ludwik

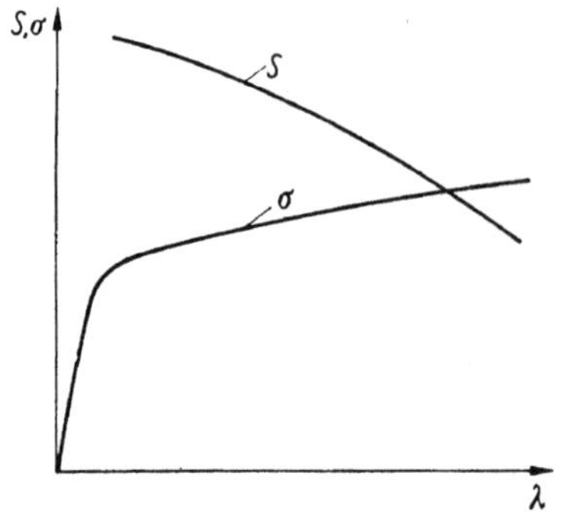

Fig. 2. Stress σ and brittle stress S as functions of plastic deformation.

strengh curve generalised for an arbitrary path of loading, the derivation of the curve being given.

A. A. Griffith [13], in his paper for the First International Congress On Applied Mechanics (Delft, 1924), extended the brittle fracture condition to cover the case of arbitrary states of stress. He pointed out that, in the general case, when brittle fracture ($\sigma \geq S_0$) occurs, σ should be taken as follows

$$\sigma = \begin{cases} \sigma_1 & \text{when} \quad 3\sigma_1 + \sigma_3 > 0\,, \\ -\dfrac{(\sigma_1 - \sigma_3)^2}{8(\sigma_1 + \sigma_3)} & \text{when} \quad 3\sigma_1 + \sigma_3 < 0\,, \end{cases} \tag{15}$$

with $\sigma_1 \geq \sigma_2 \geq \sigma_3$ being the main stresses. This well-known result may be employed in formulas (13) and (14) when applying them to estimate the states of stress other than uniaxial extension. An alternative possibility is to take in (14) as σ the largest positive reduced stress, this being equal to using the so called second "theory of strength". Although the two possibilities mentioned lead to similar results, the differences between them may be also cited.

For example, the generalised theory of *Griffith* gives the same value of limiting stress both for isotropic and linear extensions, thus introducing an error on the safe side when dealing with a state of nearly isotropic extension (this result follows from the experiments [14]). The second theory of strength, which considers distinctions between isotropic and linear extensions, underrates the value of the fracture stress under compression. These results should be kept in mind when practically exploiting the theory in question.

If the equation of failure accumulation and the fracture condition are applied in forms (9) and (13), respectively, to two types of loading which are of the utmost importance for practical purposes and which, consequently, received the most experimental study, i.e., to the proportional and to the cyclic symmetric loadings, the following relationships can be obtained:

proportional loading

$$e_p^2 = \frac{2}{K}\left(\frac{S_0^2}{\sigma^2} - 1\right), \tag{16}$$

rigid symmetric low number cyclic loading

$$N\varepsilon_p^2 = \frac{1}{2K}\left(\frac{S_0^2}{\sigma^2} - 1\right). \tag{17}$$

These two types of loading can be used to determine experimentally the parameters K and S involved in the theory. The first of these parameters characterizes plasticity and the second characterizes brittleness of solid materials. With the proportional loading the results predicted by the fracture condition are very near to those predicted by the theory developed by *N. N. Davidenkov* and *Ya. B. Fridman*. The condition of low number cycle fracture in the form of Eq. (17) yields results similar to those given by the Coffin-Manson criterion.

So, the theory of fracture considered here combines the criteria most popular at the present time that were suggested for special cases of loading.

However, any theoretical concept may be thought of as being promising if it not only describes what was known before and what could have been accounted for in advance but if it does also reveal something unexpected. It is the possibility of indirect evaluation of the brittle fracture stress, resulting from the theory presented here, that is this unexpected feature in the case treated.

So far, in order to determine this important property of a material, experiments with triaxial extension were performed. The appropriate states of stress can be achieved only in the vicinity of stress concentrations, and to measure them involves grave experimental difficulties. Therefore, in spite of the fact that there exists an extensive literature on brittle fracture stress and methods to determine it, the concrete data on this physical value do not always agree with each other. The theory proposed here offers the possibility of determining the two constants, K and S_0, from experiments with long trajectories of plastic deformation, thus avoiding problems involved in the measurement of the threedimensional stresses.

Comparison between brittle fracture stresses obtained by means of this indirect but substantially less laborious approach and results obtained by means of other methods (for example, see [15]) shows rather good agreement; discrepancies are within the accuracy of the experiment. This agreement suggests that the theory under discussion correctly represents the main features of the relationship between tough fracture and brittle fracture phenomena. If the right-hand side of the equation of failure accumulation is taken as being proportional to the magnitude of the microstress tensor rather than to the magnitude of the plastic deformation, then the agreement between theory and experiment improves; this is further evidence for the considerations taken as a basis for the theory objectively representing the physics of the fracture process. Allowance for "the saturation threshold" on the curve $\sigma = f(e_p)$ has made it possible to predict the direction of the departure of the experimental curves for the low number cycle loading from the Coffin law [11].

Being able to present the essence of the problem, this simple theory obviously cannot claim to be capable of describing fracture of solid bodies in all the diversity and detail. Thus, it has been constructed under the assumption that the plastic dilatation of a solid body can be described by means of a scalar. In practice, though, microcracks and micropores are formed oriented according to the orientation of the plastic deformation which causes them. Due to this fact, lately efforts have been attempted with increasing frequency, with the object to replace, when considering plastic dilatation, the scalar measure with the tensor characteristics.

Thus, *A. A. Il'yushin* [16] has considered and systematized numerous generalisations of the failure accumulation equation as applied to tough-elastic bodies. In the work of this author both the linear and nonlinear alternatives of the theory are embraced, and not only stresses and deformations have been taken as criterion parameters but moment stresses as well, which is, essentially, one of the possible forms of taking account of microstresses. All this material will be of great use in the future. It should be remembered, however, that phenomenological theories are characterized by the inductive rather than by the deductive development. The simplest form of phenomenological theory is usually constructed to represent little more than the basic features of the phenomenon, and it is not until comprehensive experimental testing is completed that the next step, usually a short one, is taken. An example where such gradual development may clearly be seen is the theory of plasticity. The theory of fracture features a similar graduality, so we cannot expect an easy and swift advance in this field. The comprehensive study of the potentialities of the phenomenological theory of fracture, even within the framework of the scalar presentation, is a long way to go.

For instance, it remains to be solved in which way we shall treat the two-stage nature of the fracture process. Strictly speaking, all that was said before applies solely to the first stage of the fracture process, and what has been designated above as the fracture condition would for this reason be better referred to as the condition of macroscopic crack formation. As to the second stage, which does amount to spreading of this crack, it should have been dealt with by another theory. The most successful phenomenological theory of fatigue fracture, taking account of the latter being two-stage, was developed by *V. V. Bolotin* [17]. It is natural that this made the theory of fracture more complicated, and more parameters entered into it. Besides, in this case, an uncertainty appears resulting from the difficulty of clearly distinguishing between concepts of micro- and macrocracks. Due to this, up to now, the trend is toward including the two stages in a common phenomenological theory, averaging regularities inherent in each of them, even though the separate description of these stages is physically correct. Striking instances are the Coffin and the Manson criteria in the theory of low number cycle stress and investigations done by *L. M. Kachanov* and *Yu. N. Rabotnov* in the theory of long time strength. The present author holds to the same trend taking it as being the most realistic one from the standpoint of meeting the practical demands.

Allowance for the influence of plastic dilatation upon the equation of state of a solid body should be also placed among the questions that are unclear even within the scope of the scalar alternative of the theory of fracture. *Yu. N. Rabotnov* [18] was the first to attempt evaluating this feed-back in considering the problem of long time strength.

The scalar theory of fracture, however, is to be developed in order to take into consideration the effect of the stress sign upon the rate of failure accumulation as well. The associated correction can be sometimes appreciable and deserves careful attention, even though it is unclear in what form it should be introduced.

To end this communication, I would like to formulate some conclusions.

In the presentation above fracture was treated as an irreversible process whose description required the use of differential non-holonomic equations. It is the simplest, i.e., scalar, alternative of the theory of this kind that gained the largest popularity in modern practice.

Within the framework of this alternative, the general theory of fracture of elastic-plastic bodies is proposed which is the result of the synthesis of the theory of failure accumulation and the theory of brittle fracture.

This theory, heuristic in nature, comprises two basic equations:

a) the equation of failure accumulation

$$dV/d\lambda = \gamma s \approx K e_p, \qquad (*)$$

and the fracture condition

$$\sigma \geq S = S_0/\sqrt{1 + V}. \qquad (**)$$

Equation (*) is not new, if judged only by its form, but it suggests the concrete expression for the right-hand side, namely: according to the concept that plastic dilatation is chiefly caused by microstresses, the right-hand side of equation (*) is assumed to be dependent upon the magnitude of the tensor s_{ij} which serves as a quantitative characteristic of the Bauschinger effect.

As to condition (**), it is, in essence, the criterion of brittle fracture taking into consideration, however, the fact that fracture stress is not constant but rather depends upon failure accumulated under plastic deformation. (This condition is based actually on the assumption, firstly stated by *N. N. Davidenkov* [19, 20], that fracture occurs always by tearing.)

The theory presented here does not claim to describe the process of fracture in all variety and detail but, nevertheless, it seems to represent the main features of interrelationship between the characteristics of toughness and brittleness of polycrystalline solid bodies and does explain (in the framework of phenomenological concepts) the nature of this relationship.

References

1. Novozhilov, V. V.: Prikladnaya matematika i mekhanika 28 (1964) 393–400.
2. Taylor, G. I., Quinney, H.: Proc. Roy. Soc. London 143 (1934) 307–326.
3. Palmgren, A.: Z. VDI 68 (1924).
4. Langer, B. F.: J. Appl. Mech. 4 (1937) A 160–A 162.
5. Miner, M. A.: Trans. J. Appl. Mech. 12 (1945) A 159–A 164.
6. Robinson, E. L.: Trans. ASME 74 (1952) 777–781.
7. Kachanov, L. M.: Izvestiya Akademii nauk SSSR, Otdelenie tekhnicheskikh nauk 8 (1958) 26–31.
8. Kadashevich, Yu. I., Novozhilov, V. V.: Prikladnaya matematika i mekhanika 22 (1958) 78–89.
9. Talypov, G. B.: Izvestiya Akademii nauk SSSR, Mekhanika i mashinostroenie 6 (1964) 131–137.
10. Rybakina, O. G.: Problemy mekhaniki tverdogo deformirovannogo tela, Leningrad, 1970, 375–380.
11. Rybakina, O. G.: Mekhanika tverdogo tela 6 (1969) 61–66.
12. Ludwik, P.: Elemente der technischen Mechanik. Berlin 1909.
13. Griffith, A. A.: Proc. 1st Intern. Congr. Appl. Mech. 1924; 1925, 55–63.
14. McAdam, D. J., Geil, G. W., Jenkins, W. D.: Proc. Amer. Soc. Test. Materials 47 (1947) 554–572.
15. Hendrickson, J. A., Wood, D. S., Clark, D. S.: Trans. Amer. Soc. Metals 50 (1958) 656–681.
16. Il'yushin, A. A.: Mekhanika tverdogo tela 3 (1967) 21–35.
17. Bolotin, V. V.: Statistical methods in structure mechanics, 2nd ed. Moscow 1965.
18. Rabotnov, Yu. N.: Creep of elements of structures, Moscow 1966.
19. Davidenkov, N. N.: Dynamic testing of metals, 2nd ed. Moscow-Leningrad 1936.
20. Shevandin, E. M.: Zhurnal tekhnicheskoi fiziki 18 (1948) 863–874.

Ship hydrodynamics

K. Wieghardt

Hamburg, F. R. Germany

1. Introduction

Ships have crossed the seven seas long before hydrodynamicists explained how this is possible. On the other hand, for the making of airplanes aerodynamic theory was indispensable and had to be developed simultaneously. That is, perhaps, why ship hydrodynamicists are flourishing more in secrecy than aerodynamicists. Yet, as a field of application of general fluid mechanics ship hydrodynamics is also interesting. As in aerodynamics most of the main methods are needed to describe the various aspects of the flow past a ship. Of course, supersonic flow or heat transfer are of no concern here; on the other hand, there is the free surface of the sea with waves in which theorists find much pleasure. As in aeronautics, the degree to which flow phenomena are comprehensible theoretically is very different. E.g., the intricate response of a ship in an irregular seaway can be calculated by potential theory in a surprisingly satisfactory manner, whereas the resistance of a ship in calm water is still to be found out experimentally by model tests since a theory for threedimensional, turbulent boundary layers is not yet known.

In this review for the non-specialist various flow models are mentioned which are used in ship hydrodynamics. Potential flow without or with free surfaces, gravity and free vortex sheets is describing successfully e.g. oscillatory ship motions, wave resistance, cavitation, spray, hydrofoils and propellers. On the other hand, viscosity and turbulence are, of course, essential for the ship's boundary layer and, above all, for the extrapolation of test results with models to the real ship.

A few examples of topical research show the efficiency as well as the limitations of present theories.

2. Mechanical properties of water

In ship hydrodynamics water can be considered as a homogeneous continuum with constant density ϱ; stratification effects due to different salinity are only known, for example, in Norwegian fiords where internal waves can increase ship resistance. Yet, usually there are air bubbles in the water near the sea surface because of breaking waves, and it has been observed with fast ships that the water entering the propeller is almost milky white with bubbles [1]. Hence, the test water in cavitation tanks for propellers should also contain air bubbles and possibly even the right spectrum of gas nuclei of about 0.1 to 0.4 mm in diameter [2]. As a reminder Fig. 1 shows the speed of sound a in water with bubbles vs. the volume ratio ν of air to water; for $\nu \geq 10^{-3}$ the speed of sound is smaller than in pure air [3, 4]. A similar relation was given in 1910 by *A. Mallock* [5] who assumed isothermal instead of isentropic compression of the bubbles.

Since the viscosity of water is small the Reynolds number in all ship problems is very large and the boundary layer of the ship is almost completely turbulent, and this cannot be simulated by models that are too small. Viscosity can be neglected totally for ocean waves because of the absence of rigid bodies, and, surprisingly, also for oscillatory ship motions (e.g. roll) when—apparently—viscous effects have no time to accumulate.

Shiphydrodynamicists are even interested in the fact that water may become a non-Newtonian fluid by minute concentrations of "additives". They consist of long chained molecules (molecular weight of order 10^7) which can be stretched to a length comparable with the microscale of the smallest turbulent eddies (of order say 0.1 mm). Their hydraulic effect seems to be that they slow down the breaking cascade of eddies down to the smallest scale where dissipation into heat occurs and the resistance is decreased considerably. Artificial additives are polyethylenoxide, guar gum and even asbestos fibres [6]; yet also the excrements of certain algae are effective and the growth of such algae must be prevented by fishes or chemicals in towing tanks to avoid ano malous test results.

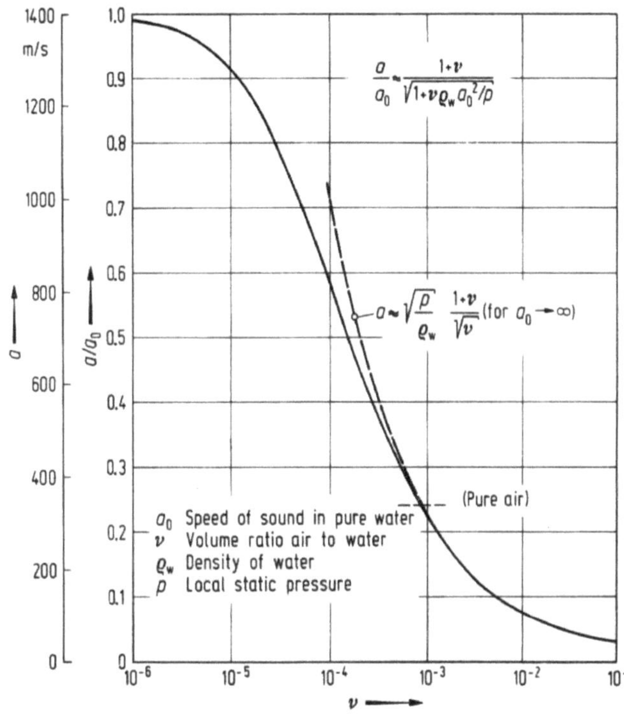

Fig. 1. Speed of sound in water containing air bubbles at atmospheric pressure [4].

3. Mathematical flow models

Just as in aerodynamics even complicate phenomena due to the compressibility of the air are quite comprehensible theoretically, there are in shiphydrodynamics good theories of the deformation and the effects of the free surface. Yet in both fields big test stations are still necessary, mainly to find out effects of viscosity such as boundary layer separation and vortex formation, wake flow and their interference with the calculable inviscid outer flow. For this, as for all turbulent flows, reliable theories are badly wanted. For example, for modern full tanker forms even the extrapolation of resistance tests with models (at Reynolds numbers R_n of the order of 10^7) to the ship ($R_n > 10^9$) is uncertain because separation is here more important than for slender ships.

The most efficient mathematical model then is still for flows in an inviscid fluid with a velocity potential. Whenever viscosity effects are negligible, yet only then, theory can describe real flows with astonishing accuracy. Figure 2 shows examples of water flows with and without gravity. The first group is without any singularities inside the flow region, i.e., without any vorticity. The classical problem here is the wave resistance of a ship, for which the first linearized theory was given by *J. H. Michell* in 1898.

For the instationary, mostly oscillatory "small" ship motions, as heaving, rolling and pitching, the concept of the virtual mass still holds to first order, which in unrestricted inviscid fluid is valid exactly.

In all these linearized theories the shape of the free surface is known to first order. In cavitating flow or in spray problems however, usually the shape of the free surface is one of the main unknowns. Such problems become even more complicate when gravity effects are considered too. Two recent nonlinear examples for twodimensional flow are the following. *G. Dagan* and *M. P. Tulin* [7] study the free surface rise in front of a blunt floating body for small and for high Froude numbers. *T. Y. Wu* [8] calculates the free surface flow past a body for large Froude number where Wagner's solution without gravity is a first order approximation near the body.

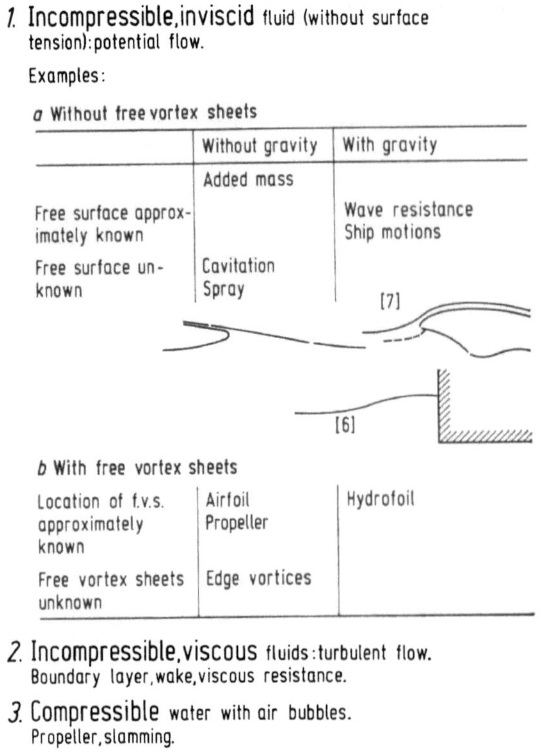

Fig. 2. Fluid models and examples of appropriate problems.

A new complication is represented by flows where the vorticity is zero except in free vortex sheets in which the potential is discontinuous. The classical example here is the wing of large aspect ratio. The shape of the free vortex sheet behind it is known a priori to first order, and the same is true for the extension of this theory to the lightly loaded propeller. With the hydrofoil free surface effects have to be considered in addition.

Not only the strength but also the shape of the free vortex sheet is unknown, for example, on a thin delta wing of small aspect ratio or for any other edge vortices. For the well known experiment with the coffee spoon—in my country named after the mathematician *F. Klein*—there is the naval architect's analogue: the effectiveness of bilge keels in damping the rolling of a ship. The kinetic energy of the spiral edge vortices produced by the bilge keels is taken from the undesirable rolling motion. The concept of a virtual mass can be used here only when it is extended to include the singularities of the potential (see Section 4.5).

The wake of a ship is a whole region with vorticity. *R. Brard* [9] describes it by an inviscid flow for the limiting case of a vanishing viscosity which acts on the local waves and on the free waves; the first effect depends on $(1/R_nF_n^2)^{1/3}$, the second one on $1/R_nF_n^4$.

Unfortunately, viscosity effects are often not negligible; for example, the resistance of a ship in calm water is almost completely viscous drag. Anyone can see that a ship's waves remove kinetic energy to infinity behind the ship, yet, their resistance component is usually less than 20% of the total resistance. Now, the ship's boundary layer is not only turbulent but also threedimensional and a theory for it is badly missing. The wake flow is, of course, also not calculable. Yet, it is of practical interest because the propeller is working in this inhomogeneous flow region which, on the other hand, is changed by the suction of the propeller in front of it.

To describe compressibility effects of sea water with air bubbles a sort of caricature of usual gasdynamics is necessary. Even when it is assumed that the air content is invariable inside the flow region there is the difficulty that the speed of sound changes locally more than in an ideal gas [4] (cf. Fig. 1: $a \sim \sqrt{p}$). Fields of application are propeller flow [3, 4] and slamming when in a heavy sea the ship's bow is falling back into the water more or less vertically. Here, short time pressures of more than 10 atm have been measured. Since a characteristic pressure is $\varrho\, a\, v$, with ϱ density, a speed of sound and v vertical velocity (see, for example, [10]) the free air content of the water should be relevant, even though the main cushioning effect is due to the entrapped air between the ship's bottom and the water surface [11].

4. Topical research examples

After this sketchy survey of theoretical models we will report on a few research examples in more details. Naturally, the choice is subjective; besides, the intention is to interest the outsider rather than the specialist.

4.1 Wave resistance

The model of a really thin "ship" with breadth to length $B/L = 1/20$ is shown in Fig. 3 [12]. Obviously, it is well fitting in the main assumption of the linearized theory, namely that the boundary condition can be applied approximately on the centerplane instead of the body's

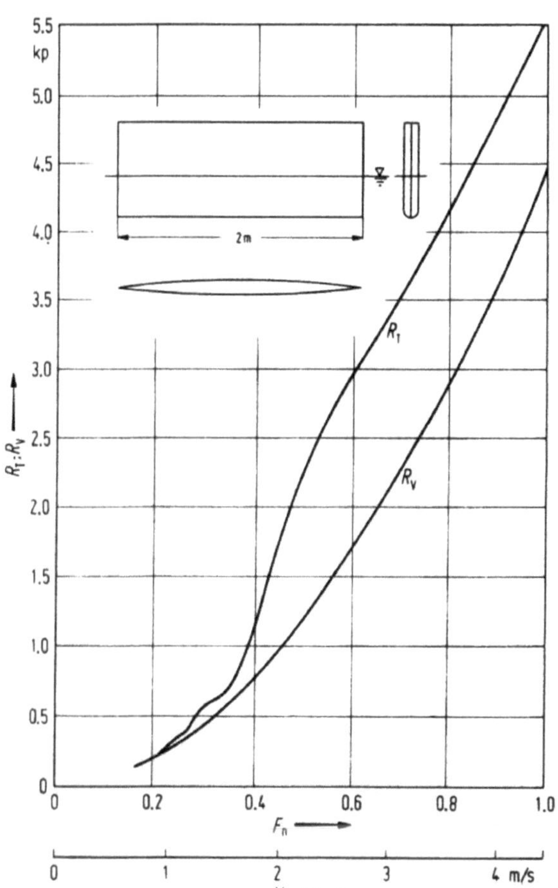

Fig. 3. Total and viscous resistance (R_T and R_V) of a thin "ship" vs. speed V or Froude number F_n [12].

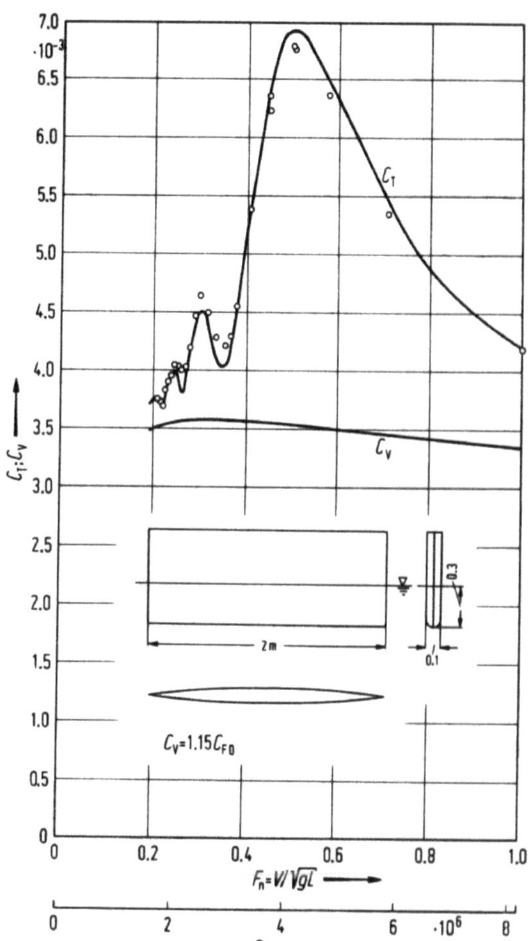

Fig. 4. Coefficients of total and viscous resistance of a thin "ship" vs. F_n or Reynolds number R_n [12].

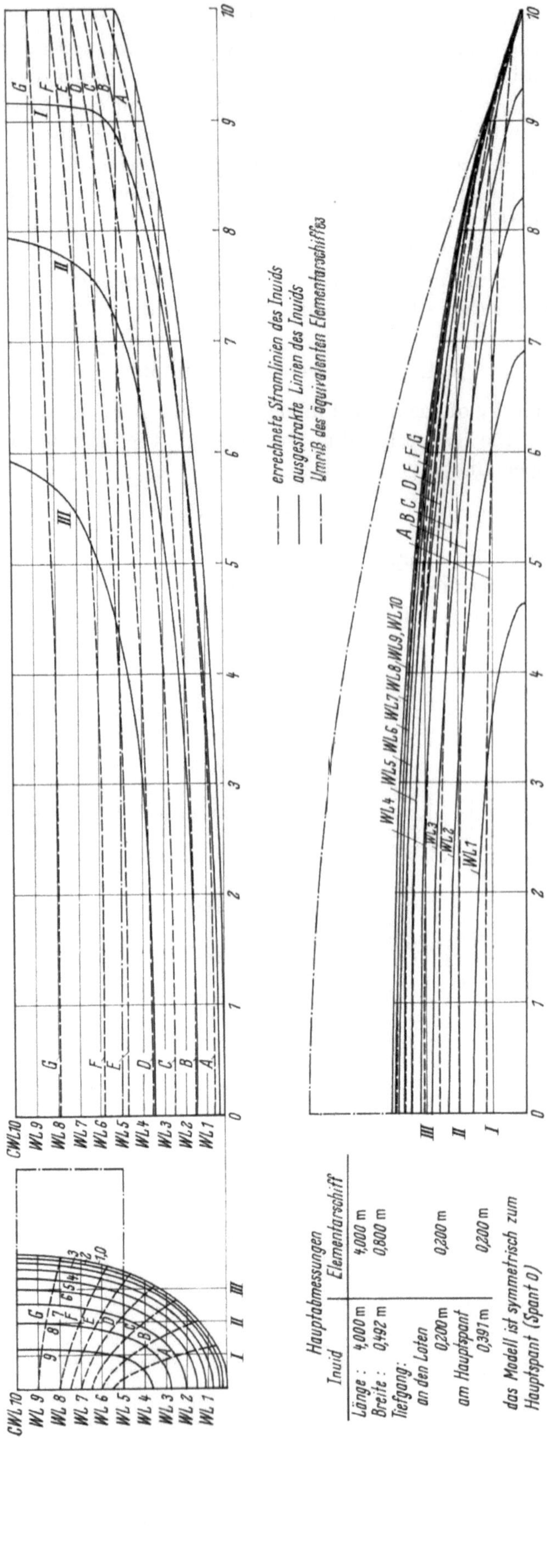

Fig. 5. Lines of a "mathematical" ship (after *Sharma* [14], Jahrb. Schiffbautechn. Ges. 59/1966, p. 478).

surface. The total resistance R_T and its viscous component R_V are plotted vs. speed V or Froude number $F_n = V/\sqrt{gL}$, g gravitational acceleration and L model length. Usually, only the resistance coefficients C_T and C_V are shown as in Fig. 4; these coefficients are R_T and R_V divided by the stagnation pressure and the wetted surface of the hull. In the test range C_V depends only little on speed or Reynolds number $R_n = VL/\nu$, with ν the kinematic viscosity, and it can be assumed proportional to the tangential frictional resistance of a plate C_{F0} with the same length (here $C_V = 1.15 \cdot C_{F0}$). When the wave resistance coefficient C_W—as calculated or as determined by an analysis of measured wave profiles—is added to this C_V thus determined one gets the C_T curve for the total resistance as drawn in Fig. 4. The test points clearly verify the theory in this case.

However, real ships are not so thin by far; their breadth to length ratio is about 1/8 or even 1/5 for big tankers. A "mathematical" ship with B/L about 1/8 is shown in Fig. 5. It is a so called "Inuid"; that is the body produced in unlimited parallel flow when—on a rectangle of the centerplane—sources and sinks are distributed whose strength changes linearly in longitudinal direction. In Fig. 6 the dimensionless wave resistance is plotted versus Froude number. Incidentally, this is a striking example of the test accuracy that can be achieved; the crosses denote tests by *T. Inui* [13] in Tokyo with a 1.75 m model, the circles tests by *S. D. Sharma* [14] in Hamburg with a 4 m model. Here, the calculated wave resistance is obviously too high.

Hence, theoreticians in various countries work on second order theories; special efforts are needed to satisfy the nonlinear dynamic boundary condition on the free surface (Bernoulli-

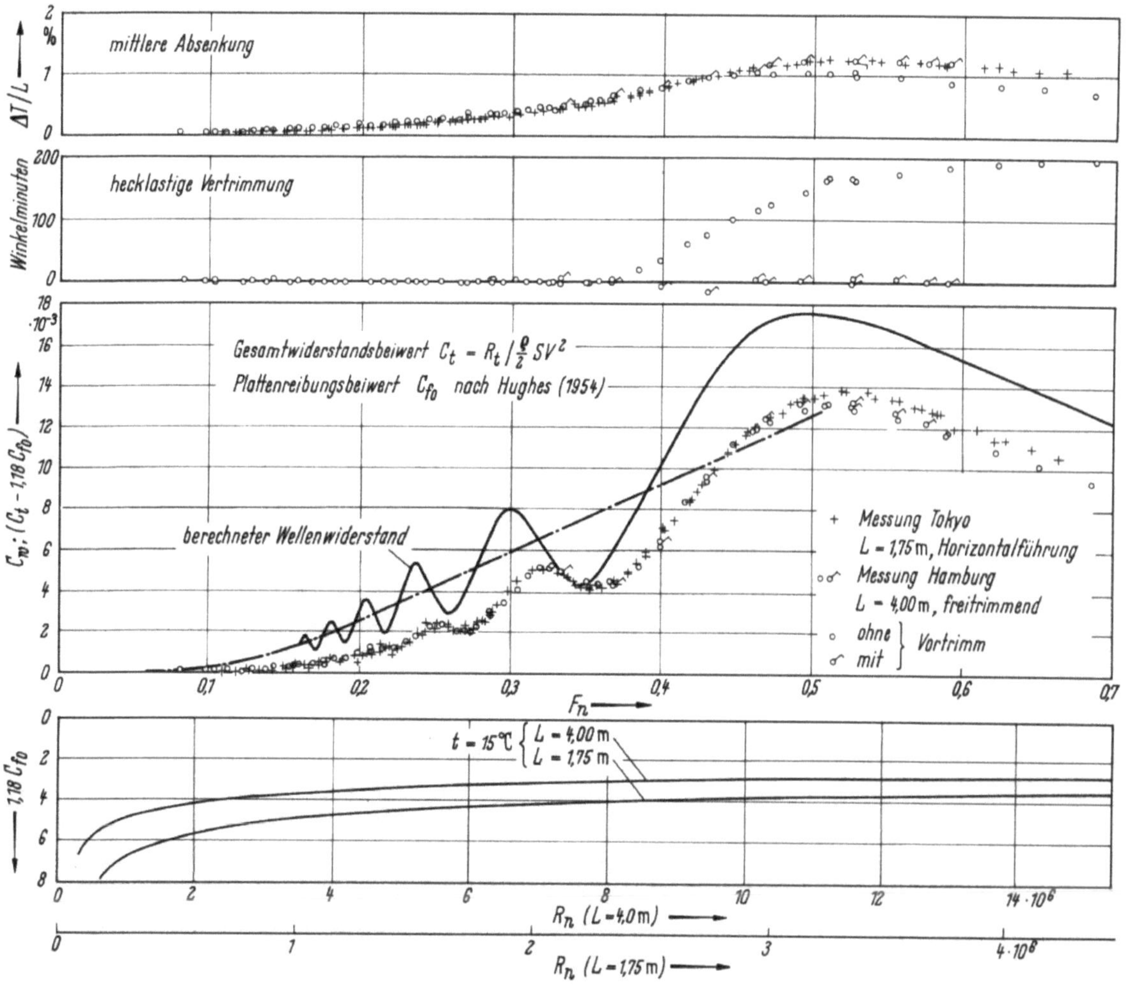

Fig. 6. Dimensionless wave resistance of an Inuid vs. F_n (after *Sharma* [14], Jahrb. Schiffbautechn. Ges. 59/1966, p. 490).

equation). Yet, there is certainly also some viscous interference. For, in real flow the boundary layer is smoothing out the pressure distribution and the displacement flow, especially near the stern, so that wavemaking there is weaker than calculated for inviscid flow.

An example of the near field wave pattern as calculated (on top) and measured (below) is shown in Fig. 7 [15] by lines of equal height; dotted lines belong to wave throughs. For the eye both patterns are rather similar. Yet, the integral for the wave resistance depends very much on the interference of the various wave systems. Therefore, sometimes slight changes of the hull form influence the wave resistance noticeably.

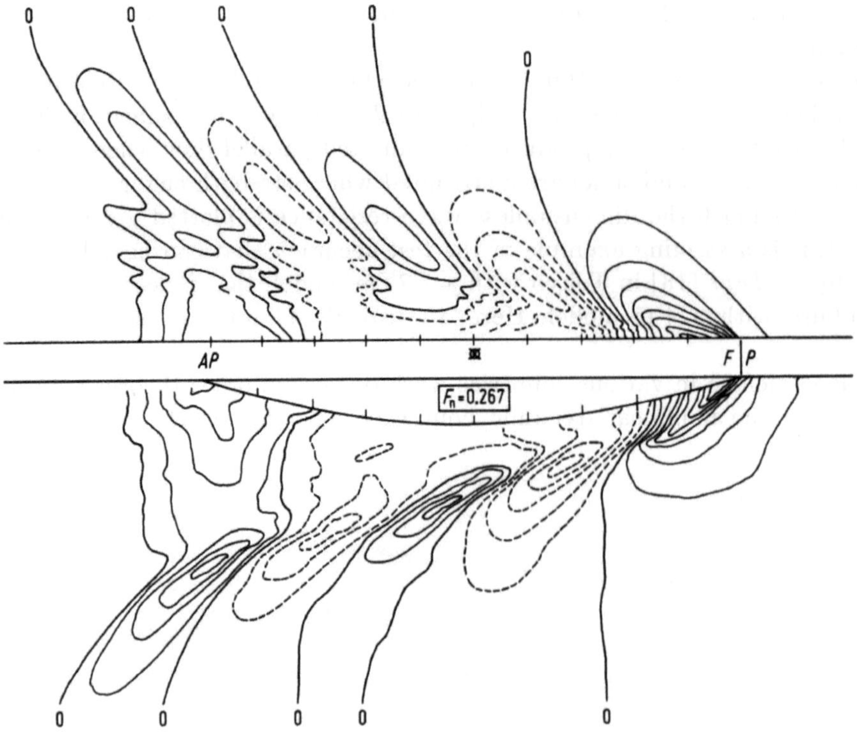

Fig. 7. Calculated (above) and measured (below) wave pattern [15].

Inspite of its shortcomings, the linearized theory—as developed mainly by *Havelock*, *Weinblum* and *Wigley*—can at least lead into the neighbourhood of optimal hulls with regard to wave resistance. E.g., with some experimental backing, it is possible to calculate good forms and positions for a bow bulb which for a fast ship can remove wave resistance at service speed almost completely.

4.2 Bow bulb for full ships at low speed

After these triumphs of theory experts were, of course, not amused when it was found out that a bow bulb can reduce the total resistance of a full form tanker in ballast condition although at these low speeds there are practically no waves at all behind the ship. An example of such model tests is given in Fig. 8: at $F_n = 0.2$ the bulb reduces the total resistance by 15%. This interests also ship owners, even when fuel cost is only about 10% of their total cost, because the speed increase, possible with the same power, yields a higher transport capacity [16].

To solve this resistance puzzle *S. D. Sharma* [17] towed deeply submerged double models of the underwater hull reflected at the ballast water line; yet, this viscous resistance was the same with or without bulb. He then measured the wave field (of floating models) in order to determine, by linearized wave theory, the energy carried away by the waves. The longitudinal cut method was used, i.e., the model was passing a probe at a fixed position which measured the water height varying with time. Yet, again, the difference in wave resistance between the models with and without bulb was very small and in the opposite sense to the difference in total resistance.

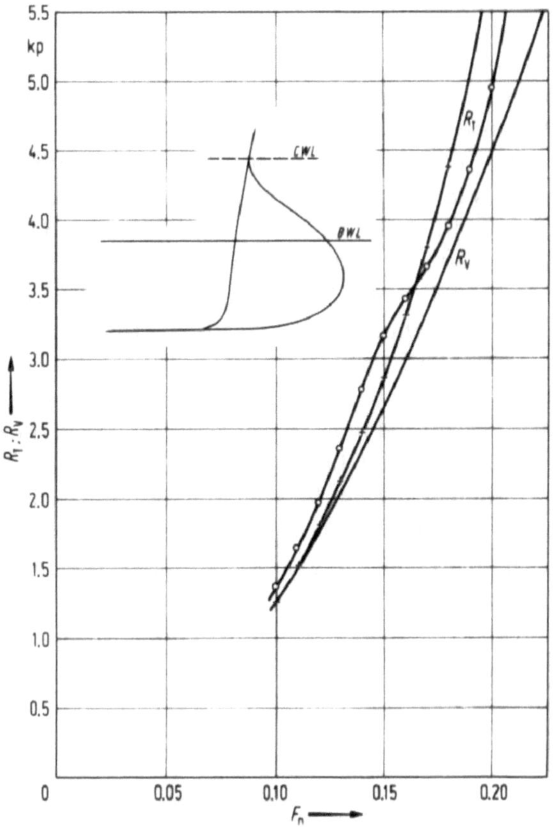

Fig. 8. Total and viscous resistance R_T and R_V (arbitrary scale) of a model without (+) and with (○) bow bulb at ballast load vs. F_n [17].

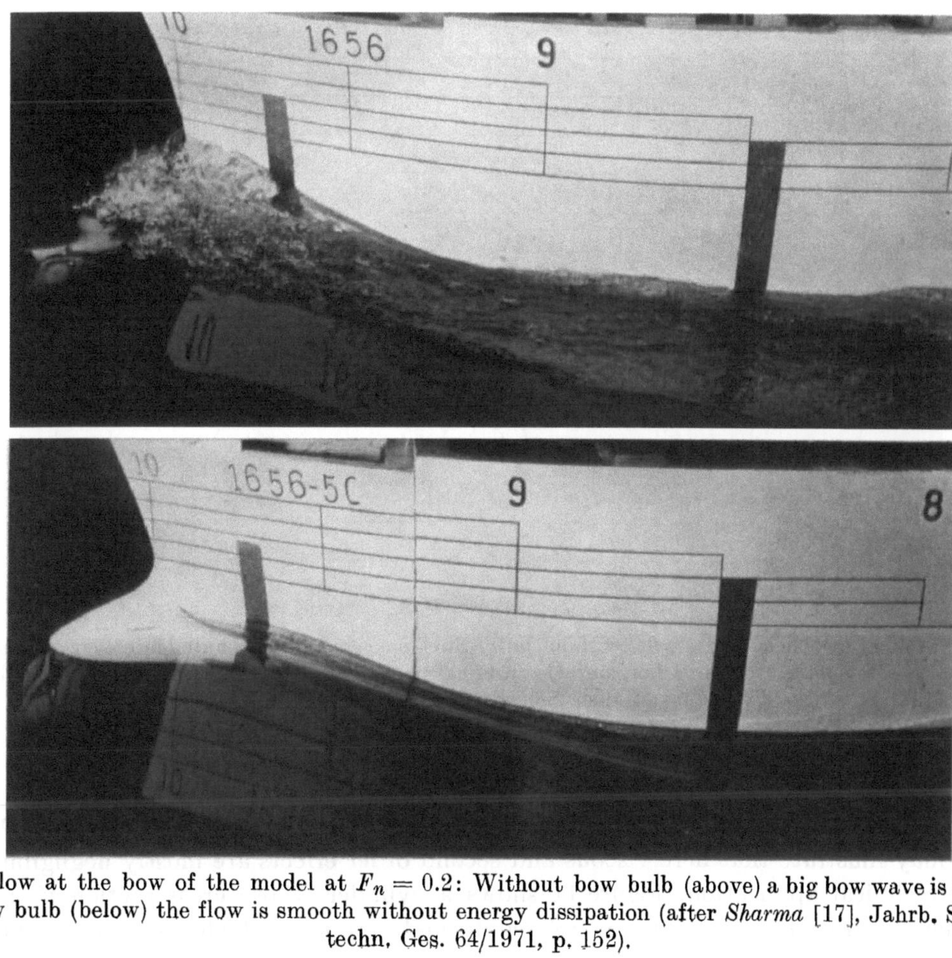

Fig. 9. Flow at the bow of the model at $F_n = 0.2$: Without bow bulb (above) a big bow wave is breaking; with bow bulb (below) the flow is smooth without energy dissipation (after *Sharma* [17], Jahrb. Schiffbautechn. Ges. 64/1971, p. 152).

However, the solution of the problem is to be seen at once at the photographs in Fig. 9. At the Froude number 0.2 a big bow wave is breaking when there is no bulb on the model, and there is energy dissipation corresponding to a spray or wave-breaking resistance, named so because it is a counterpart of the wave making resistance. Yet, on the model with bulb the water remains quite transparent and there is no wavebreaking at all. The quantitative details of this loss of mechanical energy are to be found in the wake, as far as resistance is concerned. Figure 10 shows the total head loss (divided by the stagnation pressure) versus the horizontal coordinate y (normal to model axis) over model breadth B, at various depths $-z/T$, with T the model depth. These measurements were made 2.66 m behind the model (length 6.64 m) at $F_n = 0.2$; the wake there extends to about twice the model depth. Thick lines are for the model without bulb; near the surface they indicate additional momentum loss peaks outside the main wake for $|y|/B \leq 0.5$, and this is nearly the same with or without bulb. The total momentum loss integrated over the wake region gives the viscous resistance of the model, and the difference with or without bulb is, indeed, the same as for the total towing resistance.

Hence, at least for this hull and bulb it has been proved that the beneficial effect of the bulb is due to the avoidance of wave breaking. On the other hand, sometimes a bow bulb or also a cylindrical bow might also suppress the formation of trailing bilge vortices and of separation zones at the bottom of the fore ship and hence reduce the viscous resistance [18, 19].

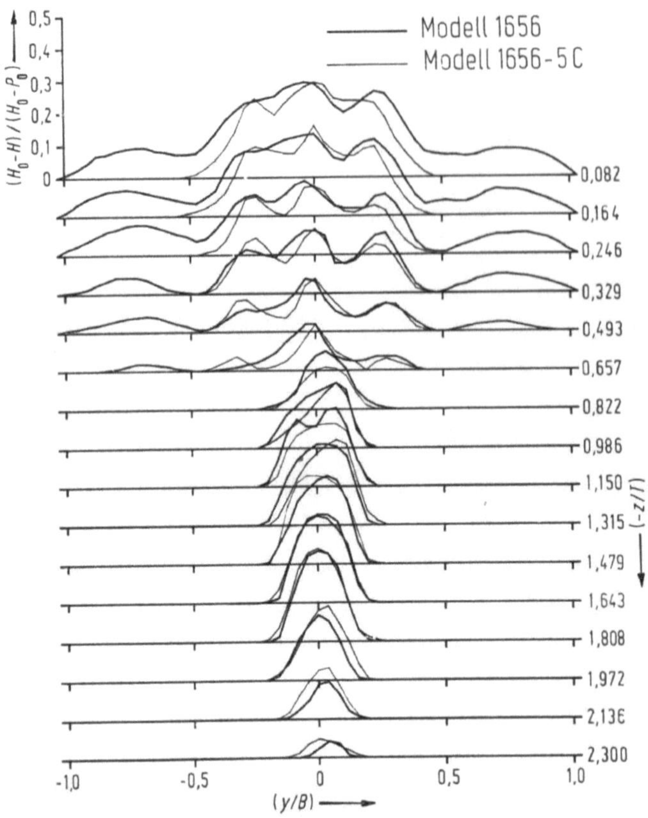

Fig. 10. The loss of total head at $F_n = 0.2$ without bulb (thick curves) and with bulb (thin curves) in a vertical plane at $0.9L$ after main section (y horizontal, z vertical coordinate, B breadth, T depth at ballast load) (after *Sharma* [17], Jahrb. Schiffbautechn. Ges. 64/1971, p. 155).

4.3 Ship motions

Theorists are most successful when calculating small oscillatory ship motions. A single example may illustrate that here viscous and second order effects are mostly negligible in the range of technical application. Figure 11 shows a cylinder whose cross section is a Lewis-profile, derived from the circle by a simple conformal mapping, which resembles a ship's section. When this cylinder has been forced to perform small heaving oscillations, $y = y_0 \cos \omega t$, then waves have been originated with the same frequency carrying energy sidewards away, and this corresponds to a damping of the motion of the body. The pressure on each point of the section is the hydrostatic pressure $\varrho g y$ plus terms proportional to $\cos \omega t$ or $\sin \omega t$, i.e. to the acceleration $\ddot{y}$ or velocity $\dot{y}$. The amplitude of the normalized total pressure fluctuation $\bar{p}$ at three points, a, b, c is plotted in Fig 11 vs. the frequency parameter $\omega^2 B/2g$, B = section breadth. The theoretical curves are confirmed by measurements [20] even for a heaving amplitude $y/(B/2)$ of 6%.

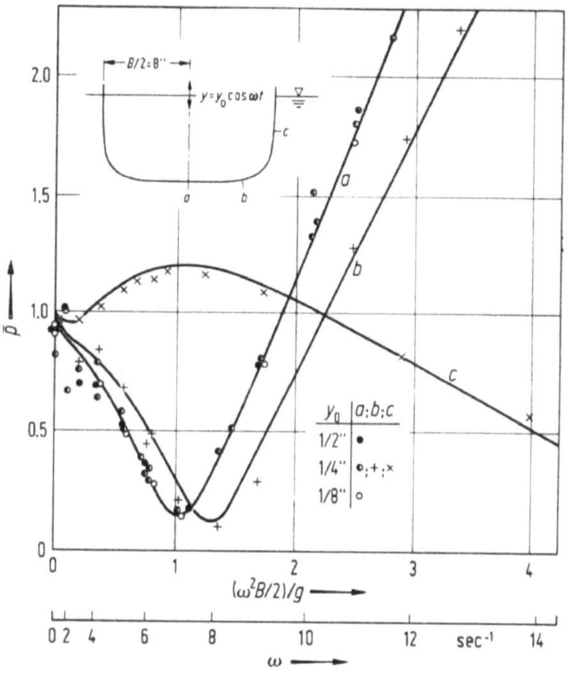

Fig. 11. Heaving oscillations of a floating Lewis profile. Pressure $p/\varrho g y_0 = (1 - p_a) \cos \omega t - p_v \sin \omega t$. Normalized total pressure $\bar{p} = [(1 - p_a)^2 + p_v^2]^{1/2}$ as measured and calculated vs. frequency parameter [20].

Such results for plane flow are then applied to slender ships (strip method). Further, *M. D. Haskind* [21, 22] has found the close relationship between these forced waves in calm water and the forces on a fixed body in incident waves, and then the motion of a floating ship in regular waves can also be calculated. Moreover, as long as linearized theory is adequate superposition is possible and with it the statistical description of ship motions in a realistic irregular seastate, as shown first by *St. Denis* and *W. J. Pierson* [23]. There are, of course, many more difficult problems still involved as, for example, the complications due to the forward motion of the ship. Yet, the goal is already in sight: oceanographical statistics for the seaway at the main ship routes shall enable one to predict the load tests to be expected in the ship's life time.

However, even the very remote goal, a nonlinear theory, will certainly be not only of academic interest, since extreme seaways and capsizing are definitely nonlinear processes. For this field an example might be cited, although—strictly speaking—it does not belong to ship hydrodynamics. *W. J. Pierson* [24] has given a nonlinear theory for wave refraction in sea areas of complex bottom topography. Hereby, he can make plausible, for example, the loss of two trawlers near the Doggerbank (North Sea) by the very high cross seas near the caustic of trains of high waves propagating at nearly right angles to each other. He suggests to locate and describe such dangerous areas for ships so that they can be avoided. Figure 12 shows waves generated at the front wall in the white shallow water part of the basin. These waves are propagated toward a portion of the tank where the depth increases linearly to deep water (black). The straight line at which the depth starts to increase is at 50° to the wave crests. After this line the waves are refracted, reflected and turned to the shallow water, at about 90° to the original waves giving rise to a high cross sea. Only short waves can go on into deep water.

Fig. 12. Shallow water waves generated at constant depth (at the front side wall) and propagated toward a portion of the tank where the depth increases linearly to a value equivalent to deep water (cf. [24]).

4.4 Edge vortices

Even for small amplitude motion agreement between theory and experiment becomes poor as soon as sharp edges are moving crosswise as, for example, with bilge keels or the keel of a triangular section in roll [25]. Often this is called a viscous effect. Yet, even if one has never heard of separation, one ought to be suspicious of the usual potential flow round an edge where

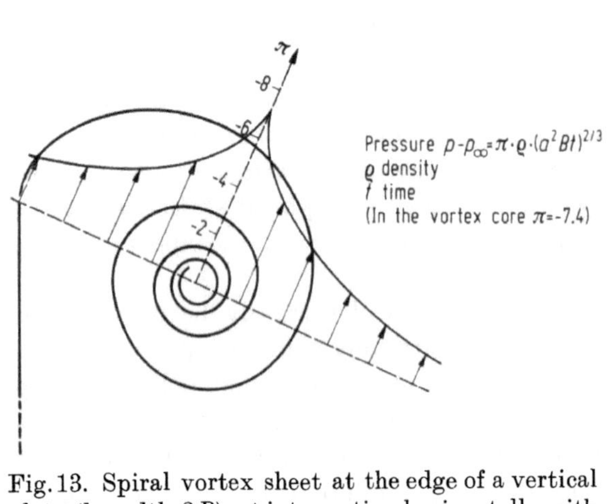

Fig. 13. Spiral vortex sheet at the edge of a vertical plate (breadth $2B$) set into motion horizontally with constant acceleration a; solution for short times $t \ll \sqrt{B/a}$. Pressure distribution along the straight line through edge and vortex core (*W. Blendermann* after [26]).

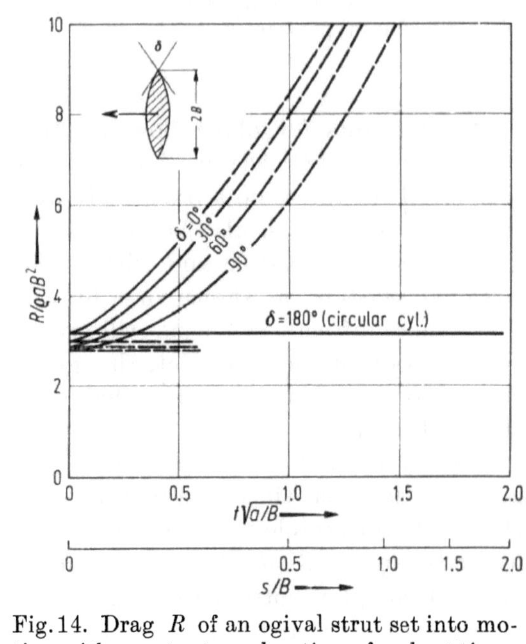

Fig. 14. Drag R of an ogival strut set into motion with constant acceleration a for short times t or distances s without mutual interferences of the vortices at both edges [26].

velocity and pressure become infinite. The only possibility to remove this singularity is to assume—in inviscid fluid—the formation of a spiral vortex, such as it is shown in Fig. 13 [26]. Here, a thin vertical plate (breadth $2B$) is accelerated from rest horizontally with constant acceleration a in a fluid at rest. At first, when the size of the spiral is small compared with the plate breadth, there is no mutual interference between the two vortices on both plate edges. The graph gives the static pressure made dimensionless for points on the straight line through the edge and the vortex core: $p - p_\infty = \pi \cdot \varrho \cdot (a^2 Bt)^{2/3}$, with ϱ density, t time $\ll \sqrt{B/a}$. Pressures and velocities are now finite everywhere in the fluid. Probably, the minimal pressure in such a vortex core determines cavitation inception.

To the kinetic energy of the spiral vortex corresponds an increase of virtual mass or resistance as shown in Fig. 14 [26] for cylinders of ogive section. Although here the acceleration a is assumed constant the resistance increases rapidly with time, beginning with the virtual mass value calculated without edge vortices [27]. In general, the virtual mass depends on time history of the motion when edge vortices appear. With time, of course, these growing spiral vortices become unstable and decay into a multitude of little vortices, yet, at least for the beginning phase this description should be realistic.

4.5 Boundary layer

At last, we must leave the happy inviscid world of the theorists to see what experimenters have found out about boundary layers of ships. For the viscous resistance already *W. Froude* proposed to use as a first approximation the skin friction of a plate with the same length as the (thin) ship. Now, turbulent friction C_{F0} of a smooth plate can be measured for Reynolds numbers of order 10^6 to 10^7 and starting from such test results one has to extrapolate to the ship's R_n of 10^8 or 10^9 and more. This "basic" line $C_{F0}(R_n)$ is still the only guide for the extrapolation from model to ship. Various formulations have been proposed as shown in Fig. 15

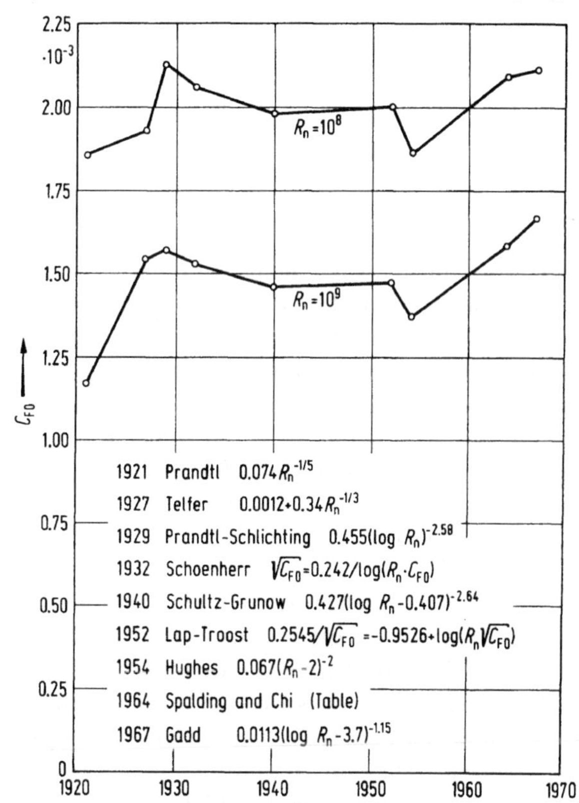

Fig. 15. Frictional resistance coefficient of the plate at two high Reynolds numbers as proposed in the last fivty years.

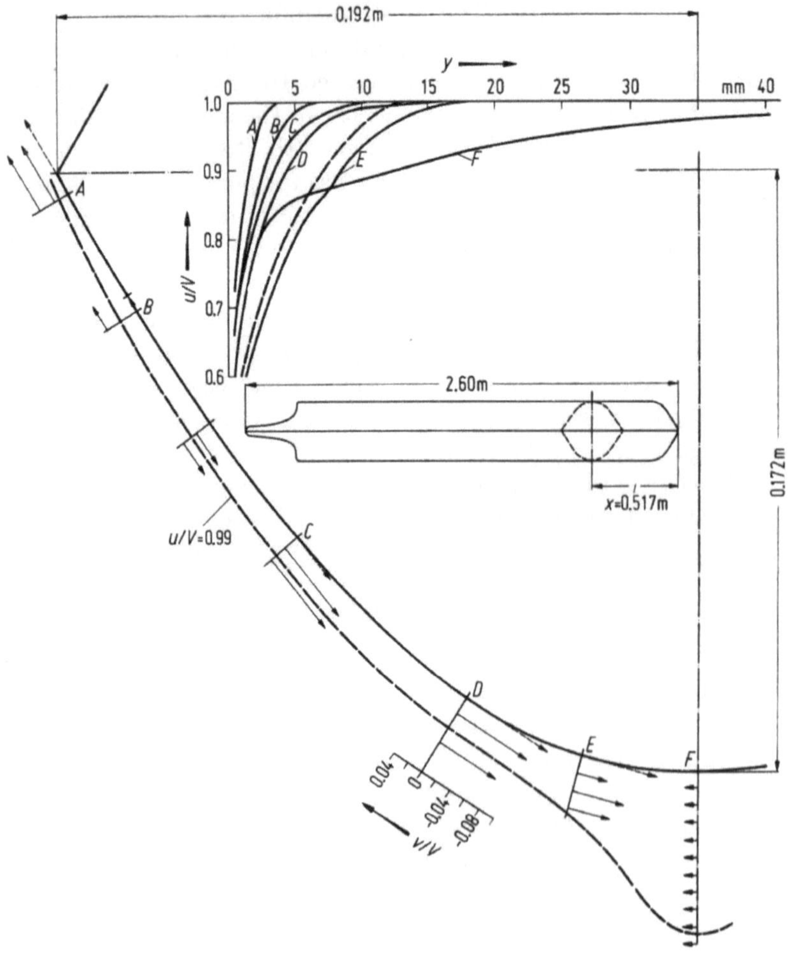

Fig. 16. Boundary layer flow at the forward end of a double model of the "meteor" [28].

where the C_{F0}-values for $R_n = 10^8$ and 10^9 are plotted versus the year when the recommendation was made. If our understanding of turbulence and the testing accuracy would increase steadily with time, both curves should look like damped oscillations approaching the unknown true values of C_{F0}. But, obviously, this final stage has not yet been reached.

The threedimensional boundary layer on a ship or model is, of course, still much more complicate, as shown in Fig. 16 [28]. The boundary layer flow past a double model of the "Meteor" has been measured in a windtunnel, i.e. without surface waves. The velocity component normal to the horizontal main direction is a few percent of the main speed, and—here at the fore ship—about the same as calculated for potential flow (dotted arrows on the hull). This secondary flow is here directed downwards and pulls the boundary layer down so that it is very thick at the keel. After Fig. 17 this secondary flow is negligible midships for inviscid flow, yet, in real flow it changes even direction there inside the boundary layer. In both figures the velocity profiles at various stations are given; the dotted profile is that of plate flow at the corresponding R_n. Near the keel these profiles have dents which are probably due to longitudinal vortices; similar profiles have been measured on a 60 m ship in Japan [29].

The wall shear stress at these two sections versus arc of length measured from keel is shown in Fig. 18 and compared with the corresponding plate values after *Prandtl-Schlichting* and *Hughes*. In spite of the strong variation of the thickness of the boundary layer the distribution of wall shear is comparatively even, resembling conditions in a tube with non-circular cross section.

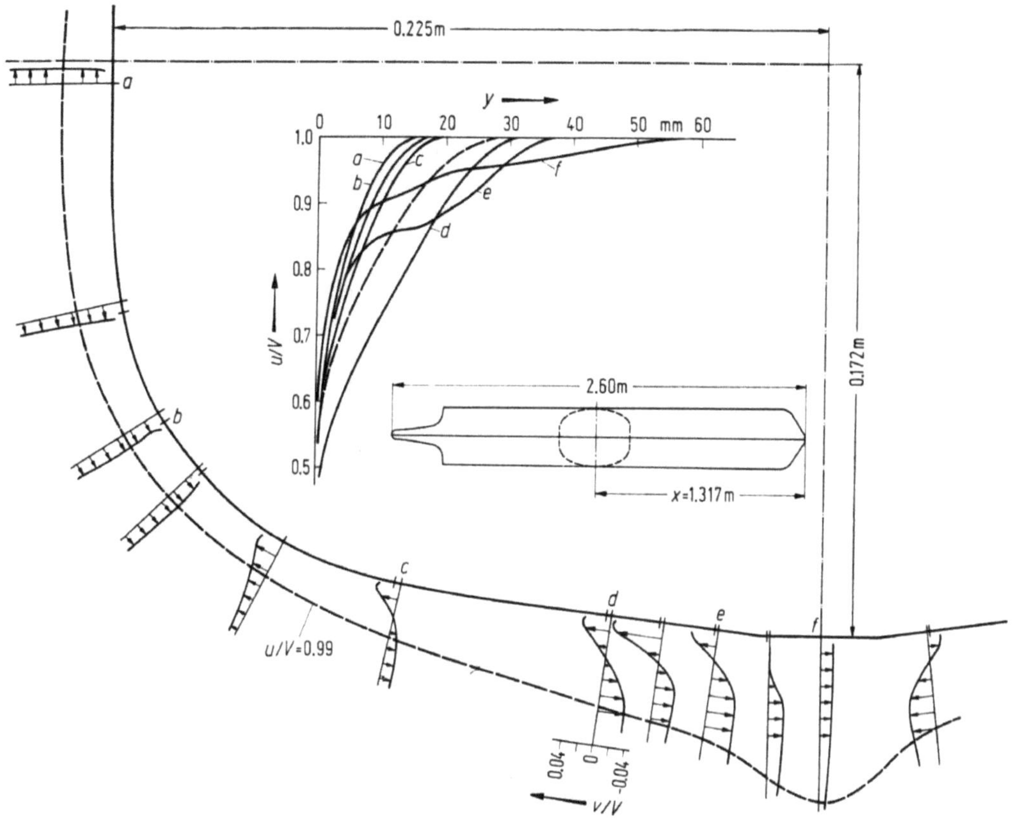

Fig. 17. Boundary layer flow amidships of a double model of the "Meteor" [28].

In Fig. 18 these wind tunnel tests are compared with measurements on the real ship. At the "Meteor" the boundary layer was measured only at one point (starboard and port) midships at $R_n = 2 \cdot 10^8$ [30]. According to the upper diagram in Fig. 19 the local skin friction of the smooth plate is indeed a good guiding line for extrapolation. In the lower diagram the ratio of twice the momentum thickness over distance from bow is compared with the total skin friction coefficient which—for the plate—equals this ratio. The full points are test results on a smaller ship, the "Lucy Ashton" [31]. For this diagram it is obviously of much greater importance to know at which point of the section measurements have been made (near the surface or under the keel) than to know the form and size of the ship!

Finally, a remark on an instationary boundary layer. *M. J. Lighthill* [32] has calculated the laminar boundary layer when the oncoming flow relative to the body oscillates in magnitude. He suggests that some of his results might be extended to turbulent flow. An application to the "coasting" of a ship whose propulsion has been stopped suddenly is shown in Fig. 20. Here, k_i is the usual virtual mass in length direction for inviscid potential flow divided by the mass of the ship, which for this estimate has been replaced by half an ellipsoid of revolution. The additional coefficient k_v is due to the turbulent boundary layer on a plate as for a thin ship. The sum of $k_i + k_v$ is plotted vs. Froude's length coefficient M, i.e. length L over cube root of displacement $\nabla^{1/3}$. (The change of wave resistance of the stopping ship is not considered here.) It is to be seen that for slender ships the longitudinal virtual mass depends much on R_n. Unfortunately however, both virtual masses are small, only a few per cent of the ship's mass, so that even an experimental verification will be difficult. In order to stop a supertanker in danger of collision it is, of course, better to rely on backing the propeller or on braking flaps or under water parachutes.

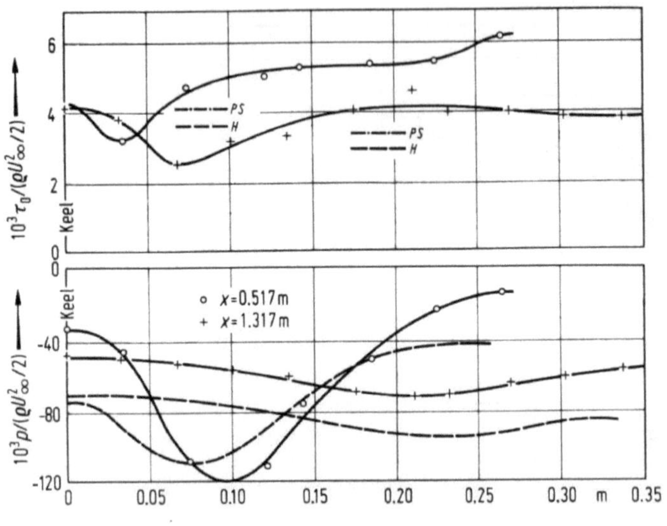

Fig. 18. Wall shear stress τ_0 and static pressure p on the two sections of Fig. 15 and 16 vs. arc of length from keel. Corresponding friction on plate after Prandtl-Schlichting (PS) or Hughes (H). Dotted curves: pressure on hull for potential flow [28].

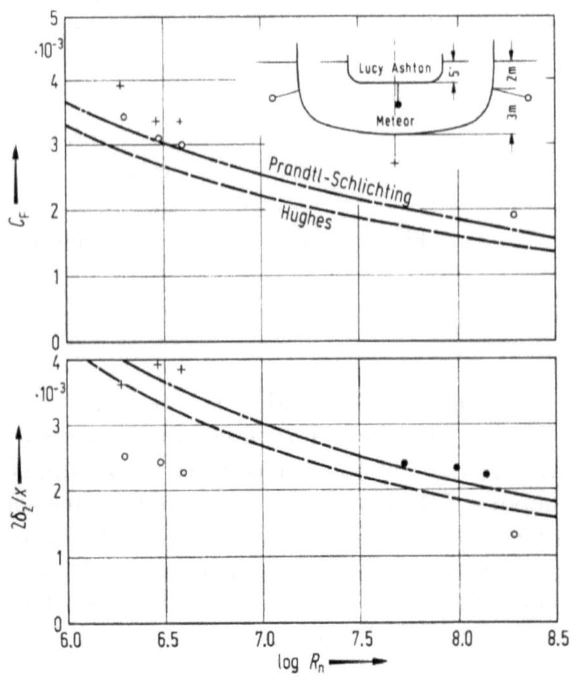

Fig. 19. Local skin friction coefficient C_f and $2\delta_2/x$ at the keel (●, +) and at position WL3 (○) for model and ship vs. R_n. Reynolds number $R_n = Vx/\nu$ with δ_2 = momentum thickness and x = longitudinal distance from bow [28, 30, 31].

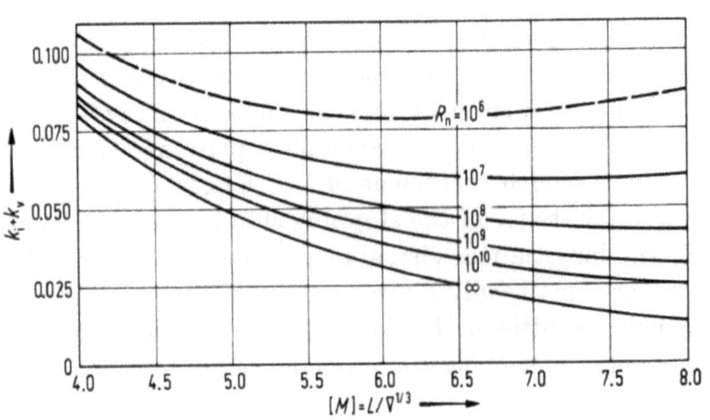

Fig. 20. "Coasting" of a ship, i.e. stopping without propeller: virtual longitudinal mass coefficient k_i (for inviscid fluid) $+k_v$ (for boundary layer) vs. length-displacement ratio. k_i calculated for half a sheroid, k_v fora plate after [32].

5. Conclusions

This bird's eye view on ship hydrodynamics has concentrated on methods of investigation without considering the practical needs of naval architects. Theoretical progress still seems possible only in fields where viscosity effects are negligible so that potential flow is adequate. Surprisingly, this is the case for the oscillatory ship motions where—apparently—viscous effects are too slow to accumulate.

The well known edge vortices can be calculated numerically in detail, nowadays. This extension of usual unsteady potential flow should give realistic results, at least for small times before the unstable spiral vortex sheet decays into many separated vortices.

Unfortunately, already for the simple flow past a ship at constant speed viscosity effects are decisive and then the flow can be studied only experimentally since a theory of three-dimensional turbulent boundary layers is badly missing. Even the extrapolation problem from model to ship, which is, obviously, fundamental for all towing tank tests, cannot be solved but by crude assumptions which become rather suspect e.g. for modern full form ships. The venerable old problem of turbulent flow is still the main stumbling-block for a rational ship hydrodynamics.

References

1. 11th Internat. Towing Tank Conference, Tokyo (1966) 176.
2. 12th Internat. Towing Tank Conference, Rome (1969) 349.
3. Wieghardt, K.: Schiffstechnik 14 (1967) 24.
4. Isay, W. H.: Schiffstechnik 18 (1971) 46.
5. Mallock, A.: Proc. Roy. Soc. A 84 (1910) 391.
6. Hoyt, J. W.: 13th Intern. Towing Tank Conf. Berlin/Hamburg 1972.
7. Dagan, G., Tulin, M. P.: J. Fluid Mech. 51 (1972) 529.
8. Wu, T. Y.: Intern. Shipbuild. Prog. 14 (1967) 88.
9. Brard, R.: J. Ship Res. 14 (1970) 207.
10. Ogilvie, T. F.: Schiffstechnik 10 (1963) 147.
11. Lewinson, G., Maclean, W. M.: J. Ship Res. 12 (1968) 116.
12. Sharma, S. D.: Hamburg. Schiffbau-Versuchsanstalt F 8/68 (1968).
13. Inui, T.: 60th Anniv. Ser. Soc. Nav. Arch. Japan 2 (1957) 173.
14. Sharma, S. D.: Jahrb. Schiffbautechn. Ges. 59 (1965) 458.
15. Kajitani, H. J.: Soc. Nav. Arch. Japan 118 (1965).
16. Eckert, E.: Jahrb. Schiffbautechn. Ges. 64 (1970) 129.
17. Sharma, S. D.: Jahrb. Schiffbautechn. Ges. 64 (1970) 140.
18. Shearer, J. R., Steele, B. N.: Roy. Inst. Nav. Arch. 112 (1970) 465.
19. Muntjewerf, J. J.: Roy. Inst. Nav. Arch. 112 (1970) 199.
20. Paulling, J. R., Porter, W. R.: Proc. 4th US Nat. Congr. Appl. Mech., Berkeley, Calif (1962) 1369.
21. Haskind, M. D.: Izv. Akad. Nauk SSSR Otd. Tekh. Nauk (1957) 65.
22. Newman, J. N.: J. Ship Res 6 (1962) 10.
23. St. Denis, M., Pierson, W. J.: Soc. Nav. Arch. Mar. Eng. 61 (1953) 280.
24. Pierson, W. J., Chao, Y. Y.: TR-61-5, New York Uni., Dep. Met. Ocean.
25. Vugts, J. H.: Intern. Shipbuild. Prog. 15 (1968) 251.
26. Blendermann, W.: Schiffstechnik 16 (1969) 3.
27. Landweber, L.: J. Ship Res. 1 (1957) 20.
28. Wieghardt, K.: van Lammeren Jubilee Memor. 1970, NSMB, Wageningen.
29. 12th Intern. Towing Tank Conference, Rome (1969) 116.
30. Wieghardt, K.: Jahrb. Schiffbautechn. Ges. 62 (1968) 189.
31. Smith, S. L.: Roy. Inst. Nav. Arch. 97 (1955) 525.
32. Lighthill, M. J.: Proc. Roy. Soc. A 224 (1954) 1.

Topical reviews on subjects of ship hydrodynamics may be found in: a) Proc. Intern. Towing Tank Conf.; b) Proc. Symp. on Naval Hydrodyn. Ed. Office of Naval Research, Arlington, USA; c) Annual Review of Fluid Mechanics, since 1969.

Some phenomena accompanying high-speed collision of solids

A. A. Deribas

Novosibirsk, USSR

1. Introduction

This report deals with some phenomena accompanying high-speed collisions of solids in the most exemplary case of metals. Explosives are mostly used in these processes to obtain high-speed collisions. Some papers are known dealing with shock compressibility of metals in which normal collisions of surfaces were used with velocities up to 15 km/sec. These experiments resulted in the creation of a gas-dynamic model of the calculation of phenomena taking place in metals up to very high pressures in the order of several megabars. This approach proved to be very successful with regard to the understanding of the process taking place in metals under shock loading. In several cases the calculations were of high practical interest.

There were also some phenomena discovered, the explanation of which could not be found in the frame of the gas-dynamic model. These phenomena could be observed under oblique collisions when the area of high pressure near the contact lines was moving along the colliding surfaces. The typical set-up of oblique collisions can be seen in Fig. 1.

1.1. Cumulation

The cumulation was the first to be investigated among these phenomena. We consider it as a process of creating a jet of metal with the help of compressing metal cones by a shaped charge of HE. The increase of the destructive effect of such charges was already discovered in the last century. A theoretical explanation of the phenomenon was given in the first post-war years (1946—1948) simultaneously in the USSR and the USA with the help of an ideal non-compressible fluid model. The authors of the hydrodynamic theory of cumulation—*M. A. Lavrentiev* in the USSR [1] and a group of authors in the USA [2] (*Birkhoff*, *Taylor* a.o.)—proposed to consider the process of the compression of cones as being similar to the well-known hydrodynamic problem of the collision of ideal non-compressible fluid jets. This method permitted to predict the creation of a high-speed cumulative jet which was later discovered during experiments.

Some effects appeared during experiments after these fundamental investigations had been carried out which could not be explained with the help of the theory of hydrodynamics.

In particular, a phenomenon was noted which consists of the disappearance of a cumulative jet for small cone angles. This effect was mainly explained by instability and destruction of the cumulative jet when it becomes too thin for small cone angles. It was also suggested that an "inner explosion" of a thin jet as a result of high gradients of pressure might be possible. These considerations were of a qualitative character and excluded the possibility of a quantitative evaluation. The work by *Walsh* and co-workers [3], published in 1954, was the first theoretical explanation of the effect of the disappearance of a jet. The authors of this paper rejected one of the main ideas of the hydrodynamic theory, i.e., the idea of incompressibility of the shell material, and investigated shock-wave situations near the point of contact. A two-dimensional scheme of collision of plates was used for this investigation. A case of supersonic velocity of the point of contact was under study. The authors of [3] succeeded in proving experimentally that the formation of a cumulative jet is incompatible with the existence of oblique shock waves connected with the point of contact.

1.2. Explosive welding

The explanation of the reasons of the disappearance of the jet proposed in [3] satisfied researchers for some period of time. However, an intensive investigation of oblique collisions of metallic surfaces was started in the early sixties, engendered by the invention of the process of explosive welding. Figure 1 shows the most exemplary scheme of welding with the velocity of the point of contact reaching the value of 2—3 km/sec and the angle of collision not exceeding 15°.

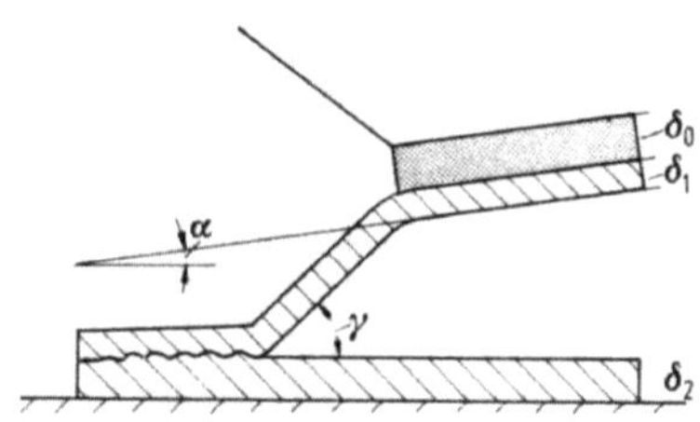

Fig. 1. Schematic diagram of an oblique collision.

The first experiments of explosive welding showed that no stable cumulative jet is formed under this process. Only small values of angles could be used for comparison with the Walsh scheme while the existence of shock waves at the point of contact was excluded for the subsonic regime of collision. Usually, a cloud of dispersed particles could be observed before the point of contact by explosive welding, and that was the reason why considerations were proposed concerning the instability of a jet at small angles γ for non-symmetrical collisions. Further investigations showed that the considerations about the decisive role of non-symmetry proved to be erroneous. Figure 1 shows the situation in which a stable plane jet was fixed at a large angle γ. Besides, a connection was established between the presence and the absence of a jet and the structure of the zone of connection behind the point of contact.

2. New phenomena

2.1. Process of wave-formation

2.1.1. The contact surface remains smooth under collisions accompanied by a cumulative jet. Periodical waves are observed on the surface of the colliding material when small angles and subsonic velocities of the point of contact are used and when a cloud of dispersed particles is observed in front of the point of contact (Fig. 2). Experimental investigations helped to establish the dependence of the length λ and the amplitude a of the waves on the parameters of collision [4]:

$$\begin{aligned} \lambda/\delta_1 &= 26 \sin^2 \gamma/2, \\ a/\lambda &\approx 0.25. \end{aligned} \tag{1}$$

The first of these formulae was obtained for the case when the thickness δ_2 of a ground plate could be considered infinite.

Taking into account the fact that no parameters are present in these formulae which could characterize the strength of the colliding materials, researchers tried to explain the process of wave-formation in terms of hydrodynamics.

2.1.2. The paper of *Hunt* [5] considers the instability of a tangential shock as the reason for wave-formation. However, consideration of this instability proved to be not enough to determine the single wave length, and *Hunt* suggested certain additional ideas which helped to achieve a "qualitative" agreement between his calculations and experiments. If we were to evaluate the essence of this conception, it should be noted that the waves are also formed in the case of symmetrical collisions with the tangential shock in the sense of [5] not being present. It is also necessary to explain the non-damping character of the oscillations of the free surface which cannot be explained by considerations of instability.

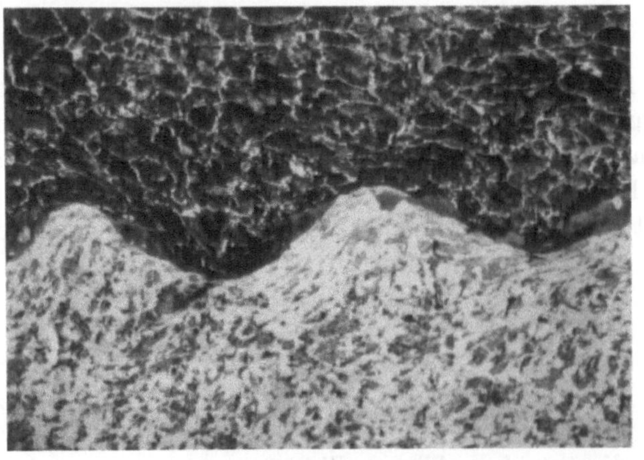

Fig. 2. Wave formation.

The paper [6] suggested an analogy between the wave-formation in the case of explosive welding and the formation of Kármán's periodical vortex-street. This idea was investigated in the paper [7] which shows that if one draws an analogy between the wave-formation and Kármán's street while using the approximation of Oseen for the Navier-Stokes equations it turns out that the Reynolds number for which waves are formed cannot exceed 4. The formation of a periodical vortex-wake in a liquid, however, was observed in the Reynolds number interval of 20—300. It is also necessary to note that periodically situated vortex areas which are characteristic for Kármán's street are not always observed and that one can observe sinusoidal waves without vortex areas (Fig. 2).

2.1.3. In the paper [8], published in 1970, the collision of plates made of identical materials was considered in terms of the theory of ideal compressible fluids. The problem for small angles of collision was solved by the theory of conformal transformation, and a complex potential, characterized by the following singularity at the point of contact, was found for the flow:

$$f(z) \sim \frac{U \sin \gamma/2}{\sqrt[4]{1 - U^2/c_0^2}} \sqrt{\frac{2\delta_1\delta_2}{\pi(\delta_1 + \delta_3)}} \frac{1}{\sqrt{z}}. \tag{2}$$

Here, U is the velocity of the point of contact, c_0 is the sound velocity. These ideas permit to determine the length scale which characterizes the size of the high pressure zone in the vicinity of the point of contact, i.e. the radius of the free surface at the origin which is determined by the following formula:

$$R = \frac{2}{\pi} \sqrt{1 - \frac{U^2}{c_0^2}} \frac{2\delta_1\delta_2}{\delta_1 + \delta_2} \sin^2 \frac{\gamma}{2}. \tag{3}$$

In the case of $\delta_2 \to \infty$, this formula becomes similar to Formula (1). A preliminary conclusion was drawn from here that the size of waves is determined by the size of the high-pressure zone in the vicinity of the point of contact.

2.1.4. A special investigation was carried out to study the mechanism of the initiation of waves. Experiments showed that waves appear at a certain distance from the point of the beginning of the collision; their size is gradually growing and reaches stationary values after several oscillations. It was supposed that to initiate the process of wave-formation it is necessary to have a certain disturbance at the point of contact; the rarefaction wave coming to the point of contact from the nearest free surface possibly serves this purpose. The distance from the start of the collision at which the point of contact meets the rarefaction wave may be easily evaluated within the following scheme of linear acoustics:

$$l = 2\delta_1 U/\sqrt{c_0^2 - U^2}. \tag{4}$$

Experiments showed that waves appear at the approximate distance l from the beginning of the collision. An estimation of the width of the rarefaction wave showed that its order of magnitude agrees with that of the length of the resulting waves [8]. Experiments with a ground plate having a step were carried out to test experimentally the role of the rarefaction wave in the initiation of waves. If the size of this step was within the amplitude of the appearing waves, the waves appeared immediately after the step independently of the distance between the step and the beginning of the collision.

In cases where the size of the step was much bigger or smaller than the size of the waves, which is determined by parameters of the collision, wave-formation behind the step did not occur.

2.1.5. These experimental results enable dus to assume that wave-formation is a self-oscillating process with a "rigid" initiation [7].

The investigation of oscillations which can exist in the system of colliding plates permits us to propose, with the help of the acoustic equations, the following general expression for the pressure:

$$p = A \sin\left[k_1 c_0 \sqrt{1 - \frac{U^2}{c_0^2}}\, t + \frac{k_1 U}{c_0 \sqrt{1 - U^2/c_0^2}} (x - x_0)\right] \sin(\nu\theta + \alpha)\, Z_\nu(k_1 r), \tag{5}$$

where

$$\operatorname{tg}\theta = \frac{y}{x - x_0}\left(1 - \frac{U^2}{c_0^2}\right)^{1/2}, \qquad r = \left[\frac{x - x_0}{\sqrt{1 - U^2/c_0^2}} + y^2\right]^{1/2}.$$

Here Z_ν is the Bessel function of order ν, while α and k_1 are arbitrary constants and x_0 is a certain arbitrary displacement of the origin where the source of disturbance is situated. We confine the investigation to the waves that spread into infinity, and we consider asymmetrical oscillations of the free surface, choosing $\alpha = 0$ for ν being an integer and $\alpha = \pi/2$ for ν being a semi-integer in order to satisfy the condition $p = 0$ at the free surface. Now one may obtain from (5) the following formula for the pressure:

$$p = A \sin\left[k_1 c_0 \sqrt{1 - U^2/c_0^2}\, t - \frac{k_1 U(x - x_0)}{c_0 \sqrt{1 - U^2/c_0^2}}\right] H_\nu^{(1)}(k_1 r) \sin\theta, \tag{6}$$

where $H_\nu^{(1)}(k_1 r)$ is the Hankel function of the first type of order ν. Oscillations of the free surface described by Formula (6), and also by corresponding equations for the velocities which could be obtained from (6) and from the equation of motion in the acoustic approximation, can describe the actual wave process at distances from the point of contact which essentially exceed the size of the high-pressure zone but which are less than the thickness δ_1 of the flying plate. The influence of free boundaries at long distances will lead to exponential damping of the amplitude. In order to determine the length of the wave or the oscillation period, which is a free parameter in Formula (6), and also to explain the non-damping character it is necessary to develop the idea of surface waves as a self-oscillating process.

2.1.6. An oscillator and a source of energy, supporting non-linear oscillations, should be present in any similar process. Let us imagine what can play the role of an oscillator in this case.

Let us consider the vicinity of the point of contact, the size of which we have estimated earlier (3). The density and, consequently, the sound velocity in this zone essentially differ, because of the high pressure, from their original values in an undisturbed material. Proceeding from this, we may consider the present vicinity as a source of oscillations with frequency ω the order of which can be determined by

$$\omega \approx c_0 R. \tag{7}$$

The characteristic time τ of the oscillation's damping may be considered as having the order of the period

$$\tau \approx 2\pi/c_0 R. \tag{8}$$

Complex frequencies $\varkappa = i\omega - \tau^{-1}$ of the oscillator may be determined as proper frequencies of the following system of equations:

$$\left.\begin{aligned} \varkappa u + U(x, y)\frac{\partial u}{\partial x} + V(x, y)\frac{\partial u}{\partial y} + \frac{1}{\varrho(x, y)}\frac{\partial p}{\partial x} &= 0, \\ \varkappa u + U(x, y)\frac{\partial v}{\partial x} + V(x, y)\frac{\partial v}{\partial y} + \frac{1}{\varrho(x, y)}\frac{\partial p}{\partial y} &= 0, \\ \varkappa p + U(x, y)\frac{\partial p}{\partial x} + V(x, y)\frac{\partial p}{\partial y} + \varrho(x, y)\, c^2(x, y)\left(\frac{\partial u}{\partial x} + \frac{\partial v}{\partial y}\right) &= 0. \end{aligned}\right\} \tag{9}$$

The velocities $U(x, y)$, $V(x, y)$, the density $\varrho(x, y)$, and the sound velocity $c(x, y)$ should be obtained by solving a stationary problem. In general, proper values cannot be obtained without complicated computations. The following is an approximate solution of this system when $U = 0$, $V = 0$ and when the distributions of the density and the sound velocity are assumed as follows:

$$\begin{aligned} \varrho(x, y) &= \begin{cases} \varrho_0, & x^2 + y^2 > R^2, \\ \varrho^* > \varrho_0, & x^2 + y^2 < R^2; \end{cases} \\ c(x, y) &= \begin{cases} c_0, & x^2 + y^2 > R^2, \\ c^* > c_0, & x^2 + y^2 < R^2. \end{cases} \end{aligned} \tag{10}$$

Since $p = 0$ at $x < 0$, $y = 0$ on the free surface (angles are assumed small) let us try to obtain the solution of the system (9) as:

$$p(x, y) = \sin\theta \cdot W(r) \tag{11}$$

while concentrating only on anti-symmetrical oscillations. From (9) we have

$$p(x, y) = \sin\theta \cdot Z_1\left(\frac{i\varkappa}{c} r\right), \qquad c = \begin{cases} c_0, & r > R_0, \\ c^*, & r < R_0. \end{cases} \tag{12}$$

Here, Z_1 means different branches of cylindrical functions of the first order depending on r being smaller or bigger than R_0. The condition at infinity is that $W(r)$ differs, at $r > R_0$, from Hankel's function $H_1^{(1)}\,(i\varkappa r/c_0)$ only by a constant. The regularity of the function to be found at point $x = 0$, $y = 0$ gives us

$$W(r) = \text{const}\; J_1(i\varkappa r/c). \tag{13}$$

The boundary conditions at the circle $r = R_0$ enable us to determine the discrete series of eigenvalues $\varkappa$. Calculations made in this case showed that the oscillation's frequency and decrement of damping depend only slightly on the parameter $\varrho^* c^{*2}/(\varrho_0 c_0^2)$. The frequency here is of the order of the radius of the high-pressure zone, the former being determined by the parameters of collision according to Formula (3).

The considerations discussed above permitted us to estimate the length of the waves appearing at explosive welding. The estimates obtained are confirmed by the existing empirical Formula (1) for the wave length, at least as far as the order of magnitude is concerned.

2.1.7. Now it is necessary to explain the non-damping character of the oscillations. The experiments to measure the temperature of the bounding zone at explosive welding are directly related to the explanation of this problem [9]. These measurements were based on the determination of the instantaneous thermo-EMF in the case when the colliding plates produce thermopaar. The thermo-EMF was registered, as a function of time, by a high-velocity oscilloscope. A typical form of the record is given in Fig. 3. Attempts to interprete such records for small values of time failed because of the instability of the records. Besides, the existence of high-pressure influences the value of the EMF (baroeffect) and engenders basic difficulties in determining the temperature by this method. This is why one had to confine the measurement to relatively large values of time after collision, when the point of contact passes entirely along the colliding surfaces. For large values of time the curve takes a smooth character (Fig. 3) and may be transformed into the curve giving the dependence of temperature on time with the help of the data of the static calibration of the thermocouple. The analysis of the obtained curve produced very interesting results. First of all, it turned out that the measured decrease of temperature with time is of a typical power-law character. Such a character of decrease suggested an idea of interpreting the obtained curve by assuming the existence, at the initial

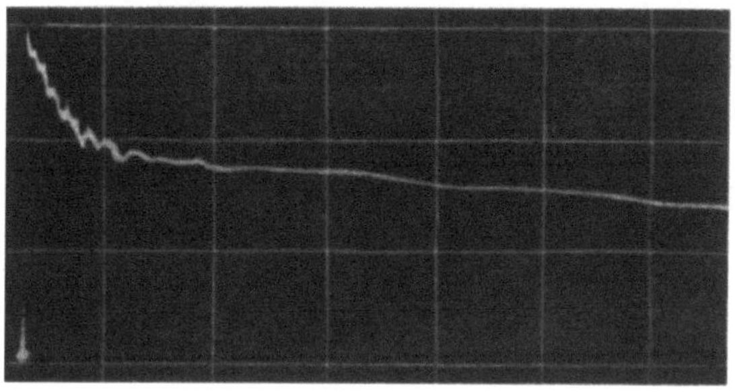

Fig. 3. Thermo-EMF vs. time.

moment, of a concentrated (linear) instant source of heating with intensity Q and also of further spreading of heat according to the mechanism of usual thermoconductivity. The formulated assumption permitted to determine, in an experimental way, the intensity Q of the hypothetical source by choosing a function from among the curves:

$$T(t) = B + A/\sqrt{t} \tag{14}$$

and determining the constants A and B in such a way that the curve (1), at its initial part, differs as little as possible from the experimental one (Fig. 4). Thus, one could determine the intensity of the instant source Q, for the solution of the thermoconductivity equation in this case is as follows:

$$T(t) = T_0 + \frac{Q}{2c\varrho\sqrt{\pi a^2 t}}. \tag{15}$$

Here a is the coefficient of thermoconductivity, c is the thermocapacity per unit of mass, T_0 is the initial temperature.

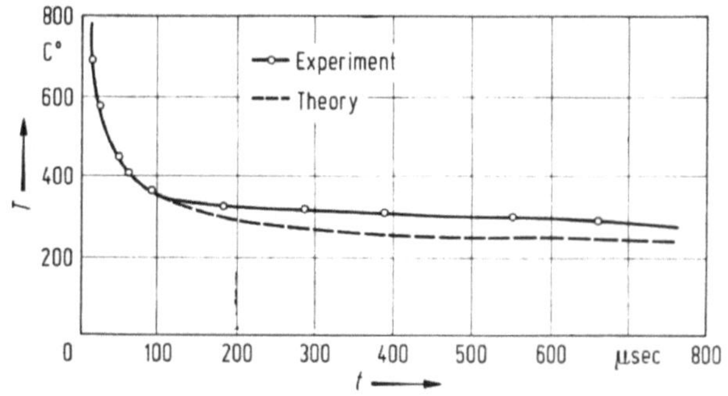

Fig. 4. Temperature vs. time.

Comparison of the experimental constants A and B with Formula (15) for the case of a collision between Ni- and Fe-plates has shown that the value of Q consists of approximately 3% of the kinetic energy of the flying plate. The main part of this energy is radiated more evenly in the entire volume of the colliding plates and it determines their final temperature which in our experiment was about 100 °C. At the same time one may clearly notice temperatures up to 700 °C at the curve in Fig. 4. This value of temperature permitted to estimate the width of the zone in which the "instant heat" Q is radiated. This zone proved to be a layer with the width not less than 10 times smaller than the amplitude of the waves formed. The most probable mechanism of the emergence of this energy is the surface friction between the layers of metals in touch.

The experiments thus have shown that a special mechanism of energy emergence exists in the surface layer. It may be supposed that it is the energy which is being spent on the supporting of non-damping self-oscillations which initiate the formation of waves.

2.2. Role of viscosity in the process of collision

The absence of the cumulative jet in the regime of oblique collisions at small angles γ has already been mentioned. It is of interest in connection with this to find out how the law of momentum conservation may be satisfied in this process. Special experiments [10] were carried out to investigate the picture of deformation in the entire volume of the colliding plates outside the zone of wave-formation. Insets of the same material were welded into the plates to mark the planes perpendicular to the colliding surfaces. The samples were cut after explosive welding and macrophotographs of the cross-section were investigated to determine the displacement, parallel to the line of welding, of the points of the marked planes depending on the distance $z(y)$

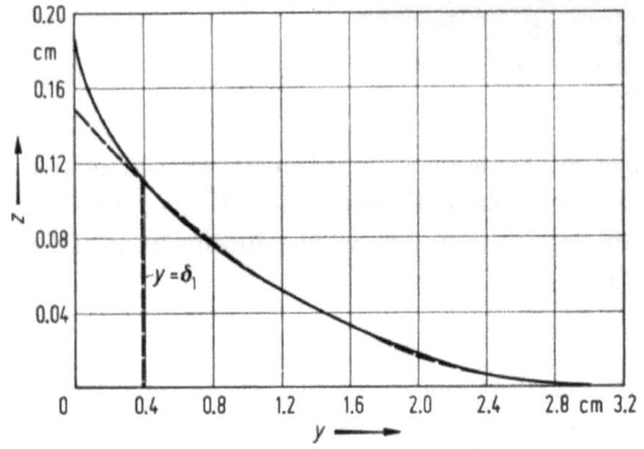

Fig. 5. Horizontal displacement vs. distance, $z(y)$.

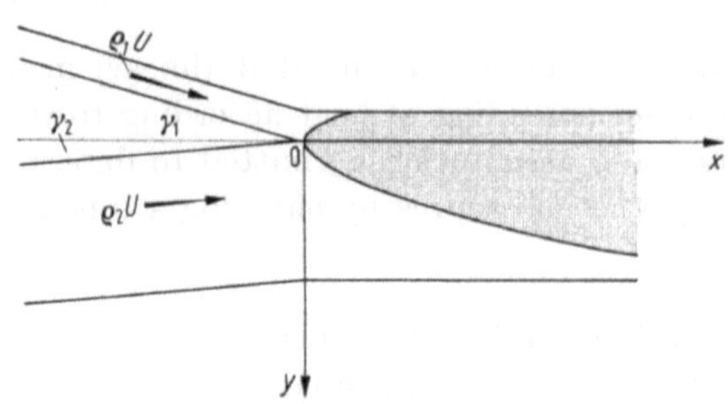

Fig. 6. Schematic diagram of the collision without reverse jet.

from the interface. A characteristic plot of the experimental dependence $z(y)$ is shown in Fig. 5. The investigation of the curves showed that at $y > \delta_1$ they may be well described by the parabola equation

$$z = a(y - \delta_2)^2, \tag{16}$$

where the coefficient a may be chosen by comparison with the experimental curve (see Fig. 5).

Let us assume that in considering the movement in a coordinate system connected with the point of contact we meet the situation in which the slowing of the colliding flows occurs at this point and an internal jet appears with a velocity smaller than that of the surrounding flow. Let us suppose that viscosity operates in the flow and a diffusion of horizontal velocity takes place and that the velocities of the flow become uniform at infinity (Fig. 6). Let us assume also that we may ignore viscosity in the non-shadowed area of the flow. Let us choose the system of coordinates in such a way that a vertical velocity component behind the point of contact would be absent. According to the law of momentum conservation we have

$$\varrho_1 \delta_1 U^2 \sin \gamma_1 = \varrho_2 \delta_2 U^2 \sin \gamma_2 . \tag{17}$$

Assuming $\gamma = \gamma_1 + \gamma_2$ for the case of small angles and using the momentum conservation law along the x-axis we may determine the velocity at infinity behind the point of contact:

$$u_\infty = \frac{U\gamma^2}{2} \frac{\varrho_1 \delta_1 \varrho_2 \delta_2}{(\varrho_1 \delta_1 + \varrho_2 \delta_2)^2} . \tag{18}$$

Let us now consider the diffusion of horizontal velocity caused by the effect of viscosity. The diffusion equation is as follows:

$$U \frac{\partial u}{\partial x} = \frac{\partial}{\partial y} \nu \frac{\partial u}{\partial y} . \tag{19}$$

Let us integrate this equation over x from $-\infty$ to $+\infty$ everywhere, with the exception of the line crossing the point of contact. We have

$$U(u_\infty - u_{-\infty}) = U\nu \, d^2z/dy^2 . \tag{20}$$

Here $z(y)$ is the horizontal displacement of the point from its original position. In our assumption $u_{-\infty} = 0$.

Let us state boundary conditions

$$\partial z/\partial y = 0 \quad \text{at} \quad y = \delta_1,$$
$$\partial z/\partial y = 0 \quad \text{at} \quad y = \delta_2,$$

which determine the absence of shear stresses at the free boundaries. Additionally, we may consider $z = 0$ at outside boundaries if we choose correspondingly the origin of displacements. The use of these conditions would enable us to obtain the formula for displacement which in the case of a ground plate would be as follows:

$$z(y) = \frac{U\,\delta_1\delta_2}{\nu(\delta_1+\delta_2)^2}\sin^2\frac{\gamma}{2}\,(y-\delta_2)^2. \tag{21}$$

The assumption of the existence of this flow is thus confirmed by the parabolic character of the experimental curve (16). Comparison of Formulae (16) and (21) permits to determine the coefficient of viscosity:

$$\mu = \frac{\varrho U\,\delta_1\delta_2}{a(\delta_1+\delta_2)^2}\sin^2\frac{\gamma}{2}. \tag{22}$$

The analysis of experimental data for different metals permitted to calculate the coefficients of viscosity. The values obtained were of the order 10^5 Poise while the difference of the values of the coefficients for different metals was within the same order.

The obtained values of the coefficients of viscosity could be compared with those obtained by other authors. Thus, during the experiments by *A. A. Il'yushin*, published in 1940 [11], the coefficient of viscosity at deformation velocities of the order of 10^3 sec^{-1} were determined and the values obtained differed only slightly from those received at explosive welding. A series of papers was published in 1964—1967 [12, 13] where the viscosity of metals under conditions of shock compression was determined by the investigation of the damping of sinusoidal shock waves. The authors of these papers obtained similar values of viscosity coefficients though these values differ several times from those which were calculated by us. It should be noted that the idea to determine the viscosity coefficient of metals at explosive welding was published in 1967 in the paper [14]. However, an error in measuring the deformation lead to a still greater mistake in the estimation of the value of the coefficient obtained.

2.3. Criterion of jet-formation

Viscosity forces considered in the previous section may be used to explain the already mentioned experimental fact of the disappearing of the cumulative jet for small angles of collision. The criterion to be found should be true for subsonic velocity of the contact point and supplement the existing criterion obtained in the paper [3]. As is known, the solution of the equations of irrotational motion of an ideal incompressible fluid is reduced to a problem of the theory of functions of complex variables by introducing the complex velocity $w(z) = u - iv$. Because of the harmonic character of u and v the solution of the problem of the motion of an ideal fluid satisfies the equations of motion of a viscous incompressible fluid. While solving specific problems it is also necessary to satisfy boundary conditions, the number of which is known to exceed, for a viscous fluid, the number of corresponding conditions for an ideal fluid. This is why the solution of the problem of the flow of an ideal fluid may describe the movement of a viscous fluid in the same area only approximately. Thus, for the viscous fluid the *stress-vector* must be equal to zero at the free boundary, while in the case of an ideal fluid the *scalar pressure* should be equal to zero at the free surface. Only when the flow components caused by viscosity approach zero at the free surface, the solution of the problem for an ideal fluid may be used to describe the flow of the viscous fluid.

Let us find out the reaction of the viscous forces on the free boundary which is compensated in the real flow by the formation of a low-velocity layer. The tensor of viscous stresses $\{\sigma_{ik}\}$ may be found with the help of the following formulae:

$$\sigma_{11} + i\sigma_{12} = i(\sigma_{21} + i\sigma_{22}) = \mu[1/2\,(u_x - u_y) + i/2(u_y + v_x)]. \tag{23}$$

According to the Cauchy-Riemann conditions we have

$$\sigma_{11} - i\sigma_{12} = \mu\, dw/dz, \tag{24}$$

or

$$\sigma_{11} + i\sigma_{12} = i\mu(U/R)\, e^{2i\phi}. \tag{25}$$

Here R is the radius of curvature of the free boundary, $Ue^{i\phi} = u - iv$ is the velocity on the free boundary whose magnitude is constant. It is necessary to determine the viscous stresses satisfying the boundary conditions. Let us assume that $\boldsymbol{n}(n_x, n_y)$ is the vector of the external normal to the free surface. Writing

$$n_x - in_y = -ie^{i\phi}, \tag{26}$$

we may determine the viscous component of the force, with which the boundary in the flow under consideration must influence the fluid, by the formula

$$\sigma_x + i\sigma_y = (\sigma_{11} + i\sigma_{12})\,(n_x - in_y). \tag{27}$$

With the help of (25) and (26) we may obtain from (27)

$$\sigma_x + i\sigma_y = \mu\,(u + iv)/R, \tag{28}$$

$$|\sigma_x + i\sigma_y| = \mu\, U/R. \tag{29}$$

The force $\sigma_x + i\sigma_y$ is thus directed along the boundary in the direction of the velocity vector and its magnitude may be found from Formula (29). The above considerations permit to suggest the criterion of jet-formation. In the problem of the symmetrical collision of plane jets one may determine the momentum of a reverse jet:

$$I_1 = 2\varrho\, \delta_1 U^2 \sin^2(\gamma/2). \tag{30}$$

For non-symmetrical collisions the problem may have several solutions. However, in this case one also may obtain an approximate formula for the horizontal component of the reverse jet's momentum:

$$I_1' = U^2 \frac{4\varrho_1\, \delta_1 \varrho_2\, \delta_2}{\varrho_1\, \delta_1 + \varrho_2\, \delta_2}\, \frac{\gamma^2}{4}. \tag{31}$$

We have already determined the force which should be applied to the free surface to have the areas of the flow of viscous and ideal fluids coincide.

Free surfaces in the ideal fluid flow turn in the vicinity of the point of contact at angles $\pi - \gamma_i$ ($i = 1, 2$) by the formation of the reverse jet. The force (29) applied to the free boundary should transmit to the flow an impulse which is directed against the flow during the first half of the turn and along the flow in its second half.

Let us assume that, if during the first half of the turn the horizontal component of the impulse of the force applied is larger than or equal to the momentum carried away by the reverse jet, there would be no reason to the formation of the latter. This assumption permits to obtain an approximate criterion of jet-formation. Let us determine the horizontal component of the momentum during the first half of the turn. We have

$$I_2' = \int_{\gamma_1}^{\pi/2} (\mu_1 U/R) \cos\phi \cdot R\, d\phi = \mu_1 U(1 - \sin\gamma_1) \tag{32}$$

and accordingly for the other jet

$$I_2'' = \int_{\gamma_2}^{\pi/2} (\mu_2 U/R) \cos\phi \cdot R\, d\phi = \mu_2 U(1 - \sin\gamma_2). \tag{33}$$

The entire horizontal momentum is determined by the relation $I_2 = I_2' + I_2''$. If we compare it with the momentum of the reverse jet we can obtain the necessary criterion of jet formation:

$$I_2 \leq I_1. \tag{34}$$

The criterion for the case of symmetrical collisions is as follows:

$$\sin\gamma/2 \geq \left(\sqrt{1 + 4\,\mathrm{Re}} - 1\right)/(2\mathrm{Re}). \tag{35}$$

Here the Reynolds number is determined, as usually, by the formula

$$\mathrm{Re} = U\,\delta/\nu\,. \tag{36}$$

It should be noted that the above result is valid only when the viscosity influences only a thin boundary layer near the surface. Special experiments were carried out to check the validity of the criterion obtained. Their results are presented in Fig. 7. The experiments have shown a satisfactory coincidence of the assumption with the data of the criterion (35) in the area of the transmittance of the jet flow into the regime of collision without formation of a reverse jet. The discrepancy between theoretical and experimental data which were noticed for small Reynolds numbers may be explained by the influence of strength and also by an incorrect determination of the Reynolds number. For a more exact experimental checkup of the criterion obtained, it is necessary to estimate the Reynolds number in the area of high pressure and temperature near the point of contact.

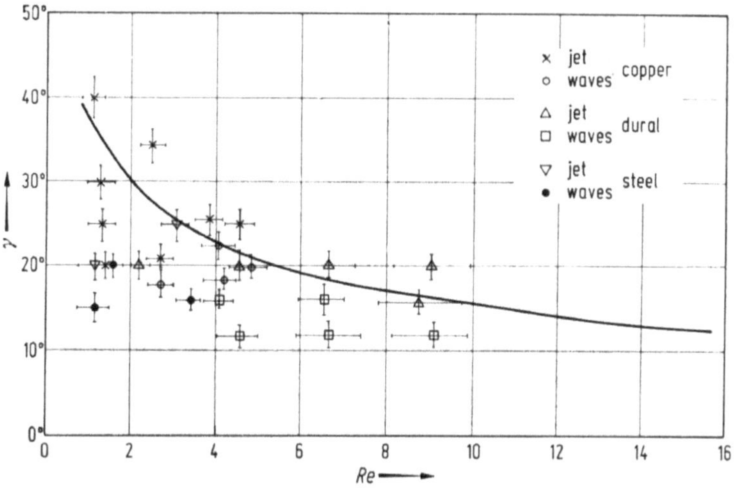

Fig. 7. Existence region of the jet.

3. On theoretical models

Some new phenomena appearing during collisions of solids were considered in the previous section. Approximate theoretical considerations were formulated which, we believe, would enable us to evaluate the estimation of the principal physical factors guiding the considered phenomena, e.g. wave formation. These theoretical approximations, however, can in no way be considered a finalized theory which would confine the investigation of the principal factors guiding the described processes to the solution of certain equations. Thus, the problem of constructing a mathematical model of the considered class of phenomena which would be essentially adequate to the subject is yet unsolved. As far as the ways of the solution of this problem are concerned we can suggest at this junction only some preliminary considerations.

3.1. As it has already been mentioned above, the explanation of the penetrative effect of the cumulative charges was obtained within the model of the ideal incompressible fluid. It is quite natural therefore that researchers tried to find, also within this simple model, the explanation of such phenomena as the disappearance of a jet at small angles and the formation of waves. Several of these attempts, e.g., Kármán's street and the instability of a tangential shock, have already been analyzed here. Similar investigations can by no means be excluded in the future.

In summing up the results of the investigations which have been carried out, we can state as the first conclusion: a theoretical model based on the assumption that metal is an ideal incompressible fluid cannot explain the main peculiarities of the phenomena which are considered in the present paper. This model cannot be essentially improved by taking into account the compressibility to allow an explanation, with the help of the model, of the main physical factors guiding the processes under consideration.

3.2. The viscosity of materials was considered in the previous section to explain some of the effects which were observed during experiments. However, this does not permit us to consider the model of a viscous fluid, described by Navier-Stockes equations, as being suitable for the calculation of these phenomena.

Let us consider some of the conditions which should be satisfied by the model necessary for the calculation of the described processes. The meaning of viscosity is known to be connected with the idea of the relaxation time of shear stresses:

$$\tau = \mu/G, \tag{37}$$

G is the shear modulus. It should be noted that according to the works of *Frenkel* [15] the time τ is considered to be the time of atom oscillations in a solid around a stationary position. The time of a single oscillation is of the order of 10^{-13} sec. The relaxation time for usual fluids including melted metals is of the order of 10^{-11} sec, i.e., an atom makes about 10^2 oscillations around thesingle stationary position within this time. Under usual conditions the lattice in a solid has essentially deaper potential depressions which seriously hampers the transition of atoms from one position to another. The relaxation time in this case is of the order of 10^4 sec, i.e. an atom makes about 10^{17} oscillations in a single position. Let us analyze from this point of view the above-mentioned value of viscosity for the processes of high-speed collisions.

It follows from Formula (37) that the relaxation time is of the order of 10^{-6} sec. Thus, metal, under high speeds of deformation which develop by collisions, behaves like a material between solid and liquid, and the relaxation time decreases by approximately ten orders of magnitude compared with the initial state. At the same time, the value of the relaxation time for collisions in liquids is less than τ by 5 orders of magnitude, and in this sense one may state that a lattice exists though being seriously deformed during collision. Such a transformation occurs with the main quantity of the colliding plates. The state of the metal on the colliding surface may evidently be considered as liquid with a relaxation time of the order of 10^{-11} sec, and one may also consider some of the intermediate layers with intermediate values of the relaxation time.

Thus, for real collisions the behaviour of the metal differs in different areas of the colliding solids. While on the external surface of the plates, which is free of stresses, metals stay elastic, in the areas close to the high-pressure zone it becomes necessary to take into account the plastic behaviour and the corresponding values of the relaxation time of the shear stress and also the transformation of the stress tensor into a spherical one. The construction of a system of equations which can take into account all these processes may enable us to calculate the collision processes with all the accompanying effects of the wave-formation type with the help of the modern computers.

3.3. Recently *Godunov* and *Romensky* [16] constructed a system of equations of a non-linear continuous medium with the help of which it may hopefully be possible in future to investigate in detail the processes similar to those which were considered in this paper. The whole proposed system of equations is as follows:

$$\left.\begin{aligned}
&\frac{\partial \varrho[E + u_i u_i/2]}{\partial t} + \frac{\partial[\varrho u_k(E + u_i u_i/2) - u_i\sigma_{ik}]}{\partial x_k} = 0,\\
&\frac{\partial \varrho u_i}{\partial t} + \frac{\partial(\varrho u_i u_k - \sigma_{ik})}{\partial x_k} = 0 \quad (i = 1, 2, 3),\\
&\frac{\partial \varepsilon_{ii}}{\partial t} + u_k\frac{\partial \varepsilon_{ii}}{\partial x_k} - \frac{\partial u_i}{\partial x_i} + 2\varepsilon_{i\alpha}\frac{\partial u_\alpha}{\partial x_i} = \phi_{ii} \quad (i = 1, 2, 3),\\
&\frac{\partial\, 2\varepsilon_{ij}}{\partial t} + u_k\frac{\partial\, 2\varepsilon_{ij}}{\partial x_k} - \frac{\partial u_i}{\partial x_j} - \frac{\partial u_j}{\partial x_i} + 2\varepsilon_{i\alpha}\frac{\partial u_\alpha}{\partial x_j} + 2\varepsilon_{ij}\frac{\partial u_\alpha}{\partial x_i} = 2\phi_{ij}\\
&\qquad (i, j) = (1, 2); (1, 3); (2, 3),\\
&\varrho = \varrho_0\sqrt{\det\|\delta_{ik} - 2\varepsilon_{ik}\|}.
\end{aligned}\right\} \tag{38}$$

Here x_i, t are independent variables, ε_{ik} are the components of the tensor of deformations, u_i are the components of the velocity vector, S is the entropy per unit mass. The material is characterized by the following equation of state

$$E = E(I_1, I_2, I_3, S), \tag{39}$$

where I_i are the invariants of the deformation tensor. The constant ϱ_0 is an initial density.

From Equation (39) it is possible, by applying Murnaghan's formulae [17], to obtain a connection between the components of the deformation tensor ε_{ik} and the stress tensor σ_{ik}, which enter into the system (38):

$$\sigma_{ki} = \varrho(\delta_{k\alpha} - 2\varepsilon_{k\alpha})\, E_{\varepsilon_{\alpha i}}. \tag{40}$$

In the proposed system the Formulae (40) are essentially a determination of the tensor ε_{ik} through the equation of state and through the stress tensor σ_{ki}.

Equations for ε_{ik} in the system (38) have, in the classic theory of elasticity, zeros on the right hand side. In this case the right hand sides ϕ_{ij} are determined as plastic components of the tensor of deformations velocities:

$$d\varepsilon_{ij}^p/dt = \phi_{ij}. \tag{41}$$

Under such conditions for ϕ_{ij} a usual derivation of Murnaghan's formulae [17] becomes invalid, and they are taken for the determination of ε_{ik}. The tensor ϕ_{ij} is supposed to be co-axial with the tensor ε_{ij} and the main components of ϕ_i are determined by the following relation:

$$\phi_i = -\frac{1}{\tau_i}\left[\varepsilon_i - \frac{\dfrac{\varepsilon_1}{\tau_1}\varrho_{\varepsilon_1} + \dfrac{\varepsilon_2}{\tau_2}\varrho_{\varepsilon_2} + \dfrac{\varepsilon_3}{\tau_3}\varrho_{\varepsilon_3}}{\dfrac{1}{\tau_1}\varrho_{\varepsilon_1} + \dfrac{1}{\tau_2}\varrho_{\varepsilon_2} + \dfrac{1}{\tau_3}\varrho_{\varepsilon_3}}\right]. \tag{42}$$

Here τ_i are Kelvin's periods of relaxation time of shear stresses along the main axes [18].

The following relation proves to be satisfied:

$$\varrho_{\varepsilon_{ij}}\phi_{ij} = 0 \tag{43}$$

which enables us to satisfy the equation of continuity and which in fact is the formalisation of the condition that plastic deformations do not lead to any change in volume.

Periods of the relaxation time τ_i should be assumed to be equal to infinity in the elastic area when shear stresses do not exceed the yield point, and for fully developed plasticity it should be assumed:

$$\tau_1 = \tau_2 = \tau_3 = \tau = \mu/G. \tag{44}$$

In the paper [16] it is proved that the system (38) is hyperbolic if the equation of state (39) satisfies several conditions of convexity which generalize the condition $E_{vv} > 0$ in gas dynamics. The equation of the characteristic conoid of the system (38) is also calculated. It is proved that the relations at characteristics, i.e. the stream lines of the system (38), produce equations for the change of the tensor of incompatibility which describe the development of dislocations in non-linear continuous media.

In order to use the system (38) for the solution of physical problems, it is necessary to have experimental data about the values of the periods of the relaxation time τ_i and about their dependence on the state of the medium (stresses and temperature). At present, only orders of magnitude of these values for different states of the medium are known and a very approximate information is obtained concerning the yield point, i.e. concerning the character of changes in the transition zone and the boundaries of this zone.

Additional information is also necessary about the equation of state (39) which would enable us to determine the contribution of each of the invariants of the tensor of deformations. The difficulty is that the shear stresses relax at very short periods of time τ_i, and high-speed registration of the processes or some other indirect experimental methods which would enable us to obtain the necessary data are needed.

References

1. Lavrent'ev, M. A.: Uspekhi matematicheskikh nauk 12 (1957) 41—56.
2. Birkhoff, G., McDougall, D., Pugh, E., Taylor, G.: J. Appl. Phys. 19 (1948) 563—582.
3. Walsh, J., Shreffler, R., Willig, E.: J. Appl. Phys. 24 (1954) 349—359.
4. Deribas, A. A.: Physics of hardening and explosive welding, Novosibirsk 1972.
5. Hunt, J. H.: Phil. Mag. 17 (1968) 669—680.
6. Klein, W.: Z. Metallkunde 4 (1965) 261—268.
7. Godunov, S. K., Deribas, A. A., Kozin, N. S.: Zhurnal prikladnoi mekhaniki i tekhnicheskoi fiziki 3 (1971) 63—72.
8. Godunov, S. K., Deribas, A. A., Zabrodin, A. V., Kozin, N. S.: J. Comp. Phys. 5 (1970) 517—539.
9. Zakharenko, I. D.: Fizika goreniya i vzryva 2 (1971) 269—272.
10. Godunov, S. K., Deribas, A. A., Zakharenko, I. D., Mali, V. I.: Fizika gorenyia i vzryva 1 (1971) 135—141.
11. Il'yushin, A. A.: Uchenye zapiski, Moscow University 39 (1940) 3—81.
12. Sakharov, A. D., Zaidel', R. M., Mineev, V. N., Oleinik, A. G.: Doklady Acad. Sci. SSSR 159 (1964) 1019—1022.
13. Mineev, V. N., Savinov, E. V.: Zhurnal eksperimental'noi i teoreticheskoi fiziki 52 (1967) 629—636.
14. Burkhardt, A., Hornbogen, E., Keller, K.: Z. Metallkunde 58 (1967) 410—416.
15. Frenkel', Ya. I.: Introduction into theory of metals, Moscow 1958.
16. Godunov, S. K., Romensky, E. I.: Zhurnal prikladnoi mekhaniki i tekhnicheskoi fiziki 6 (1972) 124—144.
17. Murnaghan, F. D.: Amer. J. Math. 59 (1937) 236—260.
18. Richardson, E. G.: Dynamics of real fluids. London: Arnold 1961.

The microrheology of human erythrocyte suspensions

Harry L. Goldsmith

Montreal, Can.

Introduction

In the human body, a suspension consisting essentially of 40 to 45% by volume of 8.3μ diameter red blood corpuscles (erythrocytes), is subjected to pulsatile flow and circulates through a network of vessels whose diameters range from 2.5 cm to 5μ. During flow, a series of complex events, associated with the movements of the individual corpuscles and surrounding elements of the suspending plasma, takes place. One has only to consider that, due to the tiny volume of each red cell ($\sim 9 \times 10^{-11}$ cm^3), a 1.0 cm length of blood vessel having an internal diameter of 0.2 cm and volume of 0.03 cm^3 would contain 160 million erythrocytes, and it becomes clear that a quantitative description of the mechanics of cell and plasma transport is a formidable task.

It is easier, then, to treat blood as a fluid, whose overall rheological properties can be determined *in vitro* by subjecting it to shear flow in instruments of known geometry, e.g. a Couette or cone and plate viscometer measuring the fluid stresses as a function of the rate of shear, or a capillary tube in which the pressure drop across a given length is determined at various volume flow rates in steady or oscillatory flow. The relevance of such results to the circulation *in vivo* can be assessed in terms of the flow velocity and pressure pulse in a given vessel as determined by physiologists [1—4]. One also hopes that the measured changes in rheological properties of the blood in disease states can be used in diagnosis and perhaps even in the treatment of such disorders [5].

This approach is valid if one is interested in such overall properties as the energy dissipated in parts of the circulation, or the mode of propagation of the pressure pulse down a given artery, or the heat exchange across a vascular bed. Ultimately, however, an understanding of the mechanics of blood flow must come from a knowledge of the flow behavior of the individual corpuscles and the surrounding plasma. And this is also important to the understanding of the effects of hematological and cardiovascular disorders since these often manifest themselves by changing the rheological properties of the plasma or the cells [5] to say nothing of the phenomena of hemostasis and thrombosis which of course occur at the level of the corpuscles [6].

To study blood flow by investigating the motions of the suspended particles is to use the approach of microrheology [7], whose general aim is to predict the macroscopic flow properties of a material from a detailed description of the behavior of its constituent elements. Here, the materials considered are suspensions of biconcave discs, each particle with a thin membrane surrounding an interior filled with an aqueous salt solution of 33% hemoglobin having a viscosity of $\sim$0.06P. The elements are the elementary volumes of plasma, a Newtonian fluid of 0.012P, each element containing a corpuscle. Several years ago, the author embarked on such an investigation of the flow behavior of individual erythrocytes as seen under a high power microscope [8], first in isolation from each other, and then when interacting at normal volume concentrations (hematocrits) in plasma undergoing laminar shear flow in tiny glass tubes having diameters from 65 to 200μ.

This paper presents a summary account of the findings of the study with special emphasis on the observed differences between the flow behavior of red cells at very low and at normal concentrations. Since the tube diameters were much greater than those of the erythrocytes, the results are not applicable to blood flow within the capillary bed; here the reader is referred to a recent review by *Skalak* on the mechanics of the microcirculation [9]. Nevertheless, the vessels were small enough so that the tube Reynolds numbers did not exceed unity at the flow rates used in the experiments.

Wherever possible, the observed motions at the microrheological level are described with the aim of relating them to the known macroscopic rheological properties of blood. It appears that the main features of the viscometric flow of human and other mammalian blood can be qualitatively accounted for by the ability of the red cell:

(i) to form linear and branched chain aggregates, known as rouleaux [10], shown in Fig. 1, whose existence depends on the presence of the protein fibrinogen in plasma, and

(ii) to deform under the influence of fluid stresses in shear flow.

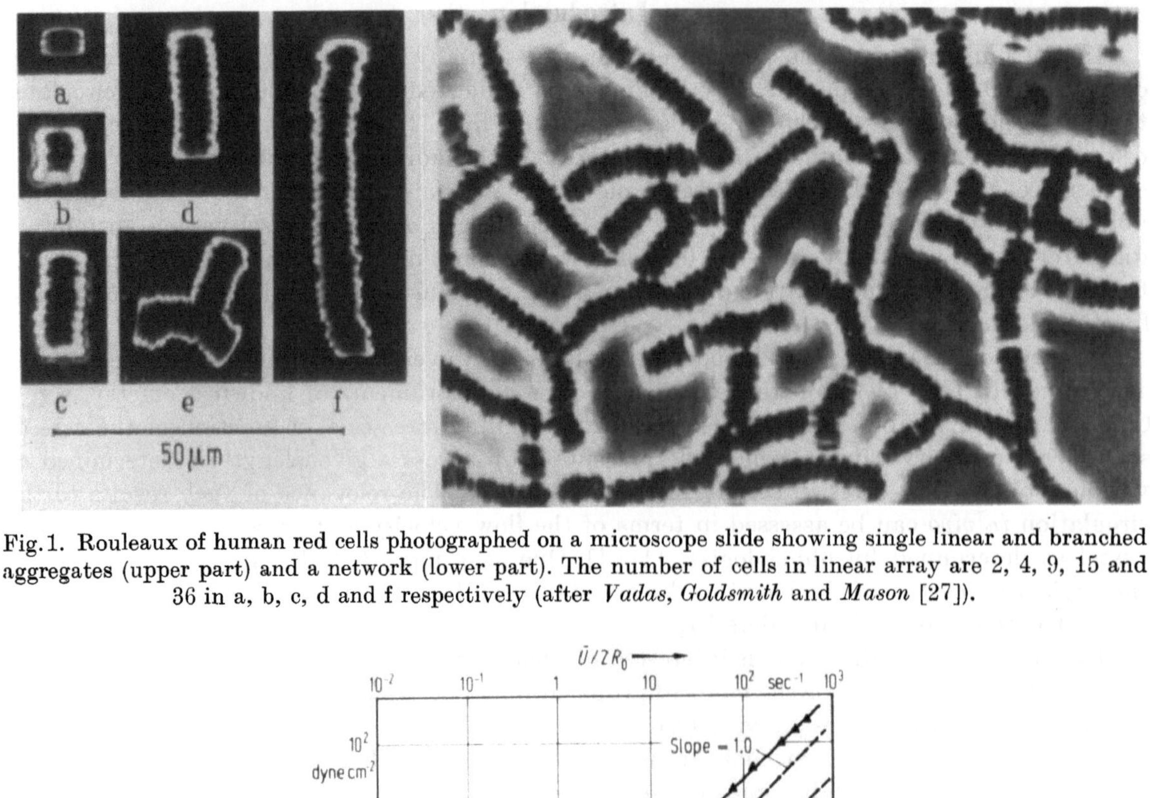

Fig. 1. Rouleaux of human red cells photographed on a microscope slide showing single linear and branched aggregates (upper part) and a network (lower part). The number of cells in linear array are 2, 4, 9, 15 and 36 in a, b, c, d and f respectively (after *Vadas*, *Goldsmith* and *Mason* [27]).

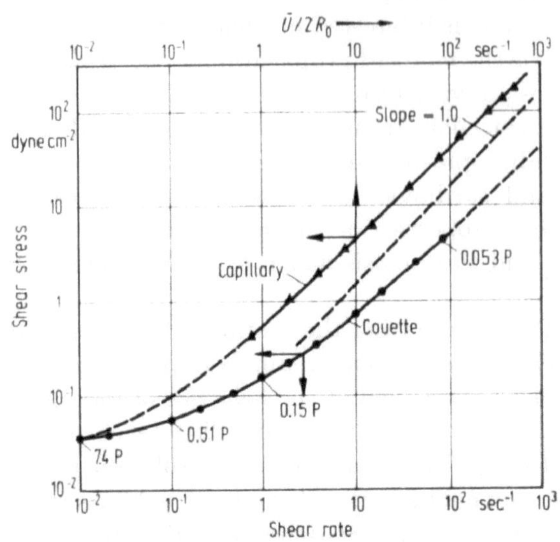

Fig. 2. Non-Newtonian properties of whole human blood at 40% hematocrit and 25 °C. A typical shear stress-shear rate plot from data obtained in a Couette viscometer and a polyethylene fiber, $R_0 = 0.041$ cm. For the capillary data the wall shear stress is plotted against $\overline{U}/2R_0$, the mean flow rate in tube diameters per sec. The arrows indicate the values of the apparent viscosity (from the work of *Merrill* and Associates [4]).

Both these properties of the cell result in the non-Newtonian, shear thinning behavior of blood, as illustrated in Fig. 2. Thus, the apparent viscosity of the suspension decreases with increasing shear rate, an effect which has been interpreted in terms of the progressive break-up of rouleaux at shear rates up to 50—60 sec^{-1} [11—13] followed at higher shear rates by deformation of the erythrocytes [14, 15], the latter being likened to that observed with emulsion droplets [16, 17]. The break-up of structures within the suspension upon the imposition of shear, and their reformation upon standing gives blood at low shear rates properties analogous to those of thixotropic suspensions of bentonite clays, metal oxide sols and printing inks to name but a few [18, 19]. At high shear rates, its viscometric behavior is somewhat similar to that of emulsions [20].

Finally, it should be emphasized that, compared to other suspensions and emulsions, human blood even at the highest concentrations is remarkably fluid. This is illustrated in Fig. 3 where the apparent relative viscosities of various systems and blood are plotted against volume concentration. The data were obtained in Couette or cone and plate viscometers at rates of shear $>100\ sec^{-1}$, at which particle aggregation ceased to be important.

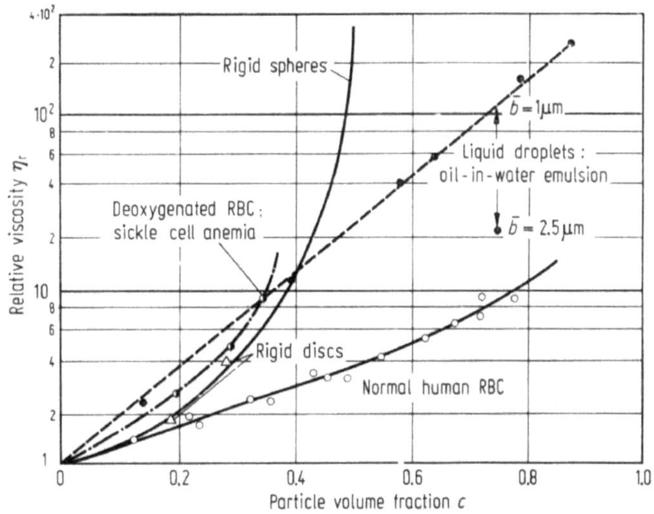

Fig. 3. Relative apparent viscosity of human blood at 25 °C as a function of volume fraction (unpublished data of the author), compared to that of suspensions of rigid latex spheres [21, 22], rigid discs [23], droplets [20] and sickled erythrocytes which are virtually non deformable [24] (after *Goldsmith* [6]).

Experimental techniques

The red cells were observed and followed in Poiseuille flow through precision bore glass tubes mounted in chambers on a microscope slide. The tubes lay on a vertically mounted platform which was mechanically moved upwards as the particles being tracked flowed downward from a syringe reservoir [25, 26]. A similar device employing hydraulic instead of mechanical drives to move the platform, has recently been described [27].

The cell motions were photographed with a 16 mm cine camera at speeds from 10 to 3,000 pictures/sec and the films analyzed, frame by frame, on a projection table.

Heparinized dilute red cell suspensions were obtained by venipuncture from healthy subjects by centrifuging and redispersing to the desired concentration in plasma or in a mixture of 1 volume of plasma to 3 volumes of Ringer solution, if rouleaux formation was to be avoided.

Particle behavior at hematocrits $>10\%$ was studied by using tracer red cells in transparent suspensions of hemolysed, unpigmented red cells—so called ghost cells [28]. The ghosts were also biconcave in shape, although their mean diameter ($\sim 7.2\mu$) and volume ($7.4 \times 10^{-11}\ cm^3$) were somewhat smaller than those of the parent red cells. The apparent viscosities of these suspensions were similar to those of red cells at a given concentration and exhibited a similar decrease with increasing shear rate. The tracer pigmented red cells were clearly visible among the ghost cells and could be tracked down the tubes.

Flow behavior of single, isolated erythrocytes

(a) The cell as a rigid rotating oblate spheroid

At rates of shear $G < 20\ \mathrm{sec}^{-1}$ in Poiseuille flow

$$G = 4\overline{U}R/R_0^2, \tag{1}$$

where R is the radial distance from the axis in a tube of radius R_0 (Fig. 4 a) in which the mean fluid velocity is $\overline{U}$, the rotations of human erythrocytes were similar to those of rigid oblate spheroids and discs [7, 29].

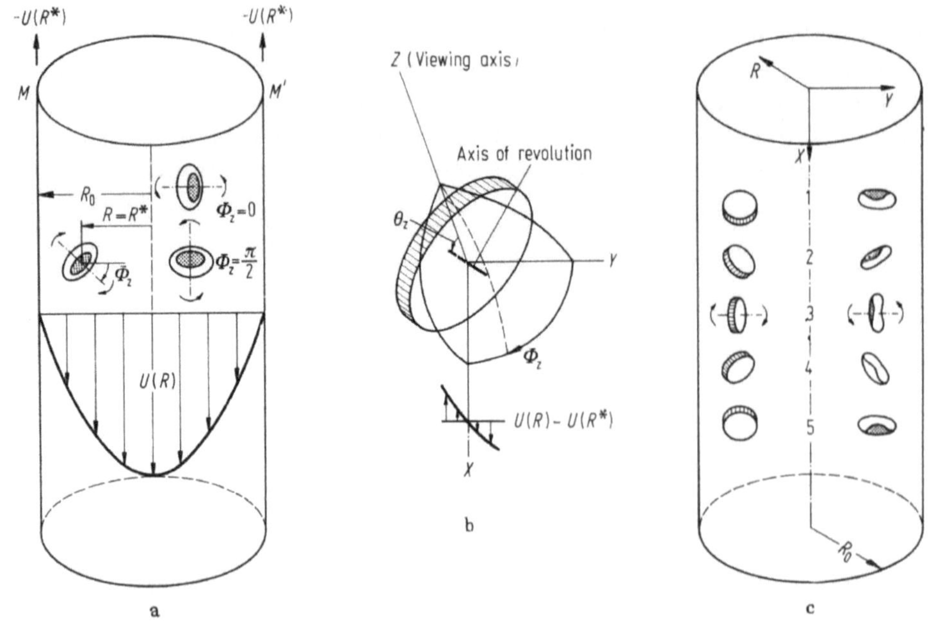

Fig. 4. a) Rigid discs and human erythrocytes observed in the median (MM') plane of a vertically positioned tube. b) Coordinate system with origin at the particle center of rotation. c) Tracings from photomicrographs of particles rotating in Poiseuille flow. For the disc, $r_p = 0.25$, $C = 2.0$.

A rigorous treatment of the rotary motion and axial spin of a single rigid spheroid with axis of revolution $2a$ and diametrical axis $2b$, and its center at the origin of an infinite field of Couette flow has been given by *Jeffery* [30]. The equations have since been shown to apply to rigid discs in Poiseuille flow [31], providing $b/R_0 \ll 1$ and hence that the variation in G across the particle is also small. The spheroid angular motions may be understood with reference to Fig. 4 in which a set of Cartesian coordinates X, Y, Z have been constructed at the cell center of rotation situated at a radial distance $R = R^*$ from the tube axis. The tube is moved upward with a velocity $-U(R^*)$ given by

$$U(R^*) = \frac{4Q}{\pi R_0^4}(R_0^2 - R^{*2}), \tag{2}$$

Q being the volume flow rate, so as to render the particle stationary—as was done in the experiments.

Assuming that the particle is non-sedimenting and inertial effects are absent, the equations for the angular velocities of the axis of revolution with respect to the spherical polar coordinates θ_z and ϕ_z having Z (the experimental optic axis) as the polar axis (Fig. 4b) are [30]:

$$\frac{d\phi_z}{dt} = \frac{G}{r_p^2 + 1}(r_p^2 \cos^2 \phi_z + \sin^2 \phi_z), \tag{3}$$

$$\frac{d\theta_z}{dt} = \frac{G(r_p^2 - 1)}{4(r_p^2 + 1)} \sin 2\theta_z \sin 2\phi_z. \tag{4}$$

These may be integrated to give

$$\tan \phi_z = r_p \tan (2\pi t/T), \tag{5}$$

where r_p is the particle axis ratio a/b, and T the period of rotation,

$$T = \frac{2\pi}{G}\left(r_p + \frac{1}{r_p}\right). \tag{6}$$

The orbit of the ends of the axis of revolution is the spherical ellipse

$$\tan \theta_z = \frac{C r_p}{(r_p^2 \cos^2 \phi_z + \sin^2 \phi_z)^{1/2}}, \tag{7}$$

C being the orbit constant.

Equations (3) to (7) have been shown to apply to rigid discs in suspensions of highly viscous Newtonian liquids *provided* an experimentally determined hydrodynamic or equivalent ellipsoidal axis ratio r_e, obtained from the measured T and G is used instead of the particle axis ratio r_p in Eq. (7).

As illustrated in Fig. 5, the angular velocities $d\phi_z/dt$ of single red cells were in accord with Eqs. (3) and (5) with maxima and minima at $\phi_z = \pi/2$ and 0 respectively. The mean values of r_e for a population of 60 erythrocytes was found to be 0.38 ± 0.07 (S.D.) [26], and as with rigid discs (7) the particle axis ratios ($\sim 2.4\mu/8.3\mu = 0.29$) were smaller than the equivalent ellipsoidal axis ratios.

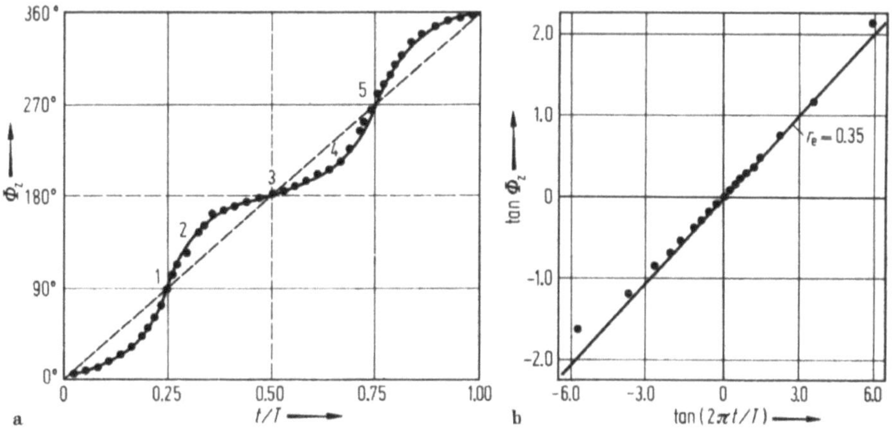

Fig. 5. a) Plot of Eq. (3) and b) plot of Eq. (5), for single rotating erythrocytes in plasma. The points are experimental, $G = 19.5$ sec^{-1}, the lines drawn were calculated using $r_e = 0.35$ obtained from the period rotation T, Eq. (6). The numbers 1—5 refer to the ϕ_z-orientations of the same particle shown in Fig. 4c.

When observing the rotations about the Z-axis in the median or diametrical plane of the tube (Fig. 4c), the projection of the upper face of a disc is an ellipse of axis ratio $s(\phi_z) \leq 1$,

$$s(\phi_z) = b'(\phi_z)/b = \cos \theta_z, \tag{8}$$

b' being the projected length of the equatorial diameter at ϕ_z. Equations (7) and (8) predict that the projected axis ratio $s(\phi_z)$ is a maximum at $\phi_z = \pi/2$ and a minimum at $\phi_z = 0$, corresponding to positions 1 and 5, and 3 respectively in Fig. 4c. In the case of the red cells there are no sharp edges to delineate the boundaries of the upper face and the time course of ϕ_z could not be followed quantitatively. Nevertheless, it was clear that the orientation of the upper face to the viewing axis was not always periodic with ϕ_z, contrary to Eq. (7). As found in experiments using aggregates of 2μ diameter latex spheres [27], this effect is due to rotary Brownian motion of the particles, which perturbs the orbits and which is more clearly noticeable when they are allowed to settle through a stationary fluid [32].

(b) The cell as a liquid droplet: deformation

The above results were obtained at flow rates considerably below those physiologically representative in arteries of, say, 100μ diameter where the mean linear velocity is probably about 1 cm sec^{-1} [1]. The observations were extended to higher shear rates and it was found that at $G > 20$ sec^{-1} the angular velocities of the erythrocytes were no longer in accord with theory, as shown by measurements of the distribution in orientations ϕ_z in a cell population in which interactions, which would interfere with the rotations, were still negligible. While it follows from Eq. (3) that, in the steady state, the fraction of particles $p(\phi_z)\, d\phi_z$ having orientations in the interval $d\phi_z$ depends only in ϕ_z and r_p [7], in the case of the red cells $p(\phi_z)$ was a function also of the shear rate. As G was increased more and more cells were found in orientations in which their major axes were aligned with the flow, i.e. ϕ_z close to 0 (Fig. 6). By contrast, rigid but still biconcave erythrocytes, produced by hardening with glutaraldehyde [26], continued to show ϕ_z-distributions independent of G and in agreement with those calculated from the measured mean r_e. The effect is thus not explicable in terms of fluid inertia at tube Reynolds numbers of the order of unity. An alternative explanation of the above behavior of normal erythrocytes at higher shear rates is that the particles, while still rotating, were attempting to align themselves in the direction of flow, as occurs in the case of non-rotating deformable liquid drops [31].

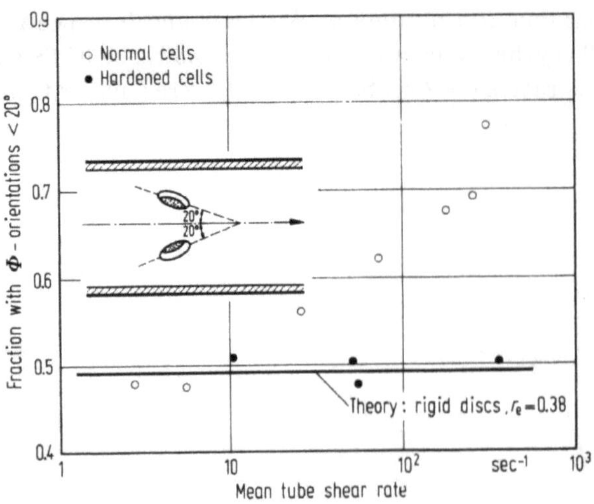

Fig. 6. The increasing fraction of normal red cells found in ϕ_z-orientations $<20°$, as the shear rate is increased. The orientation distribution of hardened cells is independent of $\bar{G}$ and agrees with that calculated for rigid discs having mean $r_e = 0.38$.

The problem of a drop of viscosity η suspended in an immiscible fluid of viscosity η_0 undergoing uniform shear flow was first treated by *Taylor* [33] and later generalized by *Cox* [34]. Under the influence of viscous stresses, the drop undergoes a change of curvature, becoming an ellipsoid whose axes are $L = 2b(1 + D)$, $2b$ and $B = 2b(1 - D)$. The first and third are the length and breadth of the respective major and minor axes which lie in the XY-plane of the flow field, $2b$ is the undistorted drop diameter and D is a deformation parameter physically given by

$$D = \frac{L - B}{L + B}. \tag{9}$$

When $D \ll 1$, it is given by the expression [34]:

$$D = \frac{5(19\lambda + 16)}{4(\gamma + 1)\sqrt{(20\gamma/Gb\eta_0)^2 + (19\lambda)^2}}, \tag{10}$$

where γ is the interfacial tension and $\lambda = \eta/\eta_0$. When interfacial tension effects are dominant, $\gamma/Gb\eta_0 \gg 1$, $\lambda = 0(1)$, Eq. (10) reduces to

$$D = \frac{Gb\eta_0}{\gamma}\,\frac{19\lambda + 16}{16\lambda + 16}, \tag{11}$$

first given by *Taylor* [34].

The orientation ϕ of the major L-axis of the drop with respect to the Y-axis of the flow field is

$$\phi = \frac{\pi}{4} + \frac{1}{2} \tan^{-1}\left(\frac{19}{20} \frac{Gb\eta_0}{\gamma}\right). \tag{12}$$

When interfacial tension effects are dominant $\phi = \pi/4$, i.e., the droplet is oriented at 45° to the streamlines.

The above equations were calculated assuming unhindered transmission of the shear stress across the droplet interface resulting in internal circulation patterns (Fig. 7), and have been shown to apply both in Couette and Poiseuille flow [31, 33, 35].

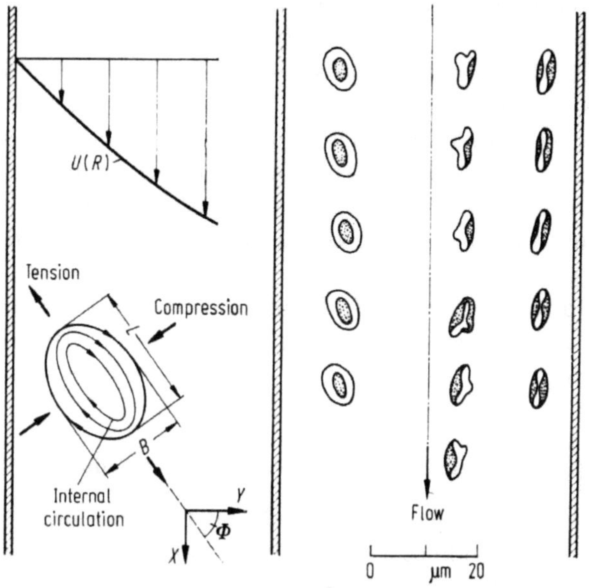

Fig. 7. a) Deformation of a liquid drop into an ellipsoid and internal circulation in Poiseuille flow. b) Deformed ellipsoid-like red cells in 30% Dextran solution, $\eta_0 = 0.51$ P. Tracings from photomicrographs at successive positions in time going from top to bottom ($G = 2.8$ sec^{-1}). c) Same suspension showing on left motion intermediate between rotation and deformation and on the right increased deformation at higher shear rate ($G = 4.7$ sec^{-1}) (after *Goldsmith* and *Marlow* [26]).

Liquid drop-like behavior of erythrocytes in plasma was visually observable at mean tube shear rates $\bar{G} = 2\bar{U}/R_0 > 150$ sec^{-1}, i.e. at shear stresses $G\eta_0 > 3$ dyne cm^{-2} when as many as 45% of the cells travelled down a tube without exhibiting angular rotation. (In a 100μ diameter artery the mean shear rate $\bar{G}$ is estimated to be 400 sec^{-1}.) As shown in Fig. 8,

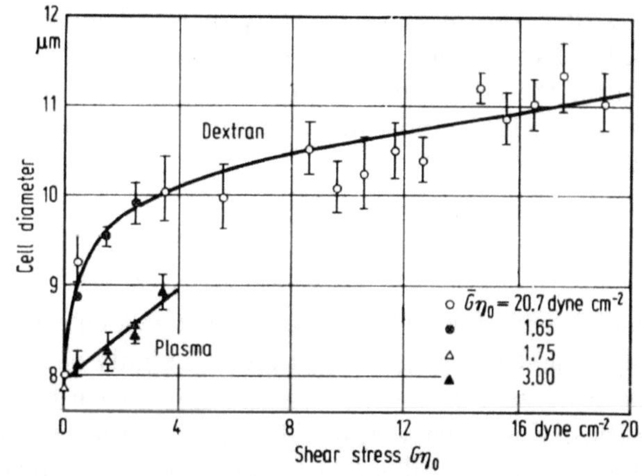

Fig. 8. Erythrocyte diameters in plasma ($\eta_0 = 0.014$ P) and Dextran ($\eta_0 = 0.51$ P) suspensions as a function of shear stress, as calculated at the particle center of rotation. The points are the mean values in each unit interval of shear stress, the bar lines indicate ± one standard error (after *Goldsmith* [6]).

measurements of the cell major diameters revealed a significant correlation with shear stress. At a given volume flow rate the cells at the periphery of a tube had larger mean diameters than those at the center, and as the shear stress increased from 0.5 to 3.5 dyne cm^{-2}, the mean cell diameters increased from 8.1 to 8.9μ.

The most striking demonstration of drop-like flow behavior, however, was obtained when the red cells were dispersed in 20 to 35% wt/vol Dextran solution [16] whose viscosities ranged from 0.15 to 0.51P. Here, as illustrated in Fig. 7b, the cells appeared as slightly irregular ellipsoids, with the dimple (biconcavity) still present and the contour continually changing, suggesting that the membrane rotated about the interior. With increasing shear stress the cells elongated and many of them were now in orientations in which the faces containing the concavity no longer lay in the median plane of the tube (Fig. 7c).

As is evident from Fig. 8, at a given shear stress the increase in cell diameter in 35% Dextran is initially several times greater than in plasma. This cannot be explained by the decrease in λ from ~5 in plasma to 0.12 in Dextran which, according to Eq. (10) would actually result in a small decrease of 22% in D. It is presumably an effect of the polymer on the cell resulting in a lowering of the effective tension at the membrane. Although the theory of drop deformation assumes a pure interface and should not therefore be applied to a fluid-filled membraneous disc, it is instructive to calculate the magnitude of the interfacial tension at the surface of a 4.1μ drop, if as shown for plasma in Fig. 8, the imposition of unit shear stress caused its major axis to increase by 0.27μ. Such a calculation (26) yields $\gamma = 1.4 \times 10^{-2}$ dyne cm^{-1}, an extremely low value, indicating great deformability of the erythrocyte in shear flow.

(c) The cell as a liquid droplet: lateral migration

In the creeping flow regime, it has been found that the deformation of a liquid drop results in its migration across the streamlines away from the tube wall [31]. Such lateral movement is not observed with rigid spherical particles at very low Reynolds numbers Re. The difference arises from the fact that the droplet is asymmetrically oriented relative to the wall (Fig. 7a). The deformation introduces terms to the flow field additional to those which exist for an undeformed sphere, and the reflection of the largest of these off the wall yields a finite migration velocity [7, 36].

A similar difference in flow behavior was found between normal (RBC) and glutaraldehyde-hardened red cells (HBC) [37], and is illustrated in the upper panels of Fig. 9 giving histograms of the number concentration distributions of cells as a function of radial distance, 1 cm downstream in a tube of 83μ diameter, Re ~0.03. Presumably, the inward migration of the normal cells arises from the greater time spent by the deformable particles in ϕ_z-orientations between 45° and 0°. Even more striking was the lateral migration shown in the Dextran solutions, $Re = 3.7 \times 10^{-3}$ and 9.1×10^{-4}, shown in the lower panels, where very few of the cells, here deformed into ellipsoid-like shapes, were still present in the outer half of the tube.

A rigorous expression for the lateral migration velocity of a drop away from a plane wall in Couette flow has been given [36], and can be applied to Poiseuille flow [7] providing the particle is close to the wall, $R/R_0 \to 1$:

$$-\frac{dR}{dt} = \frac{\overline{U}Db^3}{R_0^2} \cdot \frac{R}{(R_0 - R)^2} \cdot f(\lambda), \tag{13}$$

where $f(\lambda)$ is of the order of unity for all values of λ. It is evident that the inward migration velocity decreases rapidly with decreasing R and that it is highly dependent on b/R_0. A similar decrease in $|dR/dt|$ with decreasing R and increasing R_0 has been found with red cells [6].

Although the above theory is strictly applicable only in the creeping flow regime, rapid inward migration of deformable drops, fibers and red cells toward the tube axis in both steady [37, 38] and oscillatory flow [39], has been observed at much higher Reynolds numbers when the effects of fluid inertia are significant, and when non-deformable particles now also exhibit lateral migration [38, 40]. In the living circulation it is therefore to be expected that appreciable inward migration of the deformable red cells would occur, at least very close to the walls of vessels. However, as will become apparent in the succeeding section this is opposed by a dispersive movement due to crowding of the blood cells in the flowing suspension.

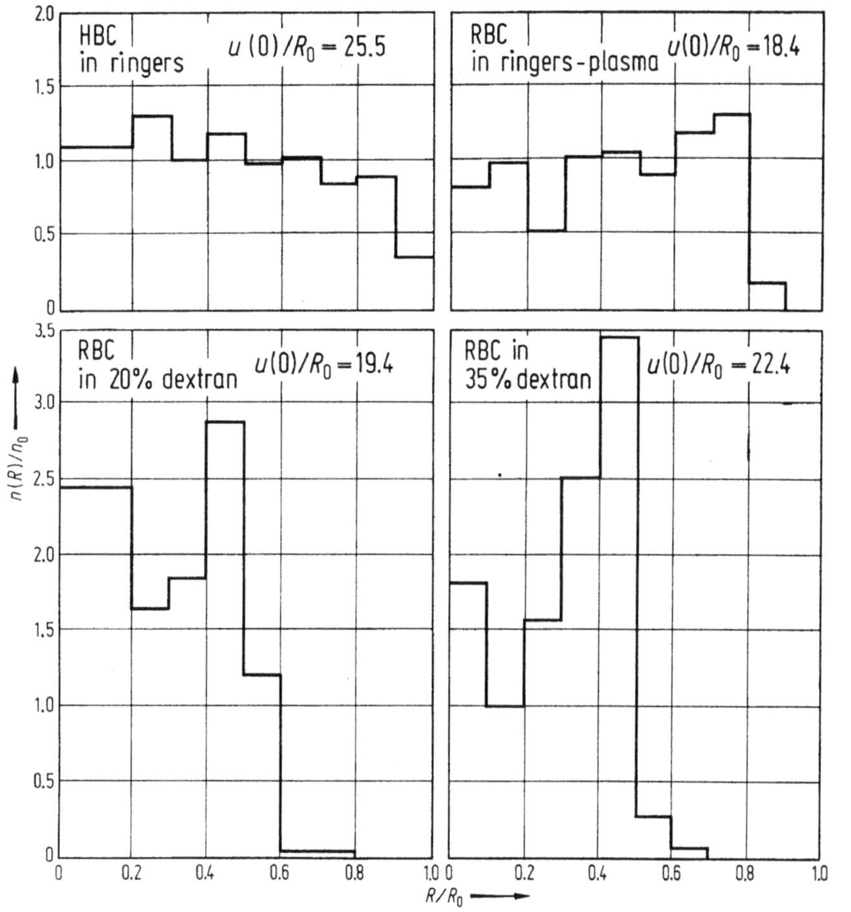

Fig. 9. Number of cells/cm³ suspension, $n(R)$, divided by the syringe reservoir concentration, n_0, at intervals of $0.1 R_0$, 1 cm downstream from the reservoir; $R_0 = 41.5\mu$, $c = 2\times10^{-3}$. The mean tube shear rates $= u(0)/R_0$ were approximately the same in each suspension. If the number distribution were uniform $n(R)/n_0 = 1$ at all R/R_0 (after *Goldsmith* [25]).

The dependence of the radial migration velocity on particle size is of significance to the flow of the disc-like platelets (thrombocytes), since, because of their smaller diameter—2.5μ—their migration rate, would, on the basis of Eq. (13) be about 200 times smaller than that of the red cells, other parameters in the expression being equal. This could lead to a relative concentration of platelets in the outer layers of vessels.

On the other hand, linear and branched rotating rouleaux of red cells, present at shear rates < 50 sec^{-1} [12], migrate to the tube axis faster than single red cells. As with erythrocytes, this was observed at very low Reynolds numbers and here is associated with their observed deformation through bending (Fig. 10). Bending sets in when the particle axis of revolution lies in the quadrant $\pi/2 < \phi_z < \pi$ of the rotational orbit where the fluid stresses are compressive [26]. Such bending of rouleaux of 6 to 10 red cells in linear array was observed at shear rates < 10 sec^{-1}, i.e., shear stresses lower than 0.2 dyne cm^{-2}. The results of the above work with model particles and blood cells thus show that red cells and their aggregates are readily deformed in shear flow at values of G much lower than those prevailing in the circulation, and that in the absence of particle interactions, erythrocytes and rouleaux would not be found at radial positions close to the wall.

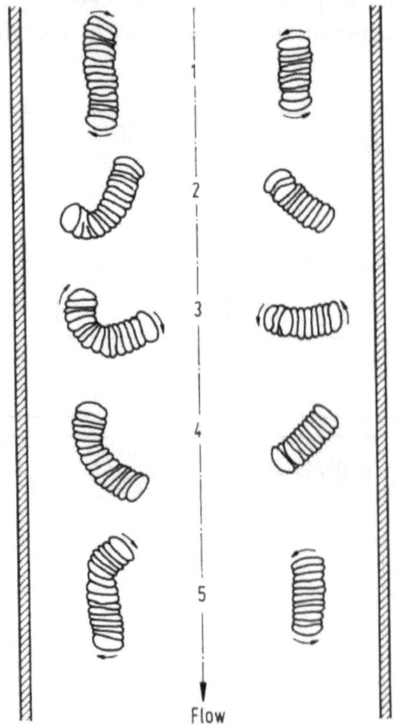

Fig. 10. Rotation of an 11- and a 16-cell rouleau of erythrocytes in Poiseuille flow, $G < 10\ \text{sec}^{-1}$. Bending commenced at position 2 where the fluid stresses are compressive. The longer particle did not straighten out in the succeeding quadrant where the stresses are tensile (after *Goldsmith* and *Marlow* [26]).

Flow behavior of erythrocytes in concentrated suspensions

(a) Collisions

Long before concentrations corresponding to the cell-crowded conditions in whole blood, in a printing ink or in an emulsion such as margarine are reached, the suspended particles will interact and collide with each other in shear flow. Already at hematocrits of 0.5% (at which low concentration there would still be 1.9 million RBC in a 1 cm length of a vessel of 0.2 cm internal diameter) there are frequent collisions as cells travelling on adjacent stream-lines are carried into close proximity to each other through the existence of a velocity gradient in the flow.

The geometry of two-body collision processes has been extensively studied in dilute suspensions of neutrally buoyant spherical particles undergoing Couette [41] or Poiseuille flow [42], and the frequency of their occurrence calculated. A discussion of most of the theoretical and experimental findings will be found in reference [7]. In a suspension containing n uniformly dispersed rigid spheres per cm^3, each of radius b, the two-body collision frequency per second per particle at a radial distance R is

$$f = 32G(R)\, b^3 n/3\,, \tag{14}$$

and since the volume fraction $c = 4\pi b^3 n/3$,

$$f = 8cG(R)/\pi\,. \tag{15}$$

These equations, which are valid providing $b/R_0 \ll 1$ and curvature of the shear field can be neglected, show that the highest collision rate occurs near the tube wall. Experimentally, it is found that when two spheres come into apparent contact they rotate as a doublet until the mirrorimage of the position of first contact is reached when they separate, as illustrated in

Fig. 11. The doublet rotates as a prolate spheroid of axis ratio $r_e = 2$ and the angular velocities are given by Eqs. (3) and (4) [41, 42]. The existence of spherical doublets for a finite time means that a steady state concentration of doublets is built up as the suspension flows. Assuming rectilinear approach of sphere centers it has been shown that the volume fraction of doublets c' existing in the tube at a given instant is

$$c' = 20c^2/3. \tag{16}$$

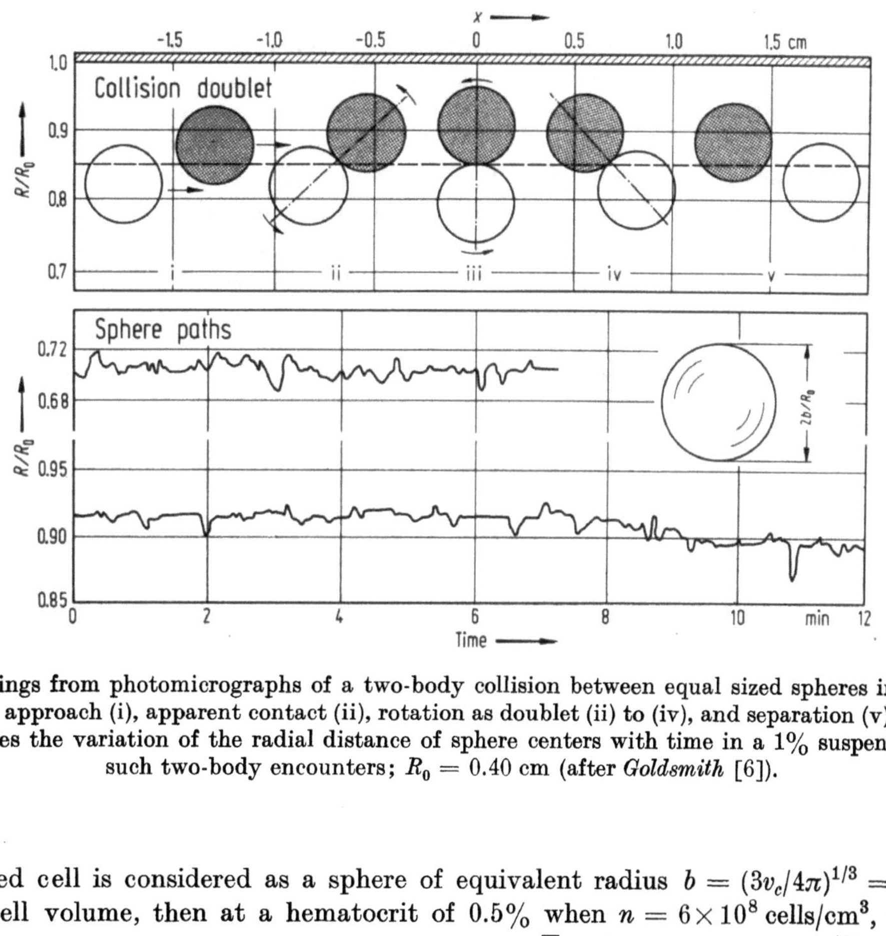

Fig. 11. Tracings from photomicrographs of a two-body collision between equal sized spheres in Poiseuille flow showing approach (i), apparent contact (ii), rotation as doublet (ii) to (iv), and separation (v). The lower part illustrates the variation of the radial distance of sphere centers with time in a 1% suspension due to such two-body encounters; $R_0 = 0.40$ cm (after *Goldsmith* [6]).

If the red cell is considered as a sphere of equivalent radius $b = (3v_c/4\pi)^{1/3} = 2.78\mu$, v^c being the cell volume, then at a hematocrit of 0.5% when $n = 6 \times 10^8$ cells/cm^3, the mean collision frequency in a 0.2 cm diameter artery in which $\overline{U} = 15$ cm sec^{-1} [1] and $\overline{G} = 300$ sec^{-1} is calculated to be $\sim$40 per second. 3.3% of the red cells would be in collision at a given instant. As c increases, the fraction of particles existing as doublets rises rapidly; at 5% volume concentration 33% of the "spherical" red cells would be in collision at any instant. However, at such concentrations, the simple theory is no longer valid since 3-, 4- and multi-body collisions now occur [42]. It is clear, then, that at hematocrits of 20%, which are still well below normal, the existence of single erythrocytes is most unlikely.

When one finally observes the corpuscles in whole blood, in which 40% or more of the available volume is occupied by cells, three main effects come to light:

(i) The velocity profile in the tube is no longer parabolic as in Poiseuille flow.

(ii) Deformation of the erythrocytes in blood occurs to a degree that is not attributable to shear alone.

(iii) The particle paths exhibit erratic displacements in a direction normal to the flow.

An understanding of the results obtained with tracer red cells in the transparent ghost cell suspensions was made much easier by the existence of data in model transparent suspensions of rigid and deformable spheres [7, 43]. In the model systems, the refractive indices of the oil and particle phases were matched and a small volume of visible spheres, having a significantly different refractive index, added to serve as tracer particles.

(b) Velocity profiles: the emulsion as a model for blood

At low Reynolds numbers, at which effects due to fluid inertia were negligible, single, isolated *rigid* spheres having radii $b \ll R_0$ translated along paths parallel to the tube axis with velocities $u(R)$ which, except very close to the wall, were equal to those of the undisturbed suspending fluid at the same radial distance given by Eq. (2) i.e., $u(R) = U(R)$ except for $R \cong R_0$ [31].

As the concentration of rigid spheres was progressively increased, the particle velocity distribution $u(R)$ remained parabolic provided $b/R_0 < 0.04$ and the volume concentrations $c < 0.2$. As c and/or b/R_0 were further increased, $u(R)$ and $U(R)$ continued to be equal, but the velocity distribution became blunted in the center of the tube with a core of radius R_c in which the particle velocities u_M were a maximum and uniform, and smaller than the centerline velocities in Poiseuille flow given by Eq. (2), $R^* = 0$ [43] (Fig. 12). This region was designated as partial plug flow but this did not necessarily mean that the profile was mathematically flat, only that within experimental error there was no *measurable* velocity gradient. However, these concentrated suspensions were still quasi-Newtonian in that the velocity profile was independent

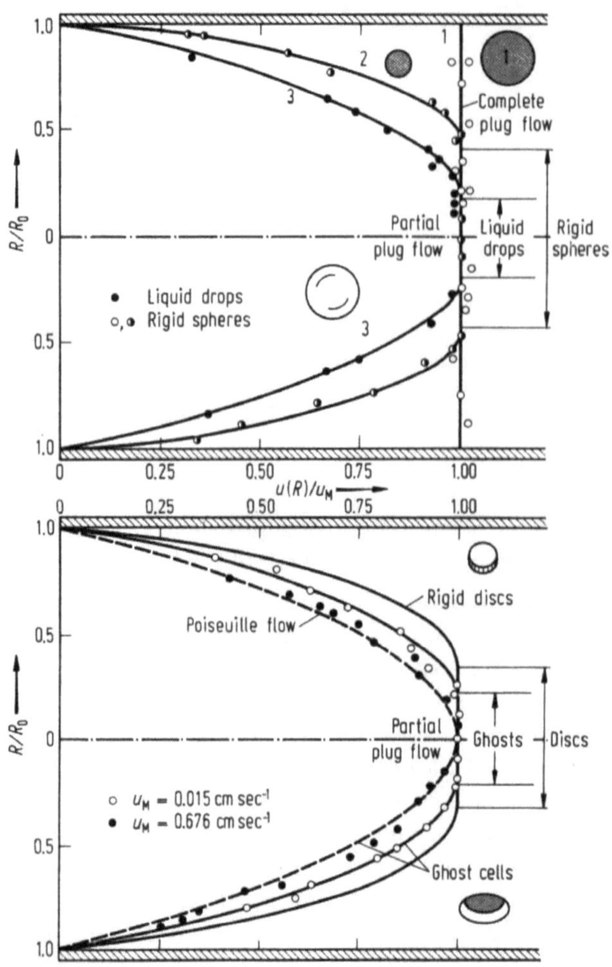

Fig. 12. Comparison of dimensionless particle velocity profiles in 32% suspensions of rigid spheres ($b/R_0 =$ 0.112 and 0.056 in curves *1* and *2* respectively) and liquid drops ($b/R_0 = 0.078$, curve *3*)—upper part, and between 32% suspensions of rigid discs ($b/R_0 = 0.078$) and ghost cells ($b/R_0 = 0.105$)—lower part. The lines drawn are the best fit of the experimental points, the dashed line is calculated from equation (2) in the form $U(R)/U(0) = 1 - R^2/R_0^2$. Note that by complete plug flow in curve *1*—upper part, we do not imply slip of fluid at the wall. Close to the boundary there must be a steep velocity gradient in the suspending fluid (after *Goldsmith* [6]).

of flow rate and suspending phase viscosity, and the pressure drop per unit length of tube was, in each suspension, directly proportional to the volume flow rate Q [43]. This result, which holds true provided the distribution of particles across the tube is uniform, is an effect which becomes important when the mean interparticle distance is of the same order as the sphere diameter and both are not too much smaller than the tube diameter [44].

By contrast, such quasi-Newtonian behavior was not shown by concentrated, monodisperse oil-in-oil emulsions [17] where the velocity profile at a given c and b/R_0 was appreciably less blunted with a smaller region of plug flow than in the rigid particle case (Fig. 12), and moreover was dependent on the flow rate as well as suspending phase viscosity: in a given system, the lower the flow rate, and the greater η_0, the greater was the degree of blunting.

Similar non-Newtonian behavior was exhibited by the ghost cell plasma suspensions at concentrations from 20 to 70%, in which the velocities of the tracer red cells were measured at mean suspension flow rates $\bar{U}$ from 0.005 to 1 cm sec^{-1} [28, 45]. Some results obtained at $c = 0.32$ are shown in the lower panel of Fig. 12, where it is evident that at a $\bar{U}$ of only 1.3 tube diameters per sec ($R_e = 36\mu$) the degree of blunting, as measured by the core radius R_c, was appreciably smaller in the ghost cell suspension ($R_c/R_0 = 0.20$) than in the suspension of rigid discs ($R_c/R_0 = 0.38$). In the case of the discs, the velocity distribution remained constant over a ten-fold range of volume flow rates corresponding to $\bar{U}$ from 0.09 to 0.9 tube diameters per second ($R_0 = 0.40$ cm) [43]. This result is even more significant when it is realized that the ratio of mean disc to tube diameter was 20% smaller than that of the ghost cells, and that, unlike the discs, the ghost cells form aggregates, the size of which increases with decreasing shear rate, and the presence of which at the above mean tube shear rate $\bar{G} = 5.2$ sec^{-1}, no doubt, contributed to the blunting of the velocity profile. Upon increasing $\bar{U}$ to 47 diameters/sec the velocity distribution in the ghost cell suspension became almost parabolic (Fig. 12).

Since, at comparable $\bar{U}/R_0$, c and b/R_0, the wall shear rates, obtained from the slopes of the plots of $u(R)$ against R at the tube wall, were appreciably smaller in emulsions and ghost cell suspensions than in suspensions of rigid spheres and rigid discs, the apparent relative viscosities would also be smaller. Similarly, the decrease in blunting of the velocity profile with increasing $\bar{U}$ in the emulsions and in blood leads to a decrease in the apparent viscosity.

(c) Cell deformation: combined effects of shear and particle crowding

As seen in Fig. 3, the relative apparent viscosities of concentrated emulsions are appreciably lower than those of suspensions of rigid latex spheres, but the relative viscosity of whole human blood is still much lower. The reason for this, as well as for the non-Newtonian behavior of the emulsions and ghost cells in plasma becomes evident when observations of their particle-crowded interior are made. In the emulsions, continuous distortion of the drops into shapes noticeably different from those of regular ellipsoids was noted [17]. The dependence of the velocity profile on Q and η_0 may then be qualitatively explained by deformation resulting from the ability of the drops to squeeze past each other, and this would be expected to increase with increasing shear stress and concentration. In the suspensions of rigid spheres and discs, the particles must pass around each other with greater disturbance to the streamlines, as could be observed by comparing particle paths at similar concentrations and b/R_0 [17]. Providing the particles are uniformly distributed, the energy dissipation at a given $\bar{U}/R_0$, c and b/R_0 is therefore greater.

Similar considerations apply to the ghost cell suspensions. Here, however, the deformation of the red cells and small rouleaux was much more striking than that of the liquid drops [45, 46], as shown in Fig. 13 at $c = 0.5$. Moreover, these large distorions of the resting cell and aggregate shape occurred already at very low shear rates ($\sim$5 sec^{-1}) corresponding to shear stresses, based on plasma viscosity, smaller than 0.07 dyne cm^{-2}, i.e., under conditions at which the isolated cell in plasma could not be seen to deform.

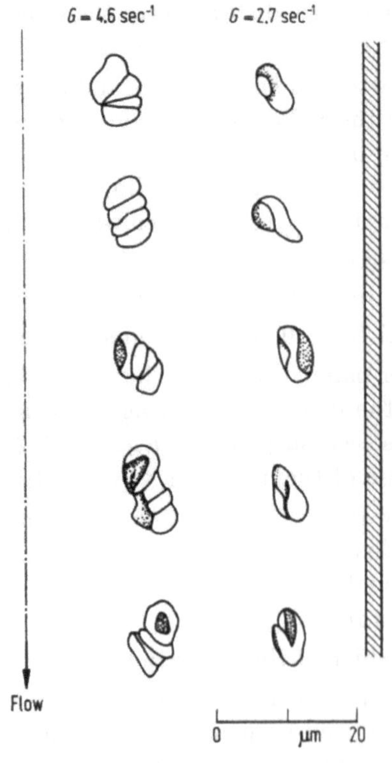

Fig. 13. Tracings from cine film showing the continuous and irregular deformation of a single erythrocyte and a 4-cell rouleaux in a ghost cell suspension, $c = 0.5$. The cell is shown at intervals of 0.4, 0.6, 2.0 and 3.2 sec respectively in which time the isolated corpuscle would excecute just over half a rotation (after *Goldsmith* [46]).

(d) Particle paths

The above comparisons of the energy dissipation in flowing systems of rigid and deformable particles were made assuming uniform distribution of the disperse phase within the vessels. However, one of the earliest questions raised by researchers in blood rheology [47, 48] and still debated today, is to what extent there is a reduction in red cell concentration near the endothelial wall of vessels and how deep into the interior does the cell-depleted zone extend. Since visual observations can only be made in smaller arteries and veins, the only *in vivo* measurements stem from examination of sections of quick-frozen rabbit femoral arteries [49]. This work, and the *in vitro* analysis of cine films of blood flow in glass tubes [50] as well as measurements of the reduction in ghost cell concentration at the vessel periphery [51] makes it appear unlikely that at hematocrits of 40—45% and normal flow rates, the plasma-rich zone can be much larger than 4μ in vessels whose diameters exceed 100μ. The resistance to crowding by the particles at these high volume concentrations will oppose any appreciable inward migration from the wall, such as would have occurred in more dilute suspensions [37, 38].

The effects of continuous collisions and interactions on the movements of particles near the wall and in the interior of tubes has been studied in suspensions of spheres [17, 43] and ghost cells [25, 45]. It has been shown that the paths of the visible tracer spheres and erythrocytes are subject to frequent erratic displacements in directions normal to the flow. Even at very low concentrations, two-body collisions result in small lateral displacements, as illustrated for rigid spheres in the lower panel of Fig. 11. As more and more particles are added to the suspension, the frequency and amplitude of these fluctuations increases.

In concentrated suspensions of rigid spheres, the magnitude and frequency of the radial displacements at a given Q is found to decrease with decreasing radial distance, and in the core of partial plug flow the fluctuations disappear.

In ghost cell suspensions, there are also pronounced radial fluctuations in the paths of the tracer red cells, but unlike the rigid particle systems these persist in the core of partial plug flow. This is illustrated in the lower panel of Fig. 14 showing tracings of the paths of red cells in a 40% ghost cell suspension in a tube of 73μ diameter. Here the velocities of the particles in a core of radius $0.27 R_0$, measured with the tube stationary and the suspension flowing past the field of view, were uniform within $\pm 1\%$ ($= u_M$), suggesting the absence of a velocity gradient. Nevertheless, it is evident from Fig. 14, curve *3*, that there were appreciable lateral movements of the red cells in this region, although their frequency was smaller than that for cells further out toward the periphery (curves *1* and *2*). When particles in the core of "partial plug flow" were followed down the tube it became apparent that movements of ghost cells and aggregates in

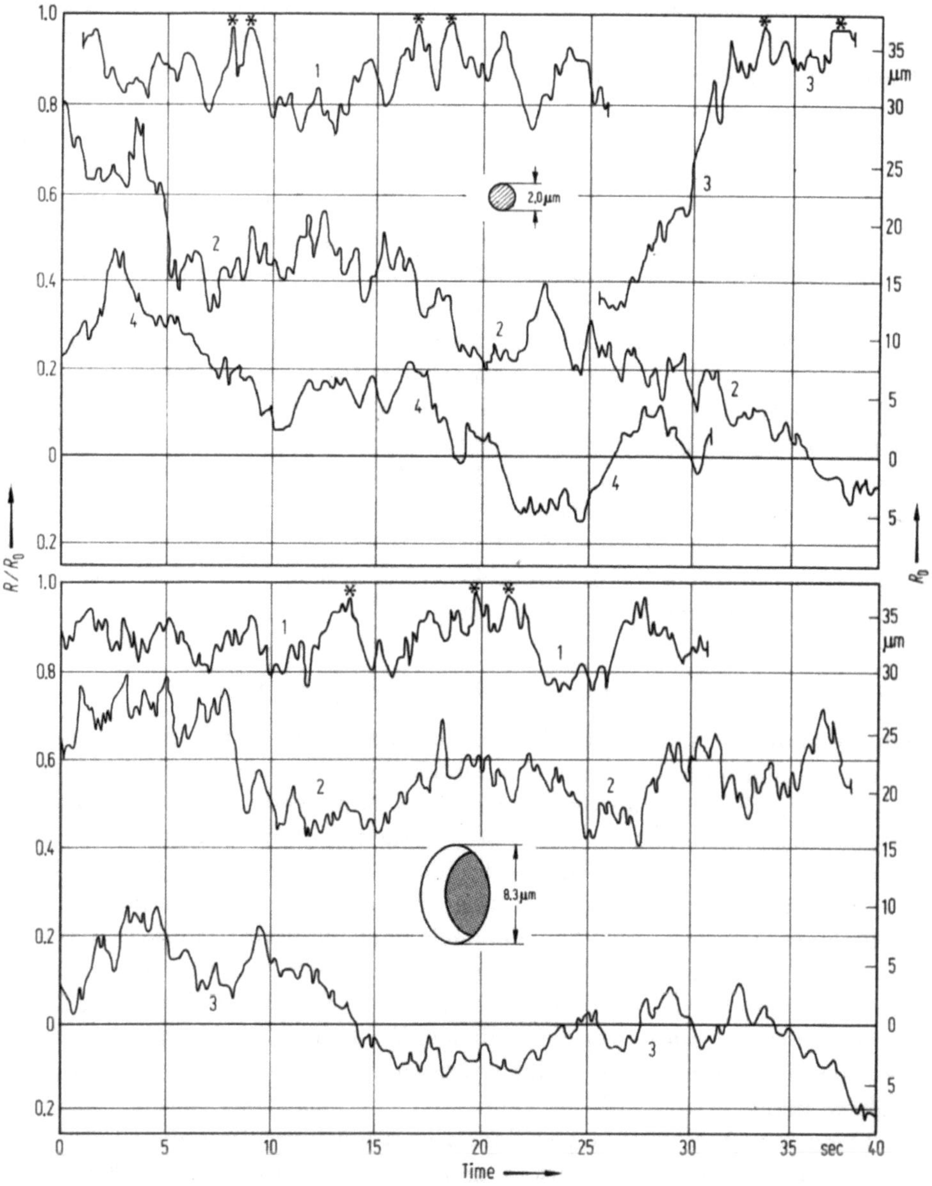

Fig. 14. Paths of tracer red cells (lower part) and 2μ polyvinyl toluene latex spheres (upper part) in a 40.2% ghost cell suspension, $R_0 = 36.5\mu$, $u(0) = 0.04$ cm sec^{-1}, $R_c/R_0 = 0.27$. Positions at which particles were observed to be in apparent contact with the tube wall are marked by asterisks.

the region outside the core were transmitted into the interior, and these resulted in differences in tracer erythrocyte translational velocities $<0.01\ u_M$. These effects became less pronounced as the tube diameter was increased.

As shown by curve *1* in the lower panel of Fig. 14, there were encounters of cells with the tube wall: often a cell moving with its center situated 3 to 6 microns from the wall would be squeezed against the boundary by interactions with its surrounding ghost cell neighbors, without sticking there. After a short time the erythrocyte moved away from the wall. Thus, due to continual particle interactions at high concentrations, cells, which if isolated, would have migrated away from the wall, were being propelled both to and from the boundary. Even if there were, on the average, a smaller concentration of erythrocytes at the periphery of a vessel, because of the multi-body interactions some particles would still be found at the wall.

The continuous sideways motions of the erythrocytes must result in a displacement of other blood corpuscles and volume elements of the suspending plasma itself. If red cells undergo wall encounters due to the radial dispersion brought on by constant interaction with their neighbors, then it is probable that platelets will also be carried to the wall in this type of mixing motion. Since platelets were difficult to photograph in ghost cell suspensions, experiments were carried out with tracer 2μ diameter polyvinyl toluene latex spheres, easily visible because of their much higher refractive index.

Some paths, typical of single latex spheres in a 40% ghost cell suspension, are shown in the upper panel of Fig. 14. It is evident that the particles undergo considerably more sideways movements than the red cells whose paths generally did not deviate more than 8μ from the mean position at this concentration. As curve *2* for the mirospheres shows, the particle in a 73μ diameter tube encompassed a total radial distance (with many fluctuations) of 32μ while travelling about 1.2 cm downstream. The path of the sphere in curve *3* covered a radial distance of 23μ before it reached the wall and adhered to the surface. Thus, wall collisions of the smaller microspheres can occur with particles starting from radial positions much further away from the surface than in the case of the red cells. Because of this, the collision rate with other microspheres and the wall would be enhanced. It may, therefore, be surmised that platelets, which are of the same size as the latex spheres, will also have greater collision rates with the wall, although this remains to be tested experimentally.

The interaction of platelets with each other and the vessel walls is of considerable interest in connection with the formation of thrombi, their adhesion to the wall and their eventual growth to produce an obstruction of the lumen [52, 53].

(e) The role of rouleaux: two-phase flow

Although the above work indicates the absence of any appreciable inward radial migration of erythrocytes, calculations show that a reduction in red cell concentration over a distance of only 2 to 4μ from the wall can result in a significant decrease in the measured apparent viscosity [54, 55]. A similar effect has been noted in the tube flow of suspensions of rigid spheres at Reynolds numbers at which the effects of fluid inertia are important and some inward migration from the wall occurs [38, 56]. This results in the two-phase flow of a particle-enriched core of relatively high viscosity surrounded by a particle-depleted peripheral layer of relatively lower viscosity. In a 15% suspension of rigid spheres, $b/R_0 = 0.04$, the initially parabolic velocity profile became blunted after particle migration had occurred [38]. But although the wall shear rate had increased, the pressure drop per unit length of the tube decreased by 20%, showing that the lowering of viscosity in the peripheral layer was sufficient to result in a net decrease in the wall shear stress.

In whole blood, however, there is, paradoxically, a much greater redistribution in tube flow at *very low* flow rates when due to marked aggregation into rouleaux, a particle-free layer is formed. This has been demonstrated *in vitro* [28, 57]: as the flow rate is progressively decreased one can observe cells being removed from the periphery of the tube and incorporated into a centrally located network of rouleaux which, at sufficiently low velocities, moves as a plug.

This results in two-phase flow of a central core where the shear rate is virtually zero surrounded by a sheared particle depleted layer of variable thickness in which a few cells and rouleaux rotate.

Concluding remarks

The motions of human erythrocytes in plasma undergoing flow through small circular tubes have been described, first for the case of the single, isolated cell and then for the cell in a particle-crowded suspension at volume concentrations normally present in whole blood.

It was shown that, but for perturbations due to rotary Brownian motion, at shear stresses <0.4 dyne cm^{-2}, the erythrocyte behaves as a rigid rotating oblate spheroid (Figs. 4c and 5). However, due to its extremely low resistance to deformation (an "interfacial tension" estimated to be only 0.014 dyne cm^{-2}), at shear stresses >1 dyne cm^{-2}, a flow behavior analogous to that of deformable fluid drops developed (Fig. 7). Thus even without considering the effects of particle interactions in the suspension, one would have anticipated considerable cell deformation at mean shear stresses of 5—7 dyne cm^{-2} (37 °C) prevailing in arteries of the same size as the tubes used here.

Observations of red cells flowing in concentrated suspensions of ghost cells then revealed that the combined effects of shear and particle crowding led to considerably greater deformation, and into a variety of irregular shapes (Fig. 13), at shear stresses of only 0.07 dyne cm^{-1}. Moreover, in such suspensions there were marked radial fluctuations in the paths of cells resulting in their collision with the wall (Fig. 14), despite the existence of a mechanism for inward particle migration at these flow rates, which was known to operate in dilute plasma suspensions (Fig. 9). Such large deformations were not seen in suspensions of liquid droplets in which the interfacial tension was ~ 4 dyne cm^{-1} [17].

These microrheological studies of normal red cell flow behavior also indicate what the effects of blood disorders can be on the overall flow properties of the system. Thus, a considerable increase in internal cell viscosity, as occurs with abnormal hemoglobins having poorer solution properties [5] (e.g., in sickle cell anemia) would lead to an almost rigid red cell and an increased apparent viscosity, especially at higher volume concentrations. When, on the other hand, the plasma viscosity increases, e.g. due to the presence of abnormally large quantities of high molecular weight proteins as in Waldenström's macroglobulanemia [58], one would then expect the increased cell deformation at a given shear rate to result in a decreased *relative* viscosity.

However, these effects cannot usually be isolated since they often occur together with other changes, particularly with an increase in the degree of aggregation of erythrocytes [58], i.ee an increase in the stability of rouleaux. One has also to consider alterations of the cell membran. in disease, which often lead to changes in cell shape and these are known to reduce the erythrocyte deformability and increase the suspension viscosity [59]. *Fung* [60] has pointed out that an important reason for the ease of distortion of the biconcave red cell is the fact that the ratio of surface area to volume is not minimized and that there must exist an infinite number of isochoric deformations which do not induce any stresses in the membrane, nor produce any change in the pressure differential between interior and exterior.

There is no doubt then, that Nature has provided a suspension, which while fulfilling its role of delivering oxygen and dissolved nutrients to the tissues, collecting CO_2 and wastes, and maintaining the body at constant heat, does so with a remarkably low expenditure of energy in both large and tiny vessels.

Acknowledgments

The author is indebted to Dr. *S. G. Mason* whose support and encouragement enabled this work to grow and develop from similar studies of model particle systems, to Dr. *C. A. Goresky* for much helpful discussion and advice, to the Medical Research Council of Canada and the Quebec Heart Foundation for grants-in-aid of this research, and to *Mary Pitteway, Katherine Davyduke* and *Jean Marlow* for their technical assistance.

References

1. McDonald, D. A.: Blood flow in arteries, Baltimore: Williams & Wilkins, 1960, 53.
2. Schultz, D. L., Tunstall-Pedoe, D. S., Lee, G. de J., Gunning, A. J., Bellehouse, B. J.: Velocity distribution and transition in the arterial system, Circulatory and Respiratory Mass Transport. Eds.: Wostenholme and Knight, London: Churchill 1969, 172—199.
3. Ling, S. C., Atabeck, H. B., Fry, D. L., Patel, D. J., Janicki, J. S.: Circulation Res. 23 (1968) 789—801.
4. Replogle, R. L., Meiselman, H. J., Merrill, E. W.: Circulation 36 (1967) 148—160.
5. Dintenfass, L.: Blood microrheology—viscosity factors in blood flow ischaemia and thrombosis, London: Butterworth 1971.
6. Goldsmith, H. L.: The flow of model particles and blood cells and its relation to thrombogenesis. Advances in hemostasis and thrombosis, I, Ed.: Spaet, New York: Grune & Stratton 1972, 97—139.
7. Goldsmith, H. L., Mason, S. G.: The microrheology of dispersions, Rheology, Theory and Applications, IV, Ed.: Eirich, New York: Academic Press 1967, 85—250.
8. Goldsmith, H. L.: Science 153 (1966) 1406—1407.
9. Skalak, R.: Mechanics of the microcirculation, Biomechanics, Eds.: Fung, Perrone and Anliker, Englewood Cliffs, N.J.: Prentice Hall 1972, 457—499.
10. Fahraeus, R.: Physiol. Rev. 9 (1929) 241—274.
11. Merrill, E. W., Cokelet, G. C., Britten, A., Wells, jr., R. E.: Circulation Res. 13 (1963) 48—55.
12. Schmid-Schönbein, H., Gaehtgens, P., Hirsch, H.: J. Clin. Invest. 47 (1968) 1447—1454.
13. Chien, S., Usami, S., Dellenback, R. J., Gregersen, M. I., Nanninga, L. B., Guest, M. M.: Science 157 (1967) 829—831.
14. Dintenfass, L.: Angiology 13 (1962) 333—344.
15. Chien, S., Usami, S., Dellenback, R. J., Gregersen, M. I.: Science 157 (1967) 827—829.
16. Schmid-Schönbein, H., Wells, R.: Science 165 (1969) 288—291.
17. Gauthier, F. J., Goldsmith, H. L., Mason, S. G.: Biorheology 9 (1972) 205—224.
18. Bauer, W. H., Collins, E. A.: Thixotropy and dilatancy; Rheology: Theory and Applications, IV, Ed.: Eirich, New York: Academic Press 1967, 423—459.
19. Casson, N.: Flow equations for pigment-oil suspensions of the printing ink type, Rheology of Disperse Systems, Ed.: Mills, New York: Pergamon Press 1959.
20. Richardson, E. G.: Emulsions, Flow properties of disperse systems, Ed.: Hermans, Amsterdam: North Holland 1953.
21. Brodnyan, J. G., Kelley, E. L.: J. Colloid Sci. 20 (1965) 7—19.
22. Taylor, H. M., Chien, S., Gregerson, M. I., Lundberg, J. L.: Nature 207 (1965) 77—78.
23. Müller, A.: Abhandlungen zur Mechanik der Flüssigkeiten, I, Univ. of Fribourg, Fribourg, Switzerland, 1936.
24. Harris, J. W., Brewster, H. A., Ham, T. H., Castle, W. B.: AMA Arch. Internal. Med. 97 (1956) 145.
25. Goldsmith, H. L.: Federation Proc. 30 (1971) 1578—1588.
26. Goldsmith, H. L., Marlow, J.: Flow behaviour of erythrocytes. I. Rotation and deformation in dilute suspensions, Proc. Roy. Soc. B 182 (1972) 351—384.
27. Vadas, E. B., Goldsmith, H. L., Mason, S. G.: The microrheology of colloidal dispersions. I. The microtube technique. J. Colloid Interface Sci. 39, 1973 (in Press).
28. Goldsmith, H. L.: Federation Proc. 26 (1967) 1813—1820.
29. Goldsmith, H. L., Mason, S. G.: J. Fluid Mech. 12 (1962) 88—96.
30. Jeffery, G. B.: Proc. Roy. Soc. A 102 (1922) 161—179.
31. Goldsmith, H. L., Mason, S. G.: J. Colloid Sci. 17 (1962) 448—476.
32. Canham, P. B., Jay, W. L., Tilsworth, E.: J. Cell Physiol. 78 (1971) 319—332.
33. Taylor, G. I.: Proc. Roy. Soc. A 146 (1934) 501—523.
34. Cox, R. G.: J. Fluid Mech. 37 (1969) 601—620.
35. Torza, S., Cox, R. G., Mason, S. G.: J. Colloid Interface Sci. 38 (1972) 395—411.
36. Chaffey, C. E., Brenner, H., Mason, S. G.: Rheol. Acta. 4 (1965) 64—72 and correction ibid 6 (1967) 100.
37. Goldsmith, H. L., Mason, S. G.: Some model experiments in hemodynamics. IV. Theoretical and clinical hemorheology, Eds.: Hartert and Copley, Berlin/Heidelberg/New York: Springer 1971, 47—59.
38. Karnis, A., Goldsmith, H. L., Mason, S. G.: Can. J. Chem. Eng. 44 (1966) 181—193.
39. Takano, M., Goldsmith, H. L., Mason, S. G.: J. Colloid Interface Sci. 27 (1968) 253—267.
40. Segré, G., Silberberg, A.: J. Fluid Mech. 14 (1962) 115—157.
41. Manley, R. St. J., Mason, S. G.: J. Colloid Sci. 7 (1962) 354—369.
42. Goldsmith, H. L., Mason, S. G.: Proc. Roy. Soc. A 282 (1964) 569—591.
43. Karnis, A., Goldsmith, H. L., Mason, S. G.: J. Colloid Interface Sci. 22 (1966) 531—553.
44. Cox, R. G., Brenner, H.: Chem. Eng. Sci. 26 (1971) 65—93.
45. Goldsmith, H. L., Mason, S. G.: Bibliotheca Anat. Karger, Basel: 10 (1969) 1—8.
46. Goldsmith, H. L.: Biorheology 7 (1971) 235—242.
47. Fahraeus, R., Lindquist, T.: Amer. J. Physiol. 96 (1931) 562—568.
48. Scott Blair, G. W.: Nature 183 (1959) 613—614.
49. Phibbs, R. H.: Orientation and distribution of erythrocytes in blood flowing through medium-sized arteries. Hemorheology, Ed.: Copley, Oxford: Pergamon Press 1969, 617—630.

50. Bugliarello, G., Sevilla, J.: Biorheology 7 (1971) 85–107.
51. Blackshear, P. L., Forstrom, R. J., Dorman, F. D., Voss, G. O.: Federation Proc. 30 (1971) 1600–1609.
52. Mustard, J. F., Murphy, E. A., Rowsell, H. C., Downie, H. G.: Amer. J. Med. 33 (1962) 621–647.
53. Madras, P. N., Morton, W. A., Petschek, H. E.: Federation Proc. 30 (1971) 1669–1676.
54. Thomas, H. W.: Biorheology 1 (1962) 41–56.
55. Charm, S. E., Kurland, G. S., Brown, S. L.: Biorheology 5 (1968) 15–43.
56. Sheshadri, V., Sutera, S. P.: Trans. Soc. Rheol. 14 (1970) 351–373.
57. Meiselman, H. J.: Some physical and rheological properties of human blood, Sc. D. Thesis, Massachusetts Institute of Technology, 1965, 270–292.
58. Wells, R. E.: New Engl. J. Med. 270 (1964) 832–839.
59. Murphy, J. R.: J. Lab. clin. Med. 69 (1967) 756–775.
60. Fung, Y. C. B.: Federation Proc. 25 (1966) 1761–1772.

Gas dynamics from the point of view of kinetic theory

J.-P. Guiraud

Paris, France

From kinetic theory to gas dynamics

Laminar gas dynamic theory is governed by the so-called compressible Navier-Stokes equations

$$\frac{\partial U}{\partial t} + A_K \frac{\partial U}{\partial x_K} = \varepsilon_1 \frac{\partial}{\partial x_J}\left(L_{JK} \frac{\partial U}{\partial x_K}\right). \tag{1}$$

The five components vector U defines the state of the gas while A_K and L_{JK} are five by five matrices functions of U. In all practical problems, there is built in the equations a small non dimensional parameter ε_1, which is the inverse of a Reynolds number. Kinetic theory, on the other hand, is governed by a quite different kind of equation

$$\frac{\partial F}{\partial t} + \xi \cdot \nabla F = \varepsilon^{-1} J(F, F), \tag{2}$$

named after *Boltzmann* who derived it exactly a century ago. The so-called distribution function $F(t, \boldsymbol{x}, \xi)$ defines the density probability of finding a molecule at $\boldsymbol{x}$, with velocity ξ. Here again, there is built in a small parameter ε, the so-called Knudsen number, which is the ratio of the mean free path (between collisions of molecules) and a typical (macroscopic) length. Knudsen and Reynolds numbers are related through a Mach number, namely $\varepsilon = M\varepsilon_1$, and it will be supposed throughout that this Mach number is of order one.

The right hand side of the Boltzmann equation, the collision operator J, is typical of kinetic theory in that it preserves mass momentum and energy, namely

$$\int \psi J \, d\xi = 0, \tag{3}$$

with ψ a five components vector, namely $\psi_0 = m$, $\psi_K = m\xi_K$, $k = 1, 2, 3$, $\psi_4 = 1/2m \, |\xi|^2$, the so-called collisionnal invariants. This remarquable property, which makes a big mathematical difference between kinetic and neutron transport theories, plays a crucial role in the most interesting problem of how to go from kinetic theory to gas dynamics. The first point is that the physical meaning of F and U implies that they are related through $U = \int \psi F \, d\xi$, a relation which is expressed by saying that the five components of U are the five gasdynamic moments of the distribution function. A second point is that there is a one-to-one connection between U and F for the distribution functions which solve $J(F, F) = 0$ and have U as (vector of) gas-dynamic moments. Such distribution functions are in the so-called Maxwell class. We set, for such a distribution function, which we call a Maxwellian, $F_M = F_M(U, \xi)$ by specifying its dependence on U. Of course, an F belonging to the Maxwell class is not a solution of the Boltzmann equation, unless U is of a very peculiar from investigated, among others, by *Lunc* (1965). The third elementary point is that, by taking the gas dynamic moments of the Boltzmann equation, one gets a conservation equation

$$\frac{\partial U}{\partial t} + A_K \frac{\partial U}{\partial x_K} = \frac{\partial \Pi_K}{\partial x_K}, \tag{4}$$

with

$$\Pi_K = \int \psi \xi_K (F_M - F) \, d\xi . \tag{5}$$

The more F is close to the Maxwell class, the more this conservation equation is close to the set of Euler equations of inviscid gas dynamics. These simple considerations are illuminating because they show that the problem under consideration is related to the smallness of ε (high Reynolds number), to the closeness of F to the Maxwell class, and to the approximation of Π_J by $\varepsilon L_{JK} \, \partial U / \partial x_K$. Stated in a different way, the last point means that the Boltzmann equation may be used to derive the constitutive relation of the gas through consideration of the microscopic processes which control it. The importance of this question is most illuminated by the remark that, up to the fifties, Boltzmann equation and its variants for polyatomic gases (*Berroir* (1970)) and mixtures, was used only for the sake of computing transport properties. All this matter is thoroughly covered by the books of *Chapman* & *Cowling* (1932) and *Hirschfelder, Curtiss* & *Bird* (1954), by the review article of *Waldman* (1958) and a number of publications on the transport properties of gases at high temperatures, which originated from engineering needs in the domain of aerothermochemistry and hypersonic flight (*Fay* (1964)).

The theory of connection between kinetic theory and gasdynamics was initiated by *Hilbert* (1912), *Chapman* (1916) and *Enskog* (1917) and it was revisited recently by *Grad* (1963), (1966), *Arseniev* (1965), *Chaxov* (1968), *Galkin* (1969), *McKean* (1969), *Scharf* (1969), *Ernst* (1970), *Hauge* (1970). A straightforward expansion in powers of ε leads to what is called a Hilbert or normal (*Scharf* (1969), *Sirovich* (1963)) "solution"

$$F^{(H)} = \sum_{n=0}^{\infty} \varepsilon^n F_n^{(H)} , \tag{6}$$

of the Boltzmann equation. The first term $F_0^{(H)}$ belongs to the Maxwell class while the following terms are obtained through succesive solution of Fredholm integral equations, each of which has for its kernel the set of five collisionnal invariants, glued together in the vector ψ. The set of five compatibility conditions, which guarantee that the integral equation is soluble, has the consequence that $\partial F_n^{(H)} / \partial t$ may be expressed through space derivatives of $F_p^{(H)}$, $p \leq n$, of order less or equal to $n - p + 1$. The Hilbert expansion implies that U and the Π_J must, similarly, be expanded, and it is checked that the set of expanded conservation equations is identical with the set of compatibility conditions.

The most noticeable property of the Hilbert expansion is that it singles out a one-to-one correspondence between $U^{(H)}$ and $F^{(H)}$. For each time t, and each position $\boldsymbol{x}$, given $U^{(H)}$ as well as all its space derivatives, $F^{(H)}$ is then completely determined. The Hilbert process is able to generate "solutions" of the Boltzmann equation, the so-called Hilbert class of solutions or of distribution functions. All along the Hilbert class, the distribution function is functionally related to its gasdynamic moments, a property which belongs also to the Maxwell class, but which covers, here, a much wider range. As a consequence, all along the Hilbert class, $F^{(H)}$ is functionally related to $U^{(H)}$, which means that the Hilbert process provides a constitutive relation, a very complex one indeed, which closes the system of five conservation equations. The Navier-Stokes constitutive relation is closely connected to the Hilbert one in that one can prove that it amounts to

$$\Pi_{J,1} = L_{J,K}(U_0) \frac{\partial U_0}{\partial x_K} , \tag{7}$$

where

$$U = U_0 + 0(\varepsilon) , \quad \Pi_J = \varepsilon \Pi_{J,1} + 0(\varepsilon^2) . \tag{8}$$

The above result is illuminating as far as the problem of going from kinetic theory to gas dynamics is concerned but it shows also that the Navier-Stokes equations do not result as a definite system of equations arising at a precise step in the process of expansion. Indeed, as soon as one expands F one cannot avoid expanding U as well (*Grad* (1963)). Then U_0 satisfies the Euler inviscid equations, while U_1 is a solution of linearized Euler equations with source terms coming from the dissipative terms in the Navier-Stokes equations, computed with U_0.

The equations for U_2 contain terms which cannot be derived, by expansion, from the Navier-Stokes equations but which belong to the Burnett level.

A moment's reflection suggests that the usefulness of the Hilbert expansion implies a small Knudsen number, which, under the assumption of an order one Mach number, means a high Reynolds number, and, as a consequence, that the corresponding Navier-Stokes solutions are (singular) perturbations of the Euler inviscid equations. As is well known to aerodynamicists, singularities develop near walls, within shock waves or wakes, and also for large distances or large time. As shown by *Grad* (1966) there is built in the Boltzmann equation a wide variety of singular limiting processes. From the point of view of kinetic theory, we expect that inviscid gas dynamics breaks down at least in each region where it fails from the point of view of Navier-Stokes theory. There are, roughly speaking, three sources of nonuniformity. First, the Hilbert class sets a high constraint in the way in which F depends on ξ, so that it is not able to fit with an essentially arbitrary, given initial F, or with a given process of reflection of molecules on the wall; to meet with these non uniformities we need an initial layer and a wall layer. Second, the process of constructing the Hilbert expansion calls for more and more spatial derivatives of F as one goes to higher and higher steps in the process, derivatives which may fail to exist near each of the wide variety of discontinuities one encounters in inviscid gas dynamics; to meet with these nonuniformities, we need shock and wake layers, but also various kinds of Mach layers. Third, in many circumstances, as is shown by inviscid acoustic theory, the Hilbert solution may grow indefinitely as one goes to infinity in space or time.

The initial layer was studied by *Grad* (1963), *Arseniev* (1965), *Galkin* (1969), *Scharf* (1969). The mathematical theory of the linearized problem is complete (*Grad* (1963)), while the nonlinear one is formally well understood. The first approximation to F in the initial layer solves the so-called spatially homogeneous Boltzmann equation, the time dependence entering through $\tau = \varepsilon^{-1}t$, that is $\partial F/\partial\tau = J(F, F)$, $F_{|\tau=0} = F_0$, given. *Carleman* (1933), (1957), *Wild* (1950), *Morgenstern* (1954), *Povzner* (1962), *Grad* (1965), *Arkeryd* (1972), proved a number of existence and uniqueness theorems about this problem, all of which give quite conclusive mathematical evidence that any initial F relaxes in time $0(\varepsilon)$ to a distribution function which is within $0(\varepsilon)$ to the Maxwell class. Under the assumption of a hard potential of interaction between molecules, with angular cut-off (*Grad* (1963a), p. 31) or with radial cut-off (*Cercignani* (1969), p. 78—80), the ultimate stage of this relaxation process is exponential. Approximate solutions by *Mintzer* (1965), *Chaxov* (1968a), *Maslova* (1968), *Abe* (1971), *Ward* & *Mintzer* (1971), and "exact" numerical solutions by the discrete ordinate technique of *Wachman* & *Hamel* (1967) or by the Monte Carlo simulation of *Nordsieck* & *Hicks* (1967) show that the relaxation is quite fast and is mainly completed in a very few collisions. A good measure of departure from equilibrium is given by the value of the H-functional of *Boltzmann*, and it comes out of the Nordsieck & Hicks' computations that it relaxes to equilibrium almost as an exponential, even for initial data very far from equilibrium. As a result of the existence of an initial layer, the initial value of $U^{(H)}$, to be used as an initial datum for the hierarchy of conservation equations, is not equal to $U_{|t=0}$, as computed from the given initial distribution function. This difference, which is called the initial slip, is of order ε and this implies that while the Navier-Stokes equations are correct up to $0(\varepsilon)$, the unsteady Navier-Stokes solutions are in error by $0(\varepsilon)$, as a result of insufficiently accurate initial data. The precise correction to be used has never been computed. For an initial F which belongs to the Maxwell class, the initial slip is $0(\varepsilon^2)$ and may be expressed in an elegant explicit form (*Grad* (1963)), but this beautiful result has never been used because it is too hard a task to compute the Hilbert expansion in a specific problem, up to $0(\varepsilon^2)$.

The wall layer, while more complex, is easier to work out and is more familiar to aerodynamicists. *Trilling* (1964) seems to be the first in having understood its real nature, in a particular context. The general case has been treated by *Darrozes* (1969), (1972). It appears that the wall layer consists of two layers: an exterior one of width $0(\varepsilon^{1/2})$, which should be identified with the well known Prandtl boundary layer, and an interior one of width $0(\varepsilon)$ known as the Knudsen layer. Within the Prandtl layer, F goes away from the Hilbert class, but keeps the droperty of being functionally related to U; as a result, in a first approximation, the constitutive relation $U \Rightarrow \Pi_J$ is the same as in the Prandtl equations. What we may call the Prandtl class for F, sets again a constraint on the way in which the distribution function depends on ξ, so that the detailed mechanism of collisions of molecules with the wall cannot be met within the Prandtl class and an adjustment is necessary through the Knudsen layer. As a first approximation one has, there,

$$\zeta\, \partial F/\partial z = J(F, F), \tag{9}$$

with εz being the distance to the wall. As shown by *Guiraud* (1965), the only solution which meets the boundary conditions at the wall and which goes to the Maxwell class at infinity is a Maxwellian, uniform throughout the Knudsen layer. This reasoning gives a mathematical proof of the physically obvious condition of no slip at the wall which is used in Prandtl boundary layer theory.

The problem of kinetic heating during reentry, has created, in the sixties, strong needs for corrections to conventional Prandtl's boundary layer theory. Such a correction is of order $0(\varepsilon^{1/2})$ [(1)] and its computation raises a good point as far as the connection between kinetic theory and gas dynamics is concerned. The problem has been worked out within the framework of Navier-Stokes theory leading to the so-called second order boundary layer theory, as reviewed by *Van-Dyke* (1963) with the result that the $0(\varepsilon^{1/2})$ correction to the Prandtl boundary layer theory value of, say, the heat transfer coefficient, is a sum of separate contributions: displacement, vorticity interaction, curvature. When the problem is worked out within the framework of kinetic theory one finds three more contributions; two of them, the velocity slip and the temperature jump have generally been included but it is not so with the third one which is related to the fact that the correct constitutive relation to be used contains terms which belong to the Burnett level in the Chapman-Enskog hierarchy, and, as such, cannot be derived from Navier-Stokes theory. This effect was anticipated by *Grad* (1966), p. 290, and thoroughly investigated by *Darrozes* (1969), (1972). Unfortunately, no numerical computation has been made of the corresponding contribution to the heat transfer coefficient. The slip velocity [2]

$$\boldsymbol{u}_s = al(\partial \boldsymbol{u}/\partial n)_w, \tag{10}$$

is the boundary value to be given to the velocity of the gas at the wall in order that, up to second order, the Prandtl layer take care of the purely kinetic theory effects occuring within the Knudsen layer; l is the mean free path, as defined, in connection with viscosity, by

$$\mu = (5/16)\, \varrho(2\pi \Re T)^{1/2}\, l. \tag{11}$$

As usual, $(\partial/\partial n)_w$ means the interior normal derivative, evaluated at the wall. In the same way, the temperature jump

$$T_J = bl(\partial T/\partial n)_w, \tag{12}$$

is the difference between the temperature which should be given to the gas at the wall and the temperature of the wall proper. The numerical values of a and b depend on the mechanisms of collisions between molecules and of reflection on the wall. Their computation is an output of the solution of the so-called Kramers problem which was considered by *Welander* (1954), *Gröss, Jackson, Ziering* (1958), *Cercignani* (1962), *Albertoni* et al. (1962), *Willis* (1962), *Bakhanov* & *Deryagin* (1961), *Sone* (1966), *Loyalka* & *Ferziger* (1967), (1968), *Cercignani* et al. (1968), *Huang* (1968), *Reddy* (1968), *Deryagin* et al. (1968), *Gorelov* & *Kogan* (1969), *Sirovich* (1969), *Lang* & *Loyalka* (1970), *Loyalka* (1971), (1971a), *Williams* (1969), *Pao* (1971), *Lin* & *Willis* (1972), *Klinc* & *Kuscer* (1972). Related to the slip and temperature jump, but not pertaining to the same step in asymptotic theory, is the phenomenon of thermal creep considered by *Sone* (1966a), *Ivtchenko* (1969), *Loyalka* & *Cipolla* (1971), *Williams* (1971), as is the phenomenon of evaporation and condensation considered by *Pao* (1971a).

[(1)] This fractional order in ε necessitates a straightforward reassessing of the Hilbert expansion.

[2] It should be borne in mind that only tangential components are concerned.

From the wide variety of discontinuities encountered in gasdynamics, the shock wave has been the most studied in kinetic theory and we shall come back to this subject later. However, from the specific point of view of constructing an Hilbert expansion, this work is of limited value because it is concerned only with the first approximation, and it has long been known that, to this order, kinetic theory treatment merely confirms the Rankine-Hugoniot relations. Nevertheless, the high altitude heat transfer problem encountered with reentry has lead *Cheng* (1966) and *Pan & Probstein* (1966), among others, to use modified Rankine-Hugoniot conditions in order to take into account viscosity and heat conductivity effects outside of the boundary layer. *Germain & Guiraud* (1966), following *Sedov* et al. (1953), investigated, in a systematic way, modified shock conditions of higher order. Their treatment is based on Navier-Stokes, theory, and *Oberei* (1965) has shown that it should be modified already in the first set of modified conditions if one relies on kinetic theory instead. Unfortunately this interesting question has not been fully investigated because its practical value is limited to the rigorous computation of the $0(\varepsilon)$ term in the Hilbert series, and this is so difficult a task that workers in the field did not refrain from using rough approximations which are doubtless sufficient for engineering purposes. It is for the same reason that a rigorous treatment of wakes in a second approximation has never been considered from the point of wiew of kinetic theory.

The Hilbert process emphasises the Euler equations in that sense that, while vicosity and heat conductivity are taken into account, the Navier-Stokes equations never enter as such into the theory. This is somewhat of a loss because the Navier-Stokes theory is well set mathematically and also because it takes care of Prandtl layers, wakes and even, if not accurately enough, shock waves, leaving the initial layer as the main non uniformity. As such, the Chapman-Enskog process seems to be superior to the Hilbert one in that it generates the Navier-Stokes equations at a definite level of approximation. However, the traditional presentation, after *Chapman & Cowling*, in terms of a formal expansion of $\partial/\partial t$ is rather confusing. *Grad* (1963) has shown that the Chapman-Enskog process amounts to a reordering of the Hilbert series, and its analogues for the *Prandtl*, wake and shock layers, in such a way that the first approximation appears as the Navier-Stokes theory. Of course, there is much arbitrariness in the process of reordering. *Grad* (1966) proposes to devise the process in such a way that it singles out the Navier-Stokes equations for the first approximation, and higher versions of it to higher orders. This goal may be reached in the following manner. Let F_M be the local Maxwellian and set $F = F_M(1 + \varepsilon\phi)$, then $\varepsilon^{-1}J = F_M(-L\phi + \varepsilon Q(\phi, \phi))$ and, through use of $d \log F_M = G \cdot dU$, one gets, in place of the Boltzmann equation

$$L\phi + G \cdot (\xi_K I - A_K) \frac{\partial U}{\partial x_K} + \eta \left\{ \frac{\partial \phi}{\partial t} + \xi_K \frac{\partial \phi}{\partial x_K} + \phi \left[\frac{\partial \log F_M}{\partial t} + \xi_K \frac{\partial \log F_M}{\partial x_K} \right] \right.$$
$$\left. - Q(\phi, \phi) + G \cdot \frac{\partial \bar{\Pi}_K}{\partial x_K} \right\} = 0, \quad \Pi_K = \varepsilon \bar{\Pi}_K, \tag{13}$$

where η should be replaced by ε. If one considers η as an independant parameter, a straightforward expansion in powers of η, such as $U = U_0 + \eta U_1 + \cdots$ leads, for U_0, to the Navier-Stokes equations, for U_1, to the linearized Navier-Stokes equations with source terms and so on. It may be checked that the conservation equations have been taken care of in deriving (13) and that, as such, the set of compatibility conditions for the solubility of the integral equations for $\phi_0, \phi_1, \ldots$ are automatically satisfied. In the same way as the Hilbert process singles out the Euler equations, at the expense of the Navier-Stokes equations, similarly the proposed process singles out the Navier-Stokes equations at the expense of the Burnett equations. While *Hilbert* goes expanding the solution, *Chapman-Enskog* go constructing a hierarchy of equations: *Euler, Navier-Stokes, Burnett*, ... The proposed process is intermediate and probably is an optimum.

We have discussed how to go from kinetic theory to gas dynamics when $Kn = M\,\mathrm{Re}^{-1}$ goes to zero, M being fixed, but there are other situations, as indicated by *Grad* (1966). Some have been investigated by *Galkin, Kogan & Fridlender* (1970), and by *Sone* (1969, 1970). For a small M and an initial F which is close to order M from a uniform Maxwellian the Hilbert process leads to linear equations, and a non uniformity develops near infinity in time. It has been investigated by *Baum* (1972) who shows that the solution evolves in three periods corresponding to the initial layer, followed by the acoustic period, and, for large time, by a period in which dissipative and nonlinear effects become important in the first approximation, in the manner described by *Lighthill* (1956).

Mathematical theory of the Boltzmann equation

The mathematical theory of the incompressible Navier-Stokes equations is rather well advanced, in contrast to the corresponding state of affairs with the compressible version, to the mathematical theory of which is devoted but the pionneering paper by *Itaya* (1970). Curiously enough, the seemingly more difficult problem of dealing with the Boltzmann equation has a much more advanced theory. There are big differences, actually, between the mathematical techniques suited for the Navier-Stokes and Boltzmann equations. For the Navier-Stokes equations one has to deal with a parabolic (elliptic for stationary problems) system, for which one has in hand very powerful techniques of functional analysis, either with Sobolev spaces of functions with square integrable derivatives, or with Schauder spaces of functions whose derivatives are Hölder continuous. On the other hand, for the Boltzmann equation one has to deal with an integro-differential equation of a type which seems to have not interested mathematicians. In both cases there is nonlinearity coming in, but in much different ways.

One of the basic difficulties with the Boltzmann equation lies in the nonlinear collision operator $J(F, F)$. Formally it is a difference between two quadratic terms $J = \mathcal{D}(F)F - \mathcal{C}(F, F)$, but, in general, $\mathcal{D}(F)$ and $\mathcal{C}(F, F)$ are given, formally, by diverging integrals. Only for rigid spheres are $\mathcal{D}$ and $\mathcal{C}$ given by convergent integrals. The difficulty arises from grazing collisions which lead to vanishing exchanges of momentum and energy. It is hard to imagine that the physics may be sensitive to such kind of collisions but the mathematics is definitely so. Up to now, in order to build a rigorous mathematical theory, there is but the trick of eliminating grazing collisions by the process of angular cut-off devised by *Grad* (1963) or radial cut-off devised by *Cercignani* (1969, p. 77).

We leave aside the subject of model equations which has a well developed mathematical theory. For the full Boltzmann equation (with cut-off), both linear and nonlinear problems have been considered. The linearization is of the Stokes type. Starting from a Maxwellian

$$\omega = (2\pi)^{-3/2} \exp\left(-|\xi|^2/2\right), \tag{14}$$

one sets $F = \omega(1 + f)$, $J = \omega(-Lf + Q(f, f))$ ending with the linearized Boltzmann equation $\partial f/\partial t + \xi \cdot \nabla f + Lf = 0$.

As in all flow processes one encounters steady and nonsteady problems, as well as interior or exterior ones, depending whether the domain of flow goes to infinity or not. For the unsteady case one makes a distinction between the spatially homogeneous problem (for which $\nabla f = 0$) and the space dependent one. Again, for the space dependent problem one may deal with flow in the whole space or with boundary and initial boundary value problems for which the domain of flow in space is limited by walls. We shall speak of the Cauchy problem proper when there is no wall.

Carleman (1933, 1957), gave the first proof of an existence and uniqueness theorem for spatial homogeneity and rigid spheres. The conditions on the initial distribution function are weak because it is sufficient that it be integrable, non negative and bounded from above by Const $\cdot (1 + |\xi|)^{-K}$ with $K \geq 6$. At the expense of a quite sophisticated technique, Carleman proves much more than mere existence and uniqueness. The solution is positive for all positive times, and remains bounded by Const $\cdot (1 + |\xi|)^{-K}$, which guarantees the existence of all the gas dynamic moments. The H-functional of Boltzmann exists for all times, is a non increasing function of time, and goes to a limit when time goes to infinity. The (stronger) result that the distribution itself tends to a maxwellian is also proven. Of course the gas dynamic moments are found to be independant of time. *Wild* (1950) and *Morgenstern* (1954) proved again an existence and uniqueness theorem, for spatial homogeneity again, in the case of Maxwell molecules (central force $\sim r^{-5}$) with angular cut-off. But their results are weaker than Carleman's as far as tendency towards equilibrium is concerned. *Arkeryd* (1972) proved a result similar to Carleman's for yet weaker conditions on the initial F, which is supposed to be integrable and to have an H-functional.

Morgenstern (1955) and *Povzner* (1962) solved the Cauchy problem proper for Maxwell molecules with angular cut-off. But they modify the collision operator in the space variables. The (modified) collision operator no longer has collisional invariants except in the spatially homogeneous case or for spheres of variable radius. Without Morgenstern modification *Grad* (1958, p. 247) was able to prove an existence theorem for Maxwell cut-off molecules, but only for a bounded time interval, in contrast to Morgenstern and Povzner solutions which are valid for all (positive) time.

The Cauchy problem proper was revisited by *Grad* (1963a, 1965, 1965a). He succeeded in making effective the simple idea of iterating on $f(t) = T(t)\, f_0 + \int_0^t T(t-\tau)\, Q(f(\tau), f(\tau))\, d\tau$, with $T(t)\, f_0$ the solution of the linearized Boltzmann equation with f_0 as a given initial value of f. One of the main steps was to find suitable norms. It is necessary to work with both Hilbert and maximum norms. For velocity space the most convenient norms are

$$\|f\|^2 = \int \omega\, |f|^2\, d\xi, \quad \langle f\rangle_n = \sup_\xi \omega^{1/2}(1 + |\xi|^2)^{r/2} |f|, \tag{15}$$

while, when dealing with both x and ξ variables one needs

$$|||f|||^2 = \int\int \omega\, |f|^2\, dx\, d\xi, \qquad \langle\|f\|\rangle_r = \sup_\xi \omega^{1/2}(1 + |\xi|^2)^{r/2} \left(\int |f|^2\, dx\right)^{1/2}, \tag{16}$$

$$\langle\langle f\rangle\rangle_r = \sup_{x,\xi} \omega^{1/2}(1 + |\xi|^2)^{r/2}\, |f|, \quad [f]_r = \sup_{x,\xi} \omega^{1/2}(1 + |\xi|^2|)^{r/2}\, (1 + |x|^2)\, |f|. \tag{17}$$

With $\eta_r(f)$ any of these norms one sets

$$\eta_r^s(f) = \sup_{|\alpha| \le s} (D^\alpha f), \tag{18}$$

with $D^\alpha f$ standing for any derivative of order $|\alpha|$. Existence of the solution of the linear problem is first established with the estimate

$$\eta_r^s(T(t)\, f_0) \le e^{Krt} \eta_r^s(f_0). \tag{19}$$

This result is too weak because it does not take into account the dissipative character of the collision operator, namely that $(Lf, f) \ge \mu\, \|Pf\|^2$, with Pf the orthogonal projection onto the subspace of collisional invariants, according to the Hilbert product $(f, g) = \int \omega f \bar{g}\, d\xi$. The dissipative inequality above, with a strictly positive μ, makes use of the cut-off and of the positiveness of the collision frequency $\nu(|\xi|)$ in the decomposition $Lf = \nu(|\xi|)\, f - Af$ of the linearized collision operator. The technique for using dissipativity is to multiply the Boltzmann equation by f, according to the Hilbert product corresponding to the norm $|||\cdot|||$, namely $((\partial f/\partial t + \xi \cdot \nabla f + Lf), f) = 0$. One has to use the vanishing of the, a priori non dissipative, term $((\xi \cdot \nabla f, f))$. According to Grad, this occurs in three cases: (i) f is periodic in space variables; (ii) f is specularly reflected on the boundary of a parallelipipedic box; (iii) f goes to zero at infinity in such a way that $[f]_r$ is bounded. Under any of these conditions Grad proves a global (in time) existence theorem as well as a uniform estimate

$$\langle\|\, T(t)\, f_0\|\rangle_r^s \le a_r \langle\| f_0\|\rangle_r^s. \tag{20}$$

This is not yet the tendency towards equilibrium which was proven separately by *Grad* (1965b), using both the decrease of $|||f(t)|||$ and the evolution of the gas dynamic moments. For an initial f_0 which is orthogonal to the collisional invariants, the tendency is exponential

$$\langle\|\, T(t)\, f_0\|\rangle_{r-1}^{s-1} \le a_r(1 + b_r t + c_r)\, e^{-\mu t} \langle\| f_0\|\rangle_r^s, \quad r \ge 3, \tag{21}$$

qut this result is valid only for cases (i) and (ii). By the way, for case (iii) the tendency towards ebuilibrium may, at most, be algebraic, a result proven by *Arseniev* (1965) through use of

Fourier transform techniques and a preliminary study of the spectral problem. *Scharf* (1967, 1969) uses again Fourier transform to prove a global existence theorem and a weak form of tendency towards equilibrium, under weak conditions on the initial distribution function. While the iterative process used to solve the nonlinear problem is simple, the proof of its effectiveness is far from trivial because of the unboundedness of the quadratic operator Q. The result is rather weak in that the existence and uniqueness is obtained in the class of functions having $\langle\|f\|\rangle_3^3$ bounded, on a finite time interval

$$t_0 = \sigma(\langle\|f_0\|\rangle_3^3)^{-1} - \tau. \tag{22}$$

As a byproduct of the analysis, *Grad* (1965) proves a deep result on the nonlinear spatially homogeneous problem. Setting $F = \omega(1 + f)$, with ω the equilibrium maxwellian, Grad proves $\langle f(t)\rangle_3 \leq 2a_3 e^{-\alpha t}\langle f_0\rangle_3$ with an α, depending of $\langle f_0\rangle_3$, which may be as close to μ as we want, provided that $\langle f_0\rangle_3$ be sufficiently small. Thanks to the dissipative inequality $(Lf, f) \geq \mu \| Pf\|^2$, this is the best result which may be obtained.

The boundary value problem as worked out by *Cercignani* (1967, 1968) and *Guiraud* (1968) leads to a basic difficulty associated with the reflection condition on the boundary. Let $\boldsymbol{u}_w$ be the velocity of a point on the wall, $\boldsymbol{n}$ be the unit vector of the interior normal; setting F^+ (resp F^-) for the restriction of F to $(\xi - \boldsymbol{u}_w)\cdot\boldsymbol{n} > 0$ (resp < 0), we may assume that F^+ is functionally related to F^- by, say, $F^+ = KF^-$. Under the Stokes linearization the boundary condition goes to $f^+ = Gf^- + \Phi^+$ where the non homogeneous term Φ^+ vanishes when the maxwellian ω in $F = \omega(1 + f)$ is in a state of equilibrium with the wall. The operator G is obtained from K through the process of linearization. The mathematics relies on the dissipativity of the operator G. This dissipativity may be derived from simple arguments as in *Darrozes* & *Guiraud* (1966) and consists in the following:

$$f^+ = Gf^- \Rightarrow -\int \omega\xi\cdot\boldsymbol{n}\,|f|^2\,d\xi \geq 2k\int\omega\,|\xi\cdot\boldsymbol{n}|\,|\Pi f^-|^2\,d\xi, \quad k > 0, \tag{23}$$

$$\Pi f^- = f^- + \text{const.}, \quad \int\omega\xi\cdot\boldsymbol{n}\Pi f^-\,d\xi = 0. \tag{24}$$

This guarantees the dissipativity of $((\xi\cdot\nabla f, f))$ and allows to derive the following inequality (*Guiraud* (1970))

$$\mu\,|||Pf|||^2 + k\int_{\partial\Omega} dS\int\omega\,|\xi\cdot\boldsymbol{n}|\,|\Pi J^- Bf|^2\,d\xi \leq \operatorname{Re}\int_{\Omega} d\boldsymbol{x}\int\omega(\xi\cdot\nabla f + Lf)\,\bar{f}\,d\xi, \tag{25}$$

for every steady solution of the linearized Boltzmann equation with homogeneous boundary condition $\Phi^+ = 0$ on the boundary $\partial\Omega$ of the bounded domain Ω of flow. In this formula B means that we take the value on $\partial\Omega$, and J^- that we restrict to $\xi\cdot\boldsymbol{n} < 0$. The projection operator Π has been defined already.

The above basic inequality guarantees the uniqueness of the solution of the stationary boundary value problem up to a constant, whose physical meaning is that the pressure may always be changed by a constant. One may prove (*Guiraud* (1970)) existence of a solution bounded according to the norm $|||\cdot|||$ corresponding to the domain Ω, provided that this domain be convex, with a smooth boundary, and that Φ^+ be bounded in a suitable Hilbert norm. There is also existence with a nonhomogeneous term ϕ on the right hand side of the equation with $|||\phi|||$ bounded. This existence result is valid for any cut-off potential. With rigid spheres (*Guiraud* (1972)) one may prove some regularity properties of the solution as function of velocity. Provided that ϕ and Φ^+ be continuous and that $\langle\langle\phi\rangle\rangle_r$ and $\langle\langle\Phi^+\rangle\rangle_r$ be bounded, the solution, rendered unique by some constraint as $((f, 1)) = 0$, satisfies the estimate

$$\langle\langle f\rangle\rangle_r \leqq C_r\{\langle\langle\Phi\rangle\rangle_r + \langle\langle\Phi^+\rangle\rangle_r\}, \quad r > 5/2.$$

This estimate allows to solve the weakly nonlinear problem (*Guiraud* (1970a, 1972)). The physical meaning of this result is as follows: The wall slides on itself with a velocity $U_0\boldsymbol{u}_w$ and a temperature $T_0(1 + a_w)$ in such a way that for $\boldsymbol{u}_w = a_w = 0$ the wall is in equilibrium with the maxwellian ω. One can show that, for $\sup_{\text{wall}}(|a_w| + |\boldsymbol{u}_w|)$ sufficiently small, there exists a solution to the nonlinear boundary value problem giving a flow regime which is unique (in the vicinity of equilibrium) up to an additive constant in the pressure. A weaker result, as far as boundary conditions are concerned, was obtained earlier by *Pao* (1967).

The initial boundary value problem has been solved by *Fetz & Shen* (1970), in the linear case, through use of perturbation theory applied to semigroups of operators in Hilbert space. *Scharf* (1967) considered the same problem again through semi-group theory, assuming dissipativity of the boundary operator. An important point is the tendency towards equilibrium which may be shown to be exponential for rigid spheres (Guiraud, unpublished communication to the Glasgow meeting on Boltzmann equation, University of Strathclyde March 1972) and homogeneous boundary condition ($\Phi^+ = 0$).

The exterior problem leads to great difficulties as shown by the example of the incompressible Navier-Stokes equations and the work of *Cercignani* (1968a) who proves that Stokes linearization, $F = \omega(1 + f)$, leads to a Stokes paradox. These difficulties have been solved for one space dimension, in the configuration of the Kramers problem, by *Rigolot-Turbat* (1971) who solved various steady and unsteady linear and nonlinear problems in one space dimension.

Sound propagation

As in most physical processes, the problem of harmonic oscillations in gases is most illuminating and it is understandable that it had been thoroughly investigated. The Navier-Stokes theory, which amounts to look for solutions of the linearized equations, with an exponential factor exp $(\sigma t - sx)$ has been reviewed by *Sherman & Talbot* (1960) who reviewed also the experiments, mainly by *Greenspan* (1949, 1950), and *Meyer & Sessler* (1957). The experimental set up singles out the so-called forced sound waves in which $\sigma = i\omega$ is purely imaginary, so that the exponential factor is exp $(i\omega(t - c^{-1}x) - \alpha x)$ with a well defined speed of propagation c and an attenuation coefficient α. The frequency comes in through the reduced period $r = p/\omega\mu = a^2/\gamma\nu\omega$, which is an inverse Knudsen number. For high r Navier-Stokes theory shows that $c = a$ with an error of order r^{-2}; while

$$\alpha \simeq 1/2(4/3 + (\gamma - 1)/Pr)\, \nu\omega^2/a^3, \tag{26}$$

ν being the kinematic viscosity and Pr the Prandtl number. Sherman & Talbot point out that the experiments are in good agreement with Navier-Stokes theory for high r, down to a value of five. *Greenspan* (1950) and *Meyer & Sessler* (1957) have performed experiments well beyond this limit down to $r = 0.02$ and their results should be checked versus a kinetic theory treatment of the problem.

The simplest approach is to use the normal mode technique which amounts to looking for solutions of the Boltzmann equation such as $F = F_0(1 + g \exp(\sigma t - s\boldsymbol{e} \cdot \boldsymbol{x}))$ where $\boldsymbol{e}$ means the unit vector normal to the planes of constant phase. One is mostly interested in the relation between s and σ when following by continuity the so-called sound modes of Navier-Stokes theory. This approach leads to the so-called spectral problem which consists in finding the dispersion relation between σ and s in order that the equation $(\sigma - s\boldsymbol{e} \cdot \boldsymbol{\xi} + L)\, g = 0$ has a non-vanishing solution. In most physical situations encountered in fluid dynamics, plane wave solutions are readily obtained. On the other hand, in kinetic theory it is at least tedious to derive accurate approximations for the dispersion relation. Moreover, the mathematical theory of the spectral problem is far from being simple, with very few rigorous treatments available as those by *Arseniev* (1965), *Nikolaenko, Thurber & Dorning* (1971) and *Nikolaenko* (1972). For free waves, for which $s = ik$ is purely imaginary, one has to deal with perturbation of the self-adjoint spectral problem of the linearized spatially homogeneous Boltzmann equation. The difficulty lies in the fact the perturbation operator $-ik\boldsymbol{e} \cdot \boldsymbol{\xi}$ is unbounded and not self-adjoint; another difficulty comes from the existence of a continuous as well as a discrete spectrum for the unperturbed spectral problem. The continuous spectrum consists in the set of values of minus the collision frequency $\nu(|\boldsymbol{\xi}|)$ which appears in the decomposition $L = \nu - A$ of the collision operator for a cut-off potential of interaction between molecules. For the forced waves the situation is even more complex.

As a matter of fact, the mathematical or numerical solution of the spectral problem is not the whole matter and a difficulty appears when interpreting the results versus experiments, as demonstrated by the controversial nature of the literature on the subject, a topic thoroughly discussed by *Thurber* (1972). No known experiment is able to check the accuracy or even the existence of free waves, while experimentally excited forced waves are not, strictly speaking, plane waves. They correspond to the solution of a boundary value problem in a half space with prescribed oscillations at the wall. The relation of this boundary value problem to the spectral problem is rather loose and difficult to assess theoretically. However, the experiments of *Greenspan* (1950), *Meyer* & *Sessler* (1957), *Hassler* (1968) give strong evidence of the appearance of plane waves with well defined propagation speed c and attenuation coefficient α.

The experimental evidence for plane waves generates a strong demand for the accurate solution of the spectral problem. *Pekeris* et al. (1962) have used an (expectedly) highly accurate version of the method of moments of *Wang Chang* & *Uhlenbeck* (1952). This amounts to using the orthogonal projection on a finite dimensional subspace of the Hilbert space with scalar product $(f, g) = \int F_0 f \bar{g}\, d\xi$. The finite dimensional subspace which is supposed to contain the collisionnal invariants is generated by a set of ψ_i $i = 1, 2, \ldots, N$. Setting $a = (g, \psi_K)$ and $(a_1, a_2, \ldots, a_N)^T = a$, with the superscript T standing for transposition of matrices, one gets a finite dimensional spectral problem $(\sigma I - sA + \Lambda)\, a = 0$, where A and Λ stand for real symmetric matrices. By setting $a = Pa + Qa = x + Qa$, with P the projection on the gas dynamic moments and Q the complementary projection, one gets

$$(\sigma I - sPAP - s^2 PAQ(Q\Lambda Q)^{-1}QAP - s^2 PAQ[\{I + (Q\Lambda Q)^{-1}(\sigma I - sQAQ)\}^{-1} - I] \times (Q\Lambda Q)^{-1}QAP)\, x = 0, \tag{27}$$

while the Navier-Stokes spectral problem amounts to

$$(\sigma I - sPAP - s^2 PAQ(Q\Lambda Q)^{-1}QAP)\, x = 0. \tag{28}$$

This shows that the Wang Chang & Uhlenbeck dispersion relation for forced sound waves agrees, when expanded in powers of frequency, with the Navier-Stokes result up to $0(\omega^2)$. *Pekeris* et al. (1962) have used an N $(= 105)$-dimensional (resp. $N = 483$) spectral problem to compute c and α as a function of frequency ω, for rigid spheres (resp. Maxwell molecules). The amount of work involved in the computation far outweights the increase in accuracy over Navier-Stokes theory. While the agreement with experiment is good down to $r = 1$, it becomes poor for r smaller than one and definitely wrong for $r < 0.2$. For $r = 0.02$, the extreme value of r in the Meyer & Sessler experiments, Pekeris et al. find an α which is thirty times too small. On the other hand, the computed values for the propagation speed deviate much less from the measured ones.

The main idea of Wang Chang & Uhlenbeck consists in subtituting for the (numerically) untractable spectral problem set in Hilbert space another one in a finite dimensional space, which amounts to solving an algebraic equation. The main drawback in this approach is the fact that however large N may be, the approximate spectral problem has always a purely discrete spectrum, a simple evidence of the failure of this technique in dealing with the continuous spectrum, which is always present when dealing with cut-off collision operators. This may be best visualized by rewriting the spectral problem for the Boltzmann equation as $(\sigma - se \cdot \xi + \nu(|\xi|))\, g - Ag = 0$. The continuous spectrum comes from the set of values of σ and s for which the parenthesis is zero for at least one value of ξ. In a series of papers *Sirovich* and *Thurber* (*Sirovich* (1963, 1963a, 1963b), *Sirovich* & *Thurber* (1965, 1965a, 1967, 1969), *Thurber* (1972)) have devised an approximate procedure retaining the advantages of reduction to finite dimensional spaces while taking duly account of the continuous spectrum. The trick is that A being a well behaved operator may, safely, be approximated by a degenerate operator. The approximation may even preserve the kernel of the collision operator, which is a highly desirable property. It is just sufficient to approximate A by $K_N = P_N \nu^{-1} A P_N$ with P_N the orthogonal projection (according to $\langle f, h\rangle = (\nu^{1/2} f, \nu^{1/2} h)$) on a finite dimensional subspace generated by a set of ψ_i, $i = 1, 2, \ldots, N$, involving the collisional invariants. Setting $P_N g =$

$\sum_{i=1}^{N} a_i\psi_i$, $a = (a_1, \ldots, a_N)^T$, the simple idea of dividing by $\nu + \sigma - se \cdot \xi$ leads to $(I - C)\, a = 0$, which is a finite dimensional spectral problem. Each entry $C_{ij} = \langle (\nu + \sigma - se \cdot \xi)^{-1}\, \nu K_N \psi_j, \psi_i \rangle$ of the matrix C is a holomorphic function of σ and s provided that $(\nu + \sigma - se \cdot \xi)$ does not vanish for any ξ. This is most easily visualized when ν is a constant, as a consequence of $C = s^{-1} D(z)$, $z = s^{-1}(\nu + \sigma)$. It is seen that $D(z)$ is a sectionaly holomorphic function of z with $\mathrm{Re}\, z = 0$ as a cut. We have indeed to deal with two functions $D^+(z)$ (resp. $D^-(z)$) holomorphic in $\mathrm{Re}\, z > 0$ (resp. $\mathrm{Re}\, z < 0$) which are not analytic continuations of each other. The occurence of a cut is, of course, a consequence of the existence of a continuous spectrum in the full problem. The dispersion relation is expressed as an algebraic equation of degree N in s, namely $P(s, z) = 0$, with coefficients which are sectionaly holomorphic functions of z. When looking at the dispersion relation $s = \phi(\omega)$ for forced sound waves one gets a frequency cut ω_c beyond which the sound modes no longer exist. In their computations *Sirovich* & *Thurber* use $N = 11$ as the most sophisticated approximation and find that the cut-off is at $r = 0.290$ for rigid spheres and $r = 0.413$ for Maxwell molecules. On the other hand, the experiments show evidence of plane waves down to $r = 0.02$. To deal with this as it stands, Sirovich and Thurber use the trick of analytically continuing C beyond its natural cut. The trick works and gives what may be called pseudo-discrete modes imbedded in the continuous spectrum. The proper mathematical theory of the analytic continuation, which is far from being trivial, has been considered by *Nikolaenko* et al. (1971), and *Nikolaenko* (1972). The numerical work of *Sirovich* & *Thurber* (1965a) provides sound waves down to $r = 0.002$ achieving good agreement with experiments down to the lowest value of 0.02.

As interesting as it may be by itself, the spectral problem should be considered in the light of a boundary value problem. *Arseniev* (1965) and *Scharf* (1967, 1969) have considered the Cauchy problem for the whole space via the spectral problem. *Mason* (1965, 1967), *Weitzner* (1965), *Bruckner* & *Ferziger* (1966), *Ostrowski* & *Kleitman* (1966), *Richardson* & *Sirovich* (1971) have studied boundary value problems qualitatively similar to the one corresponding to forced waves produced by an oscillating piston. The full problem, even for model equations is too difficult and they had to rely on some kind of simplification. We sketch here the full problem for model equations in order to stress the difficulties. We have to solve for

$$\left(i\omega + \nu + e \cdot \xi \frac{\partial}{\partial x}\right) f - \nu K_N f = 0\,, \tag{29}$$

in $x > 0$ and $f^+ = Gf^- + \Phi^+$ on $x = 0$. The unit vector e is normal to the wall and points towards the interior $x > 0$, while f^+ (resp. f^-) means f restricted to $e \cdot \xi > 0$ (resp. $e \cdot \xi < 0$). The oscillations of the gas are forced by the wall, and this is brought in by the given Φ^+. We use Laplace transform through

$$f = (2\pi i)^{-1} \int_L \exp(-sx)\, \tilde{f}\, ds\,, \tag{30}$$

and we set $g = f_{|x=0}$, which is to be found as part of the solution. Setting, in the same way as for a in the above, $y = (y_1, \ldots, y_N)^T$, with

$$y_K = \langle (\nu + i\omega - se \cdot \xi)^{-1} e \cdot \xi g, \psi_K \rangle\,, \tag{31}$$

we get the matrix equation $(I - C)\, a = y$. This allows to represent the solution as

$$\tilde{f} = (\nu + i\omega - se \cdot \xi)^{-1} \left\{ e \cdot \xi g + \sum_{i=1}^{N} a_i \nu K_N \psi_i \right\}. \tag{32}$$

Of course, y must be determined, as must be g, as part of the solution, and this makes the difficulty of the problem. The matrices C and y are holomorphic functions of s outside the shaded area of Fig. 1. The same is true for a and f with the exception of poles coming from the roots of the dispersion equation. But the solution of the problem requires that $\tilde{f}$ be holomorphic for $\mathrm{Re}\, s > 0$. This requires removing the poles which arise from the roots of $\det(I - C) = 0$ located in $\mathrm{Re}\, s > 0$, and also the cut in the domain of analyticity of $\tilde{f}$ which consists in the part of the shaded area of Fig. 1, located in Re $s > 0$. Both requirements should, when coupled with the boundary condition $g^+ = Gg^- + \Phi^+$, allow to determine g. This is too difficult to be achieved and simplification has to be made, mainly by setting $\nu = \text{Const}$ (which reduces the shaded area to a cut along a line) and by taking $N = 1$, as was done by most authors, or, as in *Bruckner & Ferziger*, by assuming an artificial boundary condition which allows to meet the requirements of holomorphy of $\tilde{f}$ in a tractable way. Taking for granted that $\tilde{f}$ is known, the Bromwich integral for f is deformed, giving rise to a residue contribution associated with roots of the dispersion

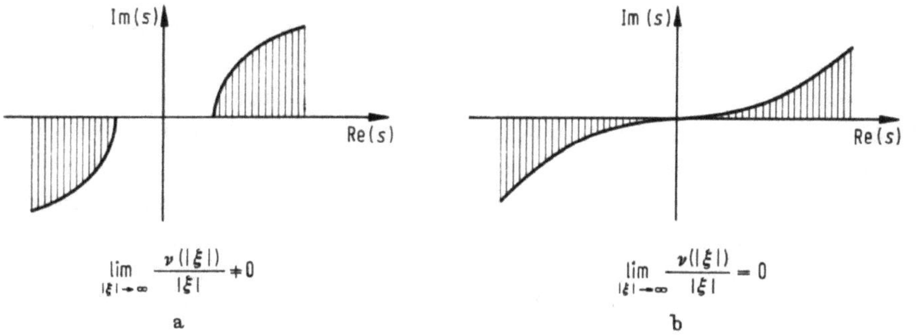

Fig. 1.

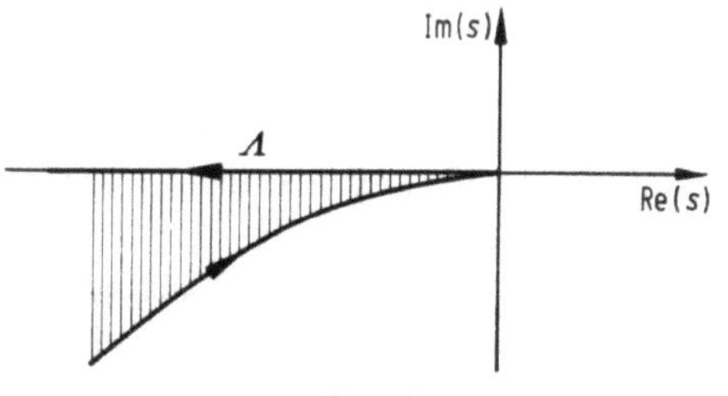

Fig. 2.

equations located in Re $s > 0$ and to a contour integral along Λ of Fig. 2. This visualizes the forced waves as a sum of normal modes coming from the discrete spectrum of the spectral problem and a continuous spectrum contribution. Bruckner & Ferziger made numerical computations of, for example, the pressure profile, which give evidence, under the conditions of experiments, that the oscillations look like well defined plane waves (for the limited distance which is considered in the experiments). These waves arise from both the discrete and continuous spectrum, both contributions showing rapid variations which compensate each other near the cut-off in frequency. As there is no analysis which fits the true conditions of the experiments, the various theoretical results are expected to remain controversial for some time.

Shock structure

The shock structure problem is a very basic one in gas dynamics and it has attracted many workers for years. There are plenty of good reasons for that. First, there is the obvious curiosity to resolve the unphysical appearance of discontinuities in gas dynamics. Second, the shock structure is the simplest basic flow in which nonlinearity and viscosity (coupled with heat conductivity) exactly balance each other as was investigated by *Lighthill* (1956). Last, but not least, the shock structure flow provides a very simple kind of flow, amenable to theoretical study and readily producible, with which a number of physical processes, relevant to the dynamics of real gases at high temperature, can be analysed both theoretically and experimentally. Rotational and vibrational relaxation, chemical kinetics in reacting gases, diffusion processes in mixtures, energy and even momentum exchanges between matter and radiation: all these processes, and others, are most easily investigated when occurring in the flow through a shock,

The first Navier-Stokes shock structure solution was given by *Taylor* in 1910, under the assumption of a very weak shock. Since then, the full problem has been completely solved and an authoritative account of the theory is given by *Gilbarg & Paolucci* (1953). One of the results is the so-called maximum slope shock thickness based on the density profile. It gives the basic length scale of the flow and is found, for moderately strong shocks, to be of the order of a few mean free paths, while, for a weak shock, this scale length thickens as the inverse of the shock strength. This result is very illuminating because it raises the problem of the adequacy of the Navier-Stokes equations for describing a flow process in which the basic length scale is not large in comparison with the mean free path. Some experiments were performed, especially at *Berkeley*, during the fifties, in order to settle this question and the results were rewiewed in *Sherman & Talbot* (1960). The conclusion was that Navier-Stokes theory is probably accurate for a shock Mach number less than 2 but that for stronger shocks it is definitely non accurate. It was then clear that the shock structure should be viewed as a basic problem in kinetic theory. At the time, the most intriguing outcome of the Sherman & Talbot investigation was the evident failure of any of the obvious kinetic improvements on Navier-Stokes theory, in the sense that with any of these, the situation was getting worse. As an example, the so-called Burnett equations fail to give any solution at all for a shock Mach number exceeding 2.2 while, for the Grad's thirteen moments equations the upper limit is 1.65 (*Grad* (1952)). This gave support to the idea that the Navier-Stokes equations provide better approximation than any of the more sophisticated ones to be derived from kinetic theory, as was suggested by an illuminating, but somewhat peculiar, example constructed by *Truesdell* (1956). It is now well understood that Hilbert's, Chapman-Enskog's, or Grad's type of improvements on the Navier-Stokes theory is only able to give better accuracy when the Navier-Stokes answer is already reasonably accurate, but that they fail completely if they are used to improve on Navier-Stokes theory when it shows evident failure.

Fortunately, as early as 1951, *Mott-Smith* devised an approximate procedure for solving the Boltzmann equation, especially in connection with the shock structure problem. While the rigorous proof of the existence of a kinetic theory shock structure solution is still lacking, there is, now, a large body of numerical as well as experimental evidence of this existence.

The weak shock raises a number of interesting theoretical points. The density, velocity and temperature profiles are given accurately by Navier-Stokes theory. Then, the distribution should be derived from the Chapman-Enskog process. A straightforward order of magnitude estimate shows the relative departure from maxwellian to be of order $(\tau/p)\ (1 + M\ c/a)\ (c/a)^2$ where τ, p, a, M are local values of viscous stress, pressure, speed of sound and Mach number while c is the molecular speed in a frame moving with the local macroscopic velocity. The maximum value of τ/p within the shock has been evaluated by *Liepman* et al. (1962) on the basis of the Navier-Stokes solution. While being fairly small for a $M = 1.2$ shock, it goes through 0.23 for a $M = 2$ shock, towards an upper limit of 1.5 for an infinitely strong shock. This is again a rough indication that $M = 2$ is a, somewhat loosely defined, upper limit of adequacy of the Navier-Stokes theory. But this is not the whole story because, however small τ/p may be, there are always sufficiently fast molecules for which the Chapman-Enskog value of the distribution function is not accurate. The mathematical consequence of this observation is that the weak shock limit is nonuniform in velocity space. The reason for this is that under the weak shock limit, while most of the molecules undergo very many collisions when processing through the shock, there are always molecules going sufficiently fast to pass through most of the shock transition almost without colliding with other molecules, or at least colliding with only a few of them, depending on the way in which the collision frequency depends on speed. This phenomenon was discovered by *Lyoubarski* (1961) and fully investigated by *Narasimha* (1968) who gave a uniformly valid representation of the distribution function, which it should be possible to test versus the direct measurements of the distribution function achieved by *Muntz & Harnett* (1969, 1972) and *Holtz, Muntz & Yen* (1971).

A considerable success has been achieved, during the sixties, in devising numerical procedures for constructing approximate or even "exact" solutions of the full Boltzmann equation as well as of model equations. An extensive study was carried out, for the Krook model, by *Liepman, Narasimha & Chahine* (1964). For a quite strong shock of $M = 10$ they find big differences with Navier-Stokes theory. The shock region is considerably wider than acording to Navier-Stokes theory, and there is marked difference between the density and temperature profiles, the last one being much wider than the other. For a viscosity going like $T^{0.816}$, and a Prandtl number one (which is inherent to the Krook model), 70% of the temperature rise in a $M = 10$ shock is accomplished in a region where only 15% of the density rise is effected. This is in contrast to the Navier-Stokes result and is a consequence of the fact that the distribution function deviates considerably from a Chapman-Enskog one. Even more illuminating is the distribution function itself, as computed by *Liepman, Narasimha & Chahine*. For the same $M = 10$ shock the distribution function has to go from a highly peaked shape on the cold side to a widely spread one on the hot side. For most of the transition the high peak appears to be convected and attenuated with relatively slight change in shape, while the downstream maxwellian progressively builds up. This remark was made by *Oguchi* (1965) the basis of an approximate solution devised for very strong shocks. The same idea was reworked in a more systematic way by *Grad* (1966), who showed that for an infinitely strong shock the distribution function should be close to

$$F = \varrho_U(1 + e^{x/\lambda})^{-1}\,\delta(\boldsymbol{\xi} - \boldsymbol{u}_U) + \varrho_D(1 + e^{-x/\lambda})^{-1}\,(2\pi\Re T_D)^{-3/2}\exp\left(-\,|\boldsymbol{\xi} - \boldsymbol{u}_D|^2/2\Re T_D\right), \quad (33)$$

the suffixes U and D referring to upstream and downstream conditions, and δ being the usual delta function. In a yet unpublished work (Glasgow[1] conference on the Boltzmann equation, March 1972) Narasimha refined the analysis. All this gives evidence that the old idea of the Mott-Smith solution being exact in the limit of infinite strength is sound, if not, strictly speaking, true.

The subject of relaxation processes in gas flows has attracted many workers and the shock tube has proven to be a quite powerful tool for such kind of investigation. Each process of relaxation may be characterized by a time scale which gives, roughly, the time necessary for the process, acting alone, to be completed. If τ is such a time scale and if τ_c is the mean collision time, we may write $\tau = Z\tau_c$ and state that Z is, roughly, the number of collisions which are necessary to complete the process. For most relaxation processes Z is quite big, and this means that the shock transition has to go in two or more steps. Rotational relaxation is an exception insofar as, at room temperature, Z may be less than five, with the consequence that rotational and translational energy jump simultaneously through the shock wave. The most interesting problem of simultaneous adjustment of rotational and translational energy within a shock wave was investigated in a pioneering work of *Talbot* and *Scala* (1961), who succeded in working out a purely continuum theory. It has long been known that the relaxation proces may be accounted for by introducing a bulk viscosity if the relaxation time is very short in comparison to the macroscopic time. According to this concept the difference between rotational and translational temperatures may be computed approximately as $T_R - T_T = Z_R\tau_c\Re T C_V^{-1}\operatorname{div}(\boldsymbol{u})$. On the other hand, when the relaxation time is comparable to the macroscopic time, a relaxation equation such as $DT_R/Dt = (Z_R\tau_C)^{-1}(T_T - T_R)$ must be used concurrently with conservation of mass momentum and (total) energy. As expected, Talbot & Scala demonstrated that the bulk viscosity model is not valid for shock waves which are not quite weak. One point which is correct with the bulk viscosity model is the qualitative fact that the rise in rotational temperature lags significantly behind the rise in translational temperature, while this last one overshoots its downstream value. For a shock Mach number of 4.6 and $Z = 5$, Tablot & Scala compute that translational temperature has risen to its downstream value when 75% of the rise in rotational temperature remains to be effected. This is evidently a consequence of the fact that the exchange of translational energy through collisions is five times more efficient than the exchange of rotational energy. The overshoot in translational temperature means that most of the available kinetic energy in the upstream flow is first converted in translational energy,

[1] University of Strathclyde

with only a small amount appearing in the form of rotational energy. As a consequence, the translational temperature reaches within the shock higher values than downstream, where the available kinetic energy has been shared in equilibrium proportions between translational and rotational energies.

Of course, from what is known about shock structure in monoatomic gases, one should expect that the continuum approach is not fully valid, and one should resort to a full kinetic theory treatment. Such a treatment was not undertaken up to the recent past. The problem is of course a very difficult one, but there is also the fact the density profile was the only one accessible to experiment, and it is known that this profile is very weakly sensitive to rotational relaxation. A breakthrough was accomplished by the demonstration, by *Robben & Talbot* (1966), that the distribution of population amongst the rotational energy levels could be determined in an even highly nonequilibrium flow. A number of approximate, but more or less sophisticated kinetic theory treatments, for moderate to quite high shock strength, in a diatomic gas, have been reported, mainly at the Boston meeting (*Turcotte*, p. 331—341, *Brau* et al. p. 343—352, *Venkataraman & Morse*, p. 353—366) and the Pisa meeting (*Haight*, *Giddens* et al.). All this confirms qualitatively the conclusions drawn from the Talbot & Scala work and give an estimate of Z of 3 to 5, at room temperature, in agreement with sound absorption experiments.

This short review of shock structure should not end without a word about diffusive separation of species, induced by shock waves in mixtures. According to continuum theory, separation is a result of baro-diffusion, thermal diffusion and ordinary diffusion. For a binary mixture of molecular masses $m_1 > m_2$, proceeding through a shock wave, the velocity of the heavy species relative to the mixture is given by

$$u_1 - u = \frac{\varrho m_1 m}{\varrho_1 m^2} D_{12} \left\{ f(1-f) \left[\frac{m_1 - m_2}{m} \frac{d \log p}{dx} - \alpha \frac{d \log T}{dx} \right] - \frac{df}{dx} \right\}, \tag{34}$$

with $f = n_1/n = \varrho_1 m/\varrho m_1$ the molar concentration, D_{12} the binary diffusion coefficient and α the coefficient of thermal diffusion. Through conservation of mass, the gas which goes faster than the mixture does so at the expense of a decrease in its concentration. *Sherman* (1960) has worked out the continuum theory, especially for an Argon-Helium mixture corresponding to a fairly high mass ratio of 10. It turns out that the heavier gas goes faster through the shock than the mixture which is getting richer in the light species. For a very small concentration of the heavier component (2% in Sherman's computation), Sherman finds that, even in a fairly weak shock, the heavy species is first accelarated, then decelerated, reaching a maximum in velocity of 18% in excess to the upstream value. While it is easy to understand that the heavy molecules are less decelerated through collisions than the light ones, it is hard to imagine any process by which the collisions might first accelerate the heavy molecules. The use of a continuum treatment under extreme conditions was considered suspect by Sherman itself, and it was evident that a kinetic treatment was necessary. The first attempt was made by *Oberei* (1966) through use of the Mott-Smith technique, with the expected result that the so-called Sherman's anomaly was, indeed, a pure continuum effect, physically not significant. The problem was revisited using various approaches, in particular at the Boston meeting (*Sirovich & Goldman*, p. 407—415, *Abe & Oguchi*, p. 425—432) without changing Oberei's conclusion. Interestingly enough, *Sirovich & Goldman* (1969) use a two-speeds one-temperature continuum model which no longer leads to the Sherman's anomaly. The diffusion equation is replaced by a relaxation one $\lambda \, d/dx \, u(u_1 - u_2) + u_1 - u_2 = (u_1 - u_2)_{\mathrm{Diff}}$ so that the one-fluid theory looks like a (singular) limit of the two-fluids theory. It is shown that, under the conditions which exhibit Sherman's anomaly, the first term on the left hand side is not a small perturbation of the second one. Full kinetic theory treatments have been given, the most significant of them being that by *Bird* (1968) who integrates the system of two Boltzmann equations for Argon-Helium through Monte Carlo simulation. An outcome of Bird's computations is the appearance, under some conditions, of an overshoot in the temperature profile of Argon, a result which is not obtained with the two fluids theory.

The sharp leading edge in hypersonic flow

The laminar flow of a gas over a very sharp flat plate, parallel to the wind, has been widely investigated by theory as well as by experiment. When the Reynolds number $\mathrm{Re} = \varrho_\infty U_\infty L/\mu_\infty$, based on plate length, is very great, the usual picture of potential flow over a boundary layer is valid except near the leading and trailing edges. The vicinity of the trailing edge was revisited, with new ideas, by *Stewartson* (1969) and *Messiter* (1970). It is found that the trailing edge region is dominated by a high coupling between potential and boundary layer flow. In incompressible flow the upstream influence of the trailing edge is felt over a distance of $L\mathrm{Re}^{-3/8}$, in a domain over the most part of which the flow is approximately inviscid, friction being restricted to a thin sublayer of thickness $L\,\mathrm{Re}^{-5/8}$. Imbedded within this sublayer is a full Navier-Stokes domain of scale $L\,\mathrm{Re}^{-3/4}$. One may show that there the Knudsen number is $0(M_\infty \mathrm{Re}^{-3/4})$ with M_∞ as the upstream Mach number. As a result, kinetic theory is not needed to investigate the main behavior near a trailing edge. For a leading edge the scale of the Navier-Stokes domain is $L\,\mathrm{Re}^{-1}$ and the Knudsen number there is $0(M_\infty)$. In supersonic or, even more so, in hypersonic flow the leading edge flow should be considered as a typical problem in kinetic theory. This has been realized for a long time and papers on the leading edge problem have appeared at each of the Rarefied Gas Meetings since 1960.

The boundary layer acts on the inviscid flow as a body whose local slope is of the order $M_\infty^2\,\mathrm{Re}_x^{-1/2}$, with $\mathrm{Re}_x = \varrho_\infty U_\infty x/\mu_\infty$ the local Reynolds number. This induces in the inviscid flow perturbations of order $M_\infty^3 \mathrm{Re}_x^{-1/2}$. On the other hand, the rarefaction parameter, as measured by the ratio of slip velocity to upstream speed, is of order $M_\infty \mathrm{Re}_x^{-1/2}$. These rough estimates show that for a very high Mach number one may well have quite strong perturbations, implying quite strong coupling between inviscid and boundary layer flows, while the Navier-Stokes equations remain valid from the point of view of kinetic theory as an approximation for small Knudsen number. Nearer to the leading edge one expects strong evidence of kinetic theory effects for a moderately large value of the rarefaction parameter. One may try to extend the strong interaction continuum theory (*Lees* (1953)) by taking into account slip, temperature jump and second order effects in the boundary layer, and even to go farther using modified Rankine-Hugoniot conditions and viscous inhomogeneous terms in the mainly inviscid layer. This amounts to an application of *Darrozes's* (1969), (1972) analysis, and has never been done.

Actually, up to the very recent past, the hypersonic leading edge problem was studied by using two extreme techniques. *Charwat* (1961), and more recently *Kogan & Degtyariev* (1965), used near free molecule flow theory, while most of the workers in the field have used, after *Oguchi* (1961), a continuum model on the basis of Navier-Stokes equations with slip and temperature jump. While relying on Navier-Stokes theory, such a treatment does not lead to a Navier-Stokes solution proper even in an asymptotic sense. It is rather a physically sensible flow model, which allows computations to be performed. Assuming that Prandtl's boundary layer equations may be used up to the shock, Oguchi succeeded in building a model of flow which shows a continuous transition from strong interaction theory, involving a boundary layer well separated from the shock by an inviscid layer, to a fully viscous regime involving a constant strength shock. Oguchi's model had the virtue to exhibit a mechanism able to put an end to the unlimited growth in surface pressure, inherent in strong interaction theory. *Pan & Probstein* (1966) refined Oguchi's analysis by including slip and temperature jump at the wall, as well as shock slip, through the application of modified Rankine Hugoniot conditions. There is no evidence of a maximum in surface pressure, an indication that Oguchi's plateau was perhaps fortuitous.

As a matter of fact, the above theories were aimed at building a model of fully viscous flow, when boundary layer and shock merge, so that one may no longer identify an inviscid region. The name of merged layer regime was given to this type of fully viscous flow. The weak point with this approach is that it relies heavily on concepts derived from asymptotic theory and uses them under circumstances where asymptotic theory fails. It is not expected that such theories be capable to go beyond the so-called incipient merged layer regime, which may be characterized by the fact that strong interaction theory, while remaining mainly correct, is felt to be insufficiently accurate.

Short of a full kinetic theory treatment or, at least, of a proper Navier-Stokes solution, the theory was in need of a sound physical idea. A breakthrough was provided by the detailed flow field surveys provided by the experimental works of *McCroskey* et al. (1967), *Harbour & Lewis* (1967), *Becker & Boylan* (1967). These show, indeed, a fully viscous layer which may be divided into two regions. Contiguous to the unperturbed flow there is a compression layer, looking much like an incomplete shock layer, with an increase in density, pressure and temperature, but with small change in longitudinal velocity. This compression layer does not go to completion as would do a shock layer, because it is limited by the shear layer whose main goal is to slow down the flow and cool it down to the values required by the wall. As a matter of fact, if one takes the density ratio as a measure of the extent to which the compression is completed one sees that at $M_\infty \mathrm{Re}_x^{-1/2} \sim 0.1$. the compression is virtually complete, while it is effected to 40% only for $M_\infty \mathrm{Re}_x^{-1/2} \sim 0.4$. *Oguchi* (1967) succeeded in using this observation to build a physically sound model of flow. He patches an incomplete shock layer with a boundary layer, again incomplete. *Shorenstein & Probstein* (1968) refined the analysis by taking into account the curvature of the shock. The new theory predicts a maximum in surface pressure, in agreement with experiment. This amounts to a quantitative statement for the physically sound idea that the unlimited rise in pressure of strong interaction theory is ended by the process of merging, which prevents the compression from being completed.

However, succesfull it may be, Oguchi's theory remains empirical, a shortcoming removed by the theories of *Rudman & Rubin* (1968), *Cheng* et al. (1969), *Kot & Turcotte* (1972). Theese authors give an (approximate) full Navier-Stokes solution of the problem. By using the thin layer approximation on the full Navier-Stokes equations they neglect $\partial/\partial x$ in comparison with $\partial/\partial y$. This approximation, which is not a boundary layer approximation in the sense of *Prandtl,* and which does not neglect the transverse pressure gradient, leads to parabolic equations which are integrated by marching ahead, down the flow. This time, the structure of the viscous layer is derived, by patching of an incomplete shock layer with an incomplete shear layer, from first principles, without reference to experiments. Kot & Turcotte take care of kinetic theory effects by using the molecular beam approximation of *Turcotte & Scholnick* (1969), devised for strong shock waves by *Oguchi* (1965) and *Grad* (1966). They obtain density profiles through the shock layer in quite good agreement with experiment.

The good agreement between Oguchi's type of theory and experiment gives new evidence of the importance of Navier-Stokes equations in gas dynamic problems when viewed from kinetic theory. Nevertheless, for a value of the rarefaction parameter approaching one, continuum theory should be given up, and one should rely on specific methods in kinetic theory. This is most convincingly recognized from the computations of Kot & Turcotte which show for the profiles nearest to the leading edge, in a region where the rarefaction parameter is precisely near unity, a slip velocity at the wall as high as 50% of the upstream velocity. Unfortunately a full kinetic theory treatment of the problem is exceedingly difficult and expensive. Very few papers deal with this important problem. *Kogan & Degtyariev* (1965) haye studied the flow over a very short flat plate, so that the whole flow is in the near free molecule regime, and it is hard to draw any valuable conclusion as far as the leading edge problem is concerned. *Charwat* (1961) assumed that the leading edge is dominated by free molecule flow which is far from being evident. If this assumption were valid, this would imply that the upstream influence of the plate is not significant on the scale of the mean free path, but this is precisely an open question, even for an infinitely sharp flat plate. *Huang* (1969) developed a discrete ordinate technique applied to the Krook model, but it was only very recently that sufficiently long flat plates (52 mean free paths) have been succesfully treated, using this technique, by *Huang* et al. (1972). They are able to show the process of building Oguchi's model of flow. *Clark* (1972) worked out a method devised by *Chu* (1967) for integrating the Krook equation in unsteady flow; he computes the flow over a semi-infinite flat plate, impulsively set into motion parallel to its plane. These works are on the borderline of the hypersonic regime. The computations of *Vogenitz, Broadwell & Bird* (1970) give a satisfactory numerical solution to the leading edge problem through use of Monte Carlo simulation; the plate length is 88 mean free paths, while the Mach number is thirty. All of these works show evidence of a shear flow region near the plate, shielded from the upstream flow by a compression region. The surface pressure profile (as a matter of fact p_{yy} is computed) shows clearly that the free molecule value is not more than half the value computed at the leading edge, which is an evidence for a significant upstream influence for Mach numbers from 5.5 to 30. On the other hand, friction and heat flux at the wall approach the free molecule value at the leading edge. Anyway, under hypersonic conditions surface pressure and heat flux grow very rapidly from the leading edge to a highly peaked maximum, located in a region between five to ten mean free paths from the edge. The experiments of *Joss & Bogdonoff* (1969) with extremely sharp flat plates give evidence of an upstream influence, for at least five mean free paths, at a Mach number of 26.

References

Arseniev, A. A.: J. Numer. Math. Phys. 5 (1965) 864.

Albertoni, S., Cercignani, C., Gotusso, L.: Phys. Fluids 6 (1963) 993—996.

Abe, K.: Phys. Fluids 14 (1971) 492—498.

Arkeryd, L.: Arch. Rat. Mech. Anal. 45 (1972) 1—34.

Bakhanov, S. P., Deryagin, B. V.: Dokl. A. N. SSSR 139 (1961) 71.

Baum, H. R.: J. Fluid Mech. 50 (1972) 271—284.

Becker, M., Boylan, D. E.: Rarefied Gas Dynamics 1966, 2 (1967) 993—1014.

Berroir, A.: Ann. Inst. H. Pancaré 22 (1970) 1—69, 117—182.

Bird, G. A.: J. Fluid Mech. 31 (1968) 657—668.

Bruckner, J. K., Ferziger, J. H.: Phys. Fluids 9 (1966) 2315—2322.

Carleman, T.: Acta Mathematica 60 (1933) 91—140.

Carleman, T.: Publications mathématiques de l'Institut Mittag-Leffler Upsala 1957.

Cercignani, C.: Annals of Physics 20 (1962) 219—233.

Cercignani, C.: J. Math. Physics 9 (1968) 633—639.

Cercignani, C.: J. Math. Phys. 8 (1967) 1653—1656.

Cercignani, C.: Phys. Fluids 11 (1968) 303—308.

Cercignani, C., Foresti, P., Sernagiorto, F.: Nuovo Cimento 47 (1968) 297—306.

Cercignani, C.: Mathematical methods in kinetic theory. Plenum Press 1969.

Chapman, S.: Phil. Trans. Roy. Soc. A 216 (1916) 279.

Chapman, S., Cowling, T. G.: The mathematical theory of non-uniform gases. (1932) Reprint: Cambridge (UK), 1958.

Charwat, A. F.: Rarefied Gas Dynamics 1960, (1961) 553—578.

Chaxov, E. M.: Prikl. Mat. Mech. 32 (1968) 1140—1142.

Chaxov, E. M.: Mekh. Zhidk. Gaza 6 (1968) 143—148.

Cheng, H. K. in Hall, J. G.: Fundamental phenomena in hypersonic flow, (1966) 90—132, Cornell Univ. Press.

Cheng, H. K., Chen, S. Y., Mobly, R., Huber, C.: Rarefied Gas Dynamics 1968, 1 (1969) 451—463.

Chu, C. K.: Rarefied Gas Dynamics 1966, 1 (1967) 589—606.

Clark, R. W.: Thesis (1971), Columbia University, personal communication of the author.

Darrozes, J. S.: Rarefied Gas Dynamics 1968, 1 (1969) 111—120.

Darrozes, J. S.: ONERA Publication No. 130 (1972).

Darrozes, J. S., Guiraud, J.-P.: Compte Rendus 262 (1966) 1368—1371.

Deryagin, B. V., Ivtchenko, I. N., Salamov, Iou. I.: Mekh. Zhidk. Gaza 4 (1968) 167—172.

Enskog, D.: Dissertation, Uppsala (1917).

Ernst, M. H.: Amer. J. Phys. 38 (1970) 908—914.

Fay, J. B. in Nelson: The high temperature aspects of hypersonic flow, Pergamon 1964, 583—605.

Fetz, B., Shen, S. F.: to appear in Rarefied Gas Dynamics (1970).

Galkin, V. S.: Mekh. Zhidk. Gaza 2 (1969) 57—62.

Galkin, V. S., Kogan, M. N., Fridlender, O. H.: Mekh. Zhidk. Gaza 3 (1970) 13—21.

Germain, P., Guiraud, J.-P.: J. Math. Pures Appl. 45 (1966) 311—358.

Gilbarg, D., Paolucci, D.: J. Rat. Mech. Analysis 2 (1953) 617—642.

Gorelov, S. L., Kogan, M. N.: Mekh. Zhidk. Gaza 2 (1969) 114—119.

Grad, H.: Comm. Pure Appl. Math. 5 (1952) 257—300.

Grad, H.: Handbuch der Physik, XII (1958) 203—294.

Grad, H.: Phys. Fluids 6 (1963) 147—181.

Grad, H.: Rarefied Gas Dynamics 1962, 1 (1963a) 26—59.

Grad, H.: Proc. Symp. Appl. Math. 17 (1965) 154—183.

Grad, H.: Comm. Pure Appl. Math. 18 (1965a) 345—354.

Grad, H.: J. Soc. Indust. Appl. Math. 13 (1965b) 259—277.

Grad, H.: Transport Theory, SIAM-AMS Proceedings 20 Abu Shumays Ed. (1966) 269—308.

Greenspan, M.: Phys. Rev. 75 (1949) 197.

Greenspan, M.: J. Acoust. Soc. America 22 (1950) 526.

Gross, E. P., Jackson, E. R., Ziering, S.: Phys. Fluids 1 (1958) 215—224.

Guiraud, J.-P.: Compte Rendus T 261 (1965) 3739—3742.

Guiraud, J.-P.: J. Mécanique 7 (1968) 171–203.
Guiraud, J.-P.: J. Mécanique 9 (1970) 443–490.
Guiraud, J.-P.: Actes Congr. Int. Math. 1970, 3 (1970a) 115–122.
Guiraud, J.-P.: J. Mécanique 11 (1972) 183–231.
Harbour, R. J., Lewis, J. H.: Rarefied Gas Dynamics 1966, 2 (1967) 1031–1046.
Hassler, H.: Acoustica 20 (1968) 271.
Hauge, E. H.: Phys. Fluids 13 (1970) 1201–1208.
Hilbert, D.: Math. Annalen 72 (1912) 562.
Hirschfelder, J. O., Curtiss, C. F., Bird, R. B.: Molecular theory of gases and liquids. Wiley (1954).
Holtz, T., Muntz, E. P., Yen, S. M.: Phys. Fluids 14 (1971) 545–548.
Huang, A. B.,: Phys. Fluids 11 (1968) 61–63.
Huang, A. B.: Rarefied Gas Dynamics 1968, 1 (1969) 529–544.
Huang, A. B., Hwang, P. F., Giddens, D. P., Srinivasan, R.: personal communication of authors (1971).
Itaya, N.: Proc. Japan Acad. Sci. 46 (1970) 379–382.
Ivtchenko, Iou. I.: Mekh. Zhidk. Gaza 6 (1969) 59–66.
Joss, W. W., Bogdonoff, S. M.: Rarefied Gas Dynamics 1968, 1 (1969) 483–492.
Klinc, T., Kuscer, I.: Phys. Fluids 15 (1972) 1018–1022.
Kot, S. S., Turcotte, D. L.: AIAA Journ. 10 (1972) 291–295.
Kogan, M. N., Degtyariev, L. M.: Astron. Acta 11 (1965) 36–42.
Lang, H., Loyalka, S. K.: Phys. Fluids 13 (1970) 1871–1873.
Lees, L.: J.A.S. 20 (1953) 143–145.
Lighthill, M. J. in: Surveys in Mechanics, Cambridge (1956), edited by Batchelor & Davies.
Liepman, H., Narasiha, R., Chahine, M. T.: Phys. Fluids 5 (1962) 1313–1324.
Liepman, H. W., Narasimha, R., Chahine, M. T.: in Görtler: Applied Mechanics, Proceedings of the Eleventh International Congress of Applied Mechanics, Springer (1964), 973–979.
Lin, J. T., Willis, D. R.: Phys. Fluids 15 (1972) 31–38.
Loyalka, S. K.: Phys. Fluids 14 (1971) 2291–2294.
Loyalka, S. K.: Phys. Fluids 14 (1971a) 21–24.
Loyalka, S. K., Cipolla, J. W.: Phys. Fluids 14 (1971) 1656–1661.
Loyalka, S. K., Ferziger, J. H.: Phys. Fluids 10 (1967) 1833–1839.
Loyalka, S. K., Ferziger, J. H.: Phys. Fluids 11 (1968) 1668–1671.
Lunc, M.: J. Mécanique 4 (1965) 455–461.
Lyuabarski, G. Ya.: JETP 13 (1961) 740–745.
McKean, H. S.: Bull, Amer. Math. Soc. 1969, 1–10.
McCroskey, W. J., Bodgonoff, S. M. McDougall, J. G.: AIAA Journ. 4 (1967) 1580–1587.
Maslova, N. B.: Vest. Leningr. Univ. 13 (1968) 88–95.
Mason, R. J.: Rarefied Gas Dynamics 1964, 1 (1965) 48–70.
Mason, R. J.: Rarefied Gas Dynamics 1966, 1 (1967) 395–410.
Meyer, E., Sessler, G.: Z. f. Phys. 149 (1957) 15.
Messiter, A. F.: SIAM J. Appl. Math. 18 (1970) 241–257.
Mintzer, D.: Phys. Fluids 8 (1965) 1076–1090.
Morgenstern, D.: Proc. Nat. Acad. Sci. USA 40 (1954) 719–721.
Morgenstern, D.: J. Rat. Mech. Anal. 4 (1955) 533–545.
Muntz, E. P., Harnett, L. N.: Phys. Fluids 12 (1969) 2027–2035.
Muntz, E. P., Harnett, L. N.: Phys. Fluids 15 (1972) 565–572.
Narasimha, R.: J. Fluid Mech. 34 (1968) 1–24.
Nikolaenko, B. in F. A. Grünbaum: Seminar on the Boltzmann equation, p. 125–210, edited by the Courant Institute of Mathematical sciences, New York (1972).
Nikolaenko, B., Thurner, J. K., Dorning, J.: J. Quant. Spectroscopy and Rad. Transfer 11 (1971) 1007–1021.
Nordsieck, A., Hicks, B. L.: Rarefied Gas Dynamics 1966, 1 (1967) 695–710.
Oberei, M. N.: Univ. Calif. Los Angeles Rept. 65-76 (1965).
Oberei, M. N.: Phys. Fluids 9 (1968) 1634–1637.
Oguchi, H.: Rarefied Gas Dynamics 1960, 2 (1961) 501–524.
Oguchi, H.: Rarefied Gas Dynamics 1964, 1 (1965) 182–193.
Oguchi, H.: Inst. of Space and Aero. Sci. Univ. Tokyo Rept. 418 (1967).
Ostrowski, H., Kleitman, D. J.: Nuovo Cimento 44 (1966) 49–67.
Pan, Y. S., Probstein, R. F. in J. G. Hall: Fundamental phenomena in hypersonic flow, Cornell Univ. Press, p. 259–309 (1966).
Pao, Y. P.: J. Math. Phys. 8 (1967) 1893–1898.
Pao, Y. P.: Phys. Fluids 14 (1971) 1340–1346.
Pao, Y. P.: Phys. Fluids 14 (1971a) 306–312.
Pekeris, C. L., Alterman, Z., Finkelstein, L., Frankowski, K.: Phys. Fluids 5 (1962) 1608–1616.
Povzner, A. Ya.: Mat. Sbornik 58 (1962) 65–86.
Reddy, K. C.: Phys. Fluids 11 (1968) 1308–1311.
Richardson, T. G., Sirovich, L.: J. Math. Phys. 12 (1971) 1784–1798.
Rigolot-Turbat, C.: Compte rendus 272 (1971) 697–700.
Rigolot-Turbat, C.: Compte rendus 272 (1971) 763–766.
Rigolot-Turbat, C.: Compte rendus 273 (1971) 58–61.
Robben, F., Talbot, L.: Phys. Fluids 9 (1966) 653–662.
Rudman, S., Rubin, S. G.: AIAA Journ. 6 (1968) 1883–1890.
Sedov, L. I., Mikhailova, M. P., Tchernii, G. G.: Vest. Mosk. Univ. Ser. 3 Fiz. Astron. No. 3 (1953) 95–100.
Scharf, G.: Helvet. Physica Acta 40 (1967) 929–945.
Scharf, G.: Helvet. Physica Acta 42 (1969) 5–22.
Sherman, F. S.: J. Fluid Mech. 8 (1960) 465–480.
Sherman, S. F., Talbot, L.: Rarefied Gas Dynamics 1958, (1960). 161–191
Shorenstein, M., Probstein, R. F.: AIAA Journ. 6 (1968) 1898–1906.

Sirovich, L.: Phys. Fluids 6 (1963) 1428—1439.
Sirovich, L.: Phys. Fluids 6 (1963a) 218—230.
Sirovich, L.: Phys. Fluids 6 (1963b) 10—20.
Sirovich, L.: Phys. Fluids 12 (1969) 586—592.
Sirovich, L., Goldman, E.: Rarefied Gas Dynamics 1968, 1 (1969) 407—415.
Sirovich, L., Thurber, J. K.: Rarefied Gas Dynamics 1964, 1 (1965) 21—47.
Sirovich, L., Thurber, J. K.: J. Acoust. Soc. Amer. 37 (1965a) 329—339.
Sirovich, L., Thurber, J. K.: J. Math. Phys. 8 (1967) 888—895.
Sirovich, L., Thurber, J. K.: J. Math. Phys. 10 (1969) 239—249.
Sone, Y.: J. Phys. Soc. Japan 21 (1966) 1620—1621.
Sone, Y.: J. Phys. Soc. Japan 21 (1966a) 1836—1837.
Sone, Y.: Rarefied Gas Dynamics 1968, 1 (1969) 243—254.
Sone, Y.: to appear in Rarefied Gas Dynamics (1970).
Stewartson, K. O.: Mathematika 16, part 1 (1969) 106—121.
Talbot, L., Scala, S. M.: Rarefied Gas Dynamics 1960, (1961a) 603.
Thurber, J.K.: See Nikolaenko (1972), p. 211—262 (1972).
Trilling, L.: Phys. Fluids 7 (1964) 1681—1691.
Truesdell, C. S.: J. Rat. Mech. Anal. 1 (1956) 55—128.
Turcotte, D. L., Scholnick, I. M.: Phys. Fluids Suppl. I, (1969) 79—82.
Van Dyke, M. D.: Rarefied Gas Dynamics 1962, 2 (1963) 212—227.
Vogenitz, F. W., Broadwell, J. E., Bird, G. A.: AIAA Journ. 8 (1970) 504—510.
Waldman, L. in: Handbuch der Physik XII (1958) 295—514.
Weitzner, H.: Rarefied Gas Dynamics 1964, 1 (1965) 1—20.
Ward, C. A., Mintzer, D.: Phys. Fluids 14 (1971) 499—509.
Wachman, M., Hamel, B.: Rarefied Gas Dynamics 1966, 1 (1967) 675—694.
Wang Chang, C. S., Uhlenbeck, G. E. (1952): Reprinted in J. de Boer. Uhlenbeck, G.: Studies in Statistical Mechanics, 5, North-Holland Elsevier, 1970, p. 43.
Wild, E.: Proc. Camb. Phil. Soc. 47 (1950) 602—609.
Welander, P.: Arkiv för Fysik 7 (1954) 507—553.
Willis, D. R.: Phys. Fluids 5 (1962) 127—135.
Williams, M. M. R.: J. Fluid Mech. 36 (1969) 145—159.
Williams, M. M. R.: J. Fluid Mech. 45 (1971) 759—768.

Creep buckling of plates and shells

Nicholas J. Hoff

Stanford, Cal. USA

a	mean radius of shell
$a_{\max}$	half-major-axis of elliptically deformed shell
b	width of rectangular flat plate
B	coefficient in Odqvist' creep law
C_1, C_2	constants
$d = h_{\text{real}}/\sqrt{3}$	distance between faces of equivalent sandwich wall determined on the basis of equal moment of inertia of section
$D = [n/(2n+1)]^{n/(n+1)} h_{\text{real}}$	Rabotnov's distance between faces of equivalent sandwich shell
E	Young's modulus
$G = E/2(1+\nu)$	shear modulus
h_{real}	real wall thickness (not thickness of equivalent sandwich wall)
k	constant in steady-state uniaxial creep law
$J_2 = (1/2)\, s_{ij}\, s_{ji}$	second invariant of the tensor of stress deviations
M	constant bending moment acting on circular cylindrical shell
M_{cr}	critical bending moment for Brazier's flattening instability in the case of purely elastic deformations
m	exponent in Odqvist's creep law
n	exponent in uniaxial creep law
P	load applied to shell (positive when compressive)
q	lateral pressure acting on wall of shell (positive when compressive)
t	time measured from instant of load application
t_{cr}	critical time
$s_{ij} = \sigma_{ij} - (1/3)\, \sigma_{kk}\, \delta_{ij}$	ijth component of tensor of stress deviations
$t_E = \varepsilon_E/\dot{\varepsilon}_{\text{nom}}$	Euler time
W	amplitude of most dangerous component of deviations or displacements (lateral for plate, and radial for shell)
$x = W/D$	nondimensional amplitude of most dangerous component of radial deviation or displacement from perfect circular cylindrical shape
$\alpha = a_{\max}/a$	ratio of major axis of ellipse after deformations to diameter of perfect circular cylindrical shell
β	exponent of time in time-hardening creep law
δ_{ij}	Kronecker's delta $(= 1$ when $i = j$; and $= 0$ when $i \neq j)$
ε	strain (positive when compressive)
$\dot{\varepsilon}$	strain rate (positive when compressive)
ε_E	Euler strain, that is the strain corresponding to the Euler stress
$\dot{\varepsilon}_{\text{nom}}$	nominal strain rate, that is the strain rate of the perfect structural element
$\dot{\varepsilon}_{ij}$	ijth component of strain rate tensor
λ	most dangerous wave length
ν	Poisson's ratio
σ	stress (positive when compressive)
σ_{ij}	ijth component of stress tensor
σ_E	Euler stress, the critical stress of the structural element in the presence of purely elastic deformations
σ_{kk}	sum of three normal stress components acting on mutually perpendicular planes
$\omega = \sqrt{3}\, W/h_{\text{real}}$	nondimensional amplitude of most dangerous component of deviations or displacements from perfectly plane configuration

Subscripts

$_0$	refers to time $t = 0_+$ at beginning of creep deformations (elastic deformations included)
$_{00}$	refers to time $t = 0_-$ at load application (before elastic deformations)

Notes

The dot · above a symbol indicates differentiation with respect to time

Repeated letter subscripts indicate summation from 1 to 3

Introduction

Definition of the problem

This paper is not meant to be the definitive publication on the creep buckling of plates and shells; it is rather a progress report on work completed in the relatively near past, mostly by the present author and his collaborators. Results obtained by other researchmen will also be discussed, particularly if there has been some interaction between these men and the present author. Omission of reference to the work of any other investigator should not be interpreted to mean that it is considered to be of an inferior quality; lack of such mention may simply indicate that the present author is unfamiliar with the work, or that he had to leave it out for lack of space.

By creep is meant a time-dependent state of deformations of the material that occurs even when the loads acting on the material remain constant. With structural metals this can be observed only at temperatures that are significant fractions of the melting temperature. Creep buckling is a phenomenon in which a structural element subject to compressive stresses collapses, and is related to the instability phenomena of perfectly elastic structures.

As the theory of creep buckling is a rather young branch of mechanics, scientific opinion has not yet settled all of its aspects. No final acceptance of any researchman's ideas is likely to take place in the near future because much more experimentation is needed to clarify all the issues. This is so particularly because all the creep laws in use today must be considered only as rough approximations to nature. The scatter among test results is great even in the case of the simplest possible experiment which is the tensile creep test under a constant load. The effect of a change in the value of the stress during the creep test is still little known quantitatively. No conclusion drawn from theory should therefore be accepted categorically unless proven by careful experimentation.

This state of affairs is not unusual in physics, but it appears to be most disconcerting to the research man in solid mechanics who was weaned on elasticity theory. In this particular field only two physical constants, Young's modulus of elasticity and Poisson's ratio, govern all the deformations of a homogeneous and isotropic body. The overwhelmingly important one of the two is Young's modulus which depends only on the average properties of the crystalline lattice, and thus it is a constant that does not change in consequence of heat treatment and cold work. But beyond the range of perfectly elastic behavior the situation is much more complicated. The curved portion of the common stress-strain curve is influenced greatly by heat treatment and cold work, and the values of the yield stress and of the so-called tangent modulus vary greatly. Although this effect is well known, it might be illustrated here by a quotation from Timoshenko's *Strength of Materials* [1]: "in tests carried out in the research laboratory of the Westinghouse Electric and Manufacturing Co. stretching by 2 percent of a die-cast aluminum rod increased the proportional limit of the material from 5600 psi to 20,000 psi." *Bauschinger* [2] showed almost a hundred years ago that stretching a rod beyond the yield stress of the material increases significantly the yield stress in tension and decreases it in compression.

One should not be astonished therefore to find that the creep properties of a metal are sensitive to heat treatment and cold work, and to small amounts of impurities. Both the plastic and the creep behavior depend greatly on the irregularities of the lattice, as opposed to the average values governing the elastic deformations, that is on the distribution of the dislocations in the material.

In tests carried out by *Sherby* and *Dorn* [3] over twenty years ago the creep strain of 4S aluminum alloy subjected for 10 hours to a stress of 28,000 psi at a temperature of 212 °F was found to be 0.004 in the case of H38 temper and 0.02 in the case of H32 temper. 2S aluminum alloy developed strains of 0.0012 in the H18 temper and 0.02 in the H12 temper when subjected to a stress of 15,000 psi for one hour at 90 °F. Yet the elastic modulus of the material was, for all practical purposes, the same for the two tempers, and, as a matter of fact, also for the two alloys.

But significant random variations in creep behavior occur even with the same material under the same conditions as was observed by *Phillips* and *Sinnott* [4]. They tested 200 specimens cut out of eight bars of 2S aluminum alloy annealed in the same manner and showing the same metallographic structure. The tests were carried out in tension at 900 °F. The scatter in the creep rupture times was the same for specimens cut out of the same bar as for specimens cut out of different bars. Out of 22 specimens subjected to a stress of 520 psi one ruptured after 3878 minutes, and one at 1168 minutes; and out of 20 specimens tested at 800 psi the maximum and minimum life times were 60.3 minutes and 19.8 minutes.

This variability of the inelastic deformations makes it necessary to determine the properties of the particular batch of material that is used in a test series. Recognition of this requirement in the case of elasto-plastic deformations enabled *von Kármán* [5] to prove the validity of the Euler formula, appropriately modified to take care of the changes in the instantaneous values of the modulus. For his famous tests of 1908 von Kármán machined specimens for the establishment of stress-strain curves and for column tests out of the same block of steel specially produced for him by the Friedrich Krupp Co.

Similarities and differences between elastic and creep buckling

The fundamental difference between elastic buckling and creep buckling is that a perfectly elastic structural element buckles when during a proportional loading process a critical value of the load is reached, while an element whose material creeps buckles under any compressive load, however smalle, when a finite critical time is reached. The latter statement is true provided that the stress-strain rate relationship is nonlinear in such a manner that doubling the stress leads to more than twice the initial strain rate; in the case of a linear relationship the critical time is indefinitely large.

It will be shown in the body of the paper that plates and shells that creep behave beyond buckling in a manner similar to the postbuckling behavior of perfectly (and linearly) elastic structural elements. In particular, plates simply supported along their four edges do not collapse when the critical time is reached but continue carrying the edge compression imposed on them, although their rate of deformation increases significantly in the neighborhood of the critical time.

On the other hand, the axisymmetric deformations of an initially slightly inaccurate circular cylindrical shell can reach such amplitudes during the creep process that the hoop compressive stresses produced destabilize the axisymmetric state of equilibrium with regard to small multilobed disturbances. When this state is reached, the shell is likely to collapse in a nonaxisymmetric manner. This kind of failure is the counterpart of the bifurcation buckling of linearly elastic shells discovered by *Koiter* [6].

Hypotheses

One approach to the determination of the stability of the equilibrium of a structural element that creeps was proposed by *Rabotnov* and *Shesterikov* in 1957 [7]. These authors assumed that the element, such as a column or a flat plate subjected to compressive loads, is initially perfect in the sense that it is perfectly straight or perfectly flat. In the absence of external disturbances the element remains perfect as it creeps. The stability of the element in the classical sense is checked by imagining that the perfect state is disturbed in such a manner that the column or the plate becomes slightly imperfect, or curved; if the dynamic equations of motion yield the result that after the disturbance the deviations from perfection increase with time, the state of equilibrium is unstable.

The analysis of the motion was carried out with the aid of a linearization of the nonlinear equation of state characterizing the behavior of the material through which linear relationships were obtained between increments in stress, strain and strain rate. Although the entire analysis of *Rabotnov* and *Shesterikov* is both correct and elegant, the results obtained by them in the case of columns are difficult to interpret physically as was pointed out in [8]. Why should a material be less resistant to buckling after strain hardening than before? In addition, *Jahsman* and *Field* [9] found that the critical times derived from the theory are much shorter than those measured in experiment. For this reason they suggested that the effect of "creep hesitation" following a decrease in the stress value, should be taken into account in the theory through a modification of the mechanical equation of state. However, the modified analysis also led to critical times far shorter than those observed in experiment.

The situation was clarified by *Shesterikov* in 1959 [10]. He discussed in detail the Rabotnov-Shesterikov criterion and he showed that it is equivalent to the following statement:

"The state of a longitudinally compressed bar will be assumed to be stable if for constant load its initial deflection does not increase faster than a linear function of time."

Obviously, stability can be defined in more than one way. The above definition is a perfectly permissible one, and it is consistent with the classical concepts of stability. But the question arises whether it leads to practically useful results. In particular, as *Shesterikov* has pointed out, every column that exhibits steady creep must be considered unstable in the sense of the Rabotnov-Shesterikov criterion. Yet it is known that columns can have useful lives of many hours and even of many years if they deform in steady creep.

An entirely different approach to the study of stability in the presence of creep starts out from the observation that no structural element can ever be manufactured in a perfect manner and that creep buckling occurs because even the smallest initial deviation from the perfect shape eventually leads to collapse in consequence of the creep deformations. This approach was followed in the first analysis of creep buckling presented by *Freudenthal* in 1946 [11]. Although a mathematical error was later discovered in this paper [12], its basic idea was adopted by many subsequent investigators, such as *Kempner* [13], *Libove* [14] and the present author [15].

Here it should be mentioned that a second independent investigation of the stability of a column subject to linear creep was published by *Rzhanitsyn* [16] simultaneously with Freudenthal's paper.

Even though the present author greatly admires the relative simplicity and the elegance of the *Rabotnov-Shesterikov* approach, he is convinced that the growth of imperfections is the physical cause of most, if not all, cases of creep buckling. For this reason he feels compelled to calculate the critical times of structural elements by following up this growth even when this can be done only in a plodding manner. He wishes to remind the reader that the problem of the buckling of linearly elastic circular cylindrical shells was solved only when *Koiter* discovered the high sensitivity of these structural elements to initial imperfections in the case of axial compressive loading [6]. In the same manner, all problems of creep buckling investigated up to now have shown great sensitivity to initial deviations.

An entirely different effect of a deterioration of the resistance of the material was proposed by *Fraeijs de Veubeke* [17] in 1958 and was explored in more detail by the present author [18]. The development of a theory based on the idea of this deterioration is a beautiful example for international cooperation. It began when *Odqvist* [19] introduced his approximate representation of primary creep into the author's analysis [15] of creep buckling. The present author followed up this work [18] by considering the effects of stress reversal during the creep buckling process. The discovery of *Fraeijs de Veubeke* was that the primary creep representation of the two last-mentioned papers is capable of a reinterpretation: the equations can characterize a material that shows instantaneous elasto-plastic and time-dependent steady creep deformations.

The physical phenomenon thus discovered is basically a very simple one: an initially slightly inaccurate column creeps steadily and increases its deviations from the straight line if a compressive load smaller than the Euler load is acting upon it. But under the constant load and increasing deformations the extreme fiber stress keeps increasing and eventually reaches values at which the instantaneous deformations are no longer purely elastic. Beyond the elastic range the instantaneous buckling of the column is governed by von Kármán's effective modulus rather than by Young's modulus of elasticity and the buckling load is smaller than the Euler load. If the compressive loading of the column is continued long enough, the bending stresses become so large that the inelastic buckling load computed with the von Kármán modulus is reduced to the value of the applied compressive load. At that moment the previous slow creep deformations of the column are replaced by a sudden snapping of the column into a collapsed state.

This physical behavior was discovered by *Fraeijs de Veubeke* while he was studying the available creep buckling theories in preparation to writing his review article [17]. The book in which the review appeared was printed only in 1958, later than three articles by the present author and his collaborators [18, 20, 21]. However, the present author was the editor of the book in which [17] was printed and he had discussed the discovery with *Fraeijs de Veubeke* before the three articles mentioned appeared.

Because of the suddenness of the collapse observed in experiments on creep buckling it appears very likely that the final cause of creep buckling can be found in instantaneous plastic deformations although the initial stage is a slow growth of the lateral displacements due to steady, or primary, creep. Good agreement between this theory and experiment was found by *Chapman* et al. in [21]. Yet this approach to the study of creep buckling has not become popular, perhaps because its application requires the establishment of instantaneous stress-strain curves at high temperature in addition to obtaining information on creep. Moreover, *Lempriere* [22], a student of the author, has shown in the case of columns that the difference between the creep buckling time of the elasto-plastic-creep theory and that obtainable on the basis of linear elastic and nonlinear steady creep deformations alone is small enough to be disregarded in most practical applications.

Method of analysis

In most of the calculations that follow, the wall of the structural element is imagined to be replaced by a sandwich wall consisting of two faces and a core. The faces are supposed to carry all the forces parallel to the tangent plane of the shell (that is, in the case of the flat plate, in the plane of the plate), and the core carries only the transverse shear force.

The purpose of the introduction of the sandwich concept in the analysis is the resulting simplification of the calculations. Without this concept the research man has to determine the distribution of the stresses across the solid wall which is a problem of considerable difficulty. When the creep deformations are governed by an nth-power creep law, the Bernoulli-Navier assumption of plane sections of a bent beam before load application remaining plane after load application leads to an nth-root distribution of the normal stresses. At the same time a tensile or compressive force is distributed uniformly over the cross section. Since the ratio of the normal stress resultant to the bending moment generally varies over the structural element in space and in time, the shape of the stress distribution curve over the wall thickness also varies in space and time; evidently, the law of superpositions is not valid in the presence of a highly nonlinear stress-strain rate relationship.

The admissibility of the sandwich approximation was checked carefully in a number of publications [23 to 26] dealing mostly with columns, and the approximation was found satisfactory for practical purposes. In some papers the distance d between the faces of the equivalent sandwich wall was determined so as to preserve the moment of inertia of the section. This requirement yields

$$d = h_{\mathrm{real}}/\sqrt{3} = 0.578 h_{\mathrm{real}}. \tag{1}$$

Rabotnov [26] suggested that the distance D between the faces of the equivalent sandwich wall be selected in such a manner as to give equal deformations for real wall and sandwich wall in pure tension and in pure bending. This condition leads to

$$D = [n/(2n + 1)]^{n/(n+1)} h_{\mathrm{real}}. \tag{2}$$

The two equations naturally agree when $n = 1$. Rabotnov's formula yields 0.528 and 0.5 for the numerical coefficient when $n = 3$ and infinity, respectively.

Most of the calculations whose results are presented here in some detail were originally carried out on the basis of the assumption that the deformations of the material were entirely caused by steady creep governed by Odqvist's law [27]:

$$\dot{\varepsilon}_{ij} = BJ_2^m s_{ij}. \tag{3}$$

In the case of uniaxial stress application this reduces to

$$\dot{\varepsilon} = k\sigma^n, \tag{4}$$

where

$$m = (1/2)\,(n - 1), \qquad B = (1/2)\,3^{(n+1)/2}k. \tag{5}$$

In a recent paper [28] it was shown that the results of an analysis based on a steady creep law can easily be extended to include the effects of simultaneous linearly elastic deformations. The constitutive equation was written in the form

$$\dot{\varepsilon}_{ij} = \frac{1}{2G}\left[\dot{s}_{ij} + \frac{1-2\nu}{3(1+\nu)}\dot{\sigma}_{kk}\,\delta_{ij}\right] + BJ_2^m s_{ij}t^\beta. \tag{6}$$

If in this equation B is set equal to zero, the problem to be solved reduces to the determination of the buckling load of a purely elastic structural element. Most of the problems of this kind already solved are represented by linear differential equations and appropriate boundary conditions. When the structural element is perfect, and the external loads give rise to membrane stresses in it whose intensity is characterized by σ, the equations are homogeneous. They have nontrivial solutions only at characteristic values of the loads; the stress σ corresponding to the lowest characteristic value will be denoted σ_E.

Let us assume that in addition to the buckling stress σ_E we also know the critical time t_{crst} of the same structural element under the condition of steady creep deformations without elastic deformations. In the latter case in the constitutive Eq. (6) the value of G is infinitely large and that of β is zero. It was shown in [28] that the critical time $t_{crst+el}$ in the presence of simultaneous linear elastic and nonlinear steady creep deformations is given by

$$t_{crst+el} = [(\sigma_E - \sigma)/\sigma_E]\,t_{crst}. \tag{7}$$

This is true provided that the calculation of the steady-creep critical time is based on the assumption that the creep deformations have the shape of the eigenfunction of the purely elastic problem corresponding to the lowest eigenvalue σ_E. But this is the usual assumption in these analyses. The effect of the eigenfunctions corresponding to the higher eigenvalues is small as was shown in [15] for columns and in [29] for plates.

All the formulas to be presented later in this paper contain the factor accounting for the effect of linear elasticity.

In creep buckling calculations the factor appeared for the first time in a paper by *Kempner* [13] published on columns in 1952; it was derived again independently by *Hult* [30] in 1955. But the factor was already known during the last century as a correction factor accounting for end-load effects in beams; in Timoshenko's *Strength of Materials* [31] its discovery is attributed to *J. Perry* [32] in 1886.

When in (6) G and β have finite values, the material of the structural element deforms simultaneously elastically and creeps in accordance with a time-hardening law. Although a time-hardening creep law cannot be well justified on physical grounds, it has often been used because in most cases it is easier to handle than a strain-hardening law. It was shown in [28] that the critical time $t_{crth+el}$ in the presence of simultaneous elastic and time-hardening creep deformations can be obtained from the critical time $t_{crst+el}$ in the presence of simultaneous elastic and steday creep deformations with the aid of the formula

$$t_{crth+el} = [(\beta + 1)\,t_{crst+el}]^{1/(\beta+1)}. \tag{8}$$

Comparison between elastic and creep buckling

The main difference between elastic and creep buckling is the following: if a suitable load system is acting on a perfectly elastic structural element and these loads are increased without changing their proportions, sooner or later a characteristic value of the loading is reached at which the structural element develops very large deformations and eventually collapses. If the same structural element subjected to the same loading exhibits nonlinear creep deformations (in the sense that doubling the loads increases the creep rate by a factor greater than 2), it collapses after the lapse of a finite time t_{cr} no matter how small the loads are. Thus in creep buckling analysis the determination of the critical time corresponding to given loads replaces the calculation of the critical value of the load acting on a perfectly elastic structure.

The postbuckling behavior of structural elements shows certain similarities. Thus *Schuman* and *Back* [33] and *von Kármán* [34] have shown that a perfectly elastic rectangular plate simply supported along its edges and subjected to edgewise compression is capable of carrying loads much higher than the critical load. In the presence of creep deformations the counterpart of this behavior is that the plate does not collapse when the critical time is reached but continues to deform steadily although the magnitude of the deformations at the critical time may be inadmissible for practical purposes.

Similarly, the development of axisymmetric deformations in an axially compressed circular cylindrical shell gives rise to both tensile and compressive hoop stresses whether the deformations are purely elastic or also due to creep. When the compressive hoop stresses are large enough, the shell can collapse into many waves around the circumference. The state when this happens is reached when the axial load reaches a critical value provided that the material is perfectly elastic. This critical value was calculated by *Koiter* [6].

In the case of creep deformations, multilobed buckling becomes noticeable after the lapse of a sufficiently long time. This was first observed by *Samuelson* [35]; the phenomenon was further investigated by the present author and his students. The results of the investigation are given later in this paper.

A recent review of creep buckling and creep rupture was presented by *Odqvist* [34] at the University of Waterloo in Canada in 1972.

Formulas for the critical time

Rectangular plate in edgewise compression

When a compressive load is acting on a thin rectangular plate as shown in Fig. 1, the deformations caused by the creep of the material increase the initial deviations of the middle surface of the plate form the theoretical plane, just as an initially slightly crooked column becomes more crooked as time passes. Theory predicts an indefinitely large amplitude of the lateral deflections after the passage of a finite time provided the creep deformations are governed by a power law with an exponent greater than one. This time is properly called a critical time.

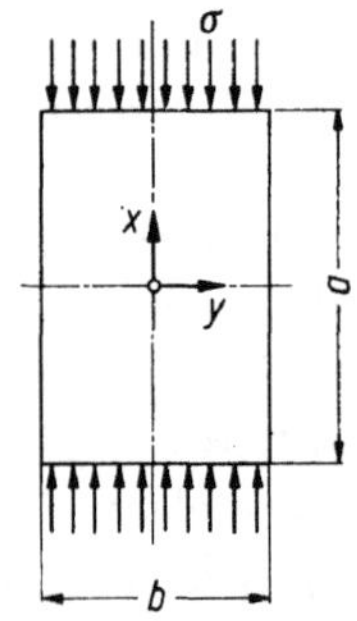

Fig. 1. Notation for flat plate.

The theory of the creep buckling of rectangular plates subjected to edgewise compression was developed in [37] in 1968. It is based on a number of simplifying assumptions:

(1) The power law is a cubic;

(2) The effect of the shear stresses on the deformations is negligibly small;

(3) The rate of change with time of the amplitude of the lateral displacements was calculated from the theory rigorously only for two extreme cases, namely for very small displace-

ments, and for very large displacements; and for the purposes of the final integration of this rate of change it was assumed that the sum of the two expressions properly represented the rate of change of the amplitude during the entire creep process;

(4) The infinite trigonometric series representing the stresses and the displacements were truncated after the second term;

(5) Simple curvature expressions valid only for small rotations were used throughout the analysis;

(6) The real plate was replaced by an equivalent sandwich wall in the calculations; and

(7) All the four edges were assumed to be simply supported.

Because of the nonlinearity of the creep law it was found impractical, if not impossible, to derive a single differential equation governing the creep deflections of the plate. For this reason the conditions of equilibrium and compatibility had to be satisfied individually.

The distance d between the face plates of the sandwich was calculated from the requirement that the second moment of the sections of real plate and sandwich plate be the same.

The result of the analysis was an expression obtained for the connection between the amplitude of the deflections and time t passed from the onset of creep:

$$t = 0.366 t_E \frac{\sigma_E - \sigma}{\sigma_E} \ln \frac{\omega^2(1.69 + \omega_0^2)}{\omega_0^2(1.69 + \omega^2)}. \tag{9}$$

In this equation σ is the applied compressive stress

$$\sigma = P/bh_{\text{real}}. \tag{10}$$

σ_E is the Euler stress

$$\sigma_E = 3.6E(h_{\text{real}}/b)^2. \tag{11}$$

ε_E the Euler strain

$$\varepsilon_E = 3.6(h_{\text{real}}/b)^2 \tag{12}$$

and

$$\dot{\varepsilon}_{\text{nom}} = k(P/bh_{\text{real}})^3 \tag{13}$$

which is the strain rate at the beginning of the test when the lateral deflections are so small that the entire plate creeps uniformly. Consequently, the Euler time is the time it would take the plate to develop a uniform compressive creep strain ε_E if the plate remained plane during the creep process and thus it crept at the uniform rate $\dot{\varepsilon}_{\text{nom}}$.

The most important quantity for the design engineer is the critical time t_{cr} at which (9) yields an indefinitely large value for ω. This is obtained from (9) if one sets ω equal to infinity and neglects 1.69 in the expression in which it is added to ω^2:

$$t_{cr} = 0.366 t_E \frac{\sigma_E - \sigma}{\sigma_E} \ln [1 + 0.564(h_{\text{real}}/W_0)^2]. \tag{14}$$

This formula shows that the critical time at which the plate buckles, that is develops indefinitely large lateral deflections, varies slowly with changes in the amplitude of the deviations at $t = 0$; similarly, it changes slowly with $(\sigma_E - \sigma)/\sigma_E$ as long as σ/σ_E is small, but this factor attains great importance when the applied stress approaches the critical stress; and it is influenced mostly by t_E in whose denominator the applied stress appears raised to the third power.

The value of W_0 at the onset of creep can be calculated if the value of the initial deviation amplitude W_{00} before load application is known:

$$W_0 = \frac{\sigma_E}{\sigma_E - \sigma} W_{00}. \tag{15}$$

The calculations presented in [37] were checked and extended in 1968 and the new information was published in [38] in 1971. Formulas were developed for the cases when n has the values of 5 and 7. The results can be written in the unified form

$$t_{cr} = C_1 \frac{\sigma_E - \sigma}{\sigma_E} t_E \ln ([1 + C_2(h_{\text{real}}/W_0)^{n-1}]. \tag{16}$$

The values of the constants C_1 and C_2 are presented in Fig. 2. The symbols in this equation have the same meaning as in (14] except that the nominal creep rate must be calculated from the expression

$$\dot{\varepsilon}_{\text{nom}} = k(P/bh_{\text{real}})^n. \tag{17}$$

In a new analysis the effect on the critical time of assumption (2) was determined and was found to be negligible. Assumptions (3) and (4) were checked by carrying out the calculations numerically with the aid of a high-speed digital computer. In this check the rate of change of the amplitude of the lateral deflections was not replaced by a simplified expression and in the trigonometric series representing the stresses and the deflections three additional terms were retained. The effect of the latter extension of the analysis was found to be negligible but that of the former proved to be noticeable at values of n higher than 3.

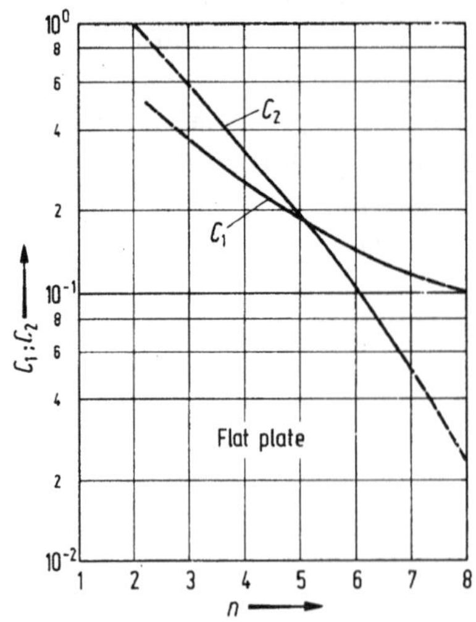

Fig. 2. Values of constants C_1 and C_2 for flat plate.

It is recommended therefore that the values of C_1 obtained through the approximate analysis be multiplied by the following correction factors based on the computer results:

Values of correction factor

ω_0 \ n	3	5	7
0.5	1	0.55	0.50
0.1	1	0.75	0.60
0.01	1	0.85	0.80
0.001	1	0.95	0.90

The most dangerous buckles that develop in consequence of creep have a wave length in a direction normal to that of the load equal to the total width of the plate. The length of the wave in the direction of the load is slightly shorter than the wave length in the normal direction. If tests are carried out with many plate specimens whose length (in the direction of the load) is two, three or many times the width a of the plate, the most dangerous buckle shape should be observed most frequently. When the shape of the plate is such that the most dangerous wave cannot develop, the critical time should be somewhat higher than that calculated from (16).

Postbuckling behavior of rectangular plate

It has been known since 1930 that thin, perfectly elastic, rectangular plates are capable of supporting compressive, inplane loads applied to them along an edge greatly in excess of the load under which they buckle. The fact was established by *Schuman* and *Back* [33] in 1930 and an explanation for the phenomenon was suggested in the same paper by *William R. Osgood*. In 1932 *von Kármán* [34] developed a very simple theory with the aid of which the ultimate load carrying capacity of the plate can be calculated.

After the buckling load of the simply supported plate has been exceeded, any further downward displacement of the upper platen of the testing machine only increases the deflections of the middle portion of the plate without noticeably increasing the average stress in the vertical fibers. But near the supported vertical edges of the plate the vertical fibers cannot deflect significantly, and along the vertical edge support they are constrained to remain perfectly straight. In this region of the plate the downward displacement of the upper platen is possible only if the vertical fibers foreshorten, and this shortening increases both the vertical normal strain and the vertical normal stress in the fibers. The result is a redistribution of the initially uniform stresses in such a manner that in the middle portion of the plate the compressive vertical normal stress is low (of the order of magnitude of the elastic buckling stress) and along the vertical edges it is high (in some cases as high as the yield stress of the material).

This is the course of events during the usual loading process of the perfectly elastic plate. There is naturally little, if any, reason to believe that similar events would not occur during a constant-load creep test of a metal plate at some elevated temperature. To test this conjecture, in 1969 *Levi* and *Hoff* [39] analyzed the creep buckling process of a plate under the condition that the vertical downward displacement of the loaded horizontal edge is constant across the width of the plate rather than the applied compressive stress.

The distribution of the vertical compressive stress across the width of the plate was represented by the first three terms of an infinite trigonometric series and the coefficients of the terms, which are functions of time, were calculated from the equilibrium and compatibility equations. The uniaxial form of the creep law employed was (4) and the exponent n was taken as 3. Because of the highly nonlinear nature of the equations, the solution was obtained with the aid of a high-speed digital computer.

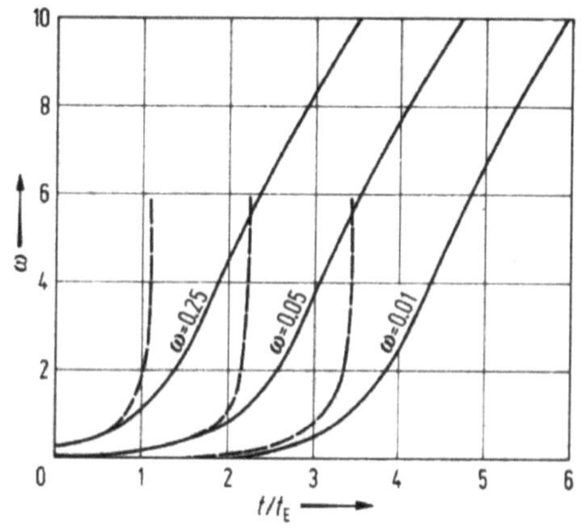

Fig. 3. Comparison of lateral displacement amplitudes according to small and large displacement theories for flat plate.

The most important results of the calculations are shown in Fig. 3. The plate was assumed the develop buckles with the ratio of the vertical wave length to the horizontal wave length (which is equal to the width of the plate) equal to 0.84. The full lines are the results when the downward displacement of the loaded edge is constant, and the dotted lines those when the load is distributed uniformly.

The significant difference between the two sets of curves is the appearance of vertical asymptotes indicating the existence of a finite critical time in the case of the dotted lines, and their complete absence in the case of the full lines. We can conclude therefore that in a rigid testing machine experiment the lateral deflections do not increase beyond all bounds when the critical time of (16) is reached. Nevertheless the critical time is a useful quantity for the designer because shortly after the critical time has been reached all the three full lines become steep indicating a rapid rise in the lateral deflections. At the critical time the plate is only slightly deformed and can safely carry the load, but soon after it develops permanent lateral deflections of such magnitude that it becomes useless for engineering purposes.

The critical time of the rectangular flat plate had been calculated earlier by *Pian* and *Johnson* [40] who used a variational theorem proposed by *Sanders, McComb* and *Schlechte* [41], rederived in [42] by *Pian*, to solve the problem of the initially slightly inaccurate plate. In the case of $n = 3$ they obtained an expression in closed form for the critical time.

A very thorough analysis of the prebuckling and postbuckling equilibrium of the plate with slight initial deviations from the plane was presented by *McComb* in [43]. The creep law was assumed as a time-hardening law and the numerical work was done with the aid of the a digital computer. The Sanders-McComb-Schlechte variational approach has, in principle, the advantage that the distribution of both the stresses and the displacements can be assumed independently and the unknown coefficients involved can be determined from the process of minimization. Unfortunately, the formulation of the problem was so complicated that McComb felt compelled to assume a linear strain distribution and a linear stress distribution through the wall thickness. Because of this assumption, his approach is not more accurate than the one in which the solid wall is replaced by a sandwich wall.

The curves obtained indicated that no finite critical time exists for the plate when the nonlinear strain-displacement relations of the Föppl-von Kármán large displacement theory are used in the calculation of the energy stored in the plate. This is in complete agreement with the findigs of [39]. However, this calculation naturally was not available to *McComb* in 1958. A comparison of his theoretical results with experiments carried out by *Mathauser* and *Deveikis* [44] led *McComb* to the conclusion that his idealization of the boundary conditions did not represent exactly the conditions existing along the V-grooves of the experiments. To the author of the present paper it appears more likely that the sudden increase in the end shortening of the plates observed by *Mathauser* and *Deveikis* was a consequence of instantaneous plastic deformations which were not included in McComb's theory.

Axisymmetric buckling of circular cylindrical shell in axial compression

Axially compressed thin-walled circular cylindrical shells (see Fig. 4) are also likely to buckle in consequence of creep. Their buckling can be induced either by initial deviations from the exact circular cylindrical shape, or by the restraint to radial displacements provided by

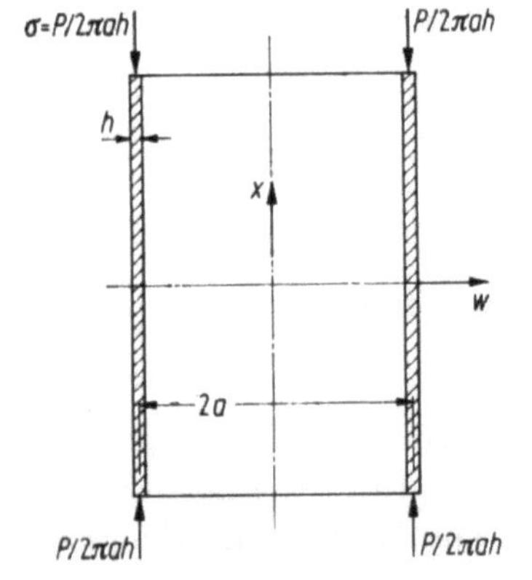

Fig. 4. Notation for circular cylindrical shell.

the support along the circular edges. The former case was investigated in [45], and the latter in [35, 46 and 47]; in both cases it was assumed that the shell deforms axisymmetrically. Since a comparison of the results of the two references shows the former case to be the more dangerous one, the expressions to be given here are all based on the existence of initial deviations from the exact shape.

As with the flat rectangular plate, here again stresses and displacements are expressed by trigonometric series truncated after the second term. Similarly, the expression for the time rate of change of the radial displacements is approximated by the sum of the two expressions calculated for very small and for very large displacements. The uniaxial form of the creep law adopted is the power law of (4) with $n = 3$. The fictitious sandwich wall was constructed in accordance with Rabotnov's proposal of 1964 [24]; the distance D between the faces of the sandwich wall is defined in (2).

A result of the analysis can be given in the form

$$t = 0.294 t_E \frac{\sigma_E - \sigma}{\sigma_E} \ln \frac{x^2(1.18 + x_0^2)}{x_0^2(1.18 + x^2)}. \tag{18}$$

This formula connects the nondimensional displacement W with the time t at which W is reached if the value of W is W_0 at the time when creep deformations commence. The critical value σ_E of the applied stress for a perfectly elastic shell is

$$\sigma_E = 0.6 E (h_{\text{real}}/a), \tag{19}$$

where a is the mean radius of the cylindrical shell. The intensity of the applied stress is

$$\sigma = P/2\pi a h_{\text{real}}. \tag{20}$$

Again, P, σ and σ_E are taken positive if they indicate compression.

The most dangerous wave length is

$$\lambda = 2.22 (aD)^{1/2}. \tag{21}$$

From (18) the critical time can be calculated by letting W, and thus x, go to infinity. One obtains

$$t_{cr} = 0.294 t_E \frac{\sigma_E - \sigma}{\sigma_E} \ln \frac{1.18 + x_0^2}{x_0^2}. \tag{22}$$

In many cases of practical interest

$$x_0^2 \ll 1.18. \tag{23}$$

Then (22) reduces to

$$t_{cr} = 0.588 t_E \frac{\sigma_E - \sigma}{\sigma_E} \ln (1.087/x_0). \tag{24}$$

It is to be remembered that these expressions are valid only when $n = 3$.

Just as with the flat rectangular plate, here again the analytical results were checked with the aid of the digital computer [48]. It was found that the agreement between analysis and computation is good when $n = 3$ and that it deteriorates rapidly when n increases beyond 3. The computations were carried out for $n = 3, 5, 7$ and 9 and the results were represented by the formula

$$t_{cr} = C_1 t_E \frac{\sigma_E - \sigma}{\sigma_E} \ln (C_2/x_0). \tag{25}$$

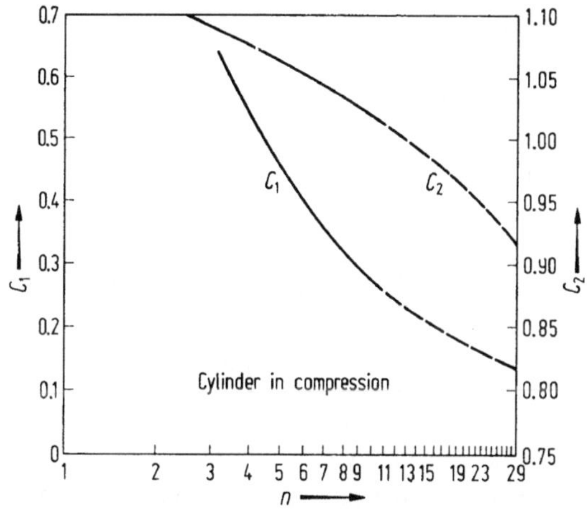

Fig. 5. Values of constants C_1 and C_2 for circular cylindrical shell in axial compression.

Through a method of extrapolation values were also found up to $n = 29$. In Fig. 5, the full lines indicate values obtained through interpolation, and the dotted lines indicate values obtained through extrapolation. They represent values calculated with the aid of the digital computer and require no correction factors.

References to other papers dealing with the axisymmetric creep buckling of circular cylindrical shells can be found in [45].

Multilobed buckling of circular cylindrical shell in axial compression

It has been known for half a century that axially compressed perfectly elastic thin-walled circular cylindrical shells buckle at stresses considerably smaller than those calculated from the linear theory of elastic stability. Although many other authors had tried to explain the discrepancy through the effect of initial deviations of the midsurface of the shell from the exact circular cylindrical shape, it was *Koiter* who succeeded in 1945 [6] in calculating the percentage reduction of the classical critical stress in function of the amplitude of the initial deviations. Koiter obtained a closed-form expression for this reduction through the device of replacing the nonlinear problem by a succession of two linear problems. In the first he calculated the growth with increasing load of the most dangerous axially symmetric initial deviations, and in the second he determined the load under which the axisymmetric mode becomes unstable and is succeeded by a multilobed form of deformations.

In many respects the buckling of axially compressed circular cylindrical shells in the presence of creep does not differ much from that accompanied by purely elastic deformations. However, the sudden jump from an initial state into the large-displacement pattern of the elastic shells is replaced by the much slower creep process. This permits the experimentalist to observe the transition from the axially symmetric shape to the multilobed shape.

Samuelson [25, 35], a student of *Odqvist*, was the first to calculate in a semi-empirical manner the time at which the hoop stresses caused by axisymmetric deformations become large enough to cause multilobed buckling. In a recent experimental study carried out at Stanford University [49] the creep tests were interrupted to measure the deformations in detail and to photograph the shell specimens. It was definitely established that buckling can begin axisymmetrically and change over later to a multilobed pattern. With the aid of the axisymmetric theory of creep deformations the time was calculated in [50] at which the hoop stresses cause elastic buckling in accordance with the Koiter theory. This happens considerably before the critical time of axisymmetric creep deformations whenever the ratio of the applied compressive load to the elastic critical load is large.

Finally, in 1970 *Levi* and *Hoff* [51] calculated the interaction between axisymmetric and multilobed components of the deformations during the creep process. The calculations were carried out with the aid of the digital computer for $n = 3$. They showed that for wave lengths greater than the length of the elastic axisymmetric buckles even very small multilobed initial deviations can significantly reduce the value of the critical time.

Flattening of circular cylindrical shell under external pressure

When a long, thin-walled circular cylindrical shell is subjected to a uniform external pressure q, the circumferential (or hoop) stress in its wall is

$$\sigma = q(a/h_{\mathrm{real}}) \,. \tag{26}$$

If the shell is perfectly elastic and perfectly shaped, the critical value of σ under which the shell buckles is given by the expression

$$\sigma_E = \frac{E}{4(1-\nu^2)}\left(\frac{h_{\mathrm{real}}}{a}\right)^2 . \tag{27}$$

Real shells are never perfect because they can be manufactured only within given tolerances. During the application of the pressure q the imperfections are exaggerated and, in particular, the first Fourier component of the deviations, which corresponds to flattening into an ellipse,

(Fig. 6), grows most noticeably as the lowest buckling load represented by (27) is approached. If α_{00} designates the ratio of the major semi axis a_{max00} of the inaccurate elliptic shell to the average radius a of the middle surface of the shell before the application of the pressure,

$$\alpha_{00} = a_{max00}/a \tag{28}$$

immediately after the application of the pressure, before creep deformations can develop, the ratio α_0 of the corresponding quantities is given by

$$\alpha_0 - 1 = (a_{max0}/a) - 1 = \sigma_E(\alpha_{00} - 1)/(\sigma_E - \sigma). \tag{29}$$

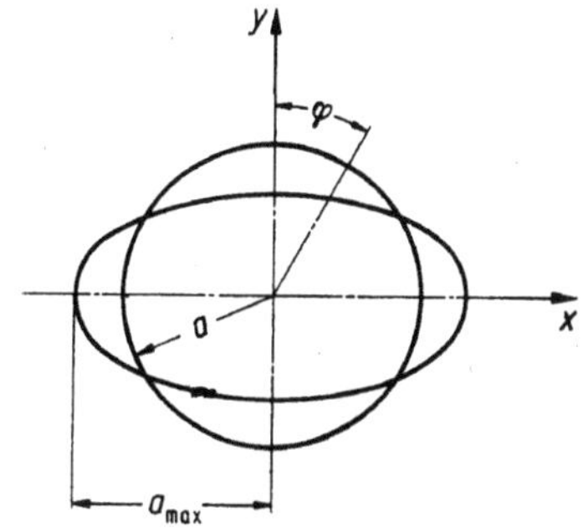

Fig. 6. Notation for flattening of circular cylindrical shell subjected to external pressure.

The further growth of α because of creep was studied in 1959 [52] by *Hoff*, *Jahsman* and *Nachbar*. The connection between α and the time t passed from the instant of the application of the pressure was found to be

$$t = 0.152\,\frac{\sigma_E - \sigma}{\sigma_E}\,t_E \ln \frac{1 + 1.365(\varepsilon_E/\alpha_0^2)}{1 + 1.365(\varepsilon_E/\alpha^2)}, \quad n = 3, \tag{30}$$

where

$$\dot{\varepsilon}_{nom} = k(qa/h_{real})^3. \tag{31}$$

The critical time prevails when α tends to infinity. Hence,

$$t_{cr} = 0.152\,\frac{\sigma_E - \sigma}{\sigma_E}\,t_E \ln\,[1 + 1.365(\varepsilon_E/\alpha_0^2)], \quad n = 3. \tag{32}$$

In [52] an expression was given for the deformation rate $\dot{\alpha}$ for arbitrary values of n, but the integration with respect to time was carried out only for $n = 3$.

The calculations of [52] were repeated by *Wah* and *Gregory* [53] who replaced the exponential creep law with the hyperbolic sine law and also carried out some experiments. In two publications [54, 55], *Bargmann* extended the theory of [52] to varying temperatures and pressures and integrated the equations with the aid of a digital computer.

Circular cylindrical shell subjected to constant bending moments

When a constant bending moment is applied to a thin-walled circular cylindrical shell which acts as a beam, as shown in Fig. 7, the initially circular cross section flattens into an ellipse in consequence of elastic deformations. As time passes, the flattening increases in consequence of creep. If the creep law is a nonlinear power law with n greater than unity, eventually the shell collapses at a finite time t_{cr} after the onset of creep.

This problem was first treated thoeretically in 1957 in [56]. Since the corresponding experiments were carried out mostly with reinforced aluminum alloy cylinders, a uniaxial creep

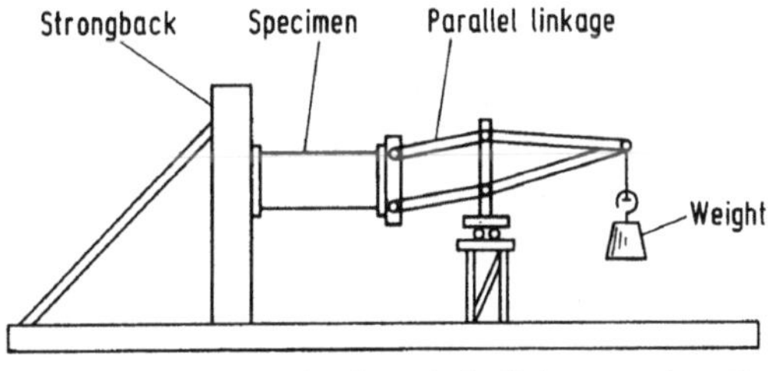

Fig. 7. Loading of cylindrical shell in pure bending.

law was used in the calculations and the material of the reinforcements was imagined to be smeared out over the surface of the shell. Two simultaneous nonlinear differential equations were derived for the curvature of the beam and for the flattening of its cross section. These equations were easily integrated numerically with the aid of a manual desk computer and the results of the computations were found to be in good agreement with experiments carried out at the Polytechnic Institute of Brooklyn during the preceding two years [57]. A report [58] on additional tests (a total of 102) and on the derivation of a formula was presented in 1960, and in 1971 a further analysis [59] was published in which the simple formulas that follow appeared.

According to these formulas the critical time can be calculated as

$$t_{cr} = [(M_{cr} - M)/M_{cr}]\,[C_1(\varepsilon_{cl}/\dot{\varepsilon}_{\text{nom}}) - C_2(\varepsilon_0/\dot{\varepsilon}_{\text{nom}})]\,. \tag{33}$$

In this expression M is the constant bending moment applied to the shell, and M_{cr} its critical value in the case of purely elastic deformations:

$$M_{cr} = 0.35\pi h_{\text{real}}^2 aE\,. \tag{34}$$

Moreover,

$$\varepsilon_{cl} = 0{,}6(h_{\text{real}}/a)\,, \tag{35}$$

$$\varepsilon_0 = M/\pi h_{\text{real}} a^2 E\,, \tag{36}$$

$$\dot{\varepsilon}_{\text{nom}} = k(M/\pi h_{\text{real}} a^2)^n\,. \tag{37}$$

The constants C_1 and C_2 depend only on the exponent n of the creep law; they are given in Fig. 8.

It is noted that in this case, the factor $(M_{cr} - M)/M_{cr}$ is a semiempirical correction factor to take care of the effects of elasticity during the creep process. It is not rigorously correct because the problem of buckling by flattening reduces to a nonlinear, and not to a linear, characteristic value problem when the deformations are governed entirely by Hooke's law. In the numerical integration process described in [56] the corresponding correction was applied rigorously. However, (33) also yields results in reasonably good agreement with experiment.

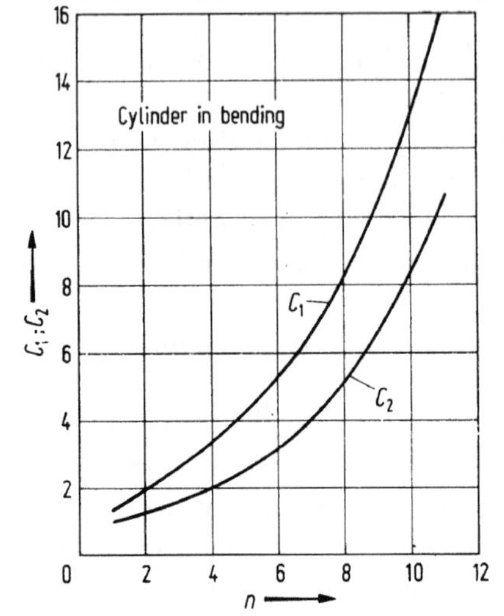

Fig. 8. Values of constants C_1 and C_2 for circular cylindrical shell subjected to pure bending.

Additional literature

The accuracy of the simplified expressions used in all analyses for the curvature was investigated by *Życzkowski* [60]. The paper shows that in the case of a column with two pinned ends the simplified expression is sufficiently accurate up to a lateral deflection of the midpoint of the column amounting to about 16 percent of the length of the column. This is 16 times the radius of gyration of a column having a slenderness ratio of 100, and 8 times the radius of gyration of a column of a slenderness ratio of 50. In other words, the simplified expression is satisfactory until very large deflections are reached.

The Rabotnov-Shesterikov synamic criterion of stability was recently investigated in detail by *Yamamoto* [61]. The same criterion was used by *Jahsman* and *Field* [62] in a study of the creep buckling of flat plates; these authors also made use of a discontinuous mechanical equation of state to account for the creep hesitation effect at unloading.

Further studies of the creep buckling of circular cylindrical shells were made by *Kurshin* [63], *Kuznetsov* and *Kurshin* [64], and by *Kuznetsov* and *Iungerman* [65].

Closing remarks

An attempt has been made to summarize the ideas that have been developed in the last fifteen years on the creep buckling of plates and shells. Formulas have been presented from which the critical time can be calculated for rectangular plates in edgewise compression, and for thin-walled circular cylindrical shells subjected to axial compression, to external radial pressure and to uniform bending. The behavior of these structural elements after buckling has been discussed in some detail.

References

1. Timoshenko, S.: Strength of Materials, Part, 2 New York: Van Nostrand 1941, 407.
2. Bauschinger, J.: Dingler's Polytechnisches Journal, 266, 1886.
3. Sherby, O. D., Dorn, J. E.: Transactions of the American Society of Metals, 43 (1951) 611.
4. Phillips, C. W., Sinnott, M. J.: Transactions of the ASM, 46 (1954) 63.
5. Kármán, Th. v.: Mitteilungen über Forschungsarbeiten des Vereins Deutscher Ingenieure, No. 81, 1910; see also Collected Works of Theodore von Kármán, 1, 90, London: Butterworth 1956.
6. Koiter, W. T.: Over de Stabiliteit van het elastisch Evenwicht, Amsterdam: Paris 1945; Proc. Roy. Netherlands Acad. Sci. Amsterdam, Series B, 66, 1963.
7. Rabotnov, G. N., Shesterikov, S. A.: Journal of the Mechanics and Physics of Solids, 6 (1957) 27.
8. Hoff, N. J.: Third U.S. Nat. Congr. Appl. Mech., Amer. Soc. Mech. Eng., New York, 1958, 29.
9. Jahsman, W. E., Field, F. A.: J. Appl. Mech. (1961).
10. Shesterikov, S. A.: PMM 23 (1959) 1101.
11. Freudenthal, A. M.: Some Time Effects in Structural Analysis, Sixth International Congress of Applied Mechanics, Paris, France, 1946, unpublished; see also A. M. Freudenthal: The Inelastic Behavior of Engineering Materials and Structures, New York: Wiley 1950, 518.
12. Kempner, J., Pohle, F. V.: Journal of the Aeronautical Sciences, 20 (Aug. 1953) 572.
13. Kempner, J.: Polytechnic Inst. Brooklyn Rep. PIBAL No. 200, November 1952; see also National Advisory Committee for Aeronautics NACA TN 3138, January 1954.
14. Libove, Ch.: J. Aeronaut. Sci. 19 (July 1952) 459.
15. Hoff, N. J.: Buckling and Stability, J. Roy. Aeronaut. Soc. 58 (Jan. 1954) 30.
16. Rzhanitsyn, A. R.: Dokl. Akad. Nauk SSSR 52, No. 25, 1946.
17. Fraeijs de Veubeke, B.: Creep Buckling, in: High Temperature Effects in Aircraft Structures, ed. N. J. Hoff, Oxford: Pergamon Press 1958, 267.
18. Hoff, N. J.: Aeronautical Quarterly, 7, Part I, (Feb. 1956) 1.
19. Odqvist, F. K. G.: J. Appl. Mech. 21 (Sept. 1954) 295.
20. Hoff, N. J.: PIBAL Report No. 356, Polytechnic Institute of Brooklyn, Brooklyn, N.Y., Aug. 1956, see also J. Roy. Aeronaut. Soc. 61 (Nov. 1957) 756.
21. Chapman, J. C., Erickson, B., Hoff, N. J.: PIBAL Report No. 406, Polytechnic Institute of Brooklyn, Brooklyn, N.Y., Oct. 1957; see also Internat. J. Mech. Sci. 1 (1960) 145.
22. Lempriere, B. M.: in: Creep in Structures, ed. N. J. Hoff, Springer: Berlin/Göttingen/Heidelberg 1962, 291.
23. Patel, S. A., Wagle, V. D.: PIBAL Report No. 358, Polytechnic Institute of Brooklyn, Brooklyn, N.Y., 1957.
24. Hoff, N. J.: The idealized column, Festschrift Richard Grammel, Ing.-Arch. 28 (1959) 89.
25. Samuelson, Å.: The Aeronautical Research Institute of Sweden, Rep. No. 108, Stockholm, 1967.
26. Rabotnov, Y. N.: Prikl. Mat. Mekh. 28 (1964) 1040.
27. Odqvist, F. K. G.: Proc. Roy. Swedish Inst. Eng. Res. No. 141 (1936) 1.
28. Hoff, N. J., Levi, I. M.: Stanford Univ. Dep. Aeronaut. and Astronaut. Rep. SUDAAR No. 425, June 1971; also in the Internat. J. Solids and Structures, 8 (1972) 1103.
29. Hoff, N. J., Berke, L., Honikman, T. C., Levi, I. M.: in Advances in Creep Design, the A. E. Johnson Memorial Volume, ed. A. I. Smith and A. M. Nicolson, Applied Sciences, Ltd., London, p. 421, 1971.
30. Hult, J. A. H.: J. Appl. Mech. 22 (Sept. 1955) 432.
31. Timoshenko, S.: Strength of Materials, 2nd Ed., Part I, New York: Van Nostrand 1941, 248.
32. Perry, J.: Engineer, December 10 and 24, 1886.
33. Schuman, L., Back, G.: Nation. Advisory Comm. Aeronaut. Rep. NACA Rep. No. 356, 1930.
34. Kármán, Th. v.: Trans. ASME 1932, 53.
35. Samuelson, Å.: Aeronaut. Res. Inst. Sweden Rep. 98, Stockholm 1964.

36. Odqvist, F. K. G.: Creep stability of bars, plates and shells, in: Stability, ed. H. H. E. Leipholz, University of Waterloo, Waterloo, Ontario, Can. 1972, 199.
37. Hoff, N. J.: Engineering Plasticity, Sir John Baker Anniversary Volume, eds. J. Heyman and F. A. Leckie, Cambridge/England: Cambridge University Press 1968, 257.
38. Hoff, N. J., Berke, L., Honikman, C., Levi, I. M.: A. E. Johnson Memorial Volume, eds.: A. I. Smith and A. M. Nicolson, London: Applied Sciences 1971, 421.
39. Levi, I. M., Hoff, N. J.: Ing.-Arch. 38, 1969, Görtler Anniversary Issue, 329.
40. Pian, T. H. H., Johnson, R. I.: MIT, Aeroelastic and Structures Research Laboratory, Technical Rep. 25—24, November 1957.
41. Sanders, J. L., McComb, Jr. H. G., Schlechte, F. R.: National Advisory Committee for Aeronautics NACA TN 4003, May 1957.
42. Pian, T. H. H.: J. Aeronaut. Sci. 24, 846, November 1957.
43. McComb, Jr., H. G.: National Advisory Committee for Aeronautics, Technical Note 4398, September 1958.
44. Mathauser, E. E., Deveikis, W. D.: National Advisory Committee for Aeronautics, Report 1308, 1957.
45. Hoff, N. J.: J. Appl. Mech., 35, Series E, (Sept. 1968) 530.
46. Pittner, E. V., Hoff, N. J.: Acta Mechanica 8 (1959) 116.
47. Samuelson, Å.: Aeronaut. Res. Inst. of Sweden Rep. 100, Stockholm 1964.
48. Honikman, T. C., Hoff, N. J.: Internat. J. Solids and Structures 7 (1971) 1685.
49. Benoit, M., Sendelbeck, R. L., Hoff, N. J.: Stanford University, Department of Aeronautics and Astronautics Report SUDAAR No. 415, Sept. 1970.
50. Hoff, N. J.: Stanford University, Department of Aeronautics and Astronautics Report SUDAAR No. 434, October 1971; also in Contributions to the Theory of Aircraft Structures, the A. van der Neut Anniversary Volume, Delft University Press, 1972, 297.
51. Levi, I. M., Hoff, N. J.: in Creep in Structures, Proc. IUTAM Symposium in Gothenburg 1970, ed. J. Hult, Berlin/Heidelberg/New York: Springer 1972, 405.
52. Hoff, N. J., Jahsman, W. E., Nachbar, W. J.: Aerospace Sci. 26 (Oct. 1959) 663.
53. Wah, T., Gregory, R. K.: J. Aerospace Sci. (March 1961) 177.
54. Bargmann, H.: European Organization for Nuclear Research, Report CERN-ISR-EN/71-32, Geneva, August 1971.
55. Bargmann, H.: 1st International Conference on Structural Mechanics in Reactor Technology, Berlin, September 1971.
56. Hoff, N. J.: J. Roy. Aeronaut. Soc. 61 (Nov. 1957) 756.
57. Erickson, B., Patel, S. A., French, F. W., Lederman, S., Hoff, N. J.: National Advisory Committee for Aeronautics Research Memorandum RM 57E17, July 23, 1957.
58. Erickson, B., French, F. W., Patel, S. A., Hoff, N. J., Kempner, J.: National Aeronautics and Space Administration Technical Note D-429, June 1960.
59. Hoff, N. J., Wang, L.: Aeronaut. J. 75 (July 1971) 471.
60. Życzkowski, M.: Creep in Structures, IUTAM Colloquium, Stanford, 1960, ed. N. J. Hoff, Berlin/Göttingen/Heidelberg: Springer 1962.
61. Yamamoto, Y.: J. Mechanics and Physics of Solids 18 (1970) 165.
62. Jahsman, W. E., Field, F. A.: Proc. Joint International Conf. on Creep, 1963, 2—43, Inst. of Mechan. Eng., London.
63. Kurshin, L. M.: Sov. Phys. Dokl. 10 (Jan. 1966).
64. Kuznetsov, A. P., Kurshin, L. M.: On the Analysis of Shell Stability under Creep Conditions by Means of Aging Theory, Stability Problems in Structural Mechanics, STROIIZDAT, 1965.
65. Kuznetsov, A. P., Iungerman, N. M.: Prikl. Mekh. i Tekh. Fiz. 128, 1965.

On the dynamics of a system of rigid bodies

A. Yu. Ishlinsky

Moscow, USSR

The existent treatises on rational mechanics do not pay attention to the dynamics of a system of rigid bodies, confining their interest to the general theorems of mechanics, the theory of rotation of a rigid body about a fixed point, some problems of analytical mechanics and the theory of stability. At the same time, the problems whose formulation is reduced precisely to the dynamics of a system of rigid bodies are quite common nowadays, at least during the last 30 years. An attempt to solve such problems on the basis of Lagrange's second method leads, as a rule, to cumbersome differential equations with a large number of terms; the mechanical interpretation of those is not at all easy. Even computers are then of little help. A solution of the problem of motion of a system of rigid bodies in terms of a linear approximation is a superposition of harmonic motions, and is, therefore, sufficiently transparent. However, in order to get the important qualitative characteristics of the motion, it is necessary to take into account the non-linear terms of initial differential equations and to integrate these equations over a time interval considerably longer than the period of partial oscillations of the associated linear system. However, such integration leads to a rapid accumulation of errors because the actual calculations by the method of finite differences involve a large number of steps.

In addition, it is rather a difficult task by itself to make sense out of the results of the numerical calculations obtained. There is a hope, though, that the new read-out techniques, allowing the solutions to be seen immediately on a television screen (the so-called displayers), will permit the computers to assist more effectively in studying characteristic properties of the motion of a system of rigid bodies.

Nevertheless, at the present time rather many general aspects and fundamental problems of the above-mentioned branch of rational mechanics have been already studied (*A. I. Lur'e, I. I. Metelitsyn, D. R. Merkin*). It has turned out that in a large number of cases the equations of motion derived by application of the general theorems of dynamics to the whole system of rigid bodies and its separate parts are more surveyable and convenient for further use than, naturally, the equivalent equations derived by using Lagrange's second method (*E. L. Nikolai, A. Yu. Ishlinsky, K. Magnus, M. Z. Litvin-Sedoi, A. I. Lur'e, I. V. Novozhilov*). Here, a right choice of basic, calculating and intermediate coordinate systems, as well as the variables to be determined, is of considerable importance.

The method of averaging which can be traced back to *Lord Rayleigh* and *B. van der Pol*, and which has been significantly developed by *B. V. Bulgakov, N. M. Krylov, N. N. Bogolybov*, and their colleagues has proved to be, from the viewpoint of mathematics, the most fruitful method allowing many phenomena of the dynamics of a system of rigid bodies to be investigated. Moreover, in some cases, the finding of Lyapunov's function of the corresponding non-linear system (*N. G. Chetaev, A. I. Lur'e*, and others), has made it possible to draw more exact conclusions concerning the stability of stationary motions of a system of rigid bodies. These conclusions had been previously obtained by solving linearized differential equations of the corresponding system.

If one omits the problem of celestial mechanics concerning the motion of two mutually gravitating rigid bodies, for example two uniform ellipsoids, then one of the first investigations in the field of dynamics of a system of rigid bodies was the study of stability conditions for two tops put one upon the other (Fig. 1). This problem, however, has been studied only in the linear formulation.

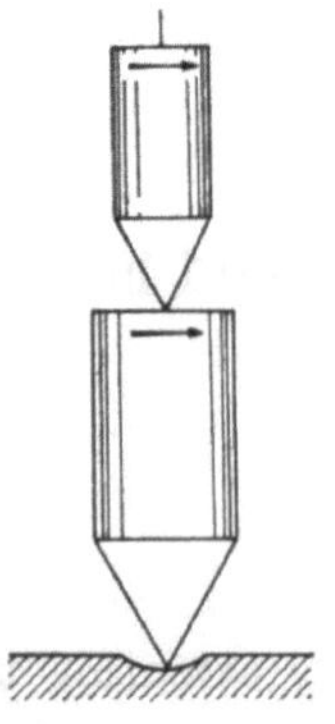

Fig. 1. Two tops put one upon another.

A classical example of a system of connected rigid bodies is provided by a gyroscope in gimbals. *E. L. Nikolai* not only compiled, by a triple application of the angular momentum theorem (to the rotor itself, to a system consisting of the rotor and the outer ring, and to the whole system of gyroscope-rotor and both rings), the equations of motion of this system but also reduced the problem to the formulae of integration. *E. L. Nikolai* seems to be the first who discovered the drift of a vibrating free gyroscope about its outer axis. In 1955 *K. Magnus* and at about the same time *B. T. Playmale* and *R. Goodstein* have derived an approximate formula for the mean angular velocity of this drift. Many scientists thought the above phenomenon paradoxical because of the widely accepted opinion that only the friction force can be the cause of the change in orientation of the rotor spin axis. However, such a statement is valid if we neglect inertia of rings (the so called precession theory of gyroscope).

Here is one of the variants of the derivation of the Magnus formula. One of the equations describing the motion of a gyroscope in gimbal, namely,

$$\Theta \frac{d^2\beta}{dt^2} - \frac{1}{2} J'(\beta) \left(\frac{d\alpha}{dt}\right)^2 - H \cos\beta \frac{d\beta}{dt} = 0$$

can be written in the form

$$\frac{d\alpha}{dt} = \frac{\Theta}{H} \frac{d}{dt} \left(\frac{1}{\cos\beta} \frac{d\beta}{dt}\right) - \frac{A_2 + C_1}{H \cos\beta} \operatorname{tg}\beta \left(\frac{d\alpha}{dt}\right)^2 + \frac{\operatorname{tg}\beta}{H \cos\beta} \left[J(\beta) \left(\frac{d\alpha}{dt}\right)^2 - \Theta \left(\frac{d\beta}{dt}\right)^2\right].$$

The gyroscope vibration through the angle β of rotation of the inner ring has a periodic character. Therefore, integrating the right-hand side of the equation over the period interval we obtain

$$\left\langle\frac{d\alpha}{dt}\right\rangle = -\frac{A_2 + C_1}{H \cos\beta^*} \operatorname{tg}\beta^* \left\langle\left(\frac{d\alpha}{dt}\right)^2\right\rangle + \frac{\operatorname{tg}\beta^{**}}{H \cos\beta^{**}} \left\langle J(\beta) \left(\frac{d\alpha}{dt}\right)^2 - \Theta \left(\frac{d\beta}{dt}\right)^2\right\rangle.$$

The mean values of the energy of vibration of the rings and the rotor of a gyroscope about the outer axis and about the inner axis are of the second order with respect to the amplitude of the angular vibration. These quantities are equal to zero in the linear treatment of the problem. This consideration enables one to draw the conclusion that in the nonlinear treatment the difference between these quantities in the right-hand side of the above-mentioned formula has a higher order, and this justifies the Magnus formula. For one particular relation between the moments of inertia of the rings and the equatorial moment of inertia of the rotor ($A_1 + A = C_1$) the problem concerning the motion of a gyroscope in gimbal was integrated in elliptic functions (*D. M. Klimov, N. P. Stepanenko*). The Magnus formula is also valid for this particular case with high accuracy.

The drift of a gyroscope in gimbals on a vehicle caused by the rings' inertia has been discovered as a particular case of the behaviour of a direction gyro during rolling and pitching of a ship as early as 1943 (*A. Yu. Ishlinsky*). Various problems concerning the mechanics of a gyroscope in gimbal have been studied by a number of Soviet authors (*D. M. Klimov, S. A. Kharlamov, N. N. Butenin, D. S. Pel'por, L. N. Slezkin*, and others). The Magnus formula itself has been made more precise both in regard to its more rigorous derivation (*Ya. A. Lunts, Kh. L. Smolitsky*, and others) and by including the terms of a still higher order with respect to the amplitude of the angular vibration of the rings.

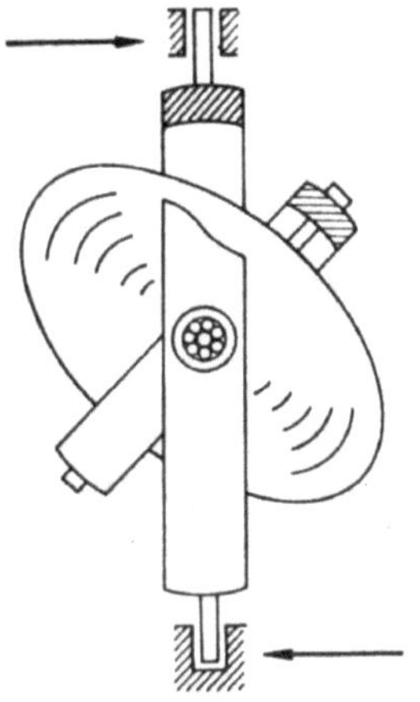

Fig. 2. Schematic diagram explaining the cause of the gyroscope drift.

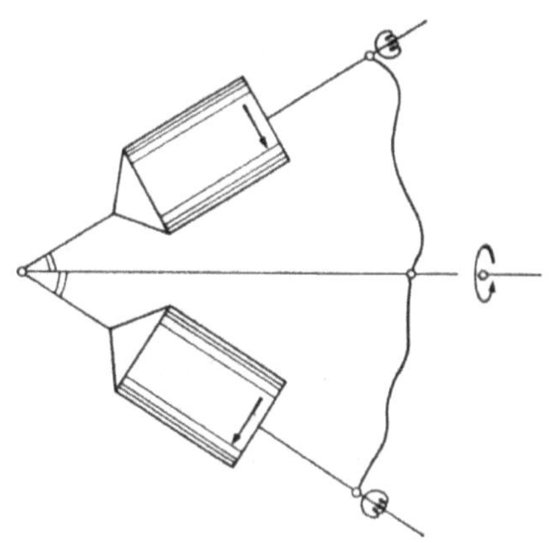

Fig. 3. The tops jointed by a spherical hinge.

The drift of a vibrating gyroscope in gimbal on an immobile base (the Nikolai-Magnus effect) is eventually the result of the influence produced on the system by the reactions of the ball bearings of its external axis (*S. S. Tikhmenev*, see Fig. 2). In this case, the base is assumed to be as if of an infinitely large mass. However, if we take the base of a finite mass, the drift will take place nevertheless, but with smaller magnitude. Here, one has to study the motion of four connected rigid bodies by using the methods of averaging, just as in the study of two tops jointed by a hinge in weightlessness (*N. P. Stepanenko*). In the last case the drift of the spin axes of the tops occurs in opposite direction. The plane containing the axes of both the tops rotates at a constant mean angular velocity about the bisector of an angle between the axes (provided the tops are identical), while the angle itself remains on average unchanged (Fig. 3). The direction and magnitude of the above mentioned angular velocity are determined by the initial conditions of the problem. More complicated problems have also been solved, in particular those concerning the resonance in the system of a gyroscope supported in gimbals (*D. M. Klimov, V. A. Filippov*), the vibration and the drift of a gyro platform having two or three gyroscopes (*A. Yu. Ishlinsky, B. I. Nazarov, V. A. Zinenko, I. V. Novozhilov*), the motion of a rigid body with an inside gyroscope (*P. V. Kharlamov*) as well as many other problems related simultaneously to electromechanics and control theory. These problems, as well as those concerned with gyrocompasses, accelerometers (newtonometers) and inertial navigation call for a special analysis and are omitted here.

Let us dwell on a treatment of the equations of dynamics of a rigid body which will be helpful for the sphere of problems under consideration. The equations of motion of a symmetrical rigid body about a fixed point may be expressed in the following "gyroscopic" form:

$$A\,dp'/dt + (C - A)\,q'r' + q'H' = M_{x'},$$

$$A\,dq'/dt + (A - C)\,r'p' - p'H' = M_{y'},$$

$$C\,dr'/dt + dH'/dt = M_{z'},$$

$$H' = C\,d\gamma/dt, \quad r = r' + d\gamma/dt, \quad \mathbf{p}' + \mathbf{q}' = \mathbf{p} + \mathbf{q}.$$

Here $x'y'z'$ is a coordinate system moving with the rotor, but not taking part in its spin rotation (Fig. 4). It is assumed that $r' \ll r$ (the case of $r' = 0$ corresponds to the axes in Bulgakov's treatment). As an example, we may take a coordinate system associated with the housing (or the inner ring of a gimbal system) inside of which the rotor of a gyroscope spins. The terms on the left-hand side of the above equations fall into two groups. One of them, namely,

$$A\,dp'/dt + (C - A)\,q'r' = M'_{x'},$$
$$A\,dq'/dt + (A - C)\,r'p' = M'_{y'},$$
$$C\,dr'/dt = M'_{z'},$$

gives the appearance of a combination of forces (moments) which should be applied to a rigid body completely identical to the given one (but "set" on the axes x', y', z' so that it would move with the coordinate system $x'y'z'$). The other group

$$q'H' = M^*_{x'} \simeq M_{x'},$$
$$-p'H' = M^*_{y'} \simeq M_{y'}, \quad (d\boldsymbol{H}'/dt = \boldsymbol{M}^* \simeq \boldsymbol{M}),$$
$$dH'/dt = M^*_{z'} \simeq M_{z'},$$

corresponds precisely to the equations of the precession theory of gyroscopes. Their left-hand sides contain the projections (upon the axes of the moving coordinate system $x'y'z'$) of the velocity of the vector point of the proper angular momentum $\boldsymbol{H}'$, whose direction is the same as that of the rotor spinning axes $z(z')$ (Fig. 5).

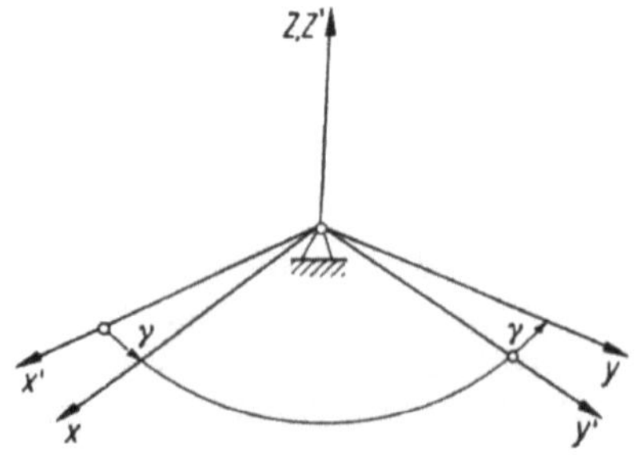

Fig. 4. Auxiliary coordinate system $x'y'z'$ moving with the rotor, but not taking part in its spin rotation.

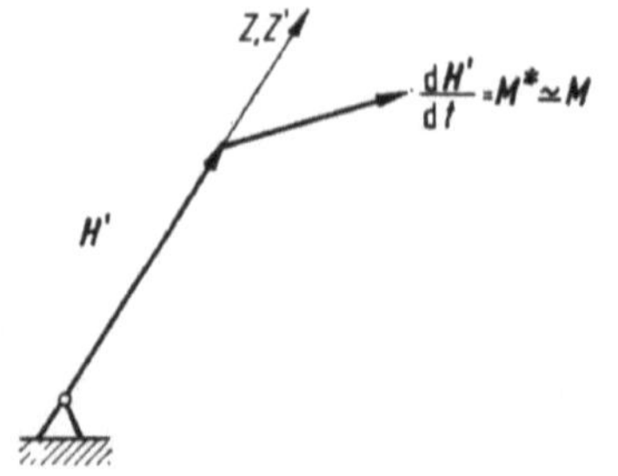

Fig. 5. Determination of the velocity of the vector point of the proper angular momentum.

The precession theory of gyroscopes ignores the forces "spent" on the motion of the above-mentioned identical body associated with the coordinate system $x'y'z'$ which does not take part in the rotor spinning. Each of the equations of motion turns then into a differential equation of the first order with respect to the generalized coordinates. This fact substantially simplifies the study of the problem, although some additional difficulties are introduced involving the correlation of initial conditions and the explanation of peculiarities of motion of certain gyroscopic systems with several rotors. Such an example is given by two vertical gyros having a common outer ring with no coupling constraint between them (Fig. 6). Professor *Ya. N. Roitenberg* and the author discussed the characteristics of such a gyroscopic system as early as in the memorable year 1941.

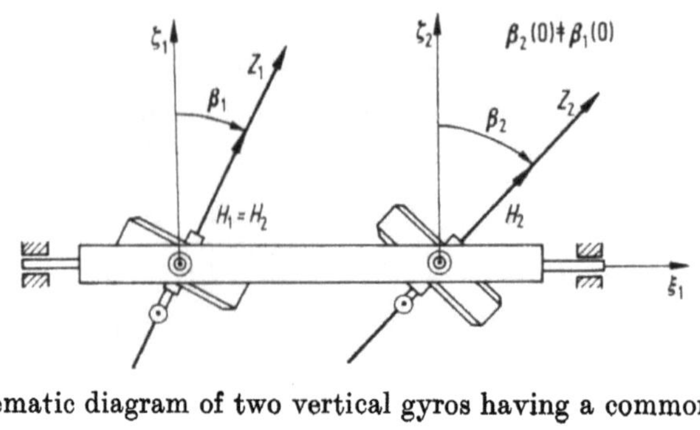

Fig. 6. Schematic diagram of two vertical gyros having a common outer ring.

In view of the fact that the gimbal rings have no large angular momentum, it is natural to assume that their action upon the rotor is of the same order of magnitude as that of the terms ignored in the equations of motion of a rigid body in the precession theory of gyroscopes. This is equivalent to the assumption that the rings have no moment of inertia. As a consequence, the

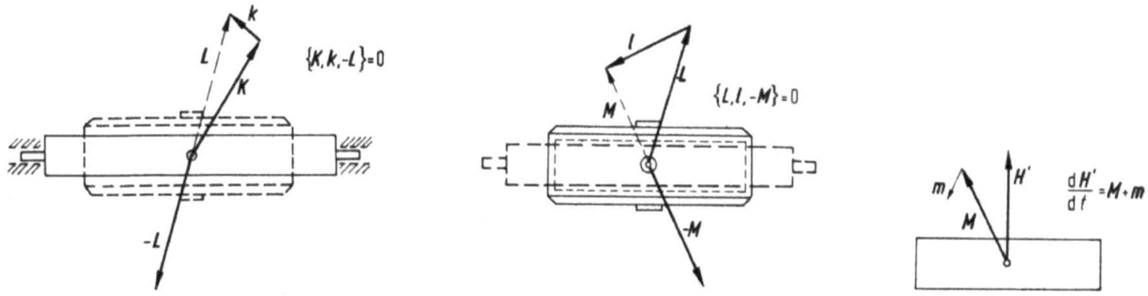

Fig.7—9. Schematic diagram for deriving equations of the precession theory of a gyroscope suspended in gimbals.

forces applied to the outer ring of the gimbal, as well as those applied to the inner ring form separately equilibrating systems of forces (Figs. 7, 8, 9). Eventually, we come to the following equations of the precession theory of a gyroscope suspended in gimbals:

$$\begin{aligned} H'q' &= (K_\xi + k_\xi) \sec\beta + (l_{z'} - M_{z'}) \operatorname{tg}\beta + l_{x'} + m_{x'}, \\ -H'p' &= L_{y'} + l_{y'} + m_{y'}, \\ dH'/dt &= M_{z'} + m_{z'}, \\ p' = \omega'_{x'}, \quad q' &= \omega'_{y'}, \quad H' = C\, d\gamma/dt, \quad d\gamma/dt \gg \omega'_{z'}. \end{aligned}$$

Without considering the meaning of individual terms in the right-hand sides of the above equations, we shall note only that here the quantity K_ξ denotes the moment applied to the outer ring about its axis ξ (Fig. 10). If the remaining moments are all zero, we obtain, in particular,

$$H'\, d\beta/dt = K_\xi/\cos\beta,$$

where β is the angular displacement of the spin axis from the perpendicular to the plane of the inner ring. Note that the commonly used method of decomposing a vector into components (Fig. 11), because of a careless handling, has led in a monograph on the gyroscope to the erro-

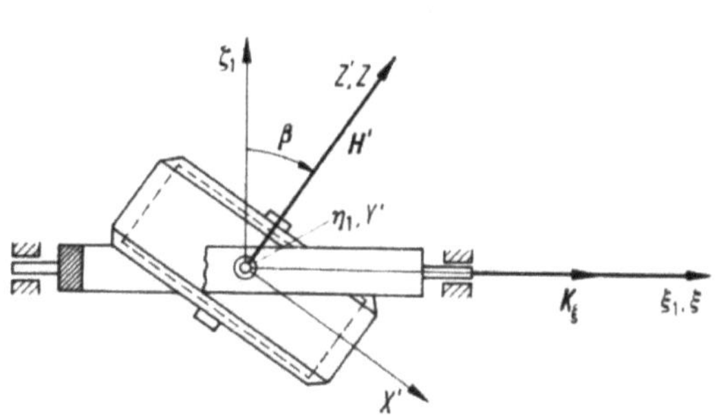

Fig. 10. Schematic diagram for solving the problem of motion of a gyroscope in gimbals influenced by the moment applied to the outer ring.

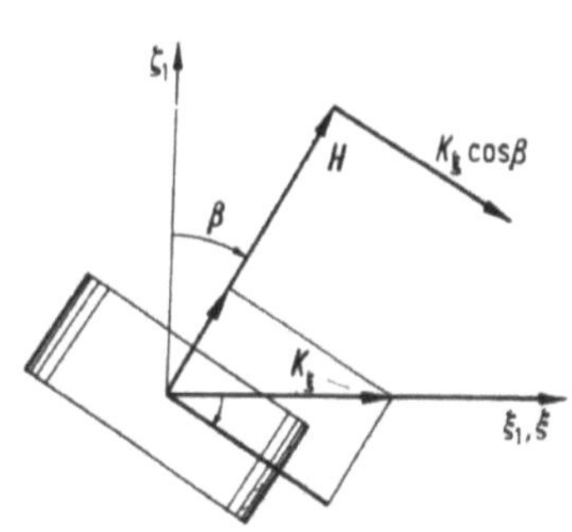

Fig. 11. Erroneous decomposition of moment K_ξ into components.

neous formula in which $\cos\beta$ is not in the denominator but in the numerator. Such mistakes are not infrequent in the mechanics of a system of rigid bodies; here one should always specify (and rigorously observe it in the subsequent treatment) just to what body one or another moment is applied. In the given case, the moment K_ξ is not applied to the rotor of the gyroscope (as shown erroneously in Fig. 11) but to the outer ring of its suspension (Fig. 12 shows a correct decomposition of the moment K_ξ into the components).

As has been mentioned above, the precession theory of gyroscopes leads to a set of differential equations whose order is twice as low as that of the equations derived in the strict theory. The equations of the precession theory are similar to the conditions of nonholonomic constraints which have restrictions placed not only on the coordinates of a mechanical system but also

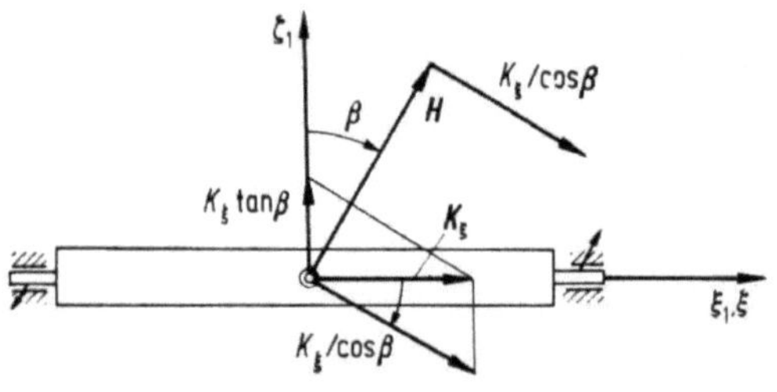

Fig. 12. Correct decomposition of moment K_ξ into components.

on the velocities. From this point of view, two examples offered by the mechanics of a system of rigid bodies are of interest, namely, a gyrohorizoncompass and a gyroscopic frame. Let us assume that the Earth is a sphere, and the field of the Earth's gravitation has a spherical symmetry. Under such assumption, the gyrocompass of Anschütz-*Geckeler* (Fig. 13), provided

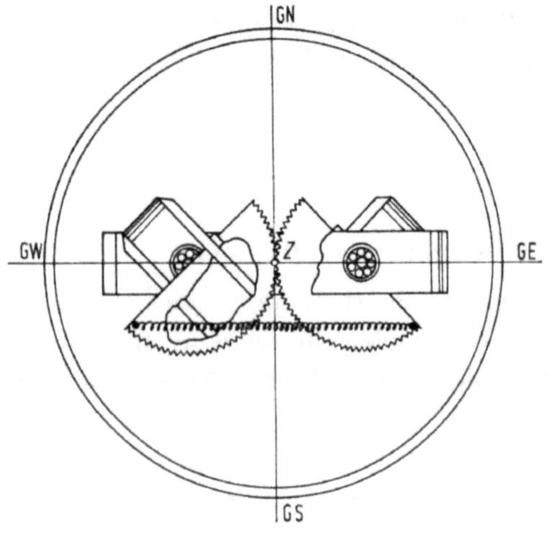

Fig. 13. Schematic diagram of the gyrocompass of Anschütz-Geckeler.

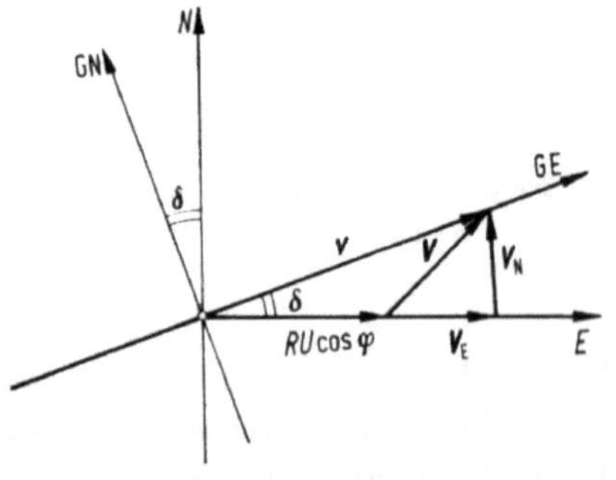

Fig. 14. Determination of the absolute velocity of the suspension point with the help of a space gyrocompass.

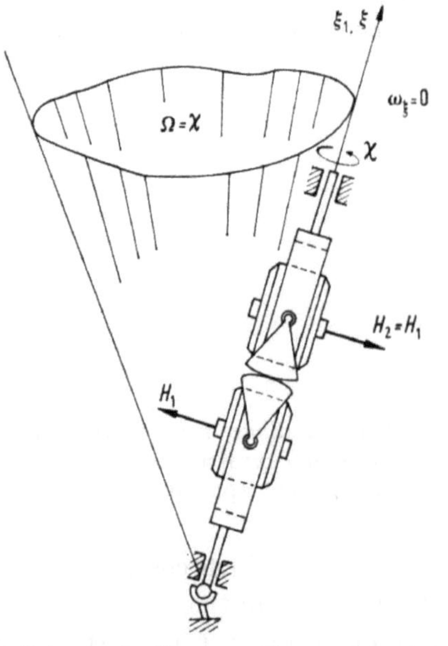

Fig. 15. Schematic diagram illustrating the accumulation of the solid angle at the motion of a two-gyroscope frame with gyroscopes connected by a linkage.

it is perfectly manufactured and certain initial conditions are exactly fulfilled, behaves as if "following" by one of its equatorial axes the absolute velocity of the suspension point (Fig. 14), while its polar axis is always directed towards the centre of the Earth in the arbitrary motion of the suspension point of the sensitive element over the sphere. The above stated facts may be, in a certain sense, interpreted as the presence of two constraints in a gyroscope, one of them being differential and the other geometric. Another example is a two-gyroscope frame with two gyroscopes connected by a linkage in such a way that they turn through equal and opposite angles (Fig. 15). If the manufacture is perfect, the projection of the angular velocity of the frame itself, that is, the common outer ring of the gyroscopes, upon its axis is zero when this axis makes an arbitrary motion. The presence of such a "nonholonomic" constraint leads to the follow-

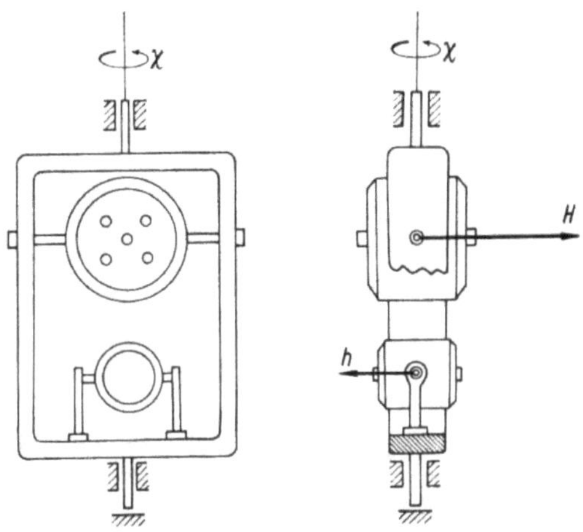

Fig. 16. Schematic diagram of the frame with two gyroscopes without linkage between them.

ing interesting consequence: If the axis of the frame describes in space an arbitrary cone and comes back to its initial position, the frame itself will not generally return to its initial position (*A. Yu. Ishlinsky*, 1952). The angle χ of its rotation turns out to be exactly equal to the measure of the solid angle Ω of the above mentioned cone. *D. M. Klimov* and *V. A. Potapenko* have considered a curious generalization of the just stated theorem for the case when gyroscopes have the angular momenta H and h of different orders of magnitude and when the second gyro is fixed on the frame (Fig. 16). In this case the angle of rotation of the frame will be less by the ratio $h:H$.

Let us come back again to the problems of dynamics and consider an interesting problem concerning the stability of rotation of two tops one of which is suspended to the other (Fig. 17a). Provided that the rotation velocities of each of these tops are sufficiently large, their rotation about the vertical axis ceases to be the only mode of stationary motion. This problem has been most thoroughly investigated for the case when one of the bodies is an inextensible string having no mass (Fig. 17b). Two other modes of stationary motion are still possible which are different from the principal one (Fig. 18) in which the thread and the spin axis of the top are vertical. It is a curious fact that all the three modes of stationary motion, including the vertical one, are stable in Lyapunov's sense. An experiment carefully carried out in vacuum (*S. V. Malashenko*) has confirmed the conclusions regarding the conservation of stability of the vertical mode of stationary motion when both bifurcational values of the angular velocity of rotation are slowly passed. The fact that unstable modes of stationary motion do not arise in the case of bifurcation, only seemingly contradicts the well-known Poincaré theorem *V. I. Vozlinsky*).

It should be noted that Lyapunov's function in the mechanics of a system of rigid bodies is mainly constructed by fitting the coefficients λ_j in the bundle of the first integrals V_j of the equations of motion

$$V = V_0 + \lambda_1 V_1 + \cdots + \lambda_n V_n.$$

This method was suggested by *N. G. Chetaev*. It has been successfully applied to a number of problems dealing with the stability of a gyroscope suspended in gimbal (*V. V. Rumyantsev*, *V. N. Skimel'*) and of the above-mentioned rotating rigid body suspended on a string (*M. E.*

Temchenko). An interesting mechanical phenomenon should be indicated in this connection: Let the gimbal rings be "added" (Fig. 19). Then the rotation of the rotor and the outer ring in the same direction would always be stable, whereas in the opposite directions it would be so only if certain inequalities were satisfied. They have been found proceeding from the linear formulation of the problem (*E. L. Nikolai*). Subsequently, the Lyapunov function has been constructed by the Chetaev method (*K. Magnus, V. V. Rumyantsev*).

I should like to conclude this outline by a few remarks concerning some problems of the mechanics of ball bearings. A ball bearing is perhaps the most used element of modern machine-building and instrument making. The need for good ball bearings compelled the engineers to overcome successfully the difficulties associated with the production of ball bearings, and the latter are now turned out by billions. Ball bearings of very high quality, the so-called precision

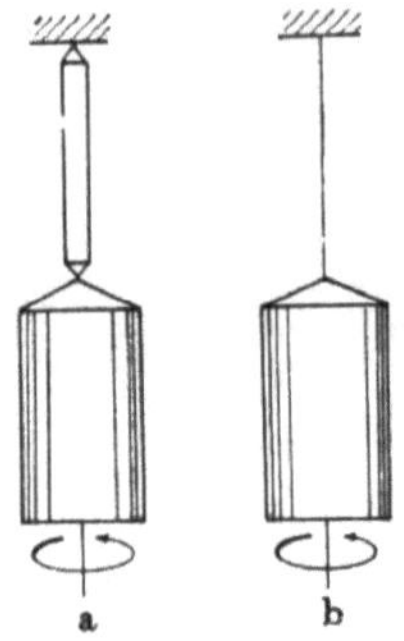

Fig. 17. The top suspended by a string.

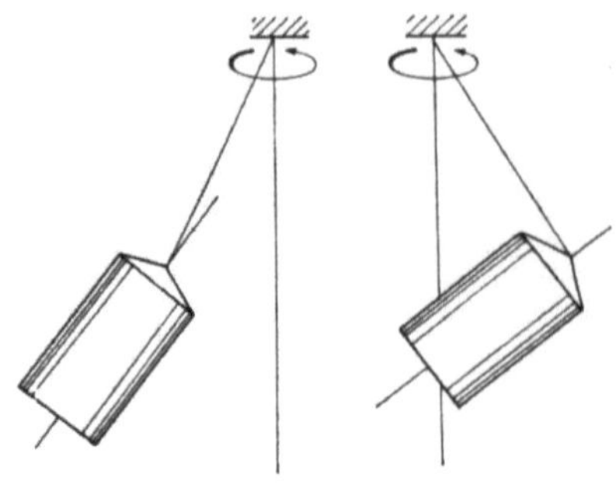

Fig. 18. Possible forms of stationary motions of the top suspended by a string.

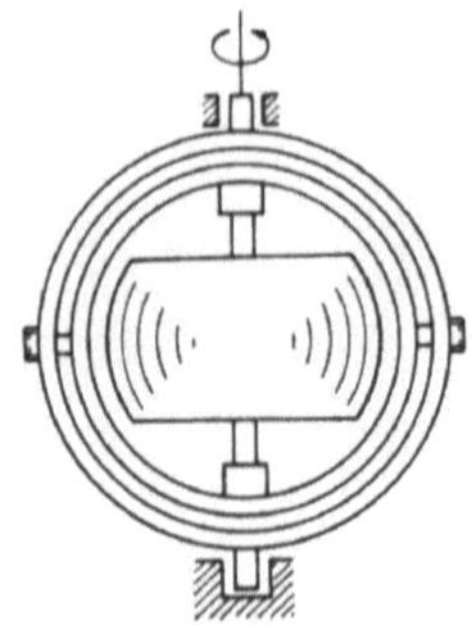

Fig. 19. The gyroscope in gimbals when the gimbal rings are "added".

ball bearings, are being used for gyroscopic devices. A ball bearing is a complicated system of contacting solids. Under the study of motion of such a system and its effect on the gyroscope it is necessary to take into account such factors as the deformation of balls and the races of the inner and outer rings, the friction force and the imperfect manufacture of the balls and the races. The works of *L. Z. Novikov, S. A. Kharlamov, V. F. Zhuravlev*, and *V. V. Filatov* have made it possible to obtain a further insight into the behaviour of ball bearings and the drift of gyroscopic systems caused by the latter. The allowance for the elastic deformation is beyond the scope of the classical problems of the dynamics of a system of rigid bodies. However, it is exactly by taking account of the elastic deformations of system elements that we have succeeded in explaining the undesirable phenomena in gyroscopic systems and, subsequently, in their eliminating. Here, in the first place, I should like to mention the drift of free gyroscopes and, moreover, the drift of a gyrocompass (*Yu. K. Zhbanov*) due to the vibration of a vehicle.

References

1. Bogolyubov, N. N., Krylov, N. M.: Introduction into non-linear mechanics, Kiev 1937.
2. Bulgakov, B. V.: Oscillations, Moscow 1954.
3. Butenin, N. V., Lunts, Ya. L.: Izvestiya vuzov, Priborostroenie 6 (1963) 75–83.
4. Chetaev, N. G.: Prikladnaya matematika i mekhanika 22 (1958) 379–381.
5. Filatov, V. V.: Mekhanika tverdogo tela 1 (1972) 30–38.
6. Grubin, C.: J. Appl. Mech. 29 (1962) 486–488.
7. Hooker, W. W., Margulies, G.: J. Astronaut. Sci. 12 (1965) 123–128.
8. Ishlinsky, A. Yu.: Prikladnaya matematika i mekhanika 21 (1957) 3–14.
9. —: ibidem 22 (1958) 359–373.
10. —: Mechanics of gyroscopic systems, Moscow 1963.
11. Kharlamov, S. A.: Prikladnaya matematika i mekhanika 26 (1962) 365–369.
12. —: ibidem 2 (1964) 50–60.
13. Klimov, D. M.: Izvestiya Acad. Sci. SSSR Mekhanika i mashinostroenie 5 (1963) 11–16.
14. —, Potapenko, V. A.: Mekhanika tverdogo tela 1 (1966) 3–5.
15. —, Stepanenko, N. P.: ibidem 6 (1967) 143–150.
16. —, Filippov, V. A.: ibidem 6 (1970) 42–54.
17. Litvin-Sedoi, M. Z.: Izvestiya vuzov, Aviatsionnaya tekhnika 2 (1966) 19–28.
18. —: Mekhanika tverdogo tela 3 (1967) 184–186.
19. Lunts, Ya. L.: Introduction into gyroscope theory, Moscow 1972.
20. Lur'e, A. I.: Trudy, Leningrad Polytechnic Institute 210 (1960) 7–22.

21. Lur'e, A. I.: Analytical mechanics, Moscow 1961.
22. Magnus, K.: Z. angew. Math. Mech. 35 (1955) 23—34.
23. —: Auswanderungserscheinungen an schwingenden Kreiseln in kardanischer Lagerung. Adv. in Aeronaut. Sci. Vol. 1, Pergamon Press 1959, 507—523.
24. —: Kreisel, Theorie und Anwendungen, Springer 1971.
25. Merkin, D. R.: Gyroscopic systems, Moscow 1956.
26. Metelitsyn, I. I.: Izvestiya Acad. Sci. SSSR Mekhanika i mashinostroenie 1 (1959) 3—9.
27. Nikolai, E. L.: Prikladnaya matematika i mekhanika 3 (1939) 3—34.
28. —: Gyroscope in gimbal., Moscow 1964
29. Novikov, L. Z.: Izvestiya Acad. Sci. SSSR Mekhanika i mashinostroenie 6 (1961) 84—90.
30. Novozhilov, I. V.: Izvestiya Acad. Sci. SSSR Mekhanika 5 (1965) 137—140.
31. —: Mekhanika tverdogo tela 2 (1968) 26—29.
32. Pel'por, D. S.: Nauchnye doklady vysshei shkoly, Mashinostroenie i priborostroenie 3 (1958) 207—211.
33. Playmale, B. T., Goodstein, R.: J. Appl. Mech. 22 (1955) 365—366.
34. Roberson, R. E.: Attitude control of satellites and space vehicles. Adv. in Space Sci. Vol. 2, Academic Press 1960, 351—436.
35. —, Wittenburg, J.: A dynamical formalism for an arbitrary number of interconnected rigid bodies, with reference to the problem of satellite attitude control. Proc. 3rd Congr. Intern. Feder. Autom. Control, London 1966.
36. Roitenberg, Ya. N.: Gyroscopes, Moscow 1966.
37. Rumyantsev, V. V.: Prikladnaya matematika i mekhanika (1958) 22, 374—378; 499—503.
38. Sandler, S. H.: J. Spacecraft and Rockets 4 (1967) 684—685.
39. Skimel', V. N.: Trudy Kazanskogo aviatsionnogo instituta 190 (1970) 70—82.
40. Slezkin, L. N.: Izvestiya Acad. Sci. SSSR Mekhanika i mashinostroenie 4 (1964) 145—147.
41. Smolitsky, Kh. L.: Mekhanika tverdogo tela 2 (1966) 17—22.
42. Stepanenko, N. P.: ibidem 1 (1969) 40—44.
43. —: ibidem 3 (1970) 155—163.
44. Temchenko, M. E.: Doklady, Acad. Sci. SSSR 117 (1957) 50—52.
45. Zhbanov, Yu. K.: Prikladnaya matematika i mekhanika 25 (1961) 933—937.
46. Zhuravlev, V. F.: Mekhanika tverdogo tela 5 (1971) 44—48.
47. Zinenko, V. A.: Izvestiya Acad. Sci. SSSR Mekhanika i mashinostroenie 3 (1964) 79—86.

Foundations of shell theory

W. T. Koiter **J. G. Simmonds**

Delft, The Netherlands Charlottesville, VA, USA

1. Introduction

Shell theory attempts the impossible: to provide a two-dimensional representation of an intrinsically three-dimensional phenomenon. A shell occupies a volume in space; indeed its relative thickness is the crucial parameter governing its response to external loads. Yet any theory of shells, virtually by definition, deals with variables defined only on a reference surface, normally the middle surface. The expectation, based on intuition and experience, is of course that the thinner the shell, the more accurately the actual three-dimensional stress and displacement fields can be inferred from a two-dimensional solution (except possibly near edges or regions of highly concentrated loading). The need of a two-dimensional theory is obvious because, even in the present computer age, such a theory is from the mathematical point of view immensely more tractable than a three-dimensional one.

To establish the foundations of shell theory in the strictest sense, it would seem that it must be shown that, given a boundary value problem in three-dimensional elasticity theory for a "thin-walled" region, there exists a set of two-dimensional shell equations with appropriate boundary conditions which predict the actual stress and displacement fields to within errors of a specified order of magnitude approaching zero with the shell thickness. To give an inkling of the enormous complexity of such a task, we note that it was only in 1970—nearly a century after the appearance of Aron's pioneering paper [1]—that the foundation question was answered affirmatively for the simplest non-trivial case imaginable. We refer here to the point-wise precise error estimates in the analysis by *Ho* and *Knowles* [2] of the linear axisymmetric torsion problem for an elastically isotropic shell of revolution of constant thickness, subject only to static end loads. The success of this analysis hinges on the fact that the exact elasticity formulation of the problem in question is actually two-dimensional, be it of an entirely different type than two-dimensional shell theory, and the associated shell problem is therefore only one-dimensional. While it seems plausible that this work can be extended to cases in which symmetry or separation of variables permits the three-dimensional elasticity problem to be decomposed into a system or sequence of two-dimensional problems, it is clear that more general situations demand new mathematical tools, if not an entirely different approach.

Perhaps the most serious obstacle in deriving a two-dimensional theory of shells is the circumstance that the stress distribution in an edge zone, of depth of the order of the shell thickness h, is nearly always essentially three-dimensional. This difficulty is even aggravated by the fact that we rarely know more about the applied edge stresses than the resulting force and moment per unit edge length on the middle surface. This state of affairs suggests that a solution of shell theory ought to be compared with a *class* of three-dimensional elasticity solutions. An essential part of the foundation problem thus becomes the identification of these "equivalent" classes of elasticity problems and the quantitative proof of Saint-Venant's principle for the difference between two solutions in the same class.

Unfortunately, a complete answer to even this relaxed form of the foundation question seems out of reach at present. For example, in nonlinear problems virtually nothing is known concerning bounds on the errors of shell solutions. The best results to date, obtained by *John* [3, 4, 5], consist of effective *error estimates for the two-dimensional differential equations* in the interior domain in the absence of surface loads, but they say nothing about the equally important question of boundary conditions. In the *linear theory* of plates *Morgenstern* [6, 7] has shown that the errors of plate theory approach zero with the plate thickness. A similar technique has been applied by *Nordgren* [8], *Danielson* [9] and the authors [10, 11, 12] to obtain various *root mean square* (L_2) *solution error estimates,* but the only point-wise error estimates available are those of *Ho* and *Knowles* [2] for the special case of torsion mentioned before. The results of *Sensenig* [13] occupy some intermediate position.

The problem of establishing a firm foundation of shell theory, as we have posed it above, is the problem of showing that two-dimensional shell theory is consistent with three-dimensional elasticity theory, and the criteria employed for assessing the validity of shell theory are *error estimates,* either for the differential equations (John's approach [3, 4, 5]) or for the solutions. Many writers, e.g., *Green* [14], *Gol'denweizer* and *Kolos* [15—24], *Rutten* [25, 26], favour a different approach for the same purpose, viz. the *asymptotic integration* of the three-dimensional equations in terms of the shell thickness *h,* divided by some characteristic length *a,* as a small parameter[1]. A distinct advantage of the asymptotic approach is its entirely systematic character which develops shell theory as a logical consequence of three-dimensional theory. At the same time, however, it is an unfortunate aspect of this systematic character (shared by John's equally systematic analysis) that the analysis is extremely complicated and tedious, much more so than is needed in the applications (cf. in particular [26]). Moreover, the asymptotic method does not enable us so far to assess the accuracy achieved. In spite of these disadvantages the asymptotic approach has contributed enormously to a deepening of our insight into the shell problem.

Finally, we mention a fundamentally different approach to shell theory in which the question of its foundation is almost irrelevant. We refer to the socalled *direct approach* in which it is postulated, ab initio, that the shell is a two-dimensional continuum subject to certain physical principles analogous to those of three-dimensional continuum mechanics. Various kinematic and static quantities are represented by three-dimensional Cartesian vectors, whose components are functions of position on the reference surface only. If a quadratic strain energy per unit area is postulated, the elastic constants (there are more than in conventional shell theory) are to be determined by laboratory tests, just as is necessary in three-dimensional elasticity. The notion of constructing a two-dimensional model of a shell, completely independent of three-dimensional considerations, originated over half a century ago with the *Cosserat* brothers [27]. They characterized the deformation of a shell by a mid-surface displacement vector plus an additional, usually independent "director" vector. This work was ignored for a long time, and it was only followed up by many writers in the past decade. We refer to a recent exhaustive exposition by *Naghdi* [28] in the "Handbuch der Physik", where a complete list of references may be found.

The elegance, economy and logical consistency of the direct approach, in its simplest form, are admittedly appealing. Even if the transverse shearing strains and the normal component of the stress couple vector are initially assumed to be non-zero, the assumption of elastic isotropy and homogeneity and an argument along the lines suggested by *Niordson* [29] lead to the field equations and boundary conditions of classical shell theory. It still seems to us, however, that the lack of a comparison, or even a connection, with three-dimensional theory is so serious a weakness that the direct approach, in spite of its considerable merits as a *model* of shell behaviour, can hardly contribute to a strengthening of the foundations of shell theory.

In this paper we shall review some recent efforts toward reinforcing the foundations of shell theory, most of them achieved in the last decade. Wherever possible we have attempted to simplify and unify or extend known results, and where it seemed appropriate to us we have drawn attention to open problems. Lack of space prevents us from discussing the significant

---

[1] Reference should also be made here to an early important paper by *Goodier* [73], overlooked by most later writers.

developments over a longer period of time and it even compels us to a selection in the review of recent advances. We ask for the reader's indulgence in connection with the personal bias involved. In addition to the references at the end of the paper we refer to some text books [30—37, 84, 85], two review articles [38, 39], and in particular to Love's famous treatise [40], for more historical information.

To keep the essential ideas in focus, we shall restrict our attention to static deformations of elastic shells which in their undeformed state are stress-free and of constant thickness h. We adopt Reissner's point of view [41, 42] that, often, the most convenient path connecting shell theory with the three-dimensional theory of elasticity is an *ascending* one which starts from shell theory as a base point. We are indeed fully confident that the equations of classical (linear or nonlinear) first-approximation shell theory, as formulated, for example, in [43—51], are entirely adequate, and this conjecture will shape our entire approach to establishing the relationship between shell theory and the three-dimensional elasticity theory.

The body of our paper thus starts with a review of the equations of classical nonlinear shell theory. There are many ways leading to Rome—or to Moscow, for that matter—and as we have seen, classical shell theory is no exception to that rule. In the main we shall follow the lines of the discussion in [47], but we shall now incorporate significant improvements due to *Danielson* [49]. Lack of space prevents even a brief discussion of the alternative approach by means of a finite rotation vector [50, 51]. This generalisation of Reissner's highly successful idea in the case of axisymmetric deformations of shells of revolution [52] holds much promise for future developments.

In Section 3 we shall exploit John's results [3] in order to obtain precise error estimates for our (approximate) constitutive equations and for the derivatives of stress resultants and bending strains along the middle surface. This discussion will enable us in Section 4 to derive a comparatively simple form of intrinsic nonlinear shell equations with *rigorous* error estimates, which we venture to call the *canonical form* of these equations. The most striking simplification in comparison with John's refined equations [4, 5] is the absence of the cubic coefficients of the strain energy density, and it is largely due to the fact that in John's lowest order interior equations [3] only the equivalent of the Mainardi-Codazzi equations and the tangential equilibrium equations are actually in need of a refinement. Additional reasons for the simplified form of the equations are the proper choice of the tensors of stress resultants and changes of curvature as dependent variables, due to *Danielson* [49], and the appropriate use of the lowest order equations (with error estimates) to simplify the refined equations, as suggested already by *John* himself [5]. We note also that the term $(1 - \nu)\, AKn_\alpha^\alpha$ in Danielson's Equation [49, (3.15)] is missing in our equations, and the omission of this term can apparently only be justified by an appeal to John's error estimates. Finally, we reduce the equations to their *canonical* form by introducing the modified tensor of changes of curvature $\varrho_{\alpha\beta}$ [43, 44, 47], defined here by (4.13). The linearized version of the equations then takes the form of the "best" linear theory in the sense of Budiansky and Sanders [44].

Whereas John's error estimates for the *equations* of shell theory represent a significant advance in the past decade, it would be even more important to have estimates for the errors of the *solutions* of the approximate equations, as compared with the solutions of the exact three-dimensional equations [39]. Some steps in this direction have been made in the case of the *linear theory*. We shall make a distinction in Section 5 between *intrinsic* error estimates, which provide a rigorous basis for establishing the equivalence of different forms of the linear shell equations [53-55], and the more fundamental *extrinsic* error estimates, which provide bounds on the difference between the three-dimensional stress and displacement fields implied by linear shell theory and those predicted by classical three-dimensional elasticity theory [8—12]. An extension of these error estimates to the case of the nonlinear theory is not immediately obvious. In fact, the possible occurrence of limit points or *snap loads* in the nonlinear case may imply *finite* errors of shell theory in the neighbourhood of such a snap load. In this case it may be more appropriate to compare shell theory and the three-dimensional theory at slightly different load levels. At present, however, we are not aware of any concrete results in this direction. On the other hand, the few available results for *bifurcation buckling* according to three-dimensional theory fully support our confidence in the predictions of shell theory [56—59].

For a closed shell, or for a shell on whose edges the tractions are distributed over the thickness in accordance with the requirements implied by shell theory, the root mean square error of the three-dimensional stress field predicted by linear shell theory is of order $\varepsilon^2 = h^2/L^2 + h/R$, where R is the smallest principal radius of curvature of the midsurface and L is the smallest local wave length of the deformation pattern [9, 11]. In case of a more "irregular" distribution of the edge tractions the resulting root mean square errors as well as the local errors in the interior domain are, in general, much larger, of order $\varepsilon^* = h/L + h/R$, whereas the local errors in the edge zone are even of order unity. The latter fact was already noted by *Friedrichs* [60], but the significant effect of the distribution of the edge loads on the accuracy of shell theory in the interior domain seems to us to have been stressed first by *Gol'denweizer* (e.g. [17, 22]). In practice we hardly ever know anything about the actual distribution of edge loads over the thickness, except in the case of a free edge to be discussed presently. In most cases we shall have to rest content with errors of order ε^*, and the significant effect of the edge load distribution also detracts much of the value of the many attempts to refine shell theory beyond a "first approximation", to be mentioned later.

In the special case of a shell with a *free edge*, discussed in Section 6, it seems again possible to approximate the actual three-dimensional stress and displacement fields to within a root mean square error of order ε^2, simply by imposing suitably *modified* Kirchhoff-type boundary conditions [61]. These modified boundary conditions are obtained by subtracting from the classical strain energy expression for shells a line integral along the free edge line on the middle surface, involving the square of the edge twist, thus accounting for the loss in torsional energy in a zone of depth $0(h)$ near a free edge. The modified boundary conditions may also be visualized by means of a fictitious edge beam with a negative torsional rigidity along the free edge, and they agree in the case of a flat plate with the boundary conditions proposed earlier by *Gol'denweizer* [17]. A rigorous proof of a root mean square accuracy of order ε^2, along the lines of [11], is somewhat tedious and is still under examination.

Even in the case of a free edge it is not equally easy to obtain an equal accuracy of order ε^2 in the entire edge zone. As pointed out by *Green* [14], and in particular in a sequence of meticulous asymptotic analyses by *Gol'denweizer* and *Kolos* [15—24], this would require the (complicated) solution of a plane strain problem for a semi-infinite strip of width h equal to the shell thickness, in addition to the (simple) solution of the torsion problem for this strip. Fortunately, the most important circumferential stresses in the edge zone, namely those at the outer and inner faces of the shell at the edge itself, may be calculated without solving the plane strain problem, again with an error of order ε^2. In the bending case of a flat plate with a cylindrical hole this analysis has been carried out by *van der Heijden* [62], and his results are confirmed by the earlier exact three-dimensional analysis of this problem by *Alblas* [63].

The formidable problems of a consistent refinement of shell theory beyond its first approximation, which may be needed for not-so-thin shells, are briefly discussed in Section 7. The paper ends with mention of some open problems in our topic which in the authors' opinion merit further research.

2. Classical nonlinear shell theory

Let $\boldsymbol{r}(x^\alpha)$, $\alpha = 1, 2$, denote the radius vector from a fixed origin in space to a generic point on the middle surface of the undeformed shell as a vector-valued function of the pair of Gaussian surface coordinates. The tangential base vectors are $\boldsymbol{a}_\alpha = \boldsymbol{r}_{,\alpha}$, where the comma preceding the subscript α denotes partial differentiation with respect to the coordinate x^α. The covariant and contravariant metric tensors of the middle surface are defined by $a_{\alpha\beta} = \boldsymbol{a}_\alpha \cdot \boldsymbol{a}_\beta$ and by $a_{\alpha\beta}a^{\beta\lambda} = \delta^\lambda_\alpha$. These tensors are employed in lowering and raising indices of all surface tensors, and the reciprocal base vectors are defined by $\boldsymbol{a}^\alpha = a^{\alpha\beta}\boldsymbol{a}_\beta$. The determinant of the covariant metric tensor is denoted by a, the altenating tensor by $\varepsilon_{\alpha\beta}$ or $\varepsilon^{\alpha\beta}$. The normal to the middle surface is defined by the unit vector $\boldsymbol{n} = \frac{1}{2}\varepsilon^{\alpha\beta}\boldsymbol{a}_\alpha \times \boldsymbol{a}_\beta$. The Christoffel symbols of the first and second kind for the undeformed middle surface are given by $\Gamma_{\lambda\alpha\beta} = \frac{1}{2}(a_{\lambda\alpha,\beta} + a_{\lambda\beta,\alpha} - a_{\alpha\beta,\lambda})$ and $\Gamma^\varkappa_{\alpha\beta} = a^{\varkappa\lambda}\Gamma_{\lambda\alpha\beta}$. Covariant surface differentiation with respect to a coordinate x^α is denoted by an additional subscript α preceded by a vertical stroke. The second fundamental tensor of the undeformed middle surface is specified by $b_{\alpha\beta} = \boldsymbol{n}\cdot\boldsymbol{r}_{,\alpha\beta}$. All derivatives in the analysis are assumed to be continuous. The equations of *Gauss* and *Mainardi-Codazzi* are written in the form [47, Eqs. (2.20), (2.22)]

$$\varepsilon^{\alpha\beta}\varepsilon^{\lambda\mu}[\Gamma_{\alpha\beta\mu,\lambda} + \Gamma^\varkappa_{\alpha\mu}\Gamma_{\varkappa\beta\lambda} + b_{\alpha\mu}b_{\beta\lambda}] = 0\,, \tag{2.1}$$

$$\varepsilon^{\alpha\beta}\varepsilon^{\lambda\mu}b_{\beta\lambda|\mu} = \varepsilon^{\alpha\beta}\varepsilon^{\lambda\mu}[b_{\beta\lambda,\mu} - \Gamma^\varkappa_{\beta\mu}b_{\varkappa\lambda}] = 0\,. \tag{2.2}$$

A point in undeformed shell space is identified by its distance z to the middle surface and by the surface coordinates of its projection on the middle surface. The shell faces $z = \pm h/2$, where h is the constant shell thickness, are surfaces parallel to the middle surface. The coordinate z is orthonormal to the surface coordinates, and the components $g_{\alpha\beta}$ of the spatial metric tensor g_{ij} (with determinant g) reduce to $a_{\alpha\beta}$ at the middle surface. The edge of the undeformed shell is assumed to be a ruled surface formed by the normals to the middle surface along an edge curve on this surface. Let $\boldsymbol{\nu}$ denote the unit vector in the tangent plane, normal to the edge curve and positive outwards. The positive sense on the edge curve is defined by the tangential unit vector $\boldsymbol{t} = \boldsymbol{n}\times\boldsymbol{\nu}$.

In a deformation of the shell the original (material) middle surface is described by a vector-valued function $'\boldsymbol{r}(x^\alpha) = \boldsymbol{r}(x^\alpha) + \boldsymbol{u}(x^\alpha)$, where $\boldsymbol{u}$ is the displacement vector. This new configuration of the original middle surface will be called the deformed middle surface, but it should be borne in mind that it does not in general coincide with the actual "middle" surface of the deformed shell. The deformed base vectors and metric tensors are defined by $'\boldsymbol{a}_\alpha = {'\boldsymbol{r}}_{,\alpha}$, $'a_{\alpha\beta} = {'\boldsymbol{a}_\alpha}\cdot{'\boldsymbol{a}_\beta}$ and $'a_{\alpha\beta}{''a^{\beta\lambda}} = \delta^\lambda_\alpha$, where the double prime at the contravariant tensor indicates that indices have *not* been raised in accordance with our previous convention. The reciprocal base vectors are defined by $''\boldsymbol{a}^\alpha = {''a^{\alpha\beta}}\,{'\boldsymbol{a}_\beta}$. The determinant of the covariant deformed metric tensor is denoted by $'a$, the alternating tensor by $'\varepsilon_{\alpha\beta} = \sqrt{'a/a}\,\varepsilon_{\alpha\beta}$, $''\varepsilon^{\alpha\beta} = \sqrt{a/'a}\,\varepsilon^{\alpha\beta}$, where the double prime indicates again that the indices have not been raised by means of $a^{\alpha\beta}$. The unit normal vector to the deformed middle surface is given by $'\boldsymbol{n} = \frac{1}{2}{'\varepsilon^{\alpha\beta}}\boldsymbol{a}_\alpha{'}\times{'\boldsymbol{a}_\beta}$, and the Christoffel symbols are $'\Gamma_{\lambda\alpha\beta} = \frac{1}{2}({'a_{\lambda\alpha,\beta}} + {'a_{\lambda\beta,\alpha}} - {'a_{\alpha\beta,\lambda}})$ and $''\Gamma^\varkappa_{\alpha\beta} = {''a^{\varkappa\lambda}}\,{'\Gamma_{\lambda\alpha\beta}}$. Covariant surface differentiation with respect to the deformed metric is denoted by a semi-colon (;) preceding the subscript of the coordinate with respect to which the differentiation is performed. The second fundamental tensor of the deformed middle surface is specified by $'b_{\alpha\beta} = {'\boldsymbol{n}}\cdot{'\boldsymbol{r}_{,\alpha\beta}}$. The equations of *Gauss* and *Mainardi-Codazzi* are the counterpart of (2.1), (2.2) with the primed quantities (double primes wherever superscripts occur).

The deformation of the middle surface, considered as a surface in space, is described completely by the pair of symmetric tensors $\gamma_{\alpha\beta} = \frac{1}{2}({'a_{\alpha\beta}} - a_{\alpha\beta})$ and $'\varrho_{\alpha\beta} = {'b_{\alpha\beta}} - b_{\alpha\beta}$, which are called the middle surface strain tensor and the tensor of changes of curvature[1]. Complete expressions for these strain measures in terms of displacement derivatives are given in [47] but will not be needed in the sequel. We do need, however, their equations of compatibility, resulting from the fact that the equations of *Gauss* and *Mainardi-Codazzi* must hold in both the undeformed and the deformed configuration of the middle surface. We introduce the abbreviation

$$\gamma_{\lambda\alpha\beta} = \gamma_{\lambda\alpha|\beta} + \gamma_{\lambda\beta|\alpha} - \gamma_{\alpha\beta|\lambda}\,, \tag{2.3}$$

[1] We have retained here the primed notation of [47] for the latter tensor in order to keep the bare symbol available for the "best" modified tensor of changes of curvature for the linear theory [43, 44].

and we note the relations ([47], Section 5)

$$ {}'\Gamma_{\lambda\alpha\beta} = {}'a_{\lambda\varkappa}\Gamma^{\varkappa}_{\alpha\beta} + \gamma_{\lambda\alpha\beta}, \qquad {}''\Gamma^{\varkappa}_{\alpha\beta} = \Gamma^{\varkappa}_{\alpha\beta} + {}''a^{\varkappa\lambda}\gamma_{\lambda\alpha\beta}. \tag{2.4} $$

The *equations of compatibility*, corresponding respectively with Gauss's equation and the Mainardi-Codazzi equations, may now be written in the form

$$ K\gamma^{\alpha}_{\alpha} + \varepsilon^{\alpha\beta}\varepsilon^{\lambda\mu}[\gamma_{\alpha\mu|\beta\lambda} + b_{\alpha\mu}{}'\varrho_{\beta\lambda} + \frac{1}{2}\,{}'\varrho_{\alpha\mu}{}'\varrho_{\beta\lambda} + \frac{1}{2}\,{}''a^{\varkappa\nu}\gamma_{\varkappa\alpha\mu}\gamma_{\nu\beta\lambda}] = 0\,, \tag{2.5} $$

$$ \varepsilon^{\alpha\beta}\varepsilon^{\lambda\mu}[{}'\varrho_{\beta\lambda|\mu} - {}''a^{\varkappa\nu}(b_{\varkappa\lambda} + {}'\varrho_{\varkappa\lambda})\,\gamma_{\nu\beta\mu}] = 0\,, \tag{2.6} $$

where $K = \frac{1}{2}\,\varepsilon^{\alpha\lambda}\varepsilon^{\beta\mu}b_{\alpha\beta}b_{\lambda\mu}$ is the Gaussian curvature of the undeformed middle surface.

The *classical equations of equilibrium* for thin shells, as formulated by *Love* [40], are *six* in number and involve *ten* unknowns, viz. four tangential stress resultants, two transverse shear forces and four stress couples. These six equations in ten unknowns may be reduced to a system of *three* equations involving no more than *six* unknowns, viz. a symmetric tensor of tangential stress resultants and a symmetric tensor of stress couples. This remarkable reduction is due to *Lur'e* [64] and *Gol'denweizer* [65]. Its physical explanation is to be found in the *principle of virtual work* [47, 66, 67], and for this reason we select from the various methods available for establishing the desired equations and boundary conditions this basic principle.

A virtual displacement of the deformed middle surface is described by a continuously differentiable infinitesimal (additional) displacement field, which may be decomposed with respect to the deformed reciprocal base vectors and the deformed unit normal vector

$$ \delta\boldsymbol{u}(x^{\alpha}) = \delta{}'u_{\alpha}{}''\boldsymbol{a}^{\alpha} + \delta{}'w{}'\boldsymbol{n}\,. \tag{2.7} $$

The associated virtual changes in the strain measures of the middle surface are given by [47, Eqs. (3.42), (3.43)]

$$ \delta\gamma_{\alpha\beta} = \frac{1}{2}\,(\delta{}'u_{\alpha;\beta} + \delta{}'u_{\beta;\alpha}) - {}'b_{\alpha\beta}\,\delta{}'w\,, \tag{2.8} $$

$$ \delta{}'\varrho_{\alpha\beta} = \delta{}'w_{;\alpha\beta} - {}'c_{\alpha\beta}\,\delta{}'w + {}''b^{\varkappa}_{\alpha}\,\delta{}'u_{\varkappa;\beta} + {}''b^{\varkappa}_{\beta}\,\delta{}'u_{\varkappa;\alpha} + {}''b^{\varkappa}_{\alpha;\beta}\,\delta{}'u_{\varkappa}\,, \tag{2.9} $$

where ${}''b^{\varkappa}_{\alpha} = {}''a^{\varkappa\beta}{}'b_{\alpha\varkappa}$ and ${}'c_{\alpha\beta} = {}''b^{\varkappa}_{\alpha}{}'b_{\varkappa\beta}$.

We restrict our attention to *spatial virtual displacement fields which obey the Kirchhoff-Love assumptions with respect to the deformed middle surface*, and we emphasize that this involves no approximation but only a restriction of our considerations. The associated virtual strains at any point in the deformed shell space x^{α}, ${}'z$, where x^{α} are the surface coordinates of its projection on the deformed middle surface and ${}'z$ is the distance to this deformed middle surface, are then homogeneous linear expressions in $\delta\gamma_{\alpha\beta}$ and $\delta{}'\varrho_{\alpha\beta}$ at this point x^{α} of the middle surface. It follows that the internal virtual work $\delta{}'V$ per unit area of the deformed middle surface is a linear form in the virtual strain measures whose coefficients are necessarily symmetric tensors of stress resultants and stress couples

$$ \delta{}'V = {}'n^{\alpha\beta}\,\delta\gamma_{\alpha\beta} + {}'m^{\alpha\beta}\,\delta{}'\varrho_{\alpha\beta}\,. \tag{2.10} $$

The external loads on the shell faces and the body forces are reduced to statically equivalent loads acting in the deformed middle surface, in accordance with our restricted Kirchhoff-Love type spatial virtual displacement field. This reduction is discussed in detail by *Naghdi* [38]. For the sake of brevity we shall assume that the reduction of loads does not introduce surface couples on the deformed middle surface. The reduced loads per unit area of the deformed middle surface may be decomposed with respect to the deformed base vectors and unit normal vector

$$ {}'\boldsymbol{p} = {}'p^{\alpha}{}'\boldsymbol{a}_{\alpha} + {}'p{}'\boldsymbol{n}\,. \tag{2.11} $$

In connection with our restricted Kirchhoff-Love type virtual displacement fields we replace, for the time being, the shell edge by a ruled surface in the deformed configuration, formed by the normals to the deformed middle surface along the (deformed) edge curve on this surface. This involves no approximation in the linear theory, where no distinction need be made between the undeformed and deformed geometries in deriving the equations of equilibrium and the dynamic boundary conditions, and the justification of this approximation in the nonlinear first-approximation theory will be discussed in Section 3. The loads on the shell edge are now likewise reduced to statically equivalent line loads along the (deformed) edge curve on the deformed middle surface, a force ${}'\boldsymbol{N}$ and a couple ${}'\boldsymbol{M}$, both per unit arc length in the deformed state, where the couple vector lies in the (deformed) tangent plane. Let ${}'\boldsymbol{\nu} = {}'\nu_\alpha {}''\boldsymbol{a}^\alpha$ denote the unit vector in the deformed tangent plane, normal to the edge curve and positive outwards. The positive sense on the edge curved is defined by the (deformed) tangential unit vector ${}'\boldsymbol{t} = {}'t_\alpha {}''\boldsymbol{a}^\alpha = {}'\boldsymbol{n} \times {}'\boldsymbol{\nu}$. We may decompose the edge loads in several ways, for example

$$ {}'\boldsymbol{N} = {}'N_{(\nu)} {}'\boldsymbol{\nu} + {}'N_{(t)} {}'\boldsymbol{t} + {}'Q {}'\boldsymbol{n} = {}'N^{\alpha} {}'\boldsymbol{a}_\alpha + {}'Q {}'\boldsymbol{n}, \tag{2.12} $$

$$ {}'\boldsymbol{M} = {}'M_{(\nu)} {}'\boldsymbol{\nu} + {}'M_{(t)} {}'\boldsymbol{t} = {}''\varepsilon_{\alpha\beta} {}'M^{\beta} {}''\boldsymbol{a}^\alpha . \tag{2.13} $$

The principle of virtual work now yields by standard methods the equations of equilibrium ([47], Eqs. (6.3), (6.4))

$$ ({}'n^{\beta\alpha} + {}''b^{\alpha}_{\varkappa} {}'m^{\beta\varkappa})_{;\beta} + {}''b^{\alpha}_{\varkappa} {}'m^{\beta\varkappa}{}_{;\beta} + {}'p^\alpha = 0, \tag{2.14} $$

$$ {}'m^{\alpha\beta}{}_{;\alpha\beta} - {}'c_{\alpha\beta} {}'m^{\alpha\beta} - {}'b_{\alpha\beta} {}'n^{\alpha\beta} - {}'p = 0, \tag{2.15} $$

as well as the dynamic boundary conditions ([47], Eqs. (6.10)—(6.12))

$$ ({}'n^{\beta\alpha} + 2 {}''b^{\alpha}_{\varkappa} {}'m^{\beta\varkappa})\, {}'\nu_\beta = {}'N^\alpha + {}''b^{\alpha}_{\varkappa} {}'M^{\varkappa}, \tag{2.16} $$

$$ {}'m^{\alpha\beta}{}_{;\alpha} {}'\nu_\beta + ({}'m^{\alpha\beta} {}'\nu_\alpha {}'t_\beta)_{,'s} = {}'M_{(\nu),'s} - {}'Q, \tag{2.17} $$

$$ {}'m^{\alpha\beta} {}'\nu_\alpha {}'\nu_\beta = - {}'M_{(t)}, \tag{2.18} $$

where the subscript ${}'s$ preceded by a comma denotes partial differentiation with respect to arc length along the deformed edge curve. We emphasize that Eqs. (2.14) and (2.15) are both necessary and sufficient to ensure the equilibrium of any shell element of finite (deformed) shell thickness ${}'h(x^\alpha)$, bounded by the shell faces and by the ruled surface formed by the normals to the deformed middle surface through any finite or infinitesimal smooth simple closed curve on his surface.

Danielson and one of the authors have proposed the introduction of modified symmetric tensors of stress resultants and stress couples [49, 86]

$$ n^{\alpha\beta} = \sqrt{{}'a/a}\,[{}'n^{\alpha\beta} + \frac{1}{2}({}''b^{\alpha}_{\varkappa} {}'m^{\varkappa\beta} + {}''b^{\beta}_{\varkappa} {}'m^{\varkappa\alpha})], \qquad m^{\alpha\beta} = \sqrt{{}'a/a}\, {}'m^{\alpha\beta}, \tag{2.19} $$

with the purpose of simplifying the equations after introduction of the (approximate) stress-strain relations. We note that their tensor of stress resultants agrees in the linear theory with the tensor employed in [43]. Employing the identity, holding for any differentiable symmetric tensor $T^{\alpha\beta}$

$$ (\sqrt{a/{}'a}\,T^{\alpha\beta})_{;\beta} = \sqrt{a/{}'a}\,[T^{\alpha\beta}|_\beta + {}''a^{\alpha\varkappa}(2\gamma_{\varkappa\lambda|\beta} - \gamma_{\lambda\beta|\varkappa})\, T^{\lambda\beta}], \tag{2.20} $$

one first obtains from (2.14) and (2.15) the modified exact equations of equilibrium [49, Eqs. (2.12), (2.13)]

$$ \begin{aligned} n^{\beta\alpha}|_\beta &+ {}''a^{\alpha\varkappa}(2\gamma_{\varkappa\lambda|\beta} - \gamma_{\lambda\beta|\varkappa})\, n^{\lambda\beta} + \frac{1}{2}({}''b^{\alpha}_{\varkappa} m^{\varkappa\beta} - {}''b^{\beta}_{\varkappa} m^{\varkappa\alpha})_{|\beta} \\ &+ {}''b^{\alpha}_{\varkappa} m^{\varkappa\beta}|_\beta + {}''b^{\alpha\varkappa}(2\gamma_{\varkappa\lambda|\beta} - \gamma_{\lambda\beta|\varkappa})\, m^{\lambda\beta} + \sqrt{{}'a/a}\,{}'p^\alpha = 0, \end{aligned} \tag{2.21} $$

$$ m^{\alpha\beta}|_{\alpha\beta} + [{}''a^{\alpha\varkappa}(2\gamma_{\varkappa\lambda|\beta} - \gamma_{\lambda\beta|\varkappa})\, m^{\lambda\beta}]_{|\alpha} - {}'b_{\alpha\beta} n^{\alpha\beta} - \sqrt{{}'a/a}\,{}'p = 0. \tag{2.22} $$

An obvious and distinct advantage of the exact equations in this form is already that all covariant differentiations, like those in the compatibility conditions (2.5), (2.6), are now performed with reference to the undeformed metric which is known a priori. We refer to ([49], Eqs. (4.6)—(4.8)) for the similar modification of the exact dynamic boundary conditions. We postpone the further (approximate) simplification of the equations and boundary conditions to Section 4 where this approximation may be argued more securely on the basis of the results achieved by *John* [3].

We emphasize once again that the equations of compatibility (2.5), (2.6) and the equations of equilibrium, (2.14), (2.15) or (2.21), (2.22), are all *fully exact*. A minor approximation has only been made in the dynamic boundary conditions (2.16)—(2.18) in the case of the nonlinear theory, and this approximation will be justified in Section 3. The foregoing theory, however, involves six equations in twelve unknowns, the symmetric tensors of middle surface strains, changes of curvature, stress resultants and stress couples. It has to be supplemented by *constitutive equations* connecting the stress resultants and stress couples on the one hand to the strain measures on the other hand. Such constitutive equations necessarily always involve certain approximations, both in the linear and nonlinear theories (unless in a direct theory all relations with three-dimensional theory are ignored). As we have mentioned before, this approximate character of all conceivable two-dimensional shell theories is an inevitable consequence of our desire to describe the behaviour of an actually always three-dimensional structure in terms of two-dimensional quantities.

We shall restrict our present discussion of the constitutive equations to the simplest possible case, based on the following assumptions:

(a) the material of the shell is elastic, homogeneous and isotropic;

(b) the strains are small everywhere in the shell;

(c) the state of stress is approximately plane and parallel to the middle surface.

We note from John's analysis in [3] that the last assumption is actually a mathematical consequence of assumptions (a) and (b) in the absence of surface loads and at a sufficient distance from an edge. In the presence of slowly varying surface loads assumption (c) is easily made plausible by a physical argument, and the solution error estimates [8—12], discussed in Section 5, confirm this argument. In the neighbourhood of a shell edge, however, the last assumption, in general, fails to hold true in a zone of depth $0(h)$ near the edge, and the required modification of the theory will be discussed in Section 6 for the case of a free edge.

Let $E^{\alpha\beta\lambda\mu}$ denote the tensor of elastic moduli for plane stress in the middle surface

$$E^{\alpha\beta\lambda\mu} = G\left[a^{\alpha\lambda}a^{\beta\mu} + a^{\alpha\mu}a^{\beta\lambda} + \frac{2\nu}{1-\nu}a^{\alpha\beta}a^{\lambda\mu}\right], \tag{2.23}$$

where G is the shear modulus and ν is Poisson's ratio. Let V denote the (approximate) expression for the strain energy per unit area of the undeformed middle surface, given by Love's formula [40, 43, 47, 29]

$$V = \frac{1}{2}hE^{\alpha\beta\lambda\mu}\left[\gamma_{\alpha\beta}\gamma_{\lambda\mu} + \frac{1}{12}h^2\,'\varrho_{\alpha\beta}\,'\varrho_{\lambda\mu}\right]. \tag{2.24}$$

The associated constitutive equations are by (2.10)

$$'n^{\alpha\beta} = \frac{1}{2}\sqrt{a/'a}\left(\frac{\partial V}{\partial\gamma_{\alpha\beta}} + \frac{\partial V}{\partial\gamma_{\beta\alpha}}\right), \quad 'm^{\alpha\beta} = \frac{1}{2}\sqrt{a/'a}\left(\frac{\partial V}{\partial'\varrho_{\alpha\beta}} + \frac{\partial V}{\partial'\varrho_{\beta\alpha}}\right). \tag{2.25}$$

The relative error in (2.24) has been estimated (qualitatively) in [43] to be of order of magnitude η, h/R or h^2/L^2, whichever of these numbers may be critical, where η is the maximum absolute principal strain in the shell, R is the minimum principal radius of curvature of the middle surface, and L is the minimum "wave length" of the deformation pattern on the middle surface. It follows that the associated constitutive equations, (2.25), involve similar errors, and we may therefore replace the square root by unity without any essential additional error. Moreover, we may also take the modified stress resultants and stress couples (2.19) to be specified by the same expressions in terms of the strain measures, viz.

$$n^{\alpha\beta} = \frac{Eh}{1-\nu^2}\left[(1-\nu)\,\gamma^{\alpha\beta} + \nu a^{\alpha\beta}\gamma^{\varkappa}_{\varkappa}\right], \quad m^{\alpha\beta} = \frac{Eh^3}{12(1-\nu^2)}\left[(1-\nu)\,'\varrho^{\alpha\beta} + \nu a^{\alpha\beta}\,'\varrho^{\varkappa}_{\varkappa}\right], \tag{2.26}$$

where $E = 2(1+\nu)\,G$ denotes Young's modulus. The inversion of these (approximate) constitutive equations reads

$$\gamma_{\alpha\beta} = \frac{1}{Eh}\left[(1+\nu)\,n_{\alpha\beta} - \nu a_{\alpha\beta}n^{\varkappa}_{\varkappa}\right], \quad '\varrho_{\alpha\beta} = \frac{12}{Eh^3}\left[(1+\nu)\,m_{\alpha\beta} - \nu a_{\alpha\beta}m^{\varkappa}_{\varkappa}\right]. \tag{2.27}$$

The (approximate) stress-strain relations (2.26) or (2.27) complement the exact equations of compatibility (2.5), (2.6) and the equally exact equations of equilibrium (2.21), (2.22) to a *complete eight-order set of twelve equations in twelve unknowns.* From a mathematical point of view it would seem immaterial whether the constitutive Eqs. (2.26) or (2.27) are employed in order to eliminate either the stress resultants and stress couples or the strain measures. From a physical standpoint, however, we shall explain in Section 3 our definite preference for retaining the tensor of stress resultants $n^{\alpha\beta}$ and the tensor of changes of curvature $'\varrho_{\alpha\beta}$ as our basic dependent variables, eliminating the stress couples and the middle surface strains by means of the second Eq. (2.26) and the first Eq. (2.27). We also postpone a discussion of the considerable simplifications which may be applied to the basic exact Eqs. (2.5), (2.6), (2.21) and (2.22) in view of the approximate nature of the stress-strain relations.

3. Reappraisal of John's results

[3, 4, 5]

John's greatest contribution to shell theory, in our opinion, was to derive rigorous and concrete pointwise estimates for all stresses and all their derivatives in the interior domain and in the absence of surface loads [3]. He accomplished this by first using the equilibrium and compatibility conditions of three-dimensional theory for a nonlinear elastic isotropic solid to derive inequalities for the L_2-norms of the derivatives, and then invoking Sobolev's lemma to infer pointwise inequalities. It will be convenient here and in the sequel to assume, without any loss in generality, that the components of the metric tensor a_{11}, a_{22} and its determinant are all of order unity. The "generalized" minimum principal radius of curvature R of the middle surface of the undeformed shell is defined in John's analysis by estimates on the order of magnitude of the second fundamental tensor $b_{\alpha\beta}$ of the undeformed middle surface and all its derivatives[1]

$$b_{\alpha\beta} = 0(R^{-1}), \quad b_{\alpha\beta|\varkappa_1\ldots\varkappa_n} = 0(R^{-1-n}). \tag{3.1}$$

Further important geometric parameters are the (constant) shell thickness h and the distance d to the shell edge. The magnitude of the strains in the shell is characterized by the maximum absolute principal extension η. Basic in John's analysis are then the dimensionless parameter

$$\theta = \max\left(\frac{h}{d}, \sqrt{\frac{h}{R}}, \sqrt{\eta}\right), \tag{3.2}$$

and the length parameter

$$\lambda = h/\theta = \min\left(d, \sqrt{hR}, h/\sqrt{\eta}\right). \tag{3.3}$$

In his three-dimensional analysis John employs the spatial stress tensor t^{ij}, defined by [3]

$$t^{ij} = \sqrt{g/'g}\,\frac{\partial W}{\partial \gamma_{ij}}, \tag{3.4}$$

where W is the strain energy density function of the isotropic elastic solid, written symmetrically with respect to γ_{ij} and γ_{ji}, and $'g$ is the determinant of the spatial metric tensor in the deformed body. With the aid of his pointwise estimates, *John* was able to show that

$$t_{\alpha\beta}(x^\varkappa, z) = \tau_{\alpha\beta}(x^\varkappa) - z\sigma_{\alpha\beta}(x^\varkappa) + 0(E\eta\theta^2), \tag{3.5}$$

$$t_{\alpha3}(x^\varkappa, z) = 0(E\eta\theta), \qquad t_{33}(x^\varkappa, z) = 0(E\eta\theta^2), \tag{3.6}$$

$$\gamma_{\alpha\beta} = \frac{1}{E}\left[(1+\nu)\,\tau_{\alpha\beta} - \nu a_{\alpha\beta}\tau^\varkappa_\varkappa\right] + 0(\eta\theta^2), \tag{3.7}$$

$$'\varrho_{\alpha\beta} = \frac{1}{E}\left[(1+\nu)\,\sigma_{\alpha\beta} - \nu a_{\alpha\beta}\sigma^\varkappa_\varkappa\right] + 0(\eta\theta^2/h), \tag{3.8}$$

$$t_{ij|\varkappa_1\ldots\varkappa_n}(x^\alpha, z) = 0(E\eta\lambda^{-n}), \tag{3.9}$$

where $\gamma_{\alpha\beta}$ and $'\varrho_{\alpha\beta}$ are again the middle surface strain tensor and the tensor of changes of curvature $('b_{\alpha\beta} - b_{\alpha\beta})$. In his first paper [3] *John* also derived his "lowest order interior shell

[1] From a physical point of view it would be worthwhile to try and relax the required estimate for the derivatives to $0(R^{-1-n/2}\,h^{-n/2})$ or $0(R^{-1}\lambda^{-n})$, where λ is defined by (3.3).

equations" as a consequence of his various estimates

$$-\frac{1}{E}\tau^{\alpha}_{\alpha}|^{\beta}_{\beta} + \varepsilon^{\alpha\beta}\varepsilon^{\lambda\mu}\left(b_{\alpha\mu}{}'\varrho_{\beta\lambda} + \frac{1}{2}{}'\varrho_{\alpha\mu}{}'\varrho_{\beta\lambda}\right) = 0(\eta\theta^2/\lambda^2), \tag{3.10}$$

$$\varepsilon^{\alpha\beta}\varepsilon^{\lambda\mu}{}'\varrho_{\beta\lambda|\mu} = 0(\eta\theta^2/h\lambda), \tag{3.11}$$

$$\frac{Eh^2}{12(1-\nu^2)}{}'\varrho^{\alpha}_{\alpha}|^{\beta}_{\beta} - (b_{\alpha\beta} + {}'\varrho_{\alpha\beta})\,\tau^{\alpha\beta} = 0(Eh\eta\theta^2/\lambda^2), \tag{3.12}$$

$$\tau^{\beta\alpha}|_{\beta} = 0(E\eta\theta^2/\lambda), \tag{3.13}$$

written as far as possible in our notation.

It was a considerable disappointment to note [47] that John's lowest order interior shell equations are evidently inadequate to cover all possible situations since they fail already to describe, for example, the simple cases of nearly inextensional bending of the linear theory. In an attempt to eliminate these shortcomings, *John* presented a refined analysis in [4, 5][1]. If his first offspring, Eqs. (3.10)—(3.13), was a bit on the lean side, his second one was depressingly ponderous.

Complexity aside, the most disturbing feature of John's refined interior shell equations, which he expresses in terms of $\gamma_{\alpha\beta}$ and $'\varrho_{\alpha\beta}$, is that the three coefficients of the cubic terms in the strain energy density function W appear sprinkled throughout. Physically, this is quite unsatisfactory, if not unacceptable. Many engineering materials such as steel or aluminium have very low yield strains, less than 0.005, making an experimental determination of the cubic elastic coefficients virtually impossible. Moreover, the variation of the constants of linear elasticity from one specimen to another can be expected to mask completely any contributions coming from higher order terms in the strain energy density. Fortunately, all can be set aright by a judicious combination of John's results in [3] and the exact equations of classical nonlinear shell theory (2.5), (2.6), (2.21), (2.22). A crucial point in this combination is the proper selection of the dependent variables in the problem, and it is also essential to restrict the refinement to those equations which are actually inadequate in John's lowest order theory.

Perhaps the most compelling reason for the elimination of the middle surface strain tensor $\gamma_{\alpha\beta}$ is to be found in the linear membrane theory of shells in the form of the simplified equations of equilibrium

$$n^{\beta\alpha}|_{\beta} + p^{\alpha} = 0, \quad b_{\alpha\beta}n^{\alpha\beta} + p = 0, \tag{3.14}$$

where the loads are now referred to the undeformed middle surface and its base vectors and unit normal vector. For shells of positive Gaussian curvature, subject to smooth surface loads and restrained at the edges so as to prevent (nearly) inextensional deformations, the linear membrane equations often yield an accurate description of the actual stresses away from narrow edge layers. The general solution of the three Eqs. (3.14) in the three unknowns of the symmetric tensor $n^{\alpha\beta}$ is *independent of the material properties*, and this feature is of course lost, if the stress resultants are eliminated in favour of the middle surface strains.

Nearly equally powerful is the argument at the other end of the scale, that the equations of the inextensional bending theory likewise ought to fall out as a simple special case of the general theory. When this theory applies, the stress resultants may be obtained, in terms of the transverse shear forces, again from the equations of equilibrium. It is obviously unreasonable to attempt their calculation from the (small) middle surface strains. In fact, in many cases of (nearly) inextensional bending it is convenient to introduce zero middle surface strains as a constraint, and the only means available to determine the stress resultants (as internal "reactions") is then the consideration of equilibrium. A similar situation is found in the theory of curved beams, where the axial force is always obtained from equilibrium considerations, even if extension of the center line are permitted.

Inspection of the equations of compatibility (2.5), (2.6) in the case of nearly inextensional bending shows immediately that it would be equally unwise to eliminate the tensor of changes of curvature $'\varrho_{\alpha\beta}$ in favour of the tensor of stress couples. Our argument to retain the pair $n^{\alpha\beta}$

[1] The equations in [4] contain some clerical errors which have been corrected in [5]. Cf. also [68].

and $'\varrho_{\alpha\beta}$ as the proper dependent variables is also supported by the well-known static-geometric analogy of linear shell theory [64, 65], which holds for a suitably defined modified tensor of changes of curvature [38, 44].

We shall now examine the consequence for the accuracy of our approximate stress-strain relations (2.26), (2.27) which may be drawn from John's analysis in [3]. It should be remembered here that the original stress resultants $'n^{\alpha\beta}$ and stress couples $'m^{\alpha\beta}$ have been defined in (2.10) by means of virtual work in spatial virtual displacement fields which obey the Kirchhoff-Love assumptions with respect to the deformed middle surface. A point in deformed shell space may be identified by the surface coordinates of its projection on the deformed middle surface and by its distance $'z$ to this surface. The virtual strains in a surface $'z = $ const. parallel to the deformed middle surface are in this new coordinate system (cf. [10], Section 4)

$$\delta'\gamma_{\alpha\beta}('z) = \delta\gamma_{\alpha\beta} - 'z\,\delta'\varrho_{\alpha\beta} + \frac{1}{2}\,'z^2({}''b^{\varkappa}_{\alpha}\,\delta'\varrho_{\varkappa\beta} + {}''b^{\varkappa}_{\beta}\,\delta'\varrho_{\varkappa\alpha} - 2\,{}''b^{\varkappa}_{\alpha}\,{}''b^{\lambda}_{\beta}\,\delta\gamma_{\varkappa\lambda}). \tag{3.15}$$

Let $'t^{ij}('z)$ denote the stress tensor in deformed shell space, referred to our new coordinate system, in a point at a distance $'z$ to the deformed middle surface. The internal virtual work per unit deformed volume is given by $'t^{ij}('z)\,\delta'\gamma_{ij}('z)$. Integrating this expression over the deformed shell thickness, say from $'z = -c_1$ to $'z = c_2$, we obtain for the internal virtual work per unit area of the deformed middle surface

$$\int_{-c_1}^{c_2} {}'t^{\alpha\beta}('z)\,\delta'\gamma_{\alpha\beta}('z)\sqrt{{}''g/{}'a}\,d'z, \tag{3.16}$$

where $''g$ is the determinant of the spatial metric tensor in the new coordinate system in a point at a distance $'z$ to the deformed middle surface. Equating the result of (3.14), (3.15) to (2.10), we obtain (exact) expressions for the original stress resultants and stress couples in terms of the stress distribution over the shell thickness

$$'n^{\alpha\beta} = \int_{-c_1}^{c_2} [{}'t^{\alpha\beta}('z) - {}'z^2\,{}''b^{\alpha}_{\lambda}\,{}''b^{\beta}_{\mu}\,{}'t^{\lambda\mu}('z)]\sqrt{{}''g/{}'a}\,d'z, \tag{3.17}$$

$$'m^{\alpha\beta} = \int_{-c_1}^{c_2} [-{}'z\,{}'t^{\alpha\beta}('z) + \frac{1}{2}\,{}'z^2\{{}''b^{\alpha}_{\varkappa}\,{}'t^{\varkappa\beta}('z) + {}''b^{\beta}_{\varkappa}\,{}'t^{\varkappa\alpha}('z)\}]\sqrt{{}''g/{}'a}\,d'z. \tag{3.18}$$

It should be borne in mind, and our present notation should also make this abundantly clear, that the stress tensor $'t^{ij}$ is not identical to John's stress tensor t^{ij}. The material points $x^{\varkappa}$, $'z$ in the new coordinate system are also distinct from a material point $x^{\varkappa}$, z (except at the middle surface where $z = 'z = 0$), and even c_1 and c_2 are (slightly) different from $h/2$. All differences, however, are small by John's estimates, and the same holds true for surface derivatives. In fact, let $'x^{\varkappa}$, $'z$ denote the coordinates in the new system of the same material point as described by $x^{\varkappa}$, z in John's coordinates. We have then the estimates $'x^{\varkappa} = x^{\varkappa} + 0(\theta\eta z)$; $'z = z + 0(\eta z)$; $c_1, c_2 = h/2 + 0(\eta h)$, and we obtain from these and John's estimates for the derivatives of the stresses $'t^{\alpha\beta}('x^{\varkappa}, 'z) = t^{\alpha\beta}(x^{\varkappa}, z) + 0(E\eta\theta^2)$. Likewise, we have for the differences between the original stress resultants (3.17), (3.18) on the one hand, and the modified quantities $n^{\alpha\beta}$, $m^{\alpha\beta}$ in (3.19) on the other hand, the estimates $Eh\eta\theta^2$ and $Eh^2\eta\theta^2$ respectively. Our result for error estimates in the stress-strain relations (2.26), (2.27) is now a consequence of (3.7), (3.8), viz.

$$\gamma_{\alpha\beta} = A[(1+\nu)\,n_{\alpha\beta} - \nu a_{\alpha\beta} n^{\varkappa}_{\varkappa}] + 0(\eta\theta^2), \tag{3.19}$$

$$m^{\alpha\beta} = D[(1-\nu)\,{}'\varrho^{\alpha\beta} + \nu a^{\alpha\beta}\,{}'\varrho^{\varkappa}_{\varkappa}] + 0(E\eta h^2\theta^2), \tag{3.20}$$

where $A = (Eh)^{-1}$ is the extensional flexibility and $D = Eh^3/12(1-\nu^2)$ is the flexural rigidity. In each (surface) differentiation of these relations the error term is divided by λ. We also note the sharper results

$$\varepsilon^{\alpha\beta}\varepsilon^{\lambda\mu}\,{}'\varrho_{\beta\lambda|\mu} = 0(\eta\theta^2/h\lambda), \qquad n^{\beta\alpha}|_{\beta} = 0(Eh\eta\theta^2/\lambda), \tag{3.21}$$

corresponding to John's Eqs. (3.11), (3.13). The verification of the second of these equations is accomplished by substitution from (3.19), (3.20) and their derivatives into the exact equation of equilibrium, (2.21), for zero surface loads.

The foregoing results based on John's analysis hold, of course, only for shells loaded by edge tractions, in the absence of surface loads. The qualitative physical argument in [43] for the validity of the stress-strain relations (3.19), (3.20), however, remains applicable in the presence of sufficiently smooth surface loads, provided that the parameter θ(3.2) is now replaced by

$$\theta = \max\left(h/L, h/d, \sqrt{h/R}, \sqrt{\eta}\right), \tag{3.22}$$

where L is the minimum "wave length" of the deformation pattern on the middle surface in the sense of [43]. Our complete confidence in this physical argument is confirmed, at least in the case of the linear theory, by the rigorous solution error estimates in [8—12], cf. also Section 5.

On the other hand, the breakdown of the error estimates in the stress-strain relations near edges, where θ grows indefinitely, seems to be irreparable, except in the case of a special distribution of edge tractions conforming to the requirements of the two-dimensional theory [10, 11]. The minor approximation made in Section 2 in the discussion of boundary conditions for the non-linear theory, involving the replacement of the actual deformed shell edge by the ruled surface of normals to the deformed middle surface along the edge curve in the middle surface, implies relative errors of at most $0(\theta^2)$, and these may certainly be neglected in comparison with the far more serious errors in the stress-strain relations in the edge zone. We cannot escape concurrence with Gol'denweizer's view (e.g. [22]) that the interaction between the three-dimensional boundary layer in the edge zone and the approximately two-dimensional interior state of stress cannot be ignored in a thin shell. It is only a poor consolation that the question is largely of academic interest because the actual distribution of the edge tractions is rarely known with any precision. The exception is, of course, the case of a free edge to be discussed in more detail in Section 6.

4. Canonical form of intrinsic equations for first-approximation nonlinear shell theory

For the sake of brevity we restrict our (approximate) simplification of the exact basic Eqs. (2.5), (2.6), (2.21), (2.22) to the case of zero surface loads where John's results, discussed in the previous section, provide rigorous error estimates for the reduced equations.

For the first term in (2.5) we have immediately the estimate

$$K\gamma^{\alpha}_{\alpha} = 0(\eta/R^2) = 0(\eta\theta^2/\lambda^2), \tag{4.1}$$

and for the last term we have equally easily the same estimate

$$''a^{\varkappa\nu}\gamma_{\varkappa\alpha\mu}\gamma_{\nu\beta\lambda} = 0(\eta^2/\lambda^2) = 0(\eta\theta^2/\lambda^2). \tag{4.2}$$

For the second term in (2.5) we obtain from (3.19) and the second Eq. (3.21) after some simple algebra

$$\varepsilon^{\alpha\beta}\varepsilon^{\lambda\mu}\gamma_{\alpha\mu|\beta\lambda} = -An^{\alpha}_{\alpha}|^{\beta}_{\beta} + 0(\eta\theta^2/\lambda^2), \tag{4.3}$$

and the final simplified equation (4.9) corresponds completely to John's lowest order Eq. (3.10).

The reduction of (2.6) requires somewhat more algebra, virtually spelt out in detail by *Danielson* [49]. Replacement of $''a^{\varkappa\nu}$ by $a^{\varkappa\nu}$ involves an error term of order $\eta\theta^4/h\lambda$, and errors of the same order are accepted in the reduction of the other terms. Use is made in particular of both estimates (3.21) and the resulting relation

$$\gamma^{\alpha}_{\beta}|_{\alpha} = -\nu An^{\alpha}_{\alpha,\beta} + 0(\eta\theta^2/\lambda), \tag{4.4}$$

and the final simplified Eq. (2.6) is presented in (4.10).

In the equation of equilibrium (2.22) we drop the second term, and we accept therefore an error term of order $Eh^2\eta\theta^2/\lambda^2$. The first term in (2.22) is reduced by means of (3.21) to

$$m^{\alpha\beta}|_{\alpha\beta} = D'\varrho^{\alpha}_{\alpha}|^{\beta}_{\beta} + 0(Eh^2\eta\theta^2/\lambda^2), \tag{4.5}$$

and our final reduced equation is (4.11), again in complete correspondence to John's lowest order Eq. (.312).

The tangential equilibrium equations, (2.21), require most algebra in their reduction [49]. In the second term we replace $''a^{\alpha\varkappa}$ by $a^{\alpha\varkappa}$, accepting an error term of order $Eh\eta\theta^4/\lambda$. By means of the stress-strain relations and the second Equation of (3.21), the second term in (2.21) is now reduced to

$$(2\gamma^{\alpha}_{\lambda|\beta} - \gamma_{\lambda\beta}|^{\alpha})\, n^{\lambda\beta} = A\,[2n^{\alpha}_{\lambda}n^{\lambda\beta} + 2\nu(n^{\alpha}_{\lambda}n^{\lambda\beta} - n^{\alpha\beta}n^{\lambda}_{\lambda})]_{|\beta} - 1/2\,A\,[(1+\nu)\,n_{\lambda\beta}n^{\lambda\beta} - \nu n^{\lambda}_{\lambda}n^{\beta}_{\beta}]|^{\alpha} + 0(Eh\eta\theta^4/\lambda). \tag{4.6}$$

For any continuously differentiable symmetric surface tensor $T^{\alpha\beta}$ we have the identity [49]

$$(T^{\alpha}_{\lambda}T^{\lambda\beta} - T^{\alpha\beta}T^{\lambda}_{\lambda})_{|\beta} = 1/2\,(T^{\lambda\beta}T_{\lambda\beta} - T^{\lambda}_{\lambda}T^{\beta}_{\beta})|^{\alpha}. \tag{4.7}$$

The second term in (2.21) is thus reduced to

$${}''a^{\alpha\varkappa}(2\gamma_{\varkappa\lambda|\beta} - \gamma_{\lambda\beta|\varkappa})\, n^{\lambda\beta} = 2A(n^{\alpha}_{\lambda}n^{\lambda\beta})_{|\beta} - 1/2\,A[(1-\nu)\,n_{\lambda\beta}n^{\lambda\beta} + \nu n^{\lambda}_{\lambda}n^{\beta}_{\beta}]|^{\alpha} + 0(Eh\eta\theta^4/\lambda). \tag{4.8}$$

The last term but one in (2.21) is of the same order as the error term already accepted, and it may therefore be omitted. In the remaining terms we replace ${}''b^{\alpha}_{\varkappa}$ by ${}'b^{\alpha}_{\varkappa}$, with again the same error term, and we reduce these terms by means of the stress-strain relations and the first equation of (3.21). The resulting reduced equation is presented in (4.12), and we have now arrived at the complete set of simplified equations

$$-An^{\alpha}_{\alpha}|^{\beta}_{\beta} + b^{\alpha}_{\beta}{}'\varrho^{\beta}_{\alpha} - b^{\alpha}_{\alpha}{}'\varrho^{\beta}_{\beta} + 1/2\,{}'\varrho^{\alpha}_{\beta}{}'\varrho^{\beta}_{\alpha} - 1/2\,{}'\varrho^{\alpha}_{\alpha}{}'\varrho^{\beta}_{\beta} = 0(\eta\theta^2/\lambda^2), \tag{4.9}$$

$${}'\varrho^{\beta}_{\alpha}|_{\beta} - {}'\varrho^{\beta}_{\beta}|_{\alpha} + (1+\nu)\,A\,[(b^{\beta}_{\alpha} + {}'\varrho^{\beta}_{\alpha})\,n^{\lambda}_{\lambda,\beta} + (b^{\beta}_{\lambda} + {}'\varrho^{\beta}_{\lambda})\,n^{\lambda}_{\beta|\alpha}] - \nu A(b^{\lambda}_{\lambda} + {}'\varrho^{\lambda}_{\lambda})\,n^{\beta}_{\beta,\alpha} = 0(\eta\theta^4/h\lambda), \tag{4.10}$$

$$D{}'\varrho^{\alpha}_{\alpha}|^{\beta}_{\beta} - (b^{\alpha}_{\beta} + {}'\varrho^{\alpha}_{\beta})\,n^{\beta}_{\alpha} = 0(Eh^2\eta\theta^2/\lambda^2), \tag{4.11}$$

$$n^{\beta}_{\alpha}|_{\beta} + 2A\,(n^{\lambda}_{\alpha}n^{\beta}_{\lambda})_{|\beta} - 1/2\,A\delta^{\beta}_{\alpha}[(1-\nu)\,n^{\varkappa}_{\lambda}n^{\lambda}_{\varkappa} + \nu n^{\varkappa}_{\varkappa}n^{\lambda}_{\lambda}]_{|\beta} + 1/2\,(1-\nu)\,D(b^{\lambda}_{\alpha}{}'\varrho^{\beta}_{\lambda} - b^{\beta}_{\lambda}{}'\varrho^{\lambda}_{\alpha})_{|\beta} + D(b^{\beta}_{\alpha} + {}'\varrho^{\beta}_{\alpha})\,{}'\varrho^{\lambda}_{\lambda,\beta} = 0(Eh\eta\theta^4/\lambda). \tag{4.12}$$

We note first of all that our Eqs. (4.9)—(4.12) agree with those of *Danielson* [49], derived by a similar, if perhaps less conclusive argument, with the single exception that our derivation on the basis of John's analysis in [3] has enabled us to omit the term (4.1) in Eq. (4.9). This represents, of course, only a small further simplification, but it is not devoid of significance since our Eqs. (4.9) and (4.11) are now formally *identical* to John's corresponding lowest order Eqs. (3.10) and (3.12). After all, this is not too much of a surprise. The occasional failure of John's lowest order equations, e.g., in the case of (nearly) inextensional bending, is indeed due to inadequate accuracy of his other Eqs. (3.11) or (3.13), in particular the latter equations in the inextensional bending case. It is no wonder therefore that a refinement only of these inadequate equations, as presented in (4.10) and (4.12), is actually sufficient in the context of a first-approximation theory.

A further argument to abstain from the refinement of Eqs. (3.10) and (3.12) in the form presented by *John* in [4, 5], at least in the context of a first-approximation theory, is to be found in a method of discussion sòlutions of the equations which is often convenient, in particular in the linear theory. Eqs. (4.10) and (4.12) are then employed to express (or at least regarded as expressing) the deviatoric parts of ${}'\varrho^{\alpha}_{\beta}$ and n^{α}_{β} in terms of the invariants ${}'\varrho^{\alpha}_{\alpha}$ and n^{α}_{α}. In the case of a slow variation of the solution over the middle surface, say with a wave length of order R (and such solutions occur in the membrane theory of shells as well as in the inextensional bending case) the errors in the deviatoric parts of ${}'\varrho^{\alpha}_{\beta}$ and n^{α}_{β} may be of order $\eta\theta^4R/h\lambda$ and $Eh\eta\theta^4R/\lambda$ respectively. Substitution of this solution into Eqs. (4.9) and (4.11) then leads to error terms of order $\eta\theta^3/\lambda^2$ and $Eh^2\eta\theta^3/\lambda^2$, one factor θ more than in the inherent errors in these equations, but one factor θ *less* than in John's refined version in [4, 5]. It is therefore questionable whether John's refinement of his Eqs. (3.10), (3.12) in [4, 5], reducing the error estimates by a factor θ^2 and resulting in an extreme complexity, is even physically meaningful in the absence of a further refinement of his Eqs. (3.11) and (3.13) beyond the already highly complicated versions in [4, 5].

It may be worthwhile to note, and readers who are familiar with the complexities of nonlinear shell theory will readily agree, that our equations are relatively simple. They are hardly more complicated than the (less accurate) simplified equations in ([47], Section 9), where the simplification had (wittingly) been carried too far to cover all possible situations. We attach particular significance to the fact that the cubic coefficients in John's strain energy density function W do not occur explicitly in our equations, although we have not made any assumption with regard to a linear stress-strain law.

A minor weakness of our Eqs. (4.9)—(4.12) is that their linearized version does not reduce to the equations of the "best" linear first-approximation shell theory in the sense of *Budiansky* and *Sanders* [44], and for this reason *Danielson* [49] has appended a note to his paper that his equations are not intended for use in the linear theory. This weakness is easily rectified, however, if the modified tensor of changes of curvature $\varrho_{\alpha\beta}$ [43, 47], defined by

$$'\varrho_{\alpha\beta} = \varrho_{\alpha\beta} + 1/2\,(b^{\varkappa}_{\alpha}\gamma_{\varkappa\beta} + b^{\varkappa}_{\beta}\gamma_{\varkappa\alpha}), \tag{4.13}$$

is introduced into our equations to replace $'\varrho_{\alpha\beta}$. Eliminating the middle surface strain tensor again by means of (3.19), and applying again (3.21) where appropriate, we arrive at our final set of equations with error estimates, which we venture to call the *canonical form of intrinsic equations* in the first-approximation nonlinear theory of shells in the absence of surface loads

$$-An^{\alpha}_{\alpha}|^{\beta}_{\beta} + b^{\alpha}_{\beta}\varrho^{\beta}_{\alpha} - b^{\alpha}_{\alpha}\varrho^{\beta}_{\beta} + 1/2\,\varrho^{\alpha}_{\beta}\varrho^{\beta}_{\alpha} - 1/2\,\varrho^{\alpha}_{\alpha}\varrho^{\beta}_{\beta} = 0(\eta\theta^2/\lambda^2), \tag{4.14}$$

$$\begin{aligned}\varrho^{\beta}_{\alpha}|_{\beta} - \varrho^{\beta}_{\beta}|_{\alpha} - 1/2(1+\nu)\,A(b^{\lambda}_{\alpha}n^{\beta}_{\lambda} - b^{\beta}_{\lambda}n^{\lambda}_{\alpha})_{|\beta} + Ab^{\beta}_{\alpha}n^{\lambda}_{\lambda,\beta}\\ + (1+\nu)\,A(\varrho^{\beta}_{\alpha}n^{\lambda}_{\lambda,\beta} + \varrho^{\beta}_{\lambda}n^{\lambda}_{\beta|\alpha}) - \nu A\varrho^{\lambda}_{\lambda}n^{\beta}_{\beta,\alpha} = 0(\eta\theta^4/h\lambda),\end{aligned} \tag{4.15}$$

$$D\varrho^{\alpha}_{\alpha}|^{\beta}_{\beta} - (b^{\alpha}_{\beta} + \varrho^{\alpha}_{\beta})\,n^{\beta}_{\alpha} = 0(Eh^2\eta\theta^2/\lambda^2), \tag{4.16}$$

$$\begin{aligned}n^{\beta}_{\alpha}|_{\beta} + 1/2(1-\nu)\,D(b^{\lambda}_{\alpha}\varrho^{\beta}_{\lambda} - b^{\beta}_{\lambda}\varrho^{\lambda}_{\alpha})_{|\beta} + Db^{\beta}_{\alpha}\varrho^{\lambda}_{\lambda,\beta} + D[\varrho^{\beta}_{\alpha}\varrho^{\lambda}_{\lambda} - 1/2\,\delta^{\beta}_{\alpha}\varrho^{\lambda}_{\lambda}\varrho^{\varkappa}_{\varkappa}]_{|\beta}\\ + 2A(n^{\lambda}_{\alpha}n^{\beta}_{\lambda})_{|\beta} - 1/2A\delta^{\beta}_{\alpha}[(1-\nu)\,n^{\varkappa}_{\lambda}n^{\lambda}_{\varkappa} + \nu n^{\varkappa}_{\varkappa}n^{\lambda}_{\lambda}]_{|\beta} = 0(Eh\eta\theta^4/\lambda).\end{aligned} \tag{4.17}$$

Within John's error estimates for the equations, it is evident from the canonical set (4.14)—(4.17) that only Eq. (4.10) is affected in its form by our change in the measure for the changes of curvature. Our canonical Eqs. (4.14)—(4.17) are equally simple (or complicated, if one likes) as Danielson's equations in our form (4.9)—(4.12), and they reduce to the "best" linear theory upon linearization. The well-known static-geometric analogy, for example, is easily recognized in the linear terms of (4.14)—(4.17).

Even if our canonical intrinsic Eqs. (4.14)—(4.17) represent perhaps the simplest possible formulation[1] of first-approximation nonlinear shell theory, they do not yield all answers to possible questions. Their actual solution poses a formidable problem, and such a direct solution seems only possible in those boundary value problems where the boundary conditions may be expressed in terms of the stress resultants $n^{\alpha\beta}$ and the changes of curvature $\varrho_{\alpha\beta}$ (or $'\varrho_{\alpha\beta}$). This is actually the case when dynamic boundary conditions are specified along the entire shell edge. The reduced form of these boundary conditions is given by *Danielson* in terms of $n^{\alpha\beta}$ and $'\varrho_{\alpha\beta}$, and since John's analysis is inherently incapable of yielding better error estimates for the edge conditions, suffice it here to refer to [49]. Whenever kinematic boundary conditions are specified along a part of the edge, we need an evaluation of the displacements from the solution of our intrinsic equations. A serious problem arises here, in principle, that our equations ensure only an *approximate* verification of the compatibility conditions. This means that the metric tensor $'a_{\alpha\beta} = a_{\alpha\beta} + 2\gamma_{\alpha\beta}$ and the second fundamental tensor $'b_{\alpha\beta} = b_{\alpha\beta} + {}'\varrho_{\alpha\beta}$ of the deformed middle surface, as obtained from the solution of our intrinsic equations, do not correspond to a surface which can actually exist in Euclidean space.

We venture to suggest a possible way out of this difficulty, even if no experience is available to show that it is practicable. The violation of the equations of *Gauss* and *Mainardi-Codazzi* for the deformed middle surface occurs in error terms in the right-hand members of order $\eta\theta^2/\lambda^2$ and $\eta\theta^4/h\lambda$ (cf. (4.14), (4.15)). We try to reduce these errors to zero by small corrections $\Delta\varrho^{\alpha}_{\beta}$ to ϱ^{α}_{β}, of order $\eta\theta^2/h$. It would seem that the three unknown corrections available should be sufficient to satisfy the equations which were violated initially. If difficulties occur, we also apply corrections $\Delta n^{\alpha}_{\beta}$ to n^{α}_{β}, of order $Eh\eta\theta^2$, subject to the restriction $\Delta n^{\beta}_{\alpha}|_{\beta} = 0$, and the associated corrections to the middle surface strains give additional flexibility in meeting the requirements of *Gauss* and *Codazzi*. It is easily verified that the error terms in (4.16), (4.17) are not affected by these corrections, and the modified solution $\varrho^{\alpha}_{\beta} + \Delta\varrho^{\alpha}_{\beta}$, $n^{\alpha}_{\beta} + \Delta n^{\alpha}_{\beta}$ is an equivalent solution which now admits the construction of an associated displacement field of the middle surface.

---

[1] An alternate epuivalent formulation may be given in terms of the finite rotation and stress function vectors in [51].

5. Error estimates for solutions

In the introduction we have already stressed the importance of estimates for the errors of the solutions of the approximate, two-dimensional shell equations, as compared with the solutions of the exact three-dimensional equations of elasticity theory. Ideally, one would like to have pointwise precise error estimates for all displacements and stresses, but this ideal has only been achieved in the simple axisymmetric torsion problem for shells of revolution in linear elasticity theory [2]. More general results have only been obtained in a more global form, as root mean square or L_2-norm error estimates, again restricted to the linear case [8—12]. It would seem that nothing is known about solution errors in the nonlinear theory. The inherent difficulties of this problem in the nonlinear case will be discussed briefly at the end of this section.

In view of the wide variety in form in wich the equations of shell theory appear in the literature, nearly every writer favouring his own equations, it is equally important to have estimates for the differences between the solutions of the equations associated with different, presumably equivalent shell theories. This simpler problem of so-called *intrinsic error estimates*, involving a comparison of solutions of the same physical dimension, has, of course, a much wider bearing. Our confidence in the predictions of the exact classical theory of elasticity for applications in engineering is based on our firm belief, that these predictions remain (approximately) valid for actual, (slightly) inhomogeneous, anisotropic and inelastic materials. A sound foundation for this belief, however, is only to be found in error estimates for the solutions of classical elasticity theory as an approximation to the solutions for actual engineering materials. In fact, *intrinsic error estimates form a necessary complement to classical elasticity theory.* Here again, concrete results have only been achieved in the linear theory, and we shall begin with a review of these [53—55][1].

Let $\boldsymbol{v}(\boldsymbol{x})$ denote an arbitrary kinematically admissible displacement field of a one-, two-, or three-dimensional elastic system. The elastic energy stored in the system is in the linear theory a positive homogeneous quadratic functional $P_2[\boldsymbol{v}]$ of the field $\boldsymbol{v}$. This functional is positive definite in the sense of *Mikhlin* [69], if the body is properly supported such that rigid-body displacements are excluded. The external loads in the linear theory of elasticity have a potential energy which is a linear functional $P_1[\boldsymbol{v}]$ of the displacement field. Let $\boldsymbol{u}[\boldsymbol{x}]$ denote the solution of the elasticity problem. It may be identified with the minimizing displacement field of the variational problem

$$P_2[\boldsymbol{v}] + P_1[\boldsymbol{v}] = \text{Min}. \tag{5.1}$$

We assume the existence of a solution $\boldsymbol{u}(\boldsymbol{x})$ [cf. e.g. [69]]. Let $\boldsymbol{w}(\boldsymbol{x})$ denote an arbitrary continuously differentiable displacement field satisfying homogeneous boundary conditions $\boldsymbol{w} = 0$ at every point where a kinematically admissible displacement field takes the zero or non-zero value of a specified displacement vector $\boldsymbol{u}_0$. Introducing the obvious and well-known notation

$$P_2[\boldsymbol{v} + \boldsymbol{w}] = P_2[\boldsymbol{v}] + P_{11}[\boldsymbol{v}, \boldsymbol{w}] + P_2[\boldsymbol{w}], \tag{5.2}$$

the variational equation, associated with the minimum problem and satisfied by the solution $\boldsymbol{u}$, reads

$$P_{11}[\boldsymbol{u}, \boldsymbol{w}] + P_1[\boldsymbol{w}] = 0, \tag{5.3}$$

holding for all admissible fields $\boldsymbol{w}$.

Some problems are simplified considerably, if the elastic energy functional $P_2[\boldsymbol{v}]$ is replaced by an approximation $P_2^*[\boldsymbol{v}]$ which is also a positive homogeneous quadratic functional of the kinematically admissible displacement field $\boldsymbol{v}$. The approximation of the original functional is expressed by the inequality

$$|P_2^*[\boldsymbol{y}] - P_2[\boldsymbol{y}]| \leq \varepsilon^2 P_2[\boldsymbol{y}], \tag{5.4}$$

holding for all continuously differentiable displacement fields $\boldsymbol{y}(\boldsymbol{x})$ without any kinematic restrictions at the boundary, and for some fixed positive number $\varepsilon^2 < 1$, normally $\varepsilon^2 \ll 1$.

[1] The discussion in [53] is restricted tacitly to homogeneous displacement boundary conditions, and we remove here this unnecessary restriction.

In three-dimensional elasticity theory we have the example of the replacement of the actual, (slightly) inhomogeneous and anisotropic elastic body by a homogeneous elastic body, in shell theory, for example, the comparison between the theory of *Flügge* and the *Sanders-Koiter* theory. The solution $\boldsymbol{u}^*(\boldsymbol{x})$ of the modified variational problem

$$P_2^*[\boldsymbol{v}] + P_1[\boldsymbol{v}] = \text{Min.}, \tag{5.5}$$

whose existence is again assumed, is now regarded as an approximation to the actual solution $\boldsymbol{u}(\boldsymbol{x})$ of the original problem. The variational equation, satisfied by the approximate solution $\boldsymbol{u}^*$,

$$P_{11}^*[\boldsymbol{u}^*, \boldsymbol{w}] + P_1[\boldsymbol{w}] = 0 \tag{5.6}$$

holds again for all admissible fields $\boldsymbol{w}$.

The basic solution error estimate obtained in [53] has a mean square character for the strains and the stresses, and it is expressed by the inequality

$$P_2[\boldsymbol{u} - \boldsymbol{u}^*] \leq 9\varepsilon^4 P_2[\boldsymbol{u}^*]. \tag{5.7}$$

If the elastic system is properly supported, we have the inequality expressing the positive definite character of the energy functional [69]

$$\|\boldsymbol{v}\|^2 \leq C^2 P_2[\boldsymbol{v}], \tag{5.8}$$

where $\|\boldsymbol{v}\|$ is the L_2-norm or root mean square of the kinematically admissible displacement field, and C some positive constant. In this case we have the solution error estimate

$$\|\boldsymbol{u} - \boldsymbol{u}^*\| \leq 3C\varepsilon^2\{P_2[\boldsymbol{u}^*]\}^{1/2}. \tag{5.9}$$

In some shell problems it appears that a modification of the strain energy functional does not yet provide the flexibility required in order to obtain the desired simplification of the field equations. In some cases, however, our purpose may be achieved by a further (small) modification of the stress-strain relations in such a way that they are no longer derivable from a strain energy function [54]. The linear boundary value problem in these cases is not self-adjoint, and it cannot be cast into the form of a minimum problem, but it may still be formulated in terms of a variational equation for the solution $\boldsymbol{u}^*(\boldsymbol{x})$

$$R_{11}^*[\boldsymbol{u}, \boldsymbol{w}] + P_1[\boldsymbol{w}] = 0, \tag{5.10}$$

where R_{11}^* is now an asymmetric bilinear functional. The approximate character of the modified Eq. (5.10) is now expressed by the inequality

$$|R_{11}^*[\boldsymbol{y}, \boldsymbol{z}] - P_{11}[\boldsymbol{y}, \boldsymbol{z}]| < 2\varepsilon^2\{P_2[\boldsymbol{y}]\}^{1/2}\{P_2[\boldsymbol{z}]\}^{1/2}, \tag{5.11}$$

holding for all pairs $\boldsymbol{y}$, $\boldsymbol{z}$ of continuously differentiable displacement fields, and again for some positive number $\varepsilon < 1$. The resulting error estimates for the approximate solution $\boldsymbol{u}^*(\boldsymbol{x})$ are identical in form to (5.7) and (5.9) except that the numerical factors 9 and 3 are replaced by 16 and 4 respectively. We refer to [54] for an application to shells whose middle surfaces have a constant value of the mean curvature H.

The foregoing intrinsic error estimates all apply in the (root) mean square sense. Pointwise estimates are now also available for the normal displacement component [55], and they are even uniform for the normal deflection under a normal point load. At present it is still an open question whether pointwise error estimates would also be possible for the tangential displacement components or, even more important, for the stresses in the shell. The Prager-Synge hypercircle method [70] may conceivably be applied to settle this question.

The problem of *extrinsic error estimates*, which apply to the comparison of the predictions of two-dimensional linear shell theory with the exact displacement and stress fields of three-dimensional linear elasticity theory, is more difficult on account of the difference in dimension of the two problems. It has only been solved rigorously in the mean square sense, and under the restriction to dynamic boundary conditions along the entire edge of the shell, with a distribution of edge tractions conforming to the requirements of shell theory [8—12]. The basic tool in the analysis is essentially a theorem of *Prager* and *Synge* [70], and we shall follow the presentation in [11] in our review. The basic theorem is formulated in terms of the complementary elastic energy of a three-dimensional elastic body $C_2[\boldsymbol{\tau}]$, essentially the elastic strain energy expressed in the stress tensor $\boldsymbol{\tau}$. Let $\boldsymbol{\sigma}^*$ denote any statically admissible stress distribution, and $\boldsymbol{\sigma}^0$ any kinematically admissible stress distribution (i.e., a stress distribution obtained, by means of Hooke's law, and the linear strain-displacement relations, from a kinematically admissible displacement field). Finally, let $\boldsymbol{\sigma}$ denote the actual solution for the stresses of the boundary value problem of three-dimensional theory. The Prager-Synge theorem is now expressed by the equation

$$C_3[\boldsymbol{\sigma} - 1/2\,(\boldsymbol{\sigma}^0 + \boldsymbol{\sigma}^*)] = C_2[1/2(\boldsymbol{\sigma}^0 - \boldsymbol{\sigma}^*)], \tag{5.12}$$

holding for all kinematically admissible stress fields $\boldsymbol{\sigma}^0$ and all statically admissible stress fields $\boldsymbol{\sigma}^*$. This equation enables us to estimate the L_2-norm or the root mean square of the error of an "approximate" solution $(\boldsymbol{\sigma}^0 + \boldsymbol{\sigma}^*)/2$ in terms of the L_2-norm of the stress difference $(\boldsymbol{\sigma}^0 - \boldsymbol{\sigma}^*)$.

The shell theory employed in [11] is the linearized version of the equations and boundary conditions (2.14)—(2.18). Following in the footsteps of *Trefftz* [71], it is shown ([11], Section 4) that a statically admissible stress distribution in the interior of the shell may be obtained from a linear distribution of pseudo-stresses $s^{\alpha\beta}(z) = \sqrt{g/a}\,\sigma^{\alpha\beta}(z)$ over the thickness. The restriction "in the interior of the shell" in the previous sentence may now be dropped, if the edge tractions in the three-dimensional problem are specified in accordance with the internal stress distribution just obtained. From the middle surface displacement field we now construct a suitable kinematically admissible three-dimensional displacement field[1]. In the absence of kinematic boundary conditions this is always easily possible [11, Section 5]. The stress difference $(\boldsymbol{\sigma}^0 - \boldsymbol{\sigma}^*)$ may now be calculated exactly from the shell solution, and we have a precise L_2-norm error estimate for $(\boldsymbol{\sigma}^0 - \boldsymbol{\sigma}^*)/2$ as an approximate solution of the three-dimensional problem. In a thin shell under sufficiently smooth surface loads, the error is always small, of order $\varepsilon^2 = h^2/L^2 + h/R$, where L is again the minimum wave length of the deformation pattern on the middle surface. A further improvement of this L_2 estimate seems to be impossible since the constitutive equations of first-approximation shell theory are known to involve errors of order ε^2. It would be of great interest, however, to obtain similar pointwise error estimates.

We emphasize again that the error estimate of order $\varepsilon^2 = h^2/L^2 + h/R$ applies only in the case where the edge tractions are specified according to the demands of shell theory. As we have mentioned before, unfortunately we have rarely accurate information about the distribution If edge loads over the thickness (except in the case of a free edge to be discussed in Section 6). on the absence of this information it is evidently impossible to obtain precise error estimates for the solutions of shell theory. The more qualitative argument in [10], however, suggests that we should expect a larger error, of order $\varepsilon^* = h/L + h/R$, in the case of more "irregular" edge load distributions, provided that the edge tractions do not exceed, in order of magnitude, the maximum value of the primary stresses in the interior of the shell. This argument is supported by the results of the next section, and a similar argument was developed in [10] for the case of kinematic boundary conditions. The somewhat depressing conclusion for most shell problems is, similar to the earlier conclusion of *Gol'denweizer* (e.g. [17, 22]), that no better accuracy of the solutions can be expected than of order $\varepsilon^* = h/L + h/R$, even if the equations of first-approximation shell theory would permit, in principle, an accuracy of order $\varepsilon^2 = h^2/L^2 + h/R$. This conclusion also explains why we shall not attempt a very detailed review of more refined interior shell theories in Section 7.

In the introduction we have already referred to the formidable difficulties encountered in attempts to extend error estimates for the solutions to nonlinear shell theory. The possible occurrence of snap loads seems even to require a different concept of solution errors, if an effective attempt at their discussion is to be made. In fact, at loads in the vicinity of a snap load for either theory, the exact or the approximate one, the other theory may not have a solution at all in the neighbourhood of the first one. This situation may occur in the comparison

[1] Substantial improvements in [11], in comparison with [10], are due to *Danielson* who graciously made his results available already before their independent publication [9].

of two different "equivalent" nonlinear shell theories as well as in the case of a comparison between shell theory and three-dimensional theory. We are not aware of any effective attempt to solve the problem.

6. Modified boundary conditions at a free edge

It will be convenient to review first the general form of the classical dynamic boundary conditions in the linear theory of shells. All quantities are now referred to the undeformed middle surface with its base and normal vectors, as well as to the undeformed edge curve with arc length s and its tangent vector $\boldsymbol{t}$ and outward normal vector $\boldsymbol{\nu}$ in the tangent plane. We recall our sign convention $\boldsymbol{t} = \boldsymbol{n} \times \boldsymbol{\nu}$. We introduce the normal curvature $\varkappa_{(\nu)}$, and the geodetic torsion and curvature $\varkappa_{(t)}$ and $\varkappa_{(n)}$ of the edge curve by the relations

$$\boldsymbol{\nu}_{,s} = \varkappa_{(n)}\boldsymbol{t} - \varkappa_{(t)}\boldsymbol{n}, \quad \boldsymbol{t}_{,s} = \varkappa_{(\nu)}\boldsymbol{n} - \varkappa_{(n)}\boldsymbol{\nu}, \quad \boldsymbol{n}_{,s} = \varkappa_{(t)}\boldsymbol{\nu} - \varkappa_{(\nu)}\boldsymbol{t}; \tag{6.1}$$

$$\varkappa_{(\nu)} = b_{\alpha\beta}t^{\alpha}t^{\beta}, \quad \varkappa_{(t)} = -b_{\alpha\beta}t^{\alpha}\nu^{\beta}. \tag{6.2}$$

The edge loads per unit arc length are decomposed with respect to the unit vectors

$$\boldsymbol{N} = N_{(\nu)}\boldsymbol{\nu} + N_{(t)}\boldsymbol{t} + Q\boldsymbol{n}, \quad \boldsymbol{M} = M_{(\nu)}\boldsymbol{\nu} + M_{(t)}\boldsymbol{t}. \tag{6.3}$$

We employ the modified tensor of changes of curvature $\varrho_{\alpha\beta}$, defined by (4.13), and the internal virtual work per unit area of the middle surface is then specified by $n^{\alpha\beta}\,\delta\gamma_{\alpha\beta} + m^{\alpha\beta}\,\delta\varrho_{\alpha\beta}$. We make use of the expression [4, 7]

$$'\varrho_{\alpha\beta} = \phi_{\alpha|\beta} - c_{\alpha\beta}w + b^{\varkappa}_{\beta}u_{\varkappa|\alpha}, \tag{6.4}$$

where ϕ_α specifies the projection of the (linear) rotation vector $\boldsymbol{\Phi}$ on the tangent plane by the relation $\phi_\alpha = \varepsilon_{\beta\alpha}\boldsymbol{\Phi}\cdot\boldsymbol{a}^{\beta}$. The virtual work of the edge loads per unit length is given by $\boldsymbol{N}\cdot\delta\boldsymbol{u} + \boldsymbol{M}\cdot\delta\boldsymbol{\Phi}$, and the boundary conditions of Kirchhoff type follow by the standard procedure of the calculus of variations in the form

$$\left[n^{\beta\alpha} + \frac{3}{2}b^{\alpha}_{\varkappa}m^{\beta\varkappa} - \frac{1}{2}b^{\beta}_{\varkappa}m^{\alpha\varkappa}\right]\nu_\alpha\nu_\beta = N_{(\nu)} - \varkappa_{(t)}M_{(\nu)}, \tag{6.5}$$

$$\left[n^{\beta\alpha} + \frac{3}{2}b^{\alpha}_{\varkappa}m^{\beta\varkappa} - \frac{1}{2}b^{\beta}_{\varkappa}m^{\alpha\varkappa}\right]t_\alpha\nu_\beta = N_{(t)} + \varkappa_{(\nu)}M_{(\nu)}, \tag{6.6}$$

$$m^{\alpha\beta}|_{\alpha}\nu_\beta + (m^{\alpha\beta}\nu_\alpha t_\beta)_{,s} = M_{(\nu),s} - Q, \tag{6.7}$$

$$m^{\alpha\beta}\nu_\alpha\nu_\beta = -M_{(t)}, \tag{6.8}$$

in view of (2.19) in agreement with the linearized version of (2.16)—(2.18), and also in agreement with [67] where a slightly different notation is employed. We emphasize again, perhaps unnecessarily, that these edge conditions are fully exact as they stand. The approximate character enters the picture when the stress-strain relations are introduced. The need of modified boundary conditions in shell theory does not arise from a defect in the Kirchhoff-type boundary conditions as such, but from a failure of the stress-strain relations to hold, even approximately, near an edge.

In the case of a *free edge* the right-hand members of (6.5)—(6.8) are all specified to be zero. This does not imply, however, that the edge tractions would also be zero. On the contrary, the non-vanishing internal edge twisting moment $m^{\alpha\beta}\nu_\alpha t_\beta$ implies tangential edge tractions of the *same order of magnitude* as occur in the interior shell domain. These tangential edge tractions are only absent in problems where the edge twisting moment happens to be zero. If we wish to consider the case of a really traction-free edge, if perhaps again only approximately, we have to investigate at least the effect of a removal of edge tractions of the same order as the internal stresses. This situation is perhaps the main reason why a number of authors, in particular *Gol'denweizer*, have devoted a major effort to analyse, by a careful examination of three-dimensional effects, the interaction between a boundary layer in the edge zone and the interior state of stress.

As we shall show, it is relatively easy to achieve the more modest purpose of a modification of the boundary conditions at a free edge such that at least the *interior* state of stress in the shell is obtained with errors of the same order $\varepsilon^2 = h^2/L^2 + h/R$ as are inherent in the first-approximation interior equations. To this end we observe that we have evidently overestimated the elastic energy in the shell by integrating (2.24) over the entire middle surface, and thus ignoring that the twisting energy decreases rapidly in an edge zone to zero at the edge itself. As happens more often, this diagnosis readily suggests a cure. From the theory of torsion for a bar with a narrow rectangular cross-section (width b and height $h \ll b$) we borrow the formula for the torsional rigidity [e.g. 72]

$$S_t = \frac{1}{3} Gbh^3 \left[1 - B\frac{h}{b} + 0\left(\frac{h}{b} e^{-\pi b/2h}\right)\right], \tag{6.9}$$

where the numerical constant B is given by[1]

$$B = \frac{192}{\pi^5} \sum_{k=0}^{\infty} (2k+1)^{-5} = 0.630\ldots \tag{6.10}$$

If classical plate theory were entirely applicable to this case, the torsional rigidity would be given simply by the first term $Gbh^3/3$. Ignoring the exponential term in (6.9), each of the two zones near a free edge of the plate is responsible for a reduction in elastic energy per unit length of the plate amounting to $BGh^4\omega^2/12$, where ω is the specific twist. Bearing in mind that shell theory is only meaningful for relatively large wave lengths L of the deformation pattern, $h/L \ll 1$, and consequently restricting our attention also to smooth edge curves, we generalize this result to plates and shells with curved free edges by *subtracting* from the classical energy expression a *line integral* along the free edges

$$\int \frac{1}{12} BGh^4 (\varrho_{\alpha\beta}\nu^\alpha t^\beta)^2 \, ds. \tag{6.11}$$

It is now a simple matter, at least in principle, to derive the associated modified boundary conditions. A difficulty, however, occurs in the circumstance that the variation $\delta\varrho_{\alpha\beta}\nu^\alpha t^\beta$ contains normal derivatives of the tangential virtual displacement components which can obviously not be removed by partial integration along the edge curve. This difficulty may be overcome by replacing the edge twist in (6.11) by

$$\omega^* = {}'\varrho_{\alpha\beta}\nu^\alpha t^\beta + \varkappa_{(t)}\gamma_{\alpha\beta}\nu^\alpha\nu^\beta - 2\varkappa_{(\nu)}\gamma_{\alpha\beta}\nu^\alpha t^\beta, \tag{6.12}$$

where ${}'\varrho_{\alpha\beta} = {}'b_{\alpha\beta} - b_{\alpha\beta}$ is the original tensor of changes of curvature, in which revised expression the terms with a factor $u_{\varkappa|\alpha}\nu^\alpha$ may be shown to cancel. Shell theory permits such a replacement even in the calculation of the surface integral of elastic energy, because the accuracy of first-approximation theory is not affected. This argument holds a fortiori for the line integral (6.11).

Let m^* denote the twisting couple at the edge, expressed in ω^* by the classical relation of shell theory $m^* = Gh^3\omega^*/6$. The line integral of internal virtual work, to be subtracted from the usual surface integral of internal virtual work (extended over the entire middle surface), is $\int Bhm^* \, \delta\omega^* \, ds$. We refer to [61] for details of the fairly lengthy algebra, and we present the result in the form of the modified boundary conditions at a free edge

$$\left[n^{\beta\alpha} + \frac{3}{2} b^\alpha_\varkappa m^{\beta\varkappa} - \frac{1}{2} b^\beta_\varkappa m^{\alpha\varkappa}\right]\nu_\alpha\nu_\beta - [\varkappa_{(\nu)}Bhm^*]_{,s} - \varkappa_{(t)}\varkappa_{(n)}Bhm^* = 0, \tag{6.13}$$

$$\left[n^{\beta\alpha} + \frac{3}{2} b^\alpha_\varkappa m^{\beta\varkappa} - \frac{1}{2} b^\beta_\varkappa m^{\alpha\varkappa}\right] t_\alpha\nu_\beta = 0, \tag{6.14}$$

$$m^{\alpha\beta}|_\alpha\nu_\beta + (m^{\alpha\beta}\nu_\alpha t_\beta)_{,s} + [\varkappa_{(n)}Bhm^*]_{,s} - \varkappa_{(\nu)}\varkappa_{(t)}Bhm^* = 0, \tag{6.15}$$

$$m^{\alpha\beta}\nu_\alpha\nu_\beta + (Bhm^*)_{,s} = 0. \tag{6.16}$$

It will be observed from a comparison with (6.5)—(6.8) that only one of the four Kirchhoff conditions remains unchanged, namely (6.6) for the tangential edge loads. We also note that

[1] Since we have employed the symbol A for the extensional flexibility $(Eh)^{-1}$, we deviate from the custom in the Soviet literature to denote this number by $1/2A$.

the additional terms involving the couple Bhm^*, after removal to the right-hand member, may be interpreted as *fictitious edge loads*, applied to the shell in the conventional theory with Kirchhoff boundary conditions, namely a force $\boldsymbol{N}^* = N^*_{(\nu)}\boldsymbol{\nu} + Q^*\boldsymbol{n}$ and a couple $\boldsymbol{M}^* = M^*_{(t)}\boldsymbol{t}$, both per unit edge length, where the components of the fictitious edge loads are specified by

$$N^*_{(\nu)} = [\varkappa_{(\nu)} Bhm^*]_{,s} + \varkappa_{(t)}\varkappa_{(n)} Bhm^*, \tag{6.17}$$

$$Q^* = [\varkappa_{(n)} Bhm^*]_{,s} - \varkappa_{(\nu)}\varkappa_{(t)} Bhm^*, \tag{6.18}$$

$$M^*_{(t)} = (Bhm^*)_{,s}. \tag{6.19}$$

It is easily verified that the fictitious loads, acting along a smooth closed free edge curve, are self-equilibrated.

It is not easily possible to compare our results with those of *Gol'denweizer* in [22] because his formulation of modified edge conditions is achieved in a completely different, in our view much more complicated form. On the other hand, in the special case of a flat plate with a free edge our results do indeed reduce to Gol'denweizer's in [17].

The simple form of our modified boundary conditions in terms of the fictitious edge loads is appealing, and it suggests that they may perhaps be obtained in a simpler way. The clue to this alternative, more elementary, if perhaps less conclusive approach is to be found in Lamb's ingenious interpretation of the Kirchhoff conditions [78]. The starting-point is here again the classical theory of torsion of a bar with a narrow rectangular cross-section (Fig. 1). The exact stress distribution *in* the cross-section is statically equivalent to a uniform distribution of torsional moments m^* per unit width (over the *entire* width b) plus a pair of concentrated forces m^* at the ends, and a pair of concentrated couples Bhm^* at the ends of opposite sign to the uniformly distriubted couples (Fig. 1). The forces m^* and couples Bhm^* are actually due to

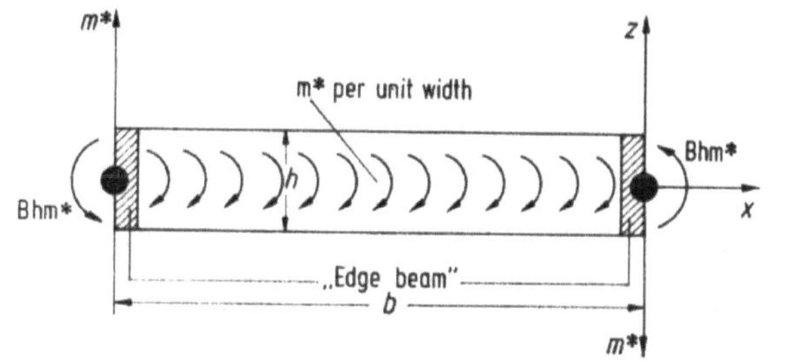

Fig. 1. Internal loads in cross-section statically equivalent to exact stress distribution in torsion.

narrow zones of width $0(h)$ near the ends. We shall neglect the width of these zones, and we shall regard them as fictitious "edge beams" of height h and negligible width, located at the ends of the cross-section, which carry the shear force m^* and the torsional couple Bhm^*. The remainder of the bar will be called the "plate" (of width b), and we shall restrict our attention to the right-hand edge beam. We take the y-axis in the longitudinal direction, perpendicular to the plane of Fig. 1, and such that xyz is a right-handed system.

The torsional moments in the plate exert uniformly distributed moments m^* per unit length on the edge beam whose vectors have the negative x-direction (right-hand screw rule!). Equilibrium of moments for the edge beam is ensured by the presence of the transverse shear force m^* indicated in Fig. 1. Even if the twist, and hence also the torsional moments m^* per unit width in the plate, are functions of the axial coordinate y, the equilibrium of moments around the x-axis is not disturbed. Equilibrium of forces in the z-direction, however, now requires a downward load $m^*_{,y}$ per unit length on the edge beam, and this load can be exerted only by the plate. The reaction on the plate is an upward load $m^*_{,y}$ per unit length, and we have arrived at Lamb's interpretation of the reduced shear force condition at a free edge of a plate.

So far it has been ignored, however, that equilibrium of moments around the y-axis is also disturbed if m^* depends on y. In order to restore this equilibrium the plate must exert on the edge beam a moment around the negative y-axis of amount $(Bhm^*)_{,y}$. Its reaction on the plate is an "external" bending moment $(Bhm^*)_{,y}$ around the positive y-axis, in agreement with

(6.19). It also agrees with (6.17)—(6.19) that the torsional moment Bhm^* in the edge beam does not result in other "external" edge loads on the plate in the case of a straight edge.

The generalisation of the fictitious edge beam concept to a curved edge of a shell is depicted in Fig. 2. On the left we have drawn an edge element of the shell with the loads exerted by the inner side of the edge beam, and on the right an element of the edge beam itself with the internal loads in one cross-section. In order to avoid a confused sketch we have omitted the reactions of

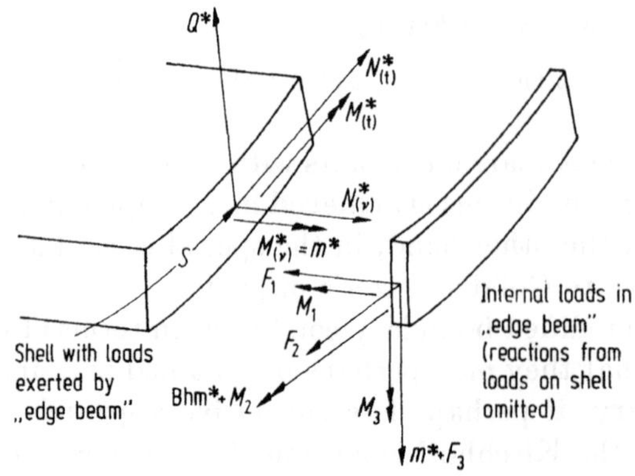

Fig. 2. Internal loads in edge beam and "external" loads exerted on shell edge.

the loads on the shell as well as the internal edge beam loads in the second cross-section. The outer side of the edge beam is, of course, free from loads in the case of a free edge. The transverse shear force m^* and the torsional moment Bhm^* are the predominant internal loads in a slightly curved edge beam, but we have to allow now for additional (small) internal loads, forces F_1, F_2, F_3 and couples M_1, M_2, M_3 in Fig. 2. The equations of equilibrium of an edge beam element are

$$M_{1,s} - \varkappa_{(n)}(Bhm^* + M_2) + \varkappa_{(t)} M_3 + F_3 = 0, \tag{6.20}$$

$$(Bhm^* + M_2)_{,s} - \varkappa_{(v)} M_3 + \varkappa_{(n)} M_1 - M^*_{(t)} = 0, \tag{6.21}$$

$$M_{3,s} - \varkappa_{(t)} M_1 + \varkappa_{(v)}(Bhm^* + M_2) - F_1 = 0, \tag{6.22}$$

$$F_{1,s} - \varkappa_{(n)} F_2 + \varkappa_{(t)}(m^* + F_3) - N^*_{(v)} = 0, \tag{6.23}$$

$$F_{2,s} - \varkappa_{(v)}(m^* + F_3) + \varkappa_{(n)} F_1 - N^*_{(t)} = 0, \tag{6.24}$$

$$(m^* + F_3)_{,s} - \varkappa_{(t)} F_1 + \varkappa_{(v)} F_2 - Q^* = 0. \tag{6.25}$$

Conclusions may only be drawn from these equations of equilibrium by taking proper account of the assumptions that the shell edge is only slightly curved, and that the stress distribution in the shell has a relatively large minimum wave length. Let ε^* denote the larger of the numbers h/L, $h\,|\varkappa_{(v)}|$, $h\,|\varkappa_{(t)}|$, $h\,|\varkappa_{(n)}|$. Our assumption that the shear force m^* and the torsional moment Bhm^* in the edge beam are the predominant internal loads is now expressed by the (qualitative) estimates

$$F_1, F_2, F_3 = 0(\varepsilon^* m^*), \; M_1, M_3, M_3 = 0(\varepsilon^* h m^*). \tag{6.26}$$

These estimates are also confirmed by a (qualitative) examination of the stress distribution in the edge zone of depth $0(h)$ in a cross-section of the shell perpendicular to the edge curve. Inspection of Eqs. (6.20)—(6.25) shows that we may now ignore M_1, M_2 and M_3. From Eqs. (6.20) and (6.22) we have then $F_1 = \varkappa_{(v)} Bhm^*$ and $F_3 = \varkappa_{(n)} Bhm^*$. The other equations may now be written in the form

$$N^*_{(v)} = \varkappa_{(t)} m^* + [\varkappa_{(v)} Bhm^*]_{,s} + \varkappa_{(t)}\varkappa_{(n)} Bhm^* - \varkappa_{(n)} F_2, \tag{6.27}$$

$$N^*_{(t)} = -\varkappa_{(v)} m^* + F_{2,s}, \tag{6.28}$$

$$Q^* = m^*_{,s} + [\varkappa_{(n)} Bhm^*]_{,s} - \varkappa_{(v)}\varkappa_{(t)} Bhm^* + \varkappa_{(v)} F_2, \tag{6.29}$$

$$M^*_{(t)} = (Bhm^*)_{,s}. \tag{6.30}$$

If all terms in the right-hand members are ignored, except the dominating first terms in (6.27)—(6.29), we arrive at the classical Kirchhoff conditions for a free edge. The correction expressed by (6.16) or (6.19) is confirmed by (6.30). In order to verify the identification of (6.17), (6.18) with (6.27), (6.29), and to confirm (6.14) by (6.28), we need an additional argument to sharpen the estimate (6.26) in the case of F_2. To this end we observe that the normal force F_2 in the edge beam arises as a resultant force of normal stresses on a cross-section perpendicular to the edge due to the presence, and in particular the variation of m^* along the edge. The order of magnitude of these stresses is thus $\varepsilon^* m^*/h^2$, and they penetrate only to a depth of order h into the shell. Their distribution over the thickness is exactly skew-symmetric in the case of a flat plate, and approximately so, with a relative error of order ε^*, in the case of a shell. Hence we have indeed the sharper estimate $F_2 = 0(\varepsilon^{*2} m^*)$, and we may omit the terms involving F_2 in (6.27), (6.28) and (6.29), thus confirming completely the previous results based on energy considerations.

The reader will have noticed that our reasoning, leading to the modified boundary conditions at a free edge, is based on qualitative error estimates, of a similar character as in [43]. Even if we have full confidence in this physical argument, in particular in the energy approach, a rigorous proof seems to require more concrete, quantitative error estimates for the solutions of the shell equations with the modified boundary conditions. We have no doubt that the Prager-Synge theorem will enable us again to derive such estimates for the L_2-norm of the errors, but the required analysis will undoubtedly be laborious, and the problem is still under investigation.

It is a more difficult problem to analyse in detail, and with the same accuracy $\varepsilon^2 = h^2/L^2 + h/R$, the stress distribution in the dege zone of depth $0(h)$. Apparently we cannot evade here the solution of one of Gol'denweizer's plane strain problems (cf. e.g. [17]). Fortunately, however, the stresses of most interest may be calculated without solving the plane strain problem. We refer here to the normal stresses on a cross-section perpendicular to the edge at the so-called "corner points", the intersections of the faces with the free edge. At these points, always in a state of (approximately) plane stress, even if the edge tractions do not vanish, the stress contribution of the plane strain problem, which is formulated in terms of edge loads to be removed, may be calculated directly from the normal edge tractions at these points. This contribution has to be added to the membrane and bending stresses of the solution of the shell problem under the modified boundary conditions (6.13)—(6.16), and the (uniaxial) warping stress associated with non-uniform torsion. We refer to van der Heijden's report [62] for details of the analysis, and we note here only the general formula in which the results may be presented

$$\sigma_{tt}\left(\text{edge}, z = \pm \frac{1}{2} h\right) = \frac{n_{tt}}{h} \mp \frac{6 m_{tt}}{h^2} \pm \left[(1+\nu)\frac{84\zeta(3)}{\pi^3} - \nu \frac{1116\zeta(5)}{\pi^5}\right] \frac{m^*_{,s}}{h}, \tag{6.31}$$

where $\zeta(n) = 2^n(2^n - 1)^{-1} \sum_{k=0}^{\infty} (2k+1)^{-n}$ is Riemann's zêta function (we note in passing that $B = (186/\pi^5)\,\zeta(5)$). The first two terms in (6.31) represent the stresses of classical shell theory for the *modified* boundary conditions (6.13)—(**6.16**). The first term between the brackets represents the uni-axial stress arising from constrained warping in non-uniform torsion, and the last term within the brackets is due to the plane strain problem.

Our remarks about the need of a rigorous proof of the accuracy achieved in the interior stresses with the aid of the modified boundary conditions also applies, of course, to the edge stresses. As a partial check, *van der Heijden* [62] has applied his approximate solution to the case of an infinite flat plate, weakened by a cylindrical hole of radius a, loaded in uniform torsion at infinity. *Alblas* had much earlier obtained the exact three-dimensional solution of this problem, as well as the first terms, those of order $(h/a)^0$ and (h/a), in the asymptotic expansion for the corner stress [63], and these exact results fully confirm van der Heijden's approximation.

Gol'denweizer has also proposed modified boundary conditions for *simply-supported* and *clamped* edges of a plate or shell (e.g. [17]). The conditions imposed, however, are unlikely to be more representative of the actual situation in engineering practice than the classical Kirchhoff conditions. We recall our earlier remarks that our lack of knowledge about the actual edge conditions is the most serious obstacle to an improved accuracy in shell theory.

7. Refined shell theories

Classical shell theory is the basic tool in the design and analysis of shell structures, and the error estimates for the nonlinear equations in section 4, and for the solutions of the linear equations in Section 5, serve only to strengthen our belief in classical theory as an adequate tool, if the basic assumptions of the first-approximation theory are satisfied, viz.

(a) the material is elastic, homogeneous and isotropic;

(b) the shell is of constant thickness h;

(c) the shell is thin, $h/R \ll 1$;

(d) the loads are smooth, $h/L \ll 1$;

(e) the strains are small, $\eta \ll 1$.

Some of these assumptions are more essential than others. For example, the theory can easily be generalized to the case of a slowly varying thickness h [10], and to anisotropic materials for which the tangent planes of surfaces parallel to the middle surface are planes of elastic symmetry [38]. The type of inhomogeneity present in sandwich or multi-layered shells requires a modification of the theory, but appropriate ad-hoc assumptions in the case of sandwich shells lead to a reasonably simple formulation with an even better accuracy than the first-approximation theory of homogeneous shells of the same thickness [76, 77, 79]. Inelastic behaviour of the material poses quite a different problem, outside the scope of our review. Nor shall we discuss in any detail large elastic deformations of thin shells or membranes, as may occur in rubber-like materials. The interested reader is referred here to some monographs on nonlinear elasticity theory [80, 81]. The distinction between two independent strain parameters, one for the extensional strains and one for the flexural strains, helpful already in the theory of small strains in the classification and further simplification of shell problems [66, 47], may here become essential to obtain a theory of manageable complexity. Sensenig's recent report [82] makes a first attempt in this direction in the case of flat plates.

Suffice it here to make a few comments on refined theories of shells in the more narrow sense, i.e. for shells which are homogeneous and isotropic, but not-so-thin, say h/R in the range $0.1-0.2$. It is a distinct advantage of the asymptotic approach that it is capable, in principle, of systematic successive refinements. In view of the labour involved already in the derivation and justification of first-approximation theory by means of the asymptotic method, it seesm questionable to us, however, whether the effort required by a systematic refinement would be justified by the results which may conceivably be achieved. It should also be borne in mind here that the rapid development of numerical techniques in three-dimensional problems, in particular the finite element technique, may obviate the need of a (complicated) refined shell theory in the near future.

It seems to us that less systematic, more or less ad-hoc, but simpler methods, based on physical intuition, may hold more promise for future developments for not-so-thin shells, where a *first* refinement beyond classical theory becomes meaningful. In fact, classical theory itself is an example of this approach, and the most significant contribution, due to the more sophisticated analyses by *Gol'denweizer*, *John* and many other writers, is a justification a posteriori, and occasionally a correction, of previously insecure results of the classical theory, rather than the independent development of the theory as a whole.

Reissner's theory of flat plates, including the effect of transverse shear deformation [74], extended to shells by *Naghdi* [75], is both a good and pioneering example of the type of theory we have in mind, as well as a typical illustration of the difficulties which have still to be overcome. Reissner's theory is a definite improvement in the interior domain, away from the edges, provided that the surface loads are still sufficiently smooth. Its primary aims, however, to allow for a better approximation of the actual boundary conditions and for a more accurate description of the state of stress near edges, have not yet been fully achieved. This somewhat unexpected shortcoming of Reissner's theory is evident from the three-dimensional analysis by *Alblas* for a plate with a traction-free circular cylindrical hole [63], and its explanation is to be found in the later more accurate asymptotic analyses of edge effects, referred to previously.

In connection with our advocacy for ad-hoc methods for the refinement of shell theory, which are already quite popular in the literature, we cannot evade making again some cautionary remarks. In the development of an approximate theory, based on apparently plausible physical assumptions, it is essential to bear in mind the physical interpretation of intermediate results at every stage in the analysis, and to apply appropriate corrections to the initial assumptions where this is required by the physics of the problem. It is indeed quite dangerous to derive a physical theory by a systematic and rigorous mathematical development of initial (approximate) assumptions unless due account is taken of the physical consequences at every step in the analysis. To physicists and engineers these remarks will look like the forcing of an open door, but experience with quite a few papers on shell theory published in the last five years shows the need of a repetition of such cautionary remarks, made on an earlier occasion [39]. We shall mention no names, but we shall cite two examples, one of them leading to absurdities, the other one to serious errors.

In the first example the usual kinematic assumption of *Kirchhoff* and *Love* (normals of the undeformed middle surface moving to normals of the deformed middle surface without any change in length) is adhered to so rigorously that the transverse shear stresses, and hence also the transverse shear forces, are concluded to be zero. This "theory" has actually found its way to print in a reputable scientific journal. The second example occurs in a paper, aiming at a more precise theory of shells by allowance for transverse shear deformations. It is assumed here that normals to the undeformed middle surface remain straight and retain their original length, although they are not required to remain normal to the deformed middle surface. Application of Hooke's law for the stresses parallel to the middle surface is equivalent to the assumption of plane strain (for these stresses), resulting in an overestimate of the extensional and flexural rigidities. In spite of the discrepancy with classical plate theory, noted in this paper, the results are presented as those of a new "more precise" plate bending theory.

We end our cursory comments on refinements in shell theory with the suggestion that the development of appropriate simplifying physical assumptions may perhaps benefit from exact three-dimensional solutions of relatively simple problems. For example, the three-dimensional analysis by *Alblas*, referred to previously, supports our approximate discussion of stresses near a free edge. In the theory of flat plates Lur'e's three-dimensional analysis for sinusoidal loading in both directions on the middle surface of an infinite plate [83] may be exploited to investigate the detailed stress distribution in a rectangular plate with some sort of simply-supported edges (zero deflection and vanishing normal stresses on the edge). We have already ascertained, in unpublished notes by one of the authors and *G. Hommel*, that Lur'e's theory confirms the validity of Reissner's theory in the interior domain. Likewise, Leibenson's effective use of three-dimensional theory in his solution of the buckling problem for a spherical shell under external pressure [56] suggests spherical shells as a second convenient test case. Finally, the cylindrical shell offers similar possibilities (cf. [59]).

8. Concluding remarks

We conclude our review with mention of a number of open problems is shell theory which, in our opinion, are in particular need of further research. Most *fundamental* among these is, perhaps, the question of error estimates for the solutions. It would be of great value, also for the applications, to have (accurate) pointwise error estimates, both for displacements and stresses. As mentioned before, the method employed by *Ho* and *Knowles* [2] may possibly be extended to more general loading conditions in the linear theory for shells of revolution. At present we have no clear idea, however, of the road to be followed for shells of arbitrary shape, not even in the linear theory, and we recall our earlier remark that the problem of solution error estimates in the case of the nonlinear theory is entirely open.

In the context of the nonlinear theory it seems desirable to compare the predictions of Danielson's Eqs. (4.9)—(4.12) and of our canonical Eqs. (4.14)—(4.17) with the results of simpler equations for special configurations, in particular Reissner's theory for axisymmetric deformations of shells of revolution [52]. In the case of rubberlike materials it will also be worthwhile to try and obtain reasonably simple and accurate stress-strain relations for finite stretching (and small bending).

As we have already emphasized repeatedly, the most critical uncertainties in shell theory arise from our ignorance of the exact nature of the actual boundary conditions, except in the case of a free edge. This problem cannot be solved by shell theory alone since it refers to the complete structure of which the shell is only an essential part. It is a regrettable consequence of this situation that the predictions of shell theory have no better accuracy than $\varepsilon^* = h/L + hR$ for structural applications, and refinements of shell theory as such are of no avail. If a solution of this problem is to be achieved at all, it will have to consider in detail the supporting structure as well as the shell itself. It would seem that a penetrating analysis of this *interaction problem between the shell and its supporting structure* is a more rewarding theme for future research than a detailed refinement of shell theory.

References

1. Aron, H.: J. Reine u. Angew. Math. 78 (1874) 136—173.
2. Ho, C. L., Knowles, J. K.: Z. A. M. P. 21 (1970) 352—377.
3. John, F.: Comm. Pure and Appl. Math. 18 (1965) 235—267.
4. John, F.: Proc. 2nd IUTAM Symp. Theory of Thin Shells, Copenhagen 1967. Berlin/Heidelberg/New York: Springer 1969, 1—14.
5. John, F.: Comm. Pure and Appl. Math. 24 (1971) 583—615.
6. Morgenstern, D.: Arch. Rat. Mech. and Anal. 4 (1959) 145—152.
7. Morgenstern, D., Szabo, I.: Vorlesungen über Theoretische Mechanik, § 9.5, Berlin/Göttingen/Heidelberg: Springer 1961.
8. Nordgren, R. P.: Quart. Appl. Math. 28 (1971) 587—595.
9. Danielson, D. A.: Proc. Kon. Ned. Ak. Wet. B 74 (1971) 294—300.
10. Koiter, W. T.: Proc. Kon. Ned. Ak. Wet B 73 (1970) 169—195.
11. Koiter, W. T.: Proc. Int. Congr. Math. Nice 1970. Paris: Gauthier-Villars, vol. 3 (1971) 123—130.
12. Simmonds, J. G.: Quart. Appl. Math. 29 (1971) 439—447.
13. Sensenig, C. B.: Int. J. Engng. Sci. 6 (1968) 435—464.
14. Green, A. E.: Proc. Roy. Soc. London A 269 (1963) 481—491.
15. Gol'denweizer, A. L.: Prikl. Mat. Mekh. 26 (1962) 668—686. English translation 1000—1025.
16. Gol'denweizer, A. L.: Prikl. Mat. Mekh. 27 (1963) 593—608. English translation 903—924.
17. Gol'denweizer, A. L.: Proc. 11th Int. Congr. Appl. Mech. München 1964. Berlin/Heidelberg/New York: Springer 1966, 306—311.
18. Kolos, A. V.: Prikl. Mat. Mekh. 28 (1964) 582—589. English translation 718—726.
19. Kolos, A. V.: Prikl. Mat. Mekh. 29 (1965) 771—781. English translation 914—925.
20. Gol'denweizer, A. L., Kolos, A. V.: Prikl. Mat. Mekh. 29 (1965) 141—155. English translation 151—166.
21. Gol'denweizer, A. L.: Prikl. Mat. Mekh. 32 (1968) 684—695. English translation 704—718.
22. Gol'denweizer, A. L.: Prikl. Mat. Mekh. 33 (1969) 996—1028. English translation 971—1001.
23. Gol'denweizer, A. L.: Proc. 2nd IUTAM Symp. Theory of Thin Shells, Copenhagen 1967. Berlin/Heidelberg/New York: Springer 1969, 31—38.
24. Gol'denweizer, A. L.: On two-dimensional equations of the general linear theory of thin elastic shells. Problems of Hydrodynamics and Continuum Mechanics (L. I. Sedov anniversary volume), Soc. Ind. Appl. Math., Philadelphia (1969) 334—352.
25. Rutten, H. S.: Proc. 2nd IUTAM Symp. on the Theory of Thin Shells, Copenhagen 1967. Berlin/Heidelberg/New York: Springer 1969, 115—134.
26. Rutten, H. S.: Asymptotic approximation in the three-dimensional theory of thin and thick elastic shells. Thesis Delft 1971, published by Netherlandse Boekdruk Industrie N.V., 's Hertogenbosch.
27. Cosserat, E., Cosserat, F.: Théorie des corps déformables. Paris: Herrmann 1909.
28. Naghdi, P. M.: The theory of shells and plates.

Flügge's Handbuch der Physik, 2nd edition, vol. VI-2. Berlin/Heidelberg/New York: Springer 1972.
29. Niordson, F. I.: Int. J. Solids and Structures 7 (1971) 1573–1579.
30. Flügge, W.: Statik und Dynamik der Schalen. Berlin: Springer 1934.
31. Flügge, W.: Stresses in shells. Berlin/Göttingen/Heidelberg: Springer 1960.
32. Wlassow, W. S.: Allgemeine Schalentheorie und ihre Anwendung in der Technik. Deutsche Ausgabe der russischen Auflage (1949) unter Redaktion von A. Kromm. Akademie-Verlag, Berlin (1958).
33. Novozhilov, V. V.: The theory of thin shells. Translated from the Russsian edition (1951) by P. G. Lowe. Groningen: Noordhoff 1959.
34. Gol'denweizer, A. L.: Theory of elastic thin shells. Translation from the Russian edition (1953) ed. G. Herrmann, Oxford: Pergamon Press (1961).
35. Girkmann, K.: Flächentragwerke, 4. Aufl. Oxford: Springer 1956.
36. Mushtari, Kh. M., Galimov, K. Z.: Nonlinear theory of thin elastic shells. Translated from the Russian edition (1957) by J. Morgenstern and J. J. Schorr-Kon. NASA-TT-F62 (1961).
37. Kraus, H.: Thin elastic shells. New York: Wiley 1967.
38. Naghdi, P. M.: Foundations of elastic shell theory. Progress in Solid Mechanics, ed. I. N. Sneddon and R. Hill. Amsterdam: North-Holland 1963, 1–90.
39. Koiter, W. T.: Proc. 2nd IUTAM Symp. on the Theory of Thin Shells, Copenhagen 1967. Berlin/Heidelberg/New York: Springer 1969, 93–105.
40. Love, A. E. H.: Mathematical theory of elasticity, 4th edition. Cambridge University Press 1927.
41. Reissner, E.: J. Math. and Phys. 42 (1963) 263–277.
42. Reissner, E.: Proc. 11th Int. Congr. Appl. Mech., München 1964. Berlin/Heidelberg/New York: Springer 1966, 20–30.
43. Koiter, W. T.: Proc. IUTAM Symp. on the Theory of Thin Elastic Shells, Delft 1959. Amsterdam: North-Holland 1960, 12–33.
44. Budiansky, B., Sanders, J. L.: On the "best" first-order linear shell theory. Progress in Applied Mechanics (Prager Anniversary Volume), New York: Macmillan 1963, 129–140.
45. Leonard, R. W.: Nonlinear first approximation thin shell and membrane theory. Thesis Virginia Polytechnic Institute 1961.
46. Sanders, J. L.: Quart. Appl. Math. 21 (1963) 21–36.
47. Koiter, W. T.: Proc. Kon. Ned. Ak. Wet. B 69 (1966) 1–54.
48. Budiansky, B.: J. Appl. Mech. 35 (1968) 393–401.
49. Danielson, D. A.: Int. J. Engng. Sci. 8 (1970) 251–259.
50. Simmonds, J. G., Danielson, D. A.: Proc. Kon. Ned. Ak. Wet. B 73 (1970) 460–478.
51. Simmonds, J. G., Danielson, D. A.: J. Appl. Mech. 39 (1972) 1085–1090.
52. Reissner, E.: Proc. 3rd Symp. Appl. Math., Ann Arbor 1949. New York: McGraw-Hill 1950, 27–52.
53. Koiter, W. T.: ZAMP 21 (1970) 534–538.
54. Simmonds, J. G.: ZAMP 22 (1971) 339–345.
55. Simmonds, J. G.: To appear in ZAMP 23 (1972).
56. Leibenson, L. S.: Matematishkii Svornik 40 (1933) 429–442.
57. Neut, A. van der: De elastische stabiliteit van den dunwandigen bol. Thesis Delft, published by H. J. Paris, Amsterdam 1932.
58. Koiter, W. T.: Proc. Kon. Ned. Ak. Wet. B 72 (1969) 40–123.
59. Newman, J. B.: To appear in J. Appl. Mech. 39 (1972).
60. Friedrichs, K. O.: Proc. 3rd Symp. Appl. Math., Ann Arbor 1949. New York: McGraw-Hill 1950, 117–124.
61. Koiter, W. T., Heijden, A. van der: Modified boundary conditions at a free edge of a thin plate or shell. To be published in Proc. Kon. Ned. Ak. Wet.
62. Heijden, A. van der: Benaderingstheorie voor het bepalen van spanningen langs de vrije schaalrand. Report 456, Laboratory of Engineering Mechanics, Dept. of Mechanical Engineering, University of Technology, Delft 1971.
63. Alblas, J. B.: Theorie van de driedimensionale spanningstoestand in een doorboorde plaat. Thesis Delft, published by H. J. Paris, Amsterdam 1957.
64. Lur'e, A. I.: Prikl. Mat. Mekh. 4, Nr. 2 (1940).
65. Gol'denweizer, A. L.: Prikl. Mat. Mekh. 4, Nr. 2 (1940).
66. Koiter, W. T.: Proc. Kon. Ned. Ak. Wet. B 64 (1961) 612–619.
67. Koiter, W. T.: Proc. Kon. Ned. Ak. Wet. B 67 (1964) 117–126.
68. Koiter, W. T.: ZAMP 20 (1969) 642–653.
69. Mikhlin, S. G.: The problem of the minimum of a quadratic functional. Translated from the Russian edition (1952) by A. Feinstein. San Francisco: Holden-Day 1965.
70. Prager, W., Synge, J. L.: Quart. Appl. Math. 5 (1947) 241–269. Cf. also J. L. Synge: The hypercircle in mathematical physics. Cambridge University Press 1957.
71. Trefftz, E.: Z. angew. Math. und Mech. 15 (1935) 101–108.
72. Sokolnikoff, I. S.: Mathematical theory of elasticity, 2nd edition, New York: McGraw Hill 1956.
73. Goodier, J. N.: Trans. Roy. Soc. Canada 32 (1938) 65–88.
74. Reissner, E.: J. Appl. Mech. 12 (1945) 69–77.
75. Naghdi, P. M.: Quart. Appl. Math. 14 (1957) 369–380.
76. Reissner, E.: Small bending and stretching of sandwich-type shells. NACA Report 975 (1950).
77. Plantema, F. J.: Sandwich construction. New York: Wiley 1966.
78. Lamb, H.: Proc. Lond. Math. Soc. 21 (1891) 119.

79. Grigoliouk, E. I., Chulkov, P. P.: On the theory of multilayer shells. Contributions to the Theory of Aircraft Structures, Delft University Press 1972, 171–183.
80. Green, A. E., Adkins, J. E.: Large elastic deformations. Oxford: Clarendon Press 1970.
81. Stoker, J. J.: Nonlinear elasticity. New York: Gordon and Breach 1968.
82. Sensenig, C. B.: Plate estimates involving two strain parameters. Report IMM 388. Courant Institute of Mathematical Sciences, New York University 1971.
83. Lur'e, A. I.: Räumliche Probleme der Elastizitätstheorie. Deutsche Augsabe der russischen Auflage (1955) von G. Landgraf. Berlin: Akademie-Verlag 1963.
84. Chernykh, K. F.: Linear theory of shells, Parts I and II, NASA TT-F 11,562. Translated from the Russian edition, Leningrad Univ. Publ. House, Part I (1962), Part II (1964).
85. Cicala, P.: Systematic approximation approach to linear shell theory. Torino: Levrotto and Bella (1965).
86. Danielson, D. A., Simmonds, J. G.: Int. J. Engng. Sci. 7 (1969) 459–468.

Drag reduction by polymer addition

Marten T. Landahl

Stockholm, Sweden, and Cambridge, Mass., USA

1. Introduction

The reduction of flow friction drag through polymer addition, a phenomenon first subjected to systematic study by *Toms* (1948) (with those name it is commonly associated) together with *Oldroyd* (1948), but of which the first documented discovery is due to *Mysels* (1949), has been the subject of very extensive research during the last decade. However, despite the combined efforts of specialists from many different fields, no satisfactory explanation of the phenomenon has as yet been found. Drag reduction has been observed with a large number of additive-fluid combinations (a few of which are shown in Fig. 1), the most effective ones being high-molecular-weight linear polymers (such as polyethyleneoxide) in water. There are clear experimental indications that drag reduction is associated with a modification of the turbulence structure, and this is most remarkable in view of the permanence and statistical stability exhibited by ordinary Newtonian shear flow turbulence. Therefore, the phenomenon of drag reduction becomes of interest not only because of the promising technical applications it offers but just as much for fundamental fluid mechanical reasons. Through the unravelling of the mechanism underlying the phenomenon one should not only improve the prospects of finding more efficient and perhaps completely novel methods of turbulence control, but should also hope to gain a substantially deepened understanding of the fundamental nature of turbulence as well.

Additive	Fluid
Polyethyleneoxide (PEO, polyox)	Water
Polyacrylamide (PAM, separan)	Water
Guar gum (GGM)	Water
Polymethylmethacrylate (PMM)	Monochlorobenzene (TOMS, 1948)
Various soaps	Water, toluene
Nylon fibres	Water (BOBKOWICZ & GAUVIN, 1965)

Fig. 1. Some additive-fluid systems exhibiting drag reduction.

In the present review of the subject, special attention is paid to the changes in turbulence structure brought about by the additives, and an effort is made to understand these in the light of recent advances in the study of shear flow turbulence in ordinary fluids. Also, a simplified theoretical model of shear flow wall turbulence is presented, with the aid of which an exploratory study of the influence of some non-Newtonian fluid properties is carried out. A particularly intriguing question is posed by the remarkable effectiveness of the extremely high-molecular-weight polymers with which only a few parts per million of weight of solvent is sufficient in some cases to lead to drag reductions of 50 per cent or more; how is it possible that such minute amounts can change the mechanical properties of the fluid enough to produce so dramatic effects? A general review of the very voluminous literature on the subject of drag reduction has not been attempted here (to date several hundred publications on the subject have appeared). Generally, the emphasis is put on the more recent contributions, particularly those having a bearing on turbulence structure, but some of the older material of interest for giving the background is also included.

2. Data on drag reduction and mean velocity profiles

Most of the data on drag reduction published to date has been obtained in pipe flows following *Toms* (1948) original procedure. However, results for flat-plate boundary layers such as those by *Kowalski* (1967) and by *Latto* & *Shen* (1970) show the same qualitative and quantitative behaviour as for pipe flow as do also those obtained in the rotating coaxial cylinder experiments of *Merrill* et al. (1966). This is perhaps not surprising in view of the similar behaviour of all turbulent wall shear flows in Newtonian fluids. We shall therefore here consider only the pipe flow data.

Systematic studies of pipe flow (e.g., *Metzner* & *Park*, 1964; *Savins*, 1964; *Elata* & *Tirosh*, 1965; *Fabula*, 1965; *Goren* & *Norbury*, 1967; *Seyer* & *Metzner*, 1967; *Virk* et al., 1967 and *Virk* et al., 1970) have identified several drag reduction characteristics common to dilute solutions of linear polymers such as polyethyleneoxide. (Other drag reducing agents such as complex soaps with micellar molecular structures produce quite different behaviour with onset of drag reduction occurring at very low shear rates, but these will not be discussed here.) At low flow rates such fluids obey the laminar friction law with a slight increase in drag due to the increased viscosity caused by the polymer. At higher velocities, the flow will go through transition at a Reynolds number which—at least if the pipe inlet disturbances are moderately high—will not be appreciably different from the Newtonian value. Drag reduction sets in at a well defined flow rate, and above this the frictional resistance depends on the polymer-solvent system, molecular weight and concentration. With increase in flow rate a maximum drag reduction independent of polymeric molecular parameters may be reached. For still higher flow rates the drag reduction effect could eventually diminish due to polymer degradation. Pipe flow drag data typical of dilute polyethyleneoxide solutions are shown in Fig. 2 taken from *Paterson* & *Abernathy* (1970).

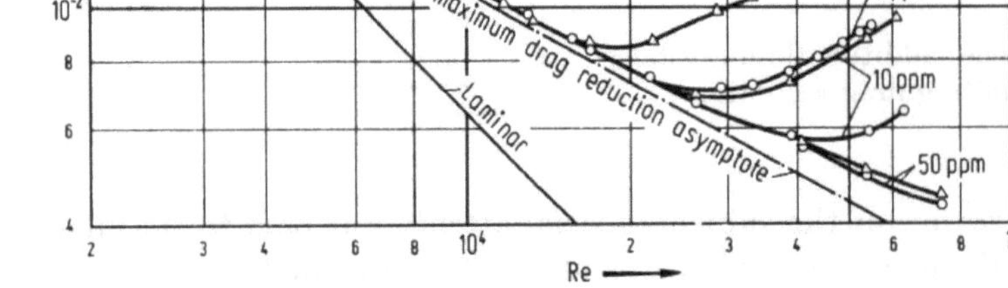

Fig. 2. Pipe flow drag data for solutions of polyethyleneoxide (from *Paterson* & *Abernathy*, 1970).

The onset of drag reduction has been shown by *Virk* et al. (1967) to be approximately independent of pipe diameter and for a given solvent-polymer combination to depend only on the wall shear stress. Furthermore, by comparing results for polymers of different molecular weight, they were able to show that the onset data correlated with the formula

$$D_M u_\tau^* / \nu = 0.015 \pm 0.005 \tag{1}$$

where D_M is the effective diameter of the randomly coiled macromolecule at rest and u_τ^* the value of the shear velocity $u_\tau = \sqrt{\tau_w/\varrho}$ at onset. This result is quite surprising because it implies that drag reduction would start when the diameter of the molecule is about 10^{-3} times the size of the smallest dynamically significant eddy in the flow. Simplified theoretical models for the macromolecule (*Peterlin*, 1966) show that a large deformation of the molecule—possibly leading to a complete extension of the random coil—would take place when the shortest time scale of motion would fall below twice the relaxation time of the molecule. However, the onset data correlate poorly on a time hypothesis. *Virk* (1971a) has also found that the onset of drag reduction in rough pipes occurs at the same shear stress as in smooth ones.

The existence of a maximum drag reduction asymptote has by now been well documented for most known solvent-polymer systems. Fig. 3, reproduced from *Virk* et al. (1970), shows data presented in Prandtl's coordinates $1/\sqrt{f}$ versus $\mathrm{Re}\sqrt{f}$. It is convenient to present results for pipe flow in this manner, because in these coordinates both the laminar drag law and the Prandtl-von Kármán law for Newtonian turbulent pipe flow

$$1/\sqrt{f} = 4.0 \log_{10}(\mathrm{Re}\sqrt{f}) - 0.4 \tag{2}$$

appear as straight lines in a semilogarithmic diagram. The drag asymptote was also shown to yield an approximately straight line of the form (*Virk* et al., 1970)

$$1/\sqrt{f} = 19.0 \log_{10}(\mathrm{Re}\sqrt{f}) - 32.4\,. \tag{3}$$

In the regime below maximum drag reduction, the friction factor varies both with polyme properties and flow variables.

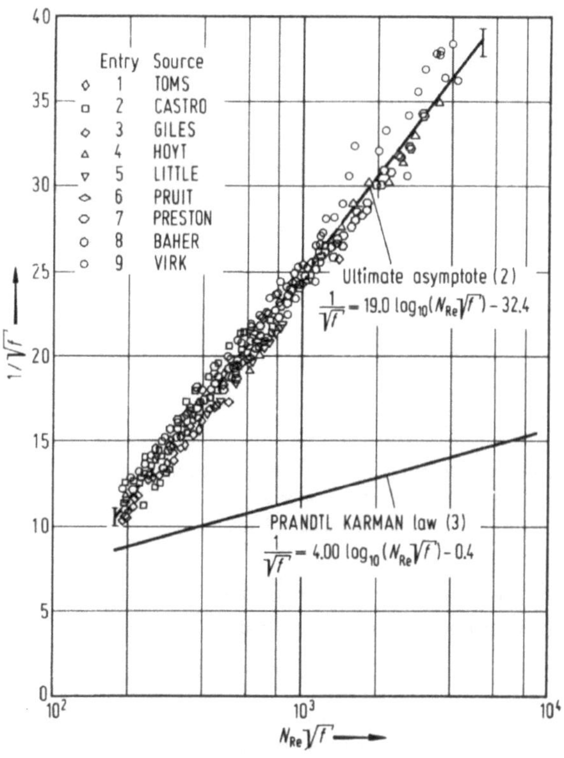

Fig. 3. The maximum drag reduction asymptote (from *Virk* et al., 1970).

Several attempts have been made to correlate friction data in the polymeric regime below maximum drag reduction, and we have singled out the recent ones of *Seyer* & *Metzner* (1969b) and of *Virk* (1971b) as being representative of two different approaches, one (Seyer & Metzner's) based on measured viscoelastic properties and the other (Virk's) working directly with molecular parameters. These are best understood in light of the behaviour of the mean velocity profiles. *Virk* (1971b) has compared published experimental data on mean velocities, and in Fig. 4 we have reproduced these (Fig. 2 of *Virk*, 1971b) showing the measured velocities plotted in the usual wall variables $U^+ = U/u_\tau$, and $y^+ = yu_\tau/\nu$. From these one can distinguish three different regions which can be represented approximately by

$$\text{(I)} \qquad U^+ = y^+; \quad 0 \leq y^+ \leq y_v^+, \tag{4}$$

$$\text{(II)} \qquad U^+ = A_2 \ln y^+ \quad B_2; \quad y_v^+ \leq y^+ \leq y_e^+, \tag{5}$$

$$\text{(III)} \qquad U^+ = A_3 \ln y^+ + B_3; \quad y_e^+ \leq y^+ \leq R^+ \tag{6}$$

respectively, where R^+ is the value of y^+ corresponding to the pipe center. The first region is the usual viscous sublayer extending to about $y^+ = 10$. The second region, called the "elastic" region, is best represented by the values $A_2 = 11.7$ and $B_2 = -17.0$. In the third region U^+ shows the same logarithmic variation with y^+ as in the Newtonian case, but the curve is displaced upwards. Thus, $A_3 \simeq 2.5$, but B_3 is a function of molecular and flow parameters. The mean velocity profile data hence indicate the existence of a Newtonian "plug" in the center of the pipe with a mixing length apparently unaffected by the polymers. At the condition of maximum drag reduction the elastic region II is assumed to extend all the way to the center of the pipe, and the Newtonian region III then disappears. *Virk* (1971b) determined the coefficients in the region II so as to make the profile give a grad variation consistent with the maximum drag asymptote under this assumption. Using (4)—(6) one can integrate the total volumetric flow rate and hence obtain a relation between friction force and flow rate.

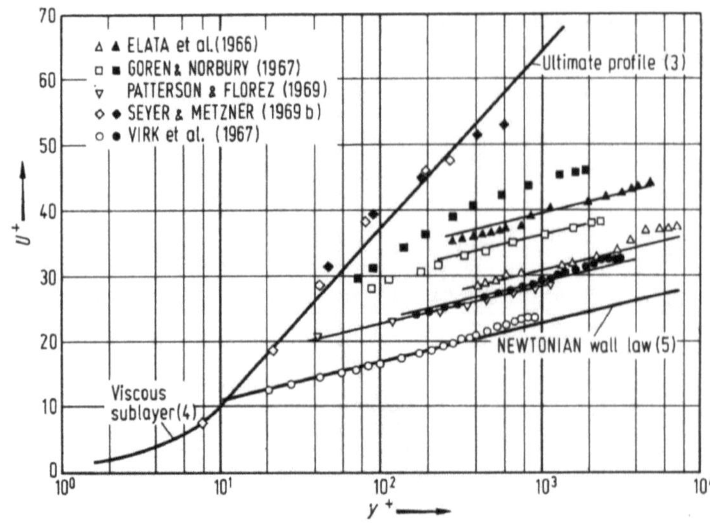

Fig. 4. Experimental mean velocity profiles during drag reduction (from *Virk*, 1971b). △ ▲ *Elata* et al. (1966); □ ■ *Goren* & *Norbury* (1967); ▽ *Patterson* & *Florez* (1969); ◇ ◆ *Seyer* & *Metzner* (1969b); ○ ● *Virk* et al. (1967).

Seyer & *Metzner* (1969b) assumed the viscous wall layer to extend all the way to the Newtonian region without an intermediate region. Thus, setting

$$U^+ = A \ln y^+ + B(\theta^+), \tag{7}$$

where $\theta^+ = \theta u_\tau^2/\nu$ and θ is the relaxation time of the fluid, they were able to correlate data for different polymer solutions. The constant A is the same as in the Newtonian case. (*Seyer* & *Metzner* took $A = 2.46$.) The nondimensional relaxation time θ^+ is the Deborah number formed by the ratio of fluid relaxation time to natural viscous time scale of the turbulence. The curve of the intercept function $B(\theta^+)$ obtained in this manner is reproduced in Fig. 5. Their data indicate an asymptotic value of B of about 30 for large relaxation times. *Virk* (1971b) made a comparison of the Seyer-Metzner model with his newer one and found that the latter is superior in representing measure profiles at conditions of large drag reduction. The Seyer-Metzner correlation is also difficult to use for highly dilute solutions, for which no reliable methods of determining the relaxation time θ are as yet available (see further the following section).

Virk (1971b) tries to relate the intercept constant B_3 in the following manner

$$B_3 = B_n + \delta \ln (R^+/R^{+*}),$$

where B_n is the value for a Newtonian flow, $B_n = 5.5$. The factor $\ln (R^+/R^{+*})$, where R^{+*} is the value of R^+ at onset, is assumed to represent an "excitation" factor which depends both on flow parameters and such polymetric parameters that may be associated with the excitation of the macromolecule. The function δ, which Virk termes a "slope increment factor" (since it gives the fractional increase due to the polymer in the slope of the curve of $1/\sqrt{f}$ versus $\log(\mathrm{Re}\sqrt{f})$), is assumed to depend solely on molecular parameters. By examining a large body of data, *Virk* (1971b) established that for each polymer-solvent system δ is approximately proportional to $c^{1/2}$, where c is the weight fraction if polymer additive. Setting

$$\delta = A_3(\bar{A}c/M)^{1/2}\, \Pi\,, \tag{8}$$

where $\bar{A}$ is Avogadros number, c the weight fraction and M the molecular weight, respectively, of the polymer (so that the quantity under square root represents the number of macromolecules per unit weight of solution) Virk finds that for all polymer species with the exception of one (guar gum) the function π (termed "intrinsic slope increment factor") can be fairly well represented by a single curve with the number of macromolecular backbone chain links, N, as a parameter. His results are reproduced in Fig. 6 (Virk's Fig. 5). Furthermore, in a logarithmic diagram all the data with the exception of that for guar gum were found to cluster around a single straight line of slope 3/2 indicating a simple power law of the form

$$\Pi = KN^{3/2}, \tag{9}$$

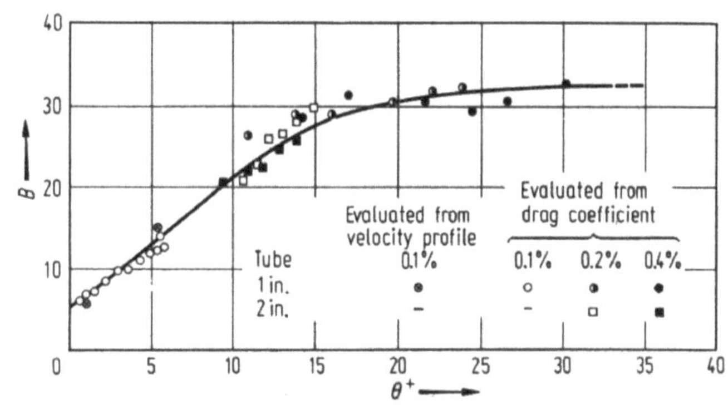

Fig. 5. Logarithmic intercept function for drag reduction velocity profiles (from *Seyer & Metzner*, 1969b). Data obtained with polyacrylamide in water.

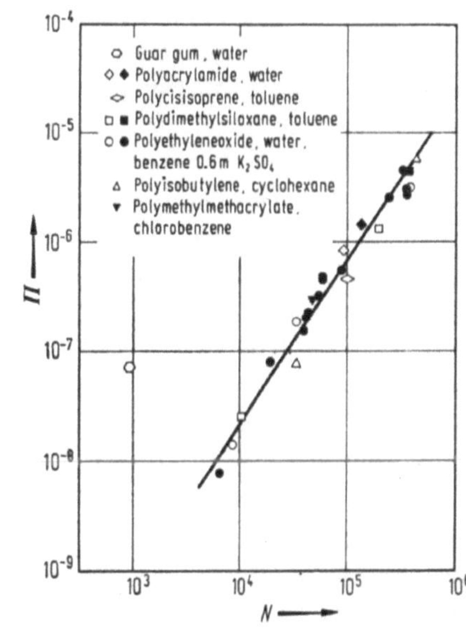

Fig. 6. Intrinsic slope increment factor Π as function of number of backbone links, N (from *Virk*, 1971b). ◇ guar gum, water; ◇ ◆ polyacrylamide, water; ◇ polycisisoprene, toluene; ■ □ polydimethylsiloxane, toluene; ○ ● polyethyleneoxide, water, bezene 0.6 m K_2SO_4; △ polyisobutylene, cyclohexane; ▼ polymethylmethacrylate, chlorobenzene.

where K is a constant found to have the approximate value $K = (2.3 \pm 0.8) \times 10^{-14}$. From a theoretical reasoning based on the estimate of the turbulent strain energy stored in the flexible macromolecule, *Virk* comes to the conclusion that Π should be proportional to b^3, where b is the bond length of the nomomer making up the macromolecule. Guar gum with a bond length of $b \sim 20$ Å, as compared to the other polymers having $b \sim 4$ Å, should therefore be about $(20/4)^3$ as effective as the others, as indeed roughly indicated by the data for guar gum which give $K \cong 280 \times 10^{-14}$.

3. Rheological properties

One of the most important and intricate questions in drag reduction research has been what specific bulk rheological property or properties of the fluid the effect might be ascribed to. It is usually stated (see, e.g., *Metzner* & *Park*, 1964) that drag reducing fluids show viscoelastic behaviour, and the suggestion that viscoelasticity is the cause of drag reduction has been almost universally accepted. Indeed, in larger concentrations the polymers effective in reducing the drag produce normal stress differences in pure shear of moderate magnitude, an effect usually associated with viscoelasticity. Direct demonstration of reversible elastic effects has also been given by *Darby* (1970), who with the aid of a specially designed transient rheometer was able to measure elastic rebound for fluids with polymer concentrations of 100 to 500 parts per million. However, such rheometric data are usually taken at low shear rates for which the normal stresses are expected to be much lower than at the high shear velocities of interest in drag reduction. Normal stress differences for drag reducing fluids are therefore usually determined from the mesaurement of the thrust of a capillary jet (see *Metzner* & *Park*, 1964), a method which does not suffer from this shortcoming. This method was the one used by *Gadd* (1966a, b) and by *Oliver* (1966), the latter also reporting results for fairly low concentration levels at which the method becomes unreliable, however. With the aid of a specially designed concentric tube rheometer, *Gadd* (1966b) was also able to detect a non-zero second normal stress difference $(\tau_{22} - \tau_{33})$ for dilute (40 p.p.m.) polyethyleneoxide solution but did not find any appreciable second normal stress difference for other good drag reducing polymers such as guar gum and polyacrylamide.

It has also been found that the viscoelastic properties exhibited depend strongly on aging and degradation of the polymer solution, especially in the case of polyethyleneoxide. *Brennen* and *Gadd* (1967) reported a much diminished total head loss with age of solution for a pitot tube in a stream of polyethyleneoxide solution indicating a diminshed first normal stress. They also stated that the second normal stress for a 70 p.p.m. solution vanished with aging. Several later investigators have noted the anomalous flow behaviour of fresh polyethyleneoxide solution; thus, *Kalashnikov* and *Kudin* (1970) found that the Strouhal number for the Kármán vortex sheet behind a circular cylinder was substantially reduced in a flow of fresh solution, whereas the same solution passed through a centrifugal pump gave a small increase in frequency as compared to water, as did a solution of guar gum. Despite such marked degradation effects, a degraded solution was still able to produce a substantial drag reduction (although somewhat diminished as compared to that of a fresh solution). Such behaviour has led *Barenblatt* et al. (1965) to propose that the viscoelastic behaviour of a polyethyleneoxide solution is a result of the formation of very large aggregates of macromolecules which in part disappear with aging. The larger aggregates will also be broken up by the shear field in the flow apparatus, so that large-scale motion such as vortex shedding would be affected by degradation but not the drag reduction.

Although standard rheometric measurements for very dilute polymer solutions (<20 p.p.m.) usually give no clear indication of departure from a purely Newtonian liquid (*Bryson* et al., 1971), there is other evidence that such fluids can show substantial non-Newtonian behaviour. *Metzner* & *Metzner* (1970) measured the stress levels developed in dilute solutions of polyacrylamide issuing from small orifices at high velocities. In this flow field the deformation is approximately that of pure rapid elongation. The ratio of elongational to shear viscosity (which is equal to three in a Newtonian fluid) determined in this way was found to be exceedingly high; a conservative estimate of the deformation rate level yielded values as high as 29,000. Possibly, the observed tendency of polymer solutions to form long threads may be related to such high values of elongational viscosity (*Lumley*, 1969).

In order to use laboratory rheological data in the equations of motion for a continuum fluid, one needs to incorporate them in a constitutive model for the fluid. A large number of such models has been proposed and studied in the rheological literature. *Spriggs* et al. (1966) made an appraisal of twenty different models for viscoelastic materials, and they concluded that in order to reproduce all essential rheological properties a model with at least four different constants was needed. In view of the uncertainty as to how viscoelastic effects come into play in the phenomenon of drag reduction, however, it would seem most appropriate to select, for exploratory studies, the simplest model incorporating viscoelasticity. Such a model is the convected Maxwell one proposed by *White* & *Metzner* (1963), which in Cartesian coordinates may be written

$$\tau_{ij} + \theta\, \delta\tau_{ij}/\delta t = 2\mu\, d_{ij}, \tag{10}$$

where τ_{ij} is the deviatoric shear stress, θ a relaxation time, $\delta/\delta t$ the Oldroyd time derivative

$$\frac{\delta\tau_{ij}}{\delta t} = \frac{\partial\tau_{ij}}{\partial t} + U_m \frac{\partial\tau_{ij}}{\partial x_m} - \tau_{mj}\frac{\partial U_i}{\partial x_m} - \tau_{im}\frac{\partial U_j}{\partial x_m},$$

and μ and d_{ij} have their usual meaning of shear viscosity and deformation rate tensor

$$d_{ij} = (1/2)\,(\partial U_i/\partial x_j + \partial U_j/\partial x_i),$$

respectively. The material coefficients θ and μ generally depend on the deformation rate. The constitutive relation (10) is a simplified version of *Oldroyd's* (1958) original one containing eight material constants. For exploratory purposes it is a great advantage to have a model with only one positive constant, namely θ, to describe departure from a Newtonian fluid. It has also some foundation in physical reality as *Lumley* (1971) has shown that a simple model for a dilute solution of polymer molecules employing a dumbbell model for the molecules leads to a constitutive relation of the Oldroyd type (albeit with three instead of two material coefficients). It is also the relation usually employed to convert data on measured normal stress differences to relaxation times; application of (10) to simple steady shear flow, $U = (U_1(x_2), 0, 0)$, gives

$$\tau_{11} - \tau_{22} = 2\theta\tau_{12}\mu\Gamma, \tag{11}$$

where $\Gamma = \partial U_1/\partial x_2$. In this model the fluid relaxation time θ is thus proportional to the first normal stress difference. On the basis of (10) one can also demonstrate the possibility of high resistance due to stretching as observed in the experiments by *Metzner* & *Metzner* (1970) quoted above.

In view of the great difficulties in measuring departures from Newtonian behaviour of very dilute drag reducing fluids, it would be natural to turn to theory to see what guidelines it may offer. In view of the large size of the macromolecules in comparison with the solvent molecules involved, a good mechanistic model for a drag reducing fluid is obtained by regarding it as a suspension of small particles in a continuum fluid. The rheology of suspensions is a field of research which has undergone a very extensive development since its inception by *Albert Einstein* in 1906 with his famous analysis of the increase in fluid viscosity due to the presence of rigid spheres. Good review articles on the subject can be found in Vols. I and IV of the series "Rheology", edited by *Eirich* (1956, 1966). In theoretical suspension mechanics one can distinguish the following three separate stages in the analysis:

i) calculation of the influence of the local flow field on the individual particle;

ii) determination of the effect of a statistical ensemble of particles on the bulk rheological properties for the suspension;

iii) analysis of the bulk motion of the suspension.

In i) fluid inertia can be ignored because of the small size of the particle, but the calculation may nevertheless become very complicated if the particle undergoes large deformations or is subjected to hydrodynamic interference from neighboring particles, situations that do arise for the fluid of interest here. Also, the effect of Brownian motion may need to be incorporated in determining the shape and average orientation of the particles. The greatest complication is perhaps caused by the coupling between stages iii) and i); the bulk motion of the suspension is influenced by the shape and orientation of the particles, which in turn depend on the local flow field created. The problem of determining the shape of the macromolecule can also become quite difficult depending on the degree of sophistication with which the molecule is modelled. The linear so-called bead- and-spring models employed by *Rouse* (1953) and by *Zimm* (1956) have been employed to calculate the relaxation time of a dilute suspension of coiled macromolecules subjected to a small oscillatory shear (*James & Acosta,* 1970). *Peterlin* (1966) employing a linear dumbbell model (i.e., two beads and one spring) demonstrated that the molecule can become infinitely extended in an irrotational straining field if the strain rate exceeds the inverse of twice the relaxation time of the molecule. In a pure shear, on the other hand, it will never become fully extended.

The infinite extension is of course a consequence of the assumption of spring linearity. With the aid of a more realistic nonlinear thread model, *Hinch** has shown recently that there is a critical strain rate above which the molecule becomes completely (at least 1/2) extended; below this the molecule is not extended at all. One would therefore expect that in a turbulent field of sufficient intensity a great proportion of the macromolecules is uncoiled. On the basis of *Peterlin's* (1966) result *Lumley* (1969) estimated that 70% of all molecules should be extended in a turbulent flow if the root mean-square strain rate exceeds the critical value of the inverse of twice the relaxation time (see also Lumley, 1972).

Such simple rheological theories point to the important conclusion that the common molecular conformation in a turbulent flow might be an extended thread rather than a random coil; therefore, the rheological properties of the polymer solution in a turbulent flow might be quite different from what they are in a laminar flow. As a theory realistically modelling coiling-uncoiling macromolecules in an unsteday flow field is likely to lead to very great mathematical difficulties; it would therefore be more useful for an exploratory study to search for a model which gives the maximum rheological effects of extended particles.

Such a model can in fact be found from the extensive recent studies of the rheological properties of suspensions of rigid elongated particles carried out by *Batchelor* (1970a, b, 1971) and his group at Cambridge University. The main concern in this group has been thes study of highly elongated rigid particles. The problem of determining the particle motion then simplifies to that of calculating the velocity and orientation giving zero net force and moment on the particle. (Inertia of the particle is neglected.) This was a problem considered by *Jeffrey* (1922), who determined the orbit of spheroids in a shearing motion and in an irrotational straining field. *Leal & Hinch* (1971) and *Hinch & Leal* (1972) have extended this analysis to incorporate Brownian motion and have shown that the orbits of highly elongated particles in shear flow tend to become such as to make them spend most of their time with their longest axes aligned with the direction of flow. *Jeffrey* (1922) also demonstrated that in an irrotational straining motion all the particles would end up in a direction with their longest axes aligned with the flow. Hence, there is a strong tendency for elongated particles to line up in the flow direction. If the particles are very slender, their contribution to the stress in pure shear will be quite small so that the suspension will behave nearly like a Newtonian fluid. On the other hand, any straining perturbation in the mean flow direction would be strongly resisted. We can use such general results, together with that for the particle stress contribution due to pure straining motion calculated by *Batchelor* (1971), to determine the perturbation stress τ_{11} arising from an unsteady twodimensioned perturbation with velocity components $u_1 = u$ and $u_2 = v$ of a parallel shear flow, a model problem that, as will be argued below, has some bearing on the dynamics of a turbulent shear flow.

---

* E. J. Hinch, private communication.

Batchelor shows that the contribution $\Delta\tau_{11}$ to the stress in pure straining motion due to the presence of slender particles, all of length l and diameter d, is given approximately by

$$\Delta\tau_{11} = \mu\overline{B}\,\partial U_1/\partial x_1, \tag{12}$$

where

$$\overline{B} = \frac{4}{3}\Phi r^2/\ln(\pi/\Phi) \equiv \frac{\pi}{3} n l^3/\ln(\pi/\Phi), \tag{13}$$

Φ is the volume fraction of particles, $r = l/d$ their aspect ratio, and n their number density. For a perturbation in the mean flow direction, the perturbation stress would therefore be obtained by the above formula with U_1 replaced by u. However, the perturbation velocity component v normal to the flow would in addition cause a misalignment of the fluid particles with the (small) angle

$$\varepsilon = \int \frac{\partial v}{\partial x} Dt, \tag{14}$$

where the symbol Dt indicates that the integration should be carried out following the particle. Since the particles are now tilted with repect to the mean flow they will experience a velocity gradient along their length proportional to $\varepsilon(dU/dy)$; hence the total first-order contribution to the stress will be

$$\Delta\tau_{11} = \mu\overline{B}\left[\frac{\partial u}{\partial x} + \frac{dU}{dy}\int^{t}\frac{\partial v}{\partial x} Dt\right]. \tag{15}$$

This result has been derived by *Hinch*[1], who also gave estimates of the error arising because not all particles are aligned. In the following we will refer to (15) as the Batchelor-Hinch model. Adding the contribution due to the solvent itself, which is obtained from the ordinary Newtonian formula, one will thus find the total perturbation stress due to the perturbation velocity field. This result will be used below with a simplified model for a turbulent shear flow exploring its large- and small-scale stability properties. *Batchelor* (1971) derived the formula (12) for rigid rods under the restriction that the suspension is dilute in the volumetric sense so that the average distance between the rods is large compared to their diameter but that the rod length is much larger than the average distance between the rods as illustrated in Fig. 7. Under such

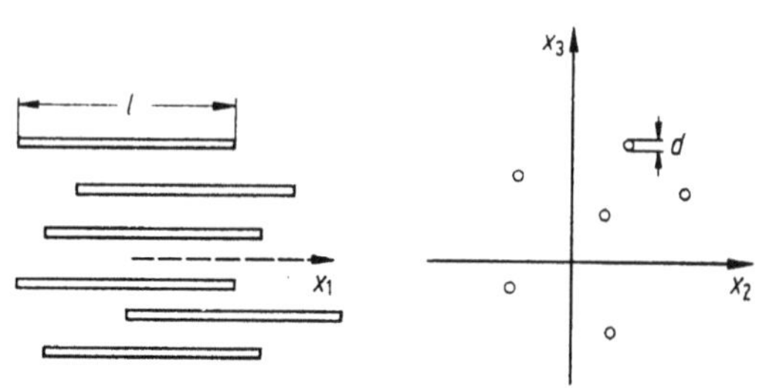

Fig. 7. *Batchelor's* (1971) model of a suspension of aligned rigid rods.

conditions, which would be satisfied for the dilute polymer solutions of present interest if the macromolecular coils were extended, the coefficient $\overline{B}$ can be very large, indeed, leading to very large ratios of extensional viscosity to shear viscosity of the order of 10^3 or more, values that order-of-magnitude-wise are consistent with the experimental results reported in the paper of *Metzner & Metzner* (1970) quoted above. Where the model obviously fails is in the assumption of rigid rods. A fluid filled with threads from uncoiled macromolecules would be able to resist extension but not compression. A more realistic model for such a fluid would have to be nonlinear and quite complicated, however, and in view of the fact that some suspensions of elongated inelastic particles have shown drag reduction (*Bobkowicz & Gauvin*, 1965), an investigation based on the simple Batchelor-Hinch model (15) should be of some value for an exploratory study.

---

[1] *E. J. Hinch*, private communication.

4. Stability and transition

In the mechanics of turbulence the phenomena of instability and transition of a laminar flow are believed to be of fundamental importance. Therefore, many theoretical and experimentai investigations have been aimed at a study of non-Newtonian effects on instability and transition.

The theory of hydrodynamic stability is concerned with whether small disturbances in a steady laminar flow field will grow or decay. We shall here discuss only two-dimensional infinitesimal disturbances of a plane parallel shear flow.

$$U_i = (U(x_2), 0, 0), \qquad (16)$$
$$(x_1 = x, x_2 = y, x_3 = z).$$

For such disturbances it is a straight-forward, if often somewhat tedious process to derive from the linearized momentum and the chosen constitutive relation a single differential equation for the stream function of the perturbation. Setting in the usual manner for the perturbation velocity components

$$u = (D\phi)\, e^{i\alpha(x-ct)}, \qquad (17)$$
$$v = -i\alpha\phi e^{i\alpha(x-ct)},$$

where α and c are the wave number and phase velocity, respectively, of the wave-like disturbance; $D = d/dy$; and the real part of $\phi(y) \exp [i\alpha(x - ct)]$ is the perturbation stream function, one obtains for the two constitutive models selected the following modified Orr-Sommerfeld equations:

a) *Convected Maxwell model*:

$$(U - c)(D^2 - \alpha^2)\phi - \phi D^2U = \frac{1}{i\alpha S\mathrm{Re}}\left[(D^2 - \alpha^2)^2\phi + \sum_{n=0}^{3} b_n D^n\phi\right], \qquad (18)$$

where Re is the Reynolds number, and

$$\begin{aligned}
S &= 1 + i\alpha\theta(U - c),\\
b_0 &= \underline{D^4S} - \alpha^2(D^2S)T - 2\alpha^2(DS)^2(1 + S^{-2})\\
&\quad - 4(DS)(D^3S)S^{-1} - 3(D^2S)^2S^{-1} + 4(DS)S^{-2}\\
&\quad - 6(DS)^2(D^2S)TS^{-1},\\
b_1 &= -2\alpha^2(DS)T + 2D^3ST + 4(DS)(D^2S)T^2\\
&\quad - 4(DS)^3TS^{-1},\\
b_3 &= 3D^2ST + 2(DS)^2T,\\
b_3 &= 2DST.\\
T &= 1 - S^{-1}.
\end{aligned}$$

This equation was given by *Porteous & Denn* (1971), who also demonstrated that for small fluid elasticity, i.e., small θ, only the underlined term in the above expressions for b_0 should be retained. The equation then becomes equivalent to the one for a second-order fluid employed by *Chun & Schwartz* (1968) and others, an equation also arrived at by *Walters* (1962) by considering a fluid of short memory.

b) *Batchelor-Hinch model*:

$$(U - c)(D^2 - \alpha^2)\phi - \phi D^2U = \frac{1}{i\alpha\,\mathrm{Re}}\left[(D^2 - \alpha^2)^2\phi - \alpha^2\bar{B}(U - c)\,D^2\left(\frac{\phi}{U - c}\right)\right]. \qquad (19)$$

All quantities appearing in (18) and (19) have been nondimensionalized in the usual manner for boundary layers employing shear layer thickness δ and free stream velocity U_1 as reference length and velocity, respectively, and δ/U_1 as reference time scale. The Batchelor parameter $\bar{B}$ is nondimensional as such and is independent of flow parameters.

Most stability results available to date have been based on the second order fluid constitutive model. For plane Poiseulle flow *Chan Man Fong* & *Walters* (1965) found that a small elasticity tended to lower the critical Reynolds number. This destabilizing effect was confirmed by *Chun* & *Schwartz* (1968) through direct numerical calculations of disturbance amplification rates, which for moderate values of the elasticity parameter were found to increase somewhat over those for a Newtonian fluid. Further demonstrations of the destabilization due to elasticity has also been given by *Porteous* & *Denn* (1971) who employed both (18) and the second order fluid model and found very little difference between the results for small values of the fluid elasticity. They also carried out an analysis for finite amplitudes. No results obtained on basis of the Batchelor-Hinch model (19) for aligned rigid rods have as yet been published, but our own preliminary calculations reported below indicate that their effect should generally be stabilizing. In the literature there have also appeared several analysis of stability of Couette and Taylor-Couette flow of viscoelastic fluid, of which perhaps the most ambitious one as far as the constitutive model employed, is that of *Karlsson* et al. (1971). Elasticity is found sometimes to act stabilizing for a Taylor-Couette flow but it is difficult to judge what possible implications such results may have for shear flow turbulence.

Experiments directly aimed at studying the stability of drag reducing fluids have been carried out by *Berman* & *Cooper* (1972). They investigated pipe flow response due to the periodic motion of a thin sleeve. Although the shear rates were not high enough for viscoelastic effects to be expected, they found that the periodic disturbance generated by the sleeve persisted longer downstream with the dilute polymer solution than with the pure solvent.

The pipe flow transition experiments that have been reported in the literature to date give rather contradictory information as to whether a drag reducing fluid will speed up or delay transition to turbulence. *Virk* et al. (1967) reported no substantial effect of polymer addition. *Castro* & *Squire* (1967) and *White* & *McEligot* (1970) found a slight increase in the transition Reynolds number, whereas *Chung* & *Graebel* (1972) obtained a large increase. *Paterson* & *Abernathy* (1972), on the other hand, observed a decrease in the transition Reynolds number of strongly drag-reducing solutions as compared to the pure solvent.

Possibly, these strongly differing results could in part have their explanation in the different methods used to detect transition. However, it is also well known from pipe flow transition measurements in Newtonian fluids (*Lindgren*, 1957) that by reducing inlet disturbances one can prolong the laminar regime almost indefinitely. Therefore, one can expect that only experiments conducted at reasonably high inlet disturbances level, such as those of *Virk* et al. (1967) will yield reproducible results for the lower transition Reynolds number. It is of interest that *Paterson* & *Abernathy* (1972), when using a squared-off pipe producing disturbed inlet flow, found little effect from the polymer on the transition Reynolds number, in agreement with the findings of *Virk* et al. (1967).

5. Turbulence structure

In recent years considerable advances have been made in the understanding of the dynamical processes taking place in a turbulent Newtonian wall shear flow. Earlier measurements of the turbulent field were focussed on determining space-time correlations of the fluctuating quantities. From such long-term averages one could hope to detect any persistent or recurring larger-scale flow features that might be of dynamical significance. Measurements of correlations such as those of *Favre* et al. (1967) and of *Willmarth* & *Tu* (1967) depict the main structure as a decaying and convected pattern which propagates with a velocity close to but not exactly equal to the local flow velocity. *Corcos* (1964) carried out a spectral analysis of the wall pressure fluctuation correlations obtained by *Willmarth* & *Wooldridge* (1962) and showed that the convection velocity had a uniquely defined value for each frequency. This was interpreted as evidence for the presence of waves in the turbulent boundary layer by *Landahl* (1967), who showed theoretically that in a system admitting propagation of waves (a wave guide) the least damped wave mode should dominate in the statistical average. *Morrison* & *Kronauer* (1969) detected such waves from measurements of frequency-wave-number spectra of the fluctuating velocities in the wall layer of a turbulent pipe flow and found that the fluctuating velocity field is dominated by waves that are highly swept. *Hussain* & *Reynolds* (1972) excited periodic disturbances in a two-dimensional turbulent channel flow and measured their amplitude distributions and propagation constants. They showed that the effect of background turbulence must be included in the equations modelling the disturbances (*Reynolds* & *Hussain*, 1972). An entirely different approach in analysing correlation data was taken by *Bakewell* & *Lumley* (1967), who decomposed the correlation functions in a set of orthogonal eigenfunctions and demonstrated that the dominating large-scale feature in the wall layer consists of a pair of randomly occurring contrarotating streamwise vortices separated in the spanwise direction of a distance of about $\lambda_z^* = 100$ expressed in wall variables.

The conceptual two-layer model proposed by *Townsend* (1956), which divides the turbulent boundary layer into an "active" part comprising the region closest to the wall and a "passive" outer one, has been a source of inspiration for many of the later invistigations. That the energetic processes in the boundary layer are mainly confined to the inner layer near the wall was given detailed confirmation by *Kline* et al. (1967) and by *Kim* et al. (1971). Their visual observations of a thick turbulent boundary layer in water with the aid of hydrogen bubble tracers revealed that most of the turbulent energy and Reynolds stress was produced during intermittently occurring localized "bursts" or turbulence. The burst is a violent instability process found to occur primarily in the zone $0 < y^+ < 100$. The observations suggested a local structure like a hair-pin shaped vortex lifting up from the lower part of the boundary layer and being stretched by the mean shear as it is transported downstream (see Fig. 8 reproduced from *Kline* et al., 1967). The average transverse distance between such events was found to be about $\lambda_z^+ \cong 100$ expressed in wall variables, strongly suggesting a connection to the statistical picture of *Bakewell* & *Lumley* (1967). Further visual observations by *Corino* & *Brodkey* (1969) revealed two distinct violent processes; an "ejection" of low-speed fluid away from the wall and a "sweep" of high-speed fluid towards the wall.

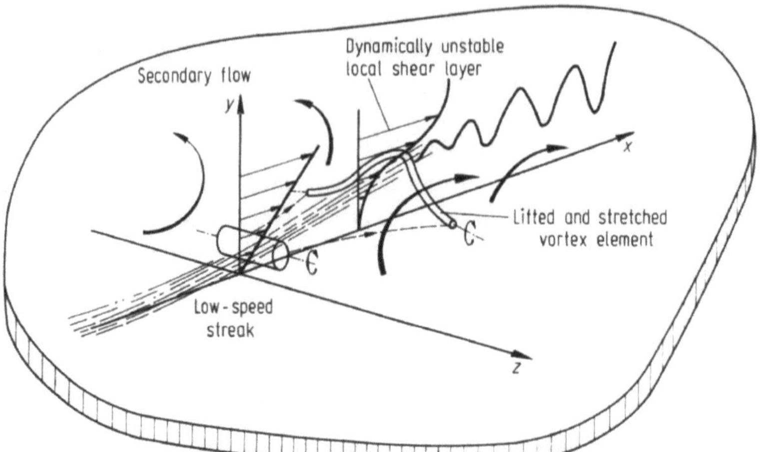

Fig. 8. Conceptual model of the structure of a low-speed streak breakup ("burst") in the wall region of a turbulent boundary layer (from *Kline* et al., 1967).

Because of their strong intermittency the presence of turbulent bursts cannot normally be detected by means of ordinary statistical methods, and their more detailed quantitative study has necessitated the developments of new experimental techniques based on selective sampling of the signal during bursts. Although different criteria have been employed by different investigators as to what should signify the occurance of an "event", all are in substantial agreement as to the main picture. *Willmarth* & *Lu* (1972) discovered that 60% of the Reynolds stress is produced when the velocity in the sublayer is less than the mean. *Blackwelder* & *Kaplan* (1972) succeeded in determining the average instantaneous velocity profile just preceding burst for different locations of the triggering probe. In Fig. 9 is shown the instantaneous profile thus obtained before burst for a setting of the triggering probe at $y^+ = 15$. The large negative velocity perturbation up to 20% of the local mean and the strong inflexional character is apparent. The inflexional character of the velocity profile preceding burst has been observed in all investigations of the phenomenon and suggests a strong similarity with the final state of breakdown of a laminar boundary layer to turbulence (*Klebanoff* et al., 1962). Other selective sampling data have been presented by *Wallace* et al. (1972), who, extending the observations by *Corino* & *Brodkey* (1969), found that the ejection and sweep stages each contributed about seventy per cent of the mean stress for a total of 140 per cent. The excess was accounted for by "interactions" towards and away from the wall, each producing a negative turbulent shear stress of about twenty per cent of the mean. Inbetween ejections, sweeps and interactions the flow was found to be of essentially laminar character but unsteady.

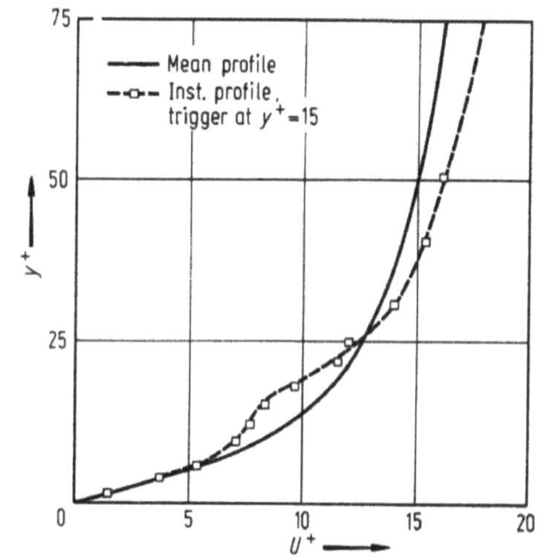

Fig. 9. Instantaneous velocity distribution just preceding burst measured by *Blackwelder* & *Kaplan* (1972). Trigger set at $y^+ = 15$.

Most of the experimental data on the turbulent structure of a non-Newtonian fluid reported in the literature so far have pertained to fluctuation intensities. From these it is immediately apparent that drag reduction is not accompanied by a corresponding decrease in fluctuation intensities. In fact, the measurements reported by *Virk* et al. (1967) actually showed an increase in the streamwise fluctuation velocity. *Seyer* & *Metzner* (1969b) on the other hand, using a hydrogen bubble technique, found a decrease. Development of laser anemometers in recent years has alleviated many of the special difficulties associated with measurements in polymer solutions employing traditional techniques. Measurements by *Goldstein* et al. (1969) gave very little difference in the u-component due to the presence of an additive. *Rudd* (1972) found a noticeable increase in the streamwise fluctuating component near the wall but hardly any change near the center of the pipe. (See Figs. 10 and 11, reproduced from *Rudd*, 1972.) The transverse (w-)component was decreased near the wall but increased above its value for the solvent at larger distances.

A few measurements of the spatial structure of non-Newtonian turbulences have been reported. For isotropic grid turbulence, *Fabula* (1966) could not detect any clear effect on the spectrum from polymer addition. *Friehe* & *Schwartz* (1970) obtained a decrease in the decay rate but no substantial change in the spectrum. Measurements in a flat plate boundary layer by *Kowalski* (1967) showed a substantial increase in the turbulence microscale near the wall by as much as a factor of two. This is consistent with the observations from mean velocity measurements which show that the thickness of the viscous wall layer increases. On the basis of data from flow in rough tubes *Virk* (1971a) concludes that the maximum increase in the viscous sublayer due to polymer additives is about 2.5 in qualitative agreement with Kowalski's result. Results from correlation measurements have only recently become available. *Fortuna* & *Hanratty* (1972) employing an electrochemical technique determined the wall-stress fluctuation in pipe flow. They discovered a strong increase in the transverse (spanwise) correlation scale (see Fig. 12) as well as a much slower longitudinal (streamwise) correlation decay (Fig. 13)

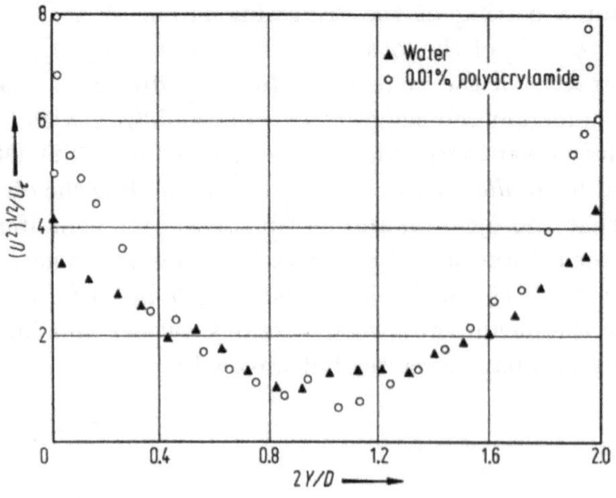

Fig. 10. Intensity of the streamwise turbulent velocities across pipe (from *Rudd*, 1972). ▲ water, ○ 0.01% polyacrylamide.

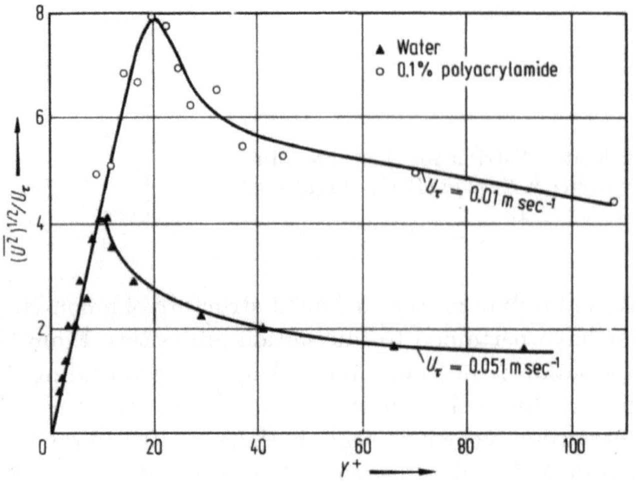

Fig. 11. Intensity of axial turbulent velocities in wall layer (from *Rudd*, 1972). ▲ water; ○ 0.1% polyacrylamide.

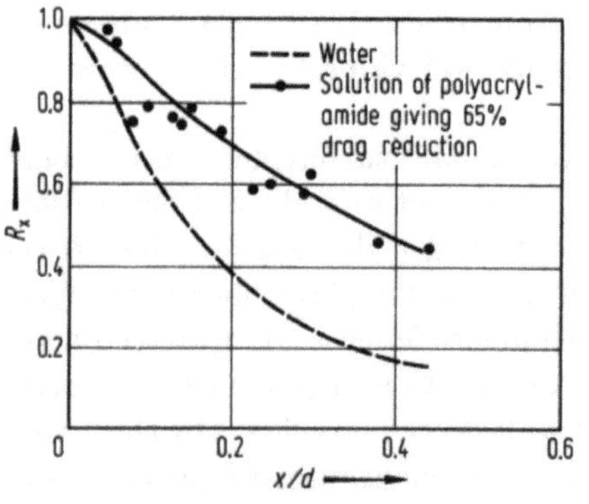

Fig. 12. Spatial correlation coefficient of streamwise fluctuating wall stresses (from *Fortuna* & *Hanratty*, 1977). — — — water; —●—●— solution of polyacrylamide giving 65% drag reduction.

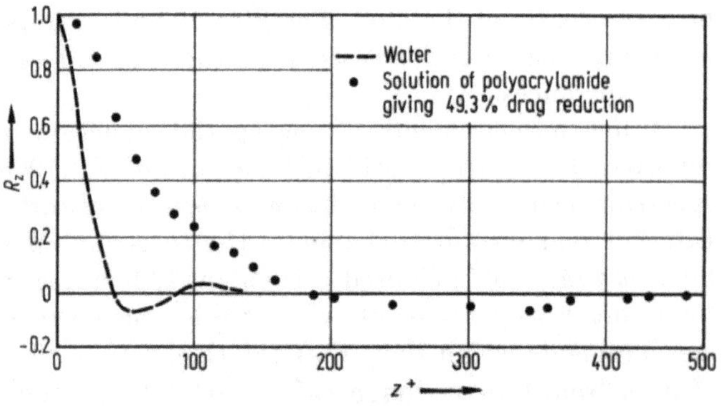

Fig 13. Spatial correlation coefficient of transverse fluctuating wall stresses (from *Fortuna* & *Hanratty*, 1972). — — — water; ● solution of polyacrylamide giving 49.3% drag reduction.

than in the Newtonian case. A reduction of the stress fluctuation amplitude for the same mean wall stress was also observed; for the transverse component up to a factor of about two, but less for the axial component. Spectra were also determined but showed no dramatic differences from their Newtonian counterparts when normalized with wall parameters. However, the low-frequency part increased somewhat and the high frequency one decreased with increasing drag reduction.

In summary, the qualitative picture that emerges from such experimental results is that polymer addition causes an increase in the scale of the smallest turbulence eddies near the wall accompanied by a reduction of Reynolds stress and an associated lowering of the wall friction. Turbulence is not supressed in the wall layer but may actually be enhanced. The most spectacular change in the turbulence structure is in the transverse scale, which can increase with up to a factor of five (*Fortuna* & *Hanratty*, 1972). The reduction of streamwise decay of the longitudinal correlation as compared to the solvent points to an increase in the life-time for the big eddies.

6. Proposed mechanisms

A number of suggestions, all of a qualitative nature, have appeared in the literature as to what mechanism might be the primary cause of the drag reduction. The key element in all of these has been the thickening of the wall layer, first observed by *Elata* et al. (1966). The explanations differ widely, however, as to how a very small amount of polymer addition could interfere with the turbulent mixing processes so as to produce a substantial increase in the scale of the smallest stress-producing eddies near the wall. There are even differing opinions on such a fundamental question as to whether an explanation on a continuum basis is sufficient. *Barenblatt* et al. (1965 and 1969) point to the observed appearance of large agglomerations of molecules and suggest that these will form viscoelastic particles at least as large as the Kolmogoroff scale of turbulence which—because they tend to become effectively rigid at very high fluctuation frequencies—will act to stabilize the small viscous eddies in the wall layer. Their effect on the inertial large-scale motion, on the other hand, is to lead to an increased intensity. The idea that entanglements of molecules would resist the motion in the wall layer was also suggested by *Fabula* et al. (1966) and discussed by *Lumley* (1967), but in his later review article (*Lumley*, 1969) he comes cautiously to the conclusion that molecular agglomerationis not an important factor in drag reduction. A continuum viewpoint was taken by *Seyer* & *Metzner* (1969a), who based their reasoning on the model of large streamwise eddies for the wall layer proposed by *Bakewell* & *Lumley* (1967) and the convected Maxwell constitutive model for the fluid. According to the latter, large values of normal stresses would be set up in a rapid elongational flow, such as would be produced when the two Bakewell-Lumley counterrotating vortices force the fluid up from the wall between them. This motion would thus be strongly resisted. *Pfenninger* (1967), following Theodorsen's (1955) idea that the turbulence could be modelled as a distribution of horseshoe vortices formed at the wall and lifting up from it as they are swept downstream, proposed that drag reduction was associated with the strong resistance to vortex stretching which experiments have shown to be exhibited by some (but not all) drag reducing fluids (see also *Gadd*, 1968). The horseshoe vortex concept has also been embraced by *Black* (1969), who, however, thinks that stretching of the more or less longitudinal vortices in the horseshoe "legs" is perhaps not as important as the direct stabilization of the wall layer. Influence on the energy budget of the turbulent fluctuations in a more unspecified manner has also been invoked. *Astarita* (1965) notes that the energy dissipation ought to decrease in a viscoelastic fluid since the dynamically process will become more elastic and hence more reversible at high frequencies. Thus, he argues, the drag must decrease in order to preserve the energy balance. Energy arguments were also used by *Savins* (1964) and are also implicit in Virk's (1971b) elastic sublayer model. Adsorption of a resilient layer of molecules on the wall layer has also been proposed as contributing to the thickened wall layer by *Bryson* et al. (1971), but no explanation has been suggested as to how this can change the turbulent processes.

7. Application of a new model

To make any quantitative or qualitative assessments of such proposed explanations of drag reduction is presently not possible because of the lack of an adequate theoretical model for the dynamics of shear flow turbulence. In view of the extreme complexity, randomness and nonlinearity of turbulence, the prospects for the discovery, in the near future, of a reasonably complete such model are very poor, indeed. However, in view of the very drastic effects produced by additives one could reasonably hope that even a fairly crude theoretical model—if retaining the essential dynamics of the turbulent processes—should be able to indicate the possibility of modifications of the turbulent field leading to drag reduction, at least qualitatively. We shall here present the first attempts at an extremely simplified such mode which, hopefully, contains enough of the essential dynamical elements of the turbulence structure to make an exploratory study of the effect of additives possible.

The new model represents an outgrowth from ideas in classical hydrodynamic stability applied to a fully turbulent flow, but it also retains *Townsend's* (1956) basic concept of a two-layer structure. The scale of wall turbulence being roughly proportional to distance from the wall, one would expect that the significant stress and energy producing eddies in the inner "active" layer close to the wall are of much smaller scale than those in the "passive" outer layer. It may therefore be meaningful to think of the turbulence conceptually as consisting of a coupled motion at two disparate scales; a primary one of large scale (of, say, the order of the boundary layer thickness) and a much smaller secondary one (of the order of the viscous wall layer thickness). The random fluctuating field is superimposed on a mean velocity field, which to a good approximation can be regarded as a parallel shear flow. Accordingly, if the fluctuation velocities are small compared to the reference free stream speed so that a linearization is allowable, a simple model equation for the fluctuating velocity field (when Fourier-transformed in x, z and t) would be of the Orr-Sommerfeld type, and a decomposition of the general unsteady motion in terms of the eigenmodes of the Orr-Sommerfeld problem should be the obvious way to proceed. However, nonlinearity would modify the picture in three important ways.

First, the primary wave field (belonging to the large-scale motion) would propagate through a flow filled with small-scale eddies, hence these will act effectively as scattering sources for the large-scale motion producing a gross effect perhaps much like that of an eddy viscosity (see *Reynolds* & *Hussain*, 1972, for efforts to model this effect). We shall here ignore this effect completely finding some justification in doing so in the calculations of *Landahl* (1967) which gave propagation and decay properties of the large-scale pressure-producing eddies in good qualitative agreement with observations without including eddy viscosity.

Secondly, nonlinearity is required to excite the primary waves since these are damped, as shown by *Landahl* (1967), and the most obvious excitation sources that have come to light so far are the turbulent bursts. As the recent experiments reviewed above indicate that these are highly localized in time and space, the excitation may be modelled as effectively point sources on the scale of the primary motion.

Thirdly, the small-scale secondary motion sees not only the parallel mean velocity field but also the primary fluctuating field, so that the secondary waves may be thought of as riding on the primary waves in very much the same way as capillary water waves on a large-scale gravity wave. On the scale of the secondary motion the mean plus the primary field will be a slowly varying one in space and time and may therefore locally be approximated as a parallel one.

For both the primary and the secondary motion we will thus have as a model equation one of the Orr-Sommerfeld type with the mean velocity replaced by mean plus primary in the latter. The secondary wave could thus be locally unstable, for example because of inflexion. Now, it has been shown recently with the aid of kinematic wave theory (*Landahl*, 1972) that there is a critical condition for a locally selfexcited secondary wave at which it will grow to very large amplitudes, given by

$$c_g = c_0, \tag{20}$$

where c_g is the group velocity of the secondary wave, and c_0 is the phase velocity of the primary wave on which it is riding. The reason for the large growth is that the group of secondary waves experiences space-time focusing at this condition (i.e., neighboring groups of waves catch up with each others and hence "compress"), and that the wave group becomes trapped in a region of the primary wave (generally near its crest), where it can continue to amplify (see Fig. 14). It is therefore hypothesized that (22) should identify the condition at which the breakdown of the primary waves into strong secondary small-scale motion

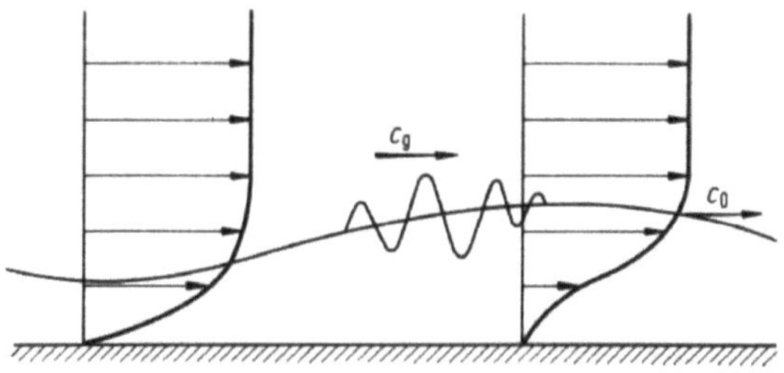

Fig. 14. Illustration of critical condition for secondary small-scale wave riding on a primary large-scale one. Breakdown occurs when $c_g = c_0$ and $c_I > 0$.

would take place. Application to the data from the transition experiments by *Klebanoff* et al. (1962) demonstrates that the condition is very closely satisfied at the appearance of the first "spikes" in the final stage of transition of a laminar boundary layer, and that it gives good agreement with the measured frequency of secondary motion. In view of the observed similarities between breakdown in the final stage of transition and turbulent bursts (as pointed out, for example, by *Willmarth & Lu*, 1972) we shall hypothesize that the bursts are the results of the attaintment of local criticality in the primary fied, and that the subsequent violent motion will cause excitation of new primary waves. (The model is illustrated conceptually in Fig. 15.) Admittedly, the model is highly speculative at this stage and contains many

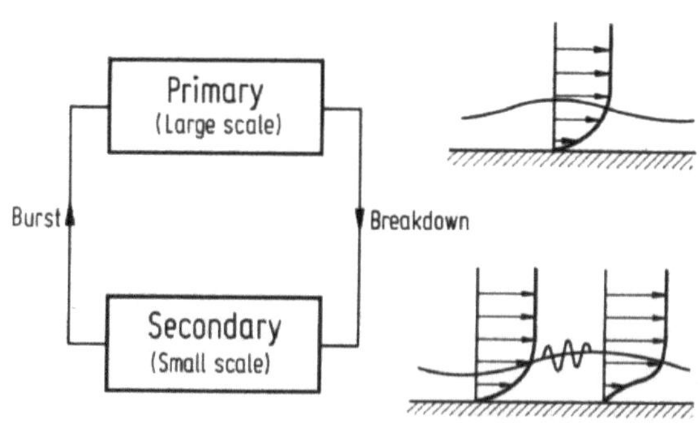

Fig. 15. Two-scale model (conceptual) of the fluctuating velocity field in a turbulent boundary layer.

difficult details which are not yet worked out and which will require a great deal more study, such as how randomness could be incorporated into the model, and how primary waves, which probably are highly oblique, could interact so as to produce a critical condition. Nevertheless, for the purpose of studying the effects of additives, a great deal could perhaps be learned from the propagation characteristics, in particular the stability characteristics, of primary and secondary motion. Accordingly, some numerical stability calculations have been carried out for velocity profiles that may be typical of the mean and instantaneous primary motion for both a Newtonian situation and one with drag reduction. These calculations were performed by *F. Bark* of the Royal Institute of Technology, Stockholm, and a detailed account of them will be given later. The three rheological models used were the ones described above: the ordinary Newtonian, the convected Maxwell and the Batchelor-Hinch model for a suspension of microscopical rigid rods. Three different velocity profiles were investigated: two qualitatively representative of the mean motion for a Newtonian and a drag-reducing fluid in a boundary layer, and one broadly similar to the inflexional profile appearing in the sublayer just before breakdown. All three are illustrated in Fig. 16. The mean profile for the Newtonian case was constructed from Reichhardt's representation for the velocity distribution in the constant-stress layer coupled with Coles' wake law profile for the outer portion. The drag-reduction velocity profile was obtained in essentially the same way, except that the viscous wall layer thickness parameter in Reichhardt's expression was increased to correspond to the maximum wall layer thickness possible according to the *Seyer-Metzner* (1969b) correlation. The same plate-length Reynolds number was employed as for the Newtonian profile. Comparing the results for the two profiles with the same constitutive model one will be able to isolate the effect on the stability characteristics of the change of profile from an ordinary turbulent one to a more laminar-looking one.

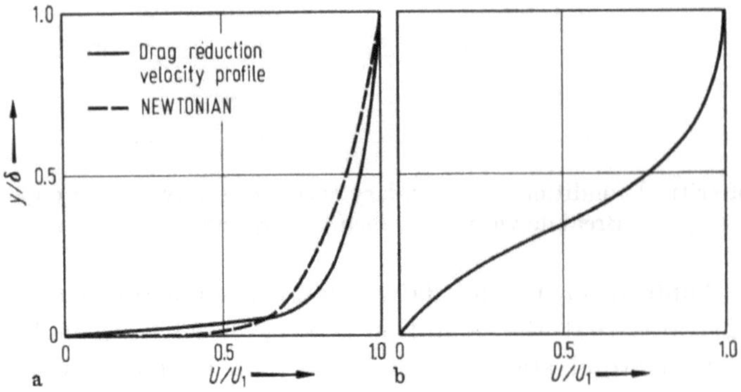

Fig. 16. Model velocity profiles used in stability studies. (a) Turbulent mean profiles; — — — Newtonian profile (1), ——— profile with drag reduction (2); (b) Instantaneous velocity profile assumed in study of secondary instability.

The velocity profile chosen to model the instantaneous wall layer breakdown profile was of the simple tanhyp-variety with the distance of the inflexion point to the wall adjusted so as to be roughly in the same ratio to the internal shear layer as in the experimental profile determined by *Blackwelder & Kaplan* (1972). This was selected in preference to a direct numerical fit to the measured profile in order to avoid undue numerical difficulties that would arise because of the fourth derivative of U appearing in the equation for the convected Maxwell model. Such simplifications, however, should not be serious at the present stage where only qualitative indications of a strong effect are being sought.

Results from the primary wave calculations are shown in Fig. 17. Following *Landahl* (1967) these were obtained in spatial form with the complex wave number $\alpha = \alpha_R = i\alpha_I$ determined as function of the real frequency ω (made dimensionless with the "outer" reference quantities δ and U_1). The imaginary part, α_I, gives a measure of the rate of decay with streamwise distance for a wave of given frequency. The results obtained with the Maxwell fluid model show that for the lower frequency, range $\omega < 1.5$, a moderate fluid elasticity (non-dimensional relaxation time θ less than unity) causes very little change in α_I relative to the Newtonian model for the same velocity profile. There is an increase of α_I relative to the value for the Newtonian profile which is entirely a profile effect. At higher frequencies a destabilizing effect is seen which at $\omega = 3$ is strong enough to lead to a complete loss of stability. Calculations for larger values of θ (not shown here) give frequency regions in which α_I becomes negative, i.e., the wave is unstable. It is difficult to judge the significance of such results as the model employed requires that all primary waves are stable, otherwise the assumption of statistical stationarity has to be given up (*Landahl*, 1967).

With the Batchelor-Hinch constitutive model, on the other hand, the decay rate α_I is increased above the value for the Newtonian case by about a factor of 2.5*. At a given non-dimensional frequency based on outer variables, one would thus expect to see a correspondingly

* However, the results for damped waves may be questioned, since the singularity in (19) produced by the second term of (15) would require special consideration for the damped case.

more rapid decay of the cross-power spectrum of the turbulent fluctuations with streamwise distance. Such a result might seem to be at variance with the experiments of *Fortuna & Hanratty* (1972), which showed a much reduced decay of the streamwise correlation of the wall stress fluctuations with streamwise distance as compared to the Newtonian case for the same mean wall stress. However, since the correlation is the inverse Fourier transform of the cross-power spectrum, the reduction in the correlation decay rate may simply be due to a shifting of the whole spectrum towards lower frequencies, and as the experiments indeed indicate, this will take place during drag reduction. Results of the stability calculations for the secondary disturbance with the model inflexional profile selected are presented in Fig. 18. In these calculations the fairly high Reynolds number of Re = 4600 was chosen in order to make the Newtonian stability characteristics become essentially the inviscid ones. For a study of the shortest scale in a Newtonian wall layer, perhaps a more realistic value of the Reynolds number would have been less than 1000; however, in order to bring out more clearly any stabilization potentialities of the other rheological models, a high Reynolds number was chosen. From the results presented in Fig. 18 it becomes clear that a moderately high elasticity of the fluid ($\theta = 0(1)$)

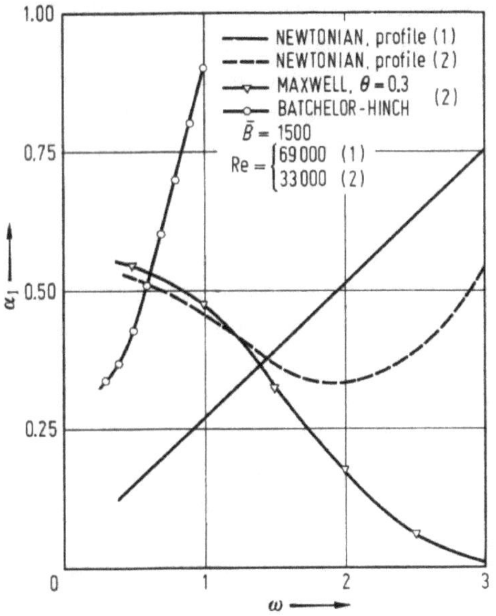

Fig. 17. Decay rates for primary waves as function of frequency for the different models. All quantities non-dimensionalized with outer variables, δ and U_1.

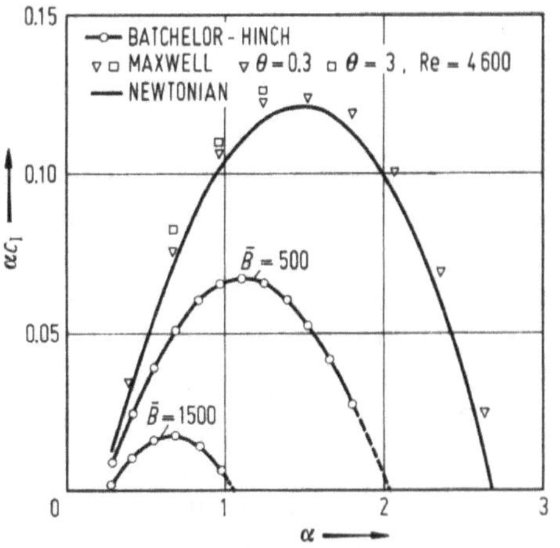

Fig. 18. Stability characteristics of secondary disturbances for the different constitutive models. Quantities made nondimensional as in Fig. 17.

has only a small destabilizing influence. However, with the Batchelor-Hinch model a strong stabilizing effect is demonstrated including the possibility of a complete stabilization of the inflexional profile for sufficiently high values of the parameter $\bar{B}$. In particular, it is observed that the strongest stabilizing effect is found for the largest wave numbers, so that the smallest scale of unstable motion is increased. This would be expected to lead to an increase in the size of the smallest stress producing eddies and a corresponding increase in the wall layer thickness. The values of $\bar{B}$ used in the calculations may perhaps seem high, but they are not unrealistic, as they give a measure of the ratio of elongational to shear viscosity of the same order as found in the experiments by *Metzner & Metzner* (1970). Furthermore, it is seen that the parameter enters into the basic stability Eq. (22) in the ratio $\bar{B}/\mathrm{Re}$, so that for a lower Reynolds number the value of $\bar{B}$ required for stabilization becomes proportionally lower.

The simplified turbulence model presented thus shows that fluid elasticity as modelled by the Maxwell constitutive equation has a destabilizing influence on both primary and secondary motion. The Batchelor-Hinch constitutive model, on the other hand, is found to produce a strong stabilization of both primary and secondary motion. The stabilizing effect on the secondary motion is probably of most significance for the drag reduction phenomenon since the associated increase in scale of the unstable secondary disturbances could be expected to lead to a thickening of the wall layer.

8. Concluding remarks

The first basic and as yet unresolved question in the puzzling phenomenon of drag reduction due to polymer addition is whether a satisfactory explanation based on continuum mechanics is possible or whether, instead, one would have to invoke processes on the non-continuum macromolecular level such as proposed by *Barenblatt* et al. (1965 & 1969). From a heuristic point of view, a theory based on non-continuum concepts will be quite difficult to quantify and the theoretician would therefore naturally be quite reluctant to part with the continuum approach. Although the initial efforts towards a simplified mechanistic theory of wall turbulence presented here have not yet led to definite conclusions in this regard, they have at least pointed to a possible continuum mechanical effect due to the polymer additive which could be highly effective even when the molecular size is much smaller than the scale of fluctuating motion. The large frictional resistance to streamwise straining perturbations caused by elongated particles aligned in the mean flow direction was shown to have a strong stabilizing effect on inflexional instability. Since there is both experimental and theoretical evidence that such instability is involved in the creation of the stress producing turbulent bursts observed to occur intermittently in the wall layer, stabilization of the small-scale motion could be expected to lead to a reduction of the turbulent stresses near the wall and an associated thickening of the wall layer. Extended polymer macromolecules mostly aligned in the mean flow direction of the turbulent velocity field could be expected to act qualitatively like the rigid rods employed in the simplified fluid model, at least during the extensional phase of the secondary fluctuation cycle. That the polymer molecules could have a rheological effect of this kind is indicated indirectly by the very high values of the ratio of elongational to shear viscosity, of the order 10^4, measured by *Metzner & Metzner* (1970). Such high values are consistent with what *Batchelor's* (1971) simplified theory would give for rods having aspect ratios like those of the extended polymer molecules of typical drag reduction additives in dilute solutions.

As yet no direct experimental demonstration of the presence of fully extended macromolecular threads during drag reduction has been given. Complete extension would not be attainable in a laminar shear flow, but the fluctuating irrotational strains present in the turbulent field could cause some coils to stay extended a large fraction of time. (*Lumley*, 1972). Intuitively, one would feel that the onset phenomenon is a manifestation of the extension of the macromolecular coils, although the observation (*Virk* et al., 1967) that onset data best correlate with a length hypothesis based on the ratio between coil radius and the smallest scale of turbulence, has not yet found a satisfactory theoretical explanation.

The calculations based on the Maxwell fluid model give quite clear indications of the effects of fluid elasticity on fluctuation field stability. Both the large-scale primary motion and the small-scale secondary one were found to be destabilized by moderate amounts of elasticity. With the model employed there is thus no indication that viscoelasticity as such could be the cause of drag reduction.

The explanation that drag reduction is associated with the high elongational viscosity produced by aligned uncoiled macromolecular threads is broadly in accord with *Virk's* (1971b) findings that the drag curve slope increment δ is approximately proportional to the square root of concentration and the three half power of the number of backbone links in the polymer. In the Batchelor-Hinch model the governing parameter $\overline{B}$ is proportional to nl^3, (ignoring the logarithmic term) where n is the number density and l the length of the aligned particles. The experimentally determined variation with concentration, i.e., of n, to the one half power would thus indicate the effect to be proportional to $(nl^3)^{1/2}$, which thus yields a variation with the bond length b and number of backbone links N as $b^{3/2}N^{3/2}$. Hence, proportionality to $c^{1/2}$ would molecular chain bond length. On the basis of an estimate of the turbulent strain energy stored in the elastic molecules, *Virk* (1971b) predicted a proportionality to b^3, in contrast to the $b^{3/2}$ be consistent with the $N^{3/2}$ variation[1]. The two models, differ however, as to the variation with dependence inherent in the $(nl^3)^{1/2}$ variation. The high drag reduction values exhibited by guar gum are more in line with the b^3 variation proposed by *Virk* than with the proportionality to $b^{3/2}$, tending to give support for Virk's hypothesis that elasticity is an important molecular property in drag reduction. In the Batchelor-Hinch model only the total length of the particle plays a role; its hydrodynamic effect extends over a sphere of diamater l and the quantity nl^3 thus measures the number of such influence spheres per unit volume. The high values of $\bar{B}$ that would be typical for dilute drag reducing fluids—of the order of 10^3 if the macromolecules were fully extended—indicate that the hydrodynamic interaction between the extended macromolecular threads should be very strong (i.e., their spheres of hydrodynamic influence would overlap strongly). In order for the solution to be dilute in the rheological sense so that the molecules take up only a small fraction of the available volume, they have to have extremely high length to diameter ratios of the order 10^3 or more.

However, there is recent experimental evidence that it may not only be the length of the extended molecule that is of importance for drag reduction. *Hand & Williams* (1971) measured the drag reduction effectiveness of polyacrylic acid and of calf thymus DNA at various pH levels of the solution. They found that the effectiveness was considerably larger for pH ranges in which the molecules are believed to have a fairly compact helical structure. For pH ranges where the molecules are supposed to be extended and behave essentially as rigid rods, the drag reduction was found to be considerably lower. Results such as these are very difficult to explain on the basis of any continuum model; but possibly entanglements and agglomerations could be an important factor.

It should be pointed out that stabilization of the instantaneous inflexional velocity profile before burst—as indicated by the calculations for a suspension of rod-like particles—is only one of several conceivable mechanisms that could produce a decrease in the rate or intensity of the bursts and thereby cause a lowering of turbulent stress. An alteration of propagation characteristics of the secondary wave group in relation to the primary disturbance field could presumably have such an effect according to the recently proposed model for breakdown (*Landahl*, 1972), but the model has not yet been sufficiently developed to allow any conclusions to be drawn regarding the possibility of an effect of this kind. Conceivably, the enhanced elongational viscosity could also have a strong influence on the structure of the longitudinal vortices appearing in the wall layer. In particular, a strong resistance to vortex stretching would be expected. However, a quantitative model to investigate this effect is likely to be quite difficult to set up.

As pointed out in the Introduction, perhaps the most interesting aspect of the research on the drag reduction effect of polymer additives is what new knowledge can be gained from it regarding the nature of turbulence. Since one now has a very effective tool with which the turbulent processes can be manipulated, new possibilities for further experimental and theoretical research on the mechanics of turbulence should open up. For example, such experiments as those by *Wells & Spangler* (1967) on the effectiveness of injection of polymer at the wall demonstrated quite unequivocally that wall shear flow turbulence is controlled by processes in a layer very close to the wall.

An important engineering motivation for turbulence research is to find new and effective methods for control of turbulence. From the research on polymer additives reviewed here it becomes apparent that to influence skin friction, it is the very small-scale stress producing eddies near the wall which may need to be interfered with directly. This may put very difficult constraints on the dynamical response characteristics of any mechanical system employed for drag modification, but the search for efficient such systems should be an extremely interesting challenge from both a scientific and engineering point of view.

---

[1] This was pointed out to me by *E. J. Hinch*.

Acknowledgements

I am much indebted to *E. J. Hinch* of Cambridge University for many interesting and useful discussions on the subject of drag reduction which, as is evident from the paper, spawned many of the new ideas presented, and for his expert guidance into the field of the rheology of suspensions.

I am also very much indebted to *Fritz Bark* of the Royal Institute of Technology for the numerical calculations presented, and for his assistance and helpful advice without which this paper would not have been possible. Mr. *Bark's* work was supported by the Swedish Board for Technical Development.

Partial support by the Air Force Office of Scientific Research under Contract No AF AFOSR F44620-71-C-0007 is also acknowledged.

References

Astarita, G.: Ind. Eng. Chem. 46 (1965) 1917.
Bakewell, H. P., Lumley, J. L.: Phys. Fluids 10 (1967) 1880.
Barenblatt, G. I., Bulina, I. G., Zeldovich, A. B., Sholomovich, G. I.: Z. Prikl. Mek. Tek. Fiz 5 (1965).
Barenblatt, G. I., Gorodsov, V. A., Kalashnikov, V. N.: Heat transfer—Soviet Research 1 (1969) 102.
Batchelor, G. K.: J. Fluid Mech. 41 (1970a) 545.
Batchelor, G. K.: J. Fluid Mech. 44 (1970b) 519.
Batchelor, G. K.: J. Fluid Mech. 46 (1971) 813.
Berman, N. S., Cooper, E. E.: A. I. Ch. E. J. 18 (1972) 312.
Black, T. J.: in: Viscous Drag Reduction (Ed. C. Wells) 383, Plenum (1969).
Blackwelder, R. F., Kaplan, R. E.: in: Turbulent Shear Flows, AGARD Conf. Proc. 93 (1972) 1—5.
Bobkowicz, A. J., Gauvin, W. H.: Can. J. Chem. Eng. 43 (1965) 87.
Brennen, C., Gadd, G. E.: Nature 215 (1967) 1368.
Bryson, A. W., Arunachalan, V. R., Fulford, G. D.: J. Fluid Mech. 47 (1971) 209.
Castro, W., Squire, W.: Appl. Sci. Res. 18 (1967) 81.
Chan Man Fong, C. F., Walters, K.: J. Méchanique 4 (1965) 29.
Chun, D. H., Schwartz, W. H.: Phys. Fluids 11 (1968) 5.
Chung, J. S., Graebel, W. P.: Phys. Fluids 15 (1972) 546.
Corcos, G. M.: J. Fluid Mech. 18 (1964) 353.
Corino, E. R., Brodkey, R. S.: J. Fluid Mech. 37 (1969) 1.
Darby, R.: Trans. Soc. Rheol. 14:2 (1970) 185.
Eirich, F. (Ed.), Rheology, Vols, I & IV, Aacdemic Press 1956, 1966.
Elata, C., Tirosh, T.: Israel J. Tech. 3 (1965) 1.
Elata, C., Lehrer, J., Kahanovitz, A.: Isreal J. Tech. 4 (1966) 84.
Fabula, A. G.: Proc. Fourth Int. Cong. Rheology, (Ed. E. H. Lee) Part 4, 455, Interscience (1965).
Fabula, A. G.: An experimental Study of Grid Turbulence in Dilute High-Polymer Solutions. Ph. D. Thesis, The Pennsylvania State University (1966).
Fabula, A. G., Lumley, J. L., Taylor, W. D.: in: Modern Developments in the Mechanics of Continua, 146, Academic Press (1966).
Favre, A., Gaviglio, J., Dumas, R.: Phys. Fluids S 138 (1967).
Fortuna, G., Hanratty, T. J.: J. Fluid Mech. 53 (1972) 575.
Friehe, C. A., Schwartz, W. H.: J. Fluid Mech. 44 (1970) 173.
Gadd, G. E.: Nature 212 (1966a) 874.
Gadd, G. E.: Nature 212 (1966b) 1348.
Gadd, G. E.: Nature 217 (1968) 1040.
Goldstein, R. J., Adrian, R. J., Kreid, D. K.: I. & E. C. Fund. 8 (1969) 498.
Goren, Y., Norbury, J. F.: ASME J. Basic Eng. 89 (1967) 814.
Hand, J. H., Williams, M. C.: Chem. Eng. Progr. Symp. Series 67 (1971) 6.
Hinch, E. J., Leal, L. G.: J. Fluid Mech. 52 (1972) 683.
Hussain, A. K. M. F., Reynolds, W. C.: J. Fluid Mech. 54 (1972) 263.
James, D. F., Acosta, A. J.: J. Fluid Mech. 42 (1970) 269.
Jeffrey, G. B.: Proc. Roy. Soc. A 102 (1922) 161.
Kalashnikov, V. N., Kudin, A. M.: Nature 225 (1970) 445.
Karlsson, S. K. F., Sokolov, M., Tanner, R. I.: Am. Inst. Chem. Eng., Chem. Eng. Progr. Symp. Ser. 67 (1971) 11.
Kim, H. T., Kline, S. J., Reynolds, W. C.: J. Fluid Mech. 50 (1971) 133.
Klebanoff, P. S., Tidstrom, K. D., Sargent, L. H.: J. Fluid Mech. 12 (1962) 1.
Kline, S. J., Reynolds, W. C., Schraub, F. A., Runstadler, P. W.: J. Fluid Mech. 30 (1967) 741.
Kowalski, T.: W 12, The Roy. Inst. Nav. Arch., London (1967).
Landahl, M. T.: J. Fluid Mech. 29 (1967) 441.
Landahl, M. T.: J. Fluid Mech. 56 (1972), 775.
Latto, B., Shen, C. H.: Can. J. Chem. Eng. 48 (1970) 34.
Leal, L. G., Hinch, E. J.: J. Fluid Mech. 46 (1971) 685.
Lindgren, E. R.: Arkiv för fysik 12 (1957) 1.
Lumley, J. L.: Appl. Mech. Rev. 20 (1967) 1139.
Lumley, L. J.: Drag reduction by additives, in Annual Review of Fluid Mech. 1 (1969), 367 Palo Alto.
Lumley, J. L.: Phys. Fluids 14 (1971) 2282.
Lumley, J. L.: in Symposia Matematica del Istituto Nazionale di Alta Matematica, Vol VI, Academic (1972).
Merrill, E. W., Smith, K. A., Shin, H., Mickley, H. S. Trans. Soc. Rheol. 10 (1966) 335.
Metzner, A. B., Park, M. G.: J. Fluid Mech. 20 (1964) 291.
Metzner, A. B., Metzner, A. P.: Rheol. Acta 9 (1970) 174.

Morrison, W. R. B., Kronauer, R. E.: J. Fluid Mech. 39 (1969) 117.
Mysels, K. J.: U.S.Pat. No. 2,492,173 (1949).
Oldroyd, J. G.: Proc. 1st Int. Congr. Rheology, Vol. 2 (1948) 130.
Oldroyd, J. G.: Proc. Roy. Soc. (London) A 245 (1958) 278.
Oliver, D. R.: Can. J. Chem. Eng. 44 (1966) 100.
Paterson, R. W., Abernathy, F. H.: J. Fluid Mech. 43 (1970) 689.
Paterson, R. W., Abernathy, F. H.: J. Fluid Mech. 51 (1972) 177.
Patterson, G. K., Florez, G. L.: in Viscous Drag Reduction (Ed. C. S. Wells) 233 Plenum (1969).
Peterlin, A.: Pure Appl. Chem. 12 (1966) 563.
Pfenninger, W.: Northrop Corp. Norair Div. Rep. BLC-179 (1967).
Porteous, K. C., Denn, M. M.: Linear and nonlinear stability of plane Poiseuille flow of viscoelastic liquids. Univ. Delaware Water Resources Center Contribution No. 16 (1971).
Reynolds, W. C., Hussain, A. K. M. F.: J. Fluid Mech. 54 (1972) 263.
Rouse, P. E. Jr.: J. Chem. Phys. 21 (1953) 1272.
Rudd, M. J.: J. Fluid Mech. 51 (1972) 673.
Savins, J. G.: Soc. Petr. Eng. J. 4 (1964) 203.
Seyer, F. A., Metzner, A. B.: Can. J. Chem. Eng. 45 (1967) 121.
Seyer, F. A., Metzner, A. B.: A. I. Ch. E. J. 15 (1969a) 426.
Seyer, F. A., Metzner, A. B.: Can. J. Chem. Eng. 47 (1969b) 525.
Spriggs, T. W., Huppler, J. D., Bird, R. P.: Trans. Soc. Rheol. 10:1 (1966) 191.
Theodorsen, T.: Article in: Fünfzig Jahre Grenzschichtforschung (Ed. H. Görtler & W. Tollmien) Braunschweig: Vieweg (1955).
Toms, B. A.: Proc. 1st Int. Congr. Rheol. Vol. 2 (1948) 135, Amsterdam: North Holland.
Townsend, A. A.: The structure of Turbulent Shear Flows, Cambridge University Press (1956).
Wallace, J. M., Eckelman, H., Brodkey, R. S.: J. Fluid Mech. 54 (1972) 29.
Walters, K.: J. Méchanique 1 (1962) 479.
Wells, C. S., Spangler, J. G.: Phys. Fluids 10 (1967) 1890.
White, J. L., Metzner, A. B.: J. Appl. Pol. Sci. 7 (1963) 1863.
White, J. L., McEligot, D. M.: J. Basic Eng., Trans. ASME (1970).
Willmarth, W. W., Wooldridge, C. E.: J. Fluid Mech. 14 (1962) 187.
Willmarth, W. W., Tu, B. J.: Phys. Fluids S 134 (1967).
Willmarth, W. W., Lu, S. S.: in Turbulent Shear Flows, AGARD Conf. Proc. 93, 3-1, Paris (1972).
Virk, P. S.: J. Fluid Mech. 45 (1971a) 225.
Virk, P. S.: J. Fluid Mech. 45 (1971b) 417.
Virk, P. S., Merrill, E. W., Mickley, H. S., Smith, K. A., Mollo-Christensen, E. L.: J. Fluid. Mech. 30 (1967) 305.
Virk, P. S., Mickley, H. S., Smith, K. A.: ASME J. Appl. Mech. 37 (1970) 488.
Zimm, B. H.: J. Chem. Phys. 24 (1956) 269.

Experimental fluid mechanics: the impact of modern instrumentation

H. W. Liepmann

Pasadena, Calif. USA

I. Introduction

The importance and, in fact, the very existence of any experimental approach had to be defended in former days against the extreme theoretical epistemologists like Eddington. Presently experimental science has to come to grips with the computer and a new set of extremists; in every field of physics, except elementary particles, the forces and governing equations are known and hence in principle all observed and applied physical phenomena are computable, given only sufficient intelligence and a sufficiently dimensioned computer. All problems in fluid mechanics as a largely classical discipline should therefore be computable from *Navier-Stokes* or similar equations or even from general molecular Boltzmann-like equations. This extreme point of view which has been put forward on occasions is clearly naive and, at least on the basis of today's facts, easily contradicted. Nobody as yet has succeeded in computing the flow past a sphere over the full range of Reynolds numbers and the chance of a purely theoretical or computational prediction of the effect of the addition of minute quantities of polymers on the turbulent skin friction of water is rather small. One of the most important numerical computations of former days is the 1937 Taylor-Green work on the vorticity cascade; an order of magnitude progress with the same problem is even today anything but easy. It is not difficult to add further counter-examples to the list even without using "strange" fluids like plasmas, ferro fluids and superfluids. Consequently, I start with the assumption that experimental fluid mechanics is going to be a vital part of mechanics and physics for some time to come. However, the impact of the extremely rapid developments in instrumentation and computers has a lasting influence on the approach to and the scope of experiments, and the attitude of an experimentalist to a problem will change with it.

I intend to use a few examples to illustrate this impact. It is interesting and significant to note that most of the progress in instrumentation is basically related to quantum effects in macroscopic systems: semi-conductors, lasers, and a large part of cryogenic techniques are all directly or indirectly related to the application of quantum mechanics to matter in bulk.

II. The problems

Engineering tests are undertaken to establish the performance of a specific device. Experiments in the terminology used here are undertaken to add to the understanding and mastery of a field. A large, sometimes crucial part of an experiment consists in the definition and realization of the simplest significant flow configuration to be studied and hence the discussion of experimental progress cannot be restricted to techniques only. In fact, the interplay between the choice of the experimental setup and the methods of measurement are most important.

The selection of a small set of fluid mechanical problems as examples is clearly biased by my own likes, dislikes and competence. One of course carries more conviction in dealing with cases where one has experienced failures and successes directly, and no attempt at an encyclopedic coverage of the subject is intended or, as a matter of fact, possible.

II.1 Shock structure

A strong shock wave traveling through a monatomic gas results in a rapid change of state from one thermodynamic equilibrium state to another. The structure of the transition layer is a classical problem in fluid mechanics and one of the very few experimentally realizable cases where kinetic theory and viscous, compressible fluid dynamics are both capable of providing solutions for the variation of the flow parameters like density, temperature, and velocity through the layer. No questionable solid-gas boundary conditions are involved and hence the shock layer represents one of the best cases for a comparison between kinetic and macroscopic theory and in particular it represents a test case for Monte Carlo techniques in kinetic theory. It is easy to estimate the thickness δ of the transition layer from the Navier-Stokes equations in terms of a^* and ν^*, the velocity of sound and kinematic viscosity evaluated at sonic conditions $\delta^* \sim 5\nu^*/a^*$. Since a strong shock in a monatomic gas travels with a velocity $C = 4a^*$, the transit time Δt is

$$\Delta t \sim \frac{5}{4} \frac{\nu^*}{a^{*2}}$$

i.e., as expected δ^* is proportional to a mean free path Λ and the transition time proportional to the mean collision time τ with a constant of proportionality of order 10 or less.

Looking at the values of Λ and τ in argon, say, we have the following picture:

p (mm Hg)	10^{-3}	10^{-2}	10^{-1}
Λ (cm)	4.5	.45	.045
τ (sec)	2×10^{-5}	2×10^{-6}	2×10^{-7}.

These numbers are significant but not the whole story: To obtain enough signal from a probe, that is a sufficiently large sample, and well defined conditions one has to produce a plane, steady shock with an area of say, $A \geq (100\Lambda)^2$. Consequently for shock tube experiments the tube diameter D has to be larger than $400a^*\tau$. Instrumentation with microsecond time resolution requires flow dimensions of from 0.1 to 1 meter. To realize this combination an advanced vacuum and shock-tube technology and reasonably sophisticated electronics are necessary. Consequently, the problem became accessible only recently.

At present the use of good shock tubes, thin-film gauges to measure shock speed, and the use of electron beams to measure the density makes it possible to measure density profiles

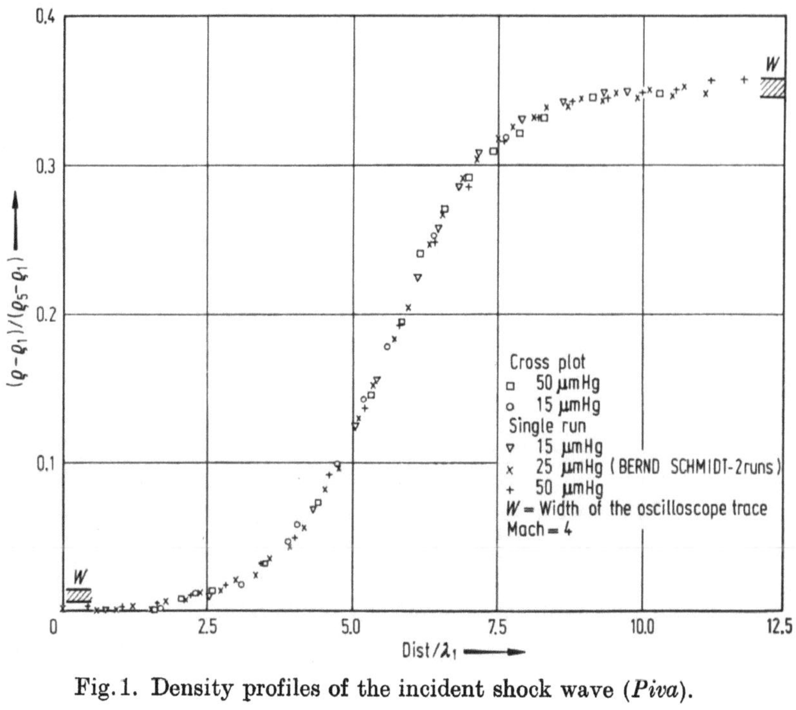

Fig. 1. Density profiles of the incident shock wave (*Piva*).

within the transition layer with good precision. Indeed, the combination of these measurements with Monte Carlo techniques based on an assumed molecular force model, make it possible to determine the molecular force parameters from the shock structure measurements (Fig. 1). The use of a steady gas flow in a rarefied wind tunnel instead of a shock tube does not make the problem overall easier to handle. The complications are shifted from a time resolution problem to the task of providing a well defined and steady high-Mach-number low-density flow and a shock holder which does not introduce significant shock curvature.

All in all, this particular problem can be considered solved, but of course it is only the simplest of a whole class of similar ones. Shock structure in more complex gases, gas mixtures and with chemical reactions is still of great interest. The technique of electron beam induced fluorescence due to *Muntz* has already been used with considerable success in some of these problems as has fast spectroscopy. Finally, very little is known as yet about shock structure in liquids and solids. It is quite likely that the time scale of the experimental device for these cases has to be lowered to the nanosecond range.

It can be expected that presently available ultra-high speed sampling techniques will ultimately completely replace the rather clumsy oscilloscope—photographic plate—digitizing process which is still almost universally used.

II.2 Coherent structures in turbulent flow

A homogeneous random field is properly describable in terms of power spectral densities, i.e., in terms of Fourier components. The granular structure of the solar surface, the Taylor vortices in rotating flows, or the large scale structure of mixing zones are convincing evidence that for some problems at least the Fourier decomposition is not natural. Indeed, all available evidence today points to the importance of large coherent structures in turbulent shear flows (Fig. 2). This picture gradually emerged from quite simple visual observations supplemented by some refinements like the hydrogen-bubble technique or say, indirect studies of intermittencies in hot-wire signals. Considerable effort is at present expended on ways and means to define flow entities, "eddies", from which the turbulent flow can be constructed by forming a statistical ensemble. Attempts to construct a sort of turbulent kinetic theory and hence to find the "molecules" of turbulence are of course not new. The Hill vortex studies of *Synge* and *Lin* as well as Theodorsen's horseshoe vortices are prominent ideas of the past. The difficulty with any such attempt is obviously the fact that since we deal with a continuum, the "forces" holding structures together and the forces of interaction between them are much more alike than in the molecular theory of matter. "Coherent structures" has to be understood in the sense of, say, solitary waves in nonlinear wave theory, i.e., flow patterns which not only have long life times but which also keep a pronounced similarity even when interacting strongly.

Large coherent structures, the Tollmien-Schlichting waves, played an important part in laminar flow as well. However, since the pattern or shapes of the laminary structures is rather simple, they could be experimentally emphasized and the pattern recognition problem reduced to a very simple phase measurement. In turbulent flow the structures are usually too complex to allow the same approach.

There is no doubt in my mind that a large part of the experimental effort in turbulence research in the near future will be productively occupied with the search for the "natural eddy". The experimental approach to the problem can follow essentially two different paths and variants thereof. Either one uses tracer techniques, i.e., marks fluid elements in a Lagrangean fashion and shifts the weight of the experimental procedure to the evaluation of the resulting pictures, or else one uses relatively large and complicated arrays of instruments fixed in the laboratory frame and electronically follows the flow by proper triggering and time delays. In both approaches a very large amount of information has to be processed, much of it being irrelevant. The whole experiment will amount to a nearly direct coupling of a computer to the measuring probes with a large degree of feedback to eliminate irrelevant information as early as possible.

Not much progress seems to have yet been made in the direct tracer technique, i.e., nothing compared to the elaborate reading devices of say, bubble chamber traces in nuclear physics.

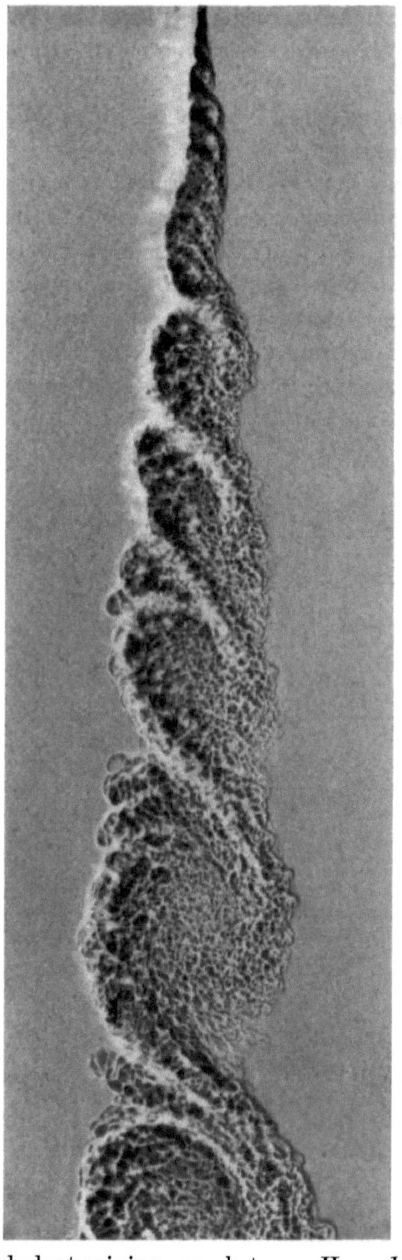

Fig. 2. Structure of a turbulent mixing zone between He and N_2 (*Roshko* & *Rebollo*).

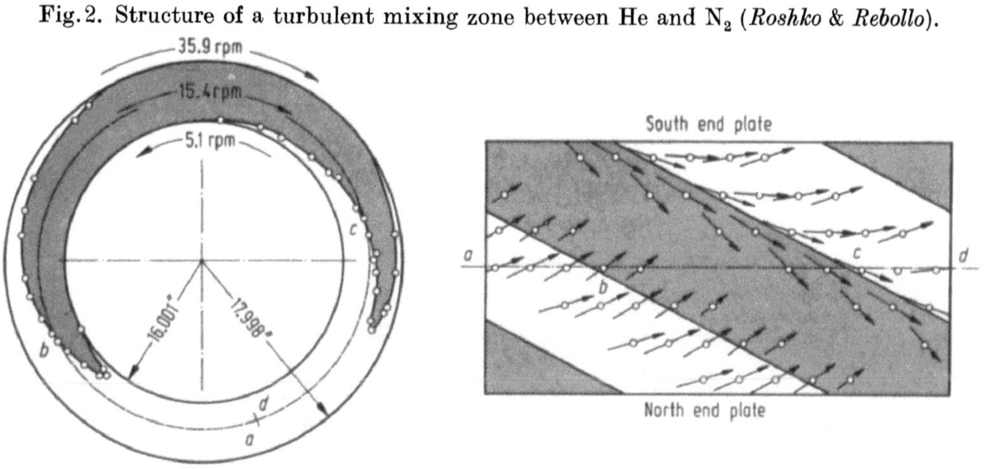

Fig. 3. Turbulent "spot" in the flow between two counter-rotating cylinders. Direction and magnitude of the flow velocity entering and leaving the spot (*Coles* & *Van Atta*).

However, the other approach, properly timing an Eulerian array, has become an indispensable tool of experimental turbulence research. Beginning with simple space-time correlation measurements using analogue tape it has now reached the state called "conditional sampling" in which an array of hot wires is triggered by a chosen event at a predetermined time. Again, only the combination of advanced solid-state electronics and digital data processing has made this progress possible. A present day hot-wire circuit is contained on a standard hard-wired card with a volume of, say, 10 cm³, compared with length dimensions at least ten times as large not long ago. This improvement of a factor of at least a thousand in volume is overshadowed by an even larger reduction in power dissipation and a large increase in stability and lifetime. Figs. 3 and 4 demonstrate typical examples of conditional sampling techniques. The first set due to *Coles* and *Van Atta* demonstrate the velocity pattern entering and leaving a turbulent spot, the second due to *Kaplan* and *Blackwelder* demonstrate the "bursting" phenomena in the turbulent boundary layer.

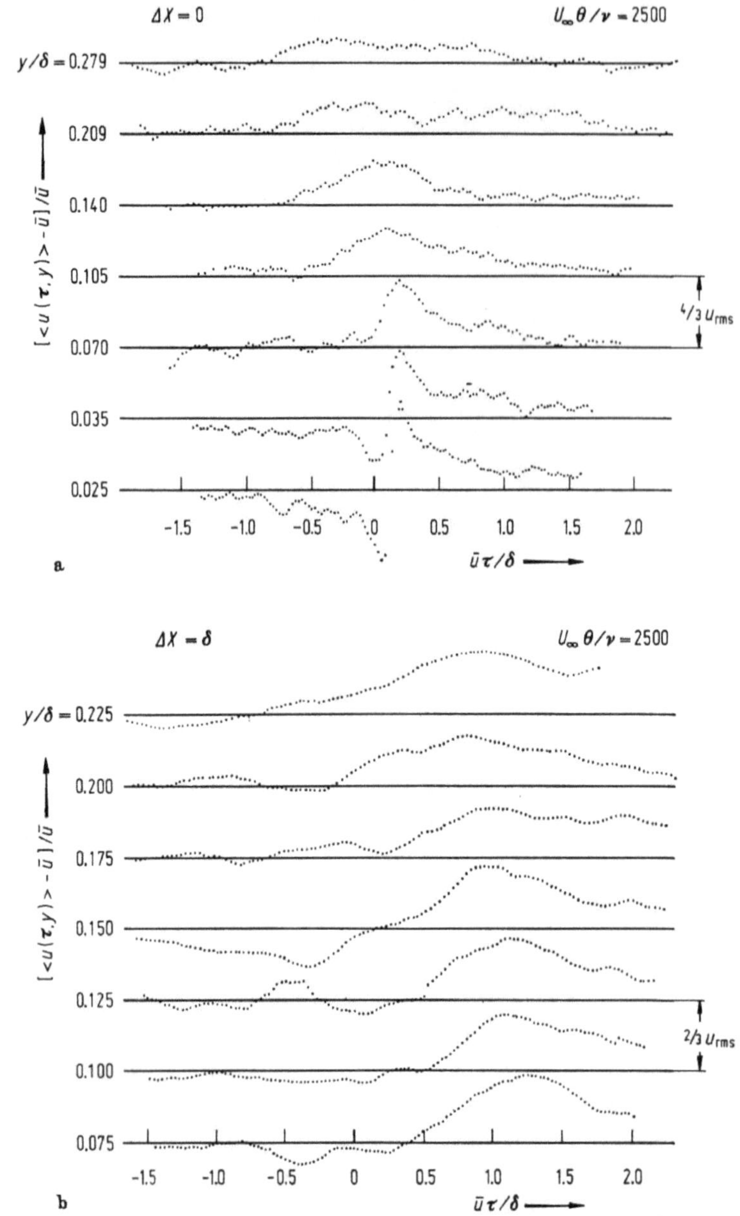

Fig. 4a. Delayed conditional averages of the velocity in a turbulent boundary layer at $y^+ = 15$ triggered by a "burst" (*Blackwelder & Kaplan*).

Fig. 4b. Delayed conditional averages of the velocity in a turbulent boundary layer at $y^+ = 15$ triggered by a "burst" (*Blackwelder & Kaplan*).

II.3 Turbulent mixing of different gases

A complete theory of turbulence in a single fluid does not yet exist. However, on the basis of many years of experimentation, a phenomenological approach is quite feasible and many technically important flow problems of today can be computed on this basis. Turbulent mixing of different gases or liquids is in a far more primitive state. A secure experimental basis for even a phenomenological approach has been missing until recently due to technical difficulties which could only be overcome by modern instrumentation. Consider a two-dimensional mixing region between two gases:

a) Due to its inherent conical or rather wedge-like similarity this configuration is the simplest significant one. The experimental problems are immediately obvious: The Reynolds number of the flow has to be sufficiently high to insure turbulent flow, the size of the facility sufficiently large to permit detailed measurements of the mixing zone, but the gas supply is severely limited by the bulk and cost of storage facilities. Indeed a standard 1 m^3 gas tank charged to some 250 atm. discharges through a sonic orifice of 1 cm^2 in only a few seconds. Consequently, for a reasonable laboratory installation the size of the working volume of the installation is of the order of 0.1 m to 1 m and the flow times of the order of seconds. To alter these dimensions by an order of magnitude or more leads immediately to a very large increase in cost, time of construction, lack of mobility, and all the other complications associated with large size.

Consequently, a probe, probe carriage and data processing system have to be devised capable of taking, say, a few hundred or more data points across a 1 cm thick layer in a few seconds.

b) The common instruments in fluid mechanical flow measurements, the pitot tube and the hot wire, measure ϱu^2 and $|\varrho u|$ respectively, and hence have to be supplemented by a velocity or density sensitive instrument capable of measuring fluctuations as well.

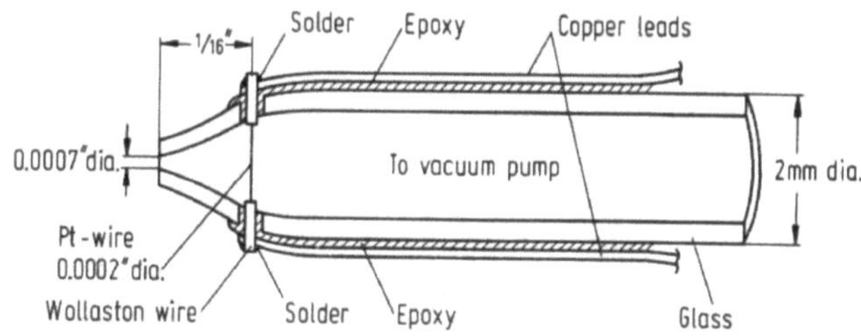

Fig. 5. Mass flow probe for density measurements (*Brown* & *Roshko*).

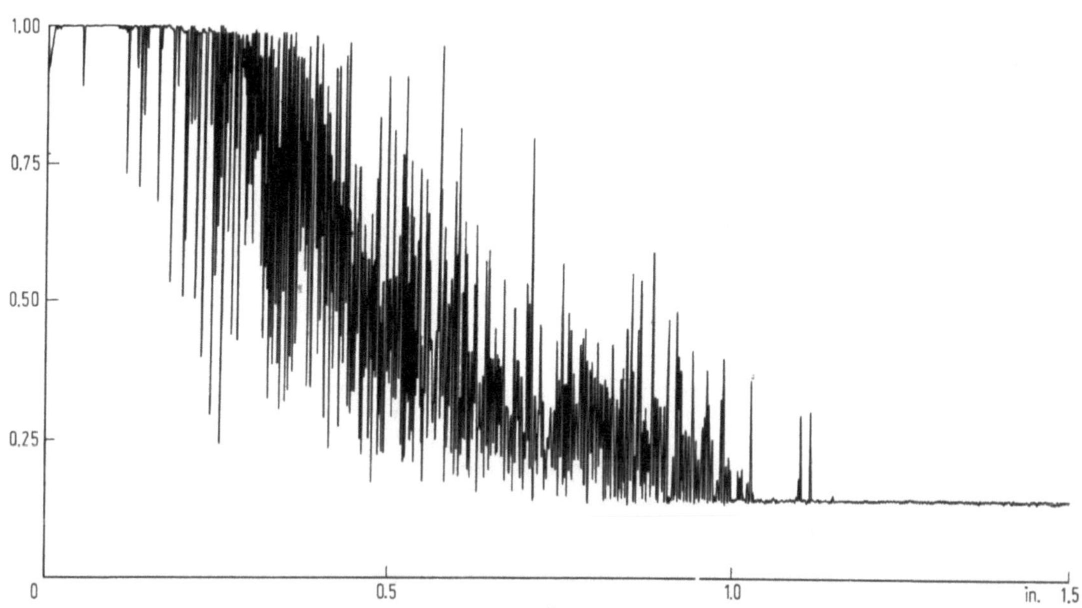

Fig. 6. Density traverse through a mixing layer between He and N_2 (*Brown* & *Roshko*).

The two requirements a and b could not have been successfully solved more than say 10 years or so ago, at least not without a very great and specialized effort in general instrument development. One way in which the problem was handled recently by *Brown, Roshko* and *Rebollo* is shown in Figs. 5, 6 and 7. A carriage driven by a pulse-controlled stepping motor traverses the mixing zone in 1500 steps, stopping a few milliseconds at each step to permit two measurements to be taken and to be digitally recorded. A mass flow probe specifically designed for this purpose measures the density and its fluctuations. The output on digital tape is ready for any type of computer processing. The details of concept and results of this work are reported at this Congress by Dr. *Roshko*.

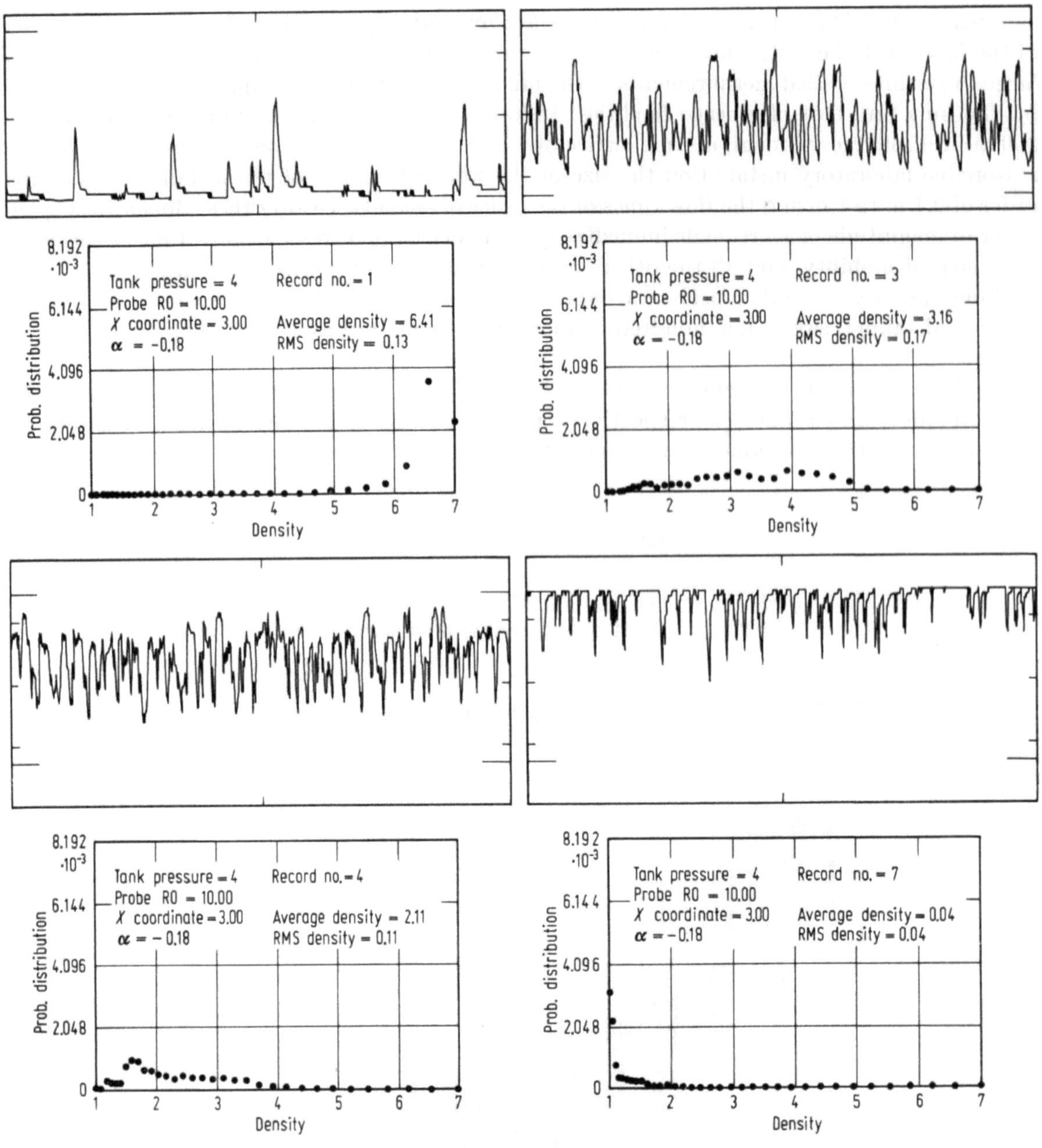

Fig. 7. Density fluctuations in a mixing layer between He and N_2 (*Roshko* & *Rebollo*).

II.4 Chemical reactions in turbulent mixing

If the previous mixing problem is extended to the technically very important case of chemically reacting gases additional problems arise: The detection of species cannot be done entirely by density measurements; instead, instrumentation sensitive to a specific molecule needs to be used. Here one can either choose a chemical reaction which destroys or creates an easily detectable species or else go all out to develop methods which, within reason, are capable of detecting any molecules. Practically all the approaches to the problem employ optical methods: For the first, extinction measurements are being used, as in the $2NO_2 \leftrightarrows N_2O_4$ reaction. The second type of instrumentation is detailed spectroscopic measurements of emission, absorption or, probably more appropriately, Raman lines.

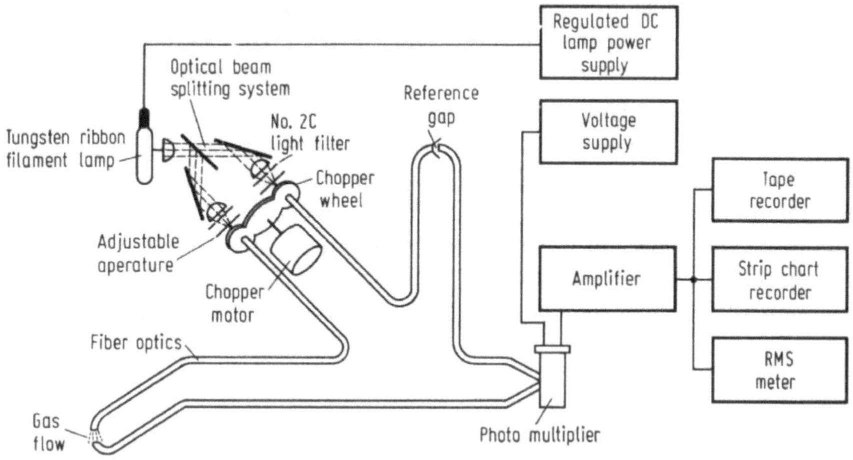

Fig. 8. Fiber optics for light absorption measurements in a reactive mixing layer (*Batt, Kubota & Laufer*).

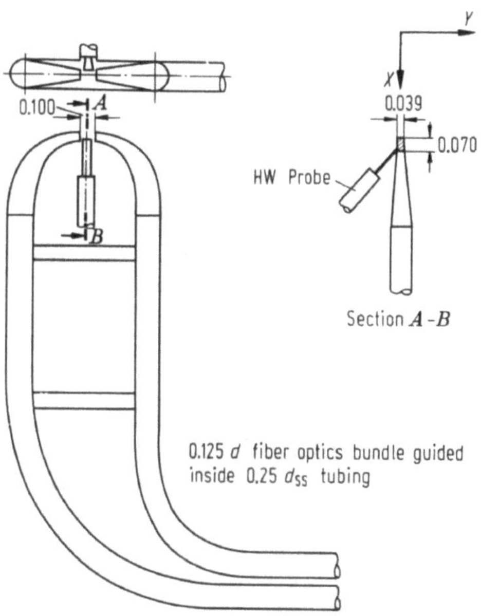

Fig. 9. Correlation probe configuration.

Again only recent technological advances make these tasks possible. Even the first, absorption measurements, need fiber optics or sophisticated multi-beam probes and correlation techniques for their full usefulness. The enormous advances in light intensity detection systems are essential for the second approach, and finally, fast Raman spectroscopy would have been impossible prior to the development of the laser and is even today not easy.

Figs. 8 and 9 illustrate some of the fiber optical and related techniques used in a specific set of experiments by *Batt, Laufer* and *Kubota*. The combination of a hot-wire and a fiber optical extinction sensor makes it possible to measure velocity-concentration correlations.

II.5 Polymer additives

The discovery of *Thoms* that small amounts of polymers added to water flow drastically change the skin friction came as a complete surprise. The quantities involved, a few parts per million, are too small to explain the phenomena by a simple change in viscosity. Consequently, the polymer additives must directly affect the turbulence creation and annihilation mechanisms and thus some extremely interesting problems in fluid mechanics are touched upon, besides the direct technical importance of the phenomenon itself.

There is hence, a clear need for detailed experimentation. However, hot-wire and hot-film anemometry, etc., are quite problematic in this medium. The advent of the laser doppler velocimeter made direct measurements of velocity, velocity fluctuation, shear, etc., possible and at present the field is being investigated and clarified.

Velocity profiles and turbulent fluctuations have been measured in pipe flow and it has already been shown, by measuring $\overline{u'v'}$, that the decrease in skin friction in a pipe due to polymer additives is reflected, as expected, in the Reynolds stress distribution, and that the acoustic radiation from a boundary layer in polymer solution is reduced (Fig. 10). On the other hand, recent measurements of the velocity profile and turbulence level in a jet and of isotropic turbulence behind grids show little or no effect of the polymers (Figs. 11 and 12). Thus the polymer addition is effective only in turbulence near solid boundaries. This in itself is very interesting indeed since all indications at present are that the turbulence boundary layer is largely controlled

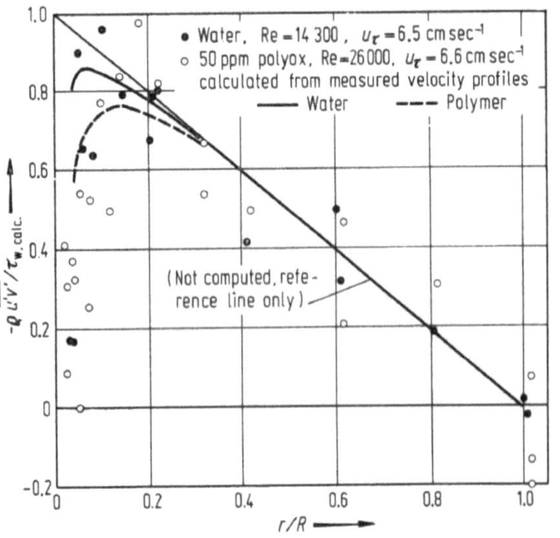

Fig. 10. Comparison of normalized Reynolds stress for water and polymer at the same average wall shear, $u_\tau = 6.6$ cm/sec (*Logan*).

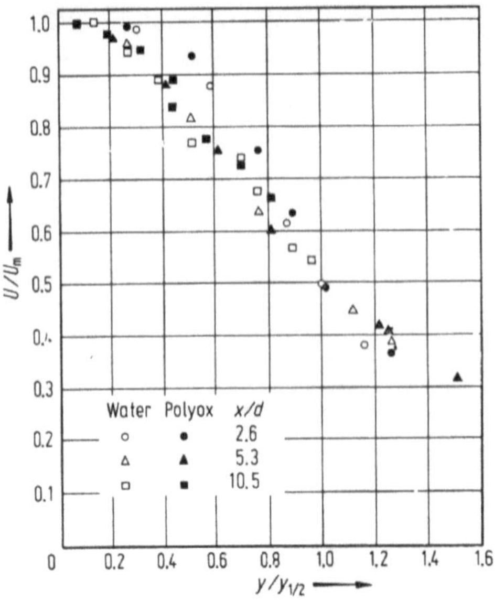

Fig. 11. Velocity profiles in a circular, turbulent jet with and without polymer additives (*Barker*).

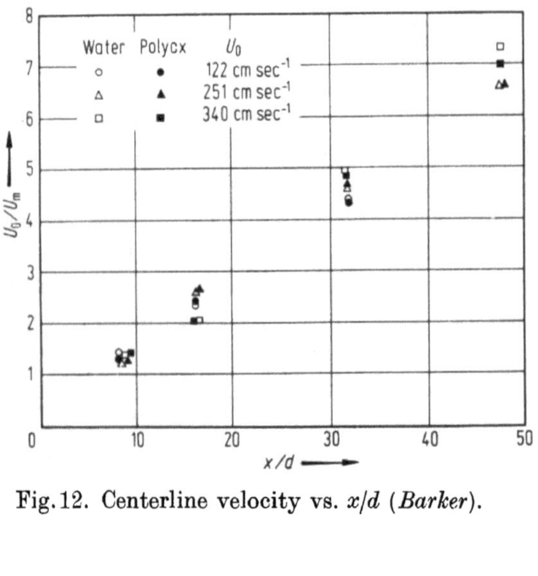

Fig. 12. Centerline velocity vs. x/d (*Barker*).

from the neighborhood of the laminar sublayer. The difference between free turbulence in mixing zones and boundary layer turbulence seems to be quite similar to the corresponding differences in laminar instability, i.e., related to a subtle u, v phase difference near the wall.

II.6 Cryogenic fluid mechanics

Liquid helium is now readily available and the technique of experiments at temperatures in the 1° range is quite accessible. Besides the study of specifically low temperature phenomena like superfluid flow, cryogenic techniques offer new possibilities for classical fluid mechanics problems.

For example, the specific heat of solids decreases at low temperatures according to the Debye T^3 law, i.e., $C_v \sim R(T/\theta)^3$ where R is the gas constant, θ a characteristic temperature. The specific heat of liquid helium near its boiling point remains about $2R$; consequently $C_{v\,\text{solid}}/C_{v\text{He}} \sim (T/\theta)^3$. The characteristic temperatures for most solids are of the order of a few hundred degrees and hence operating between 3 and 4 °K, $C_{v\,\text{solid}}/C_{v\text{He}} \sim 10^{-6}$ i.e., the heat capacity of a vessel can be made vanishingly small compared to the liquid helium it contains. This fact combined with the highly developed art of low-temperature measurements can make experiments on thermal instabilities in fluids more attractive and accurate in liquid helium than in water, as has been shown recently by *Ahlers* at the Bell Labs.

As another example, consider a simple shock tube. Elementary shock tube theory shows that the maximum shock Mach number M_s which can be reached using, say, a single monatomic perfect gas is given simply by $M_s = 4\sqrt{T_4/T_1}$ where T_4 is the driver gas and T_1 the driven gas temperature respectively. Ordinarily $T_1 \sim 300$ °K and T_4 is increased by heating or combustion. However, one can without too much trouble, keep $T_4 \sim 300$ °K and reduce T_1 to liquid helium temperatures. The result is a small shock tube capable of producing very high Mach number shocks with little deviation from ideal behavior. Figs. 13 and 14 show such a device and some of the results.

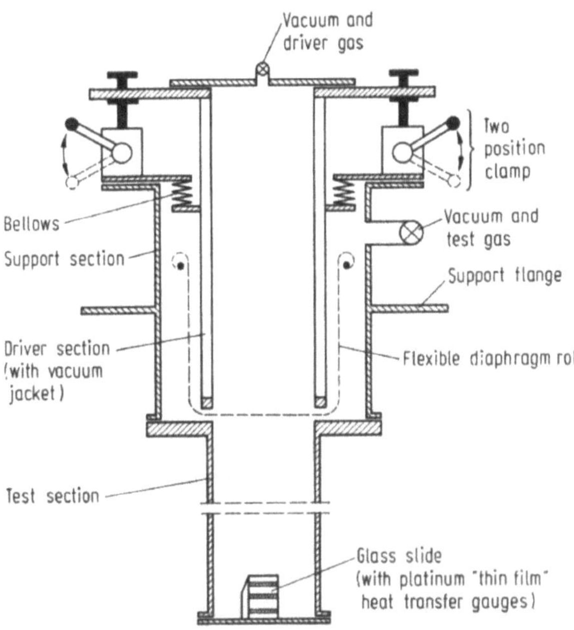

Fig. 13. Cryogenic shock tube (*Cummings*).

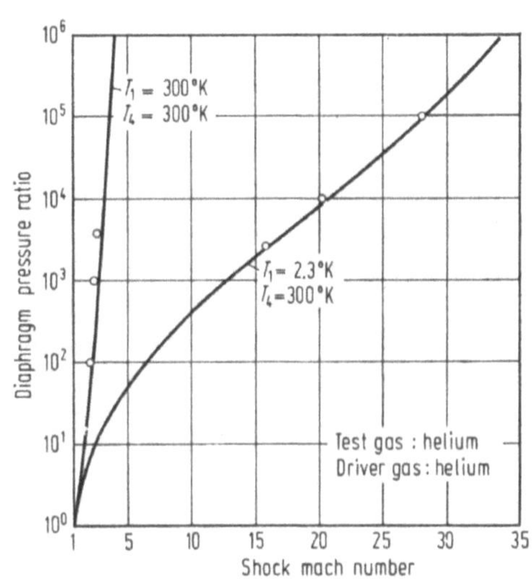

Fig. 14. Cryogenic shock tube data (*Cummings*).

Indeed, in this fashion it is possible in principle to study shock waves and shock wave reflections at Mach numbers of up to 40 or so without encountering any gas imperfection, even in the reflection zone.

Finally, at present a whole array of cryogenic instruments to measure very weak electromagnetic fields by means of the Josephson effect is being developed and I am sure will eventually become part of the experimental arsenal of fluid mechanics.

III. Sensors and processing

The ideal experimental sensor should interact with the system to be investigated only to the extent necessary to transmit the desired information and only within the space-time volume of interest. Progress in experimental methods can be measured by the degree of approximation to this asymptotic ideal. It is clearly impossible to make sweeping statements here about progress on a wide front because of the immense variety of situations. But one can, I believe, make a few general remarks about progress in principle, applicable to a wide selection of measurements. In particular it seems to me that three aspects of measuring methods have undergone major changes: 1. Time resolution. 2. Signal discriminating techniques. 3. Wave and beam methods.

1. The characteristic time τ of any electronic apparatus is $\tau \sim D/c$ i.e., the ratio of a measure of the size of the measuring apparatus, including cables, etc., divided by the speed of light. Prior to the development of solid-state electronics $D \sim 1$ m and hence $\tau \sim 3$ nanoseconds. Solid state circuitry makes it possible to decrease D by something like a factor of 100. The corresponding usable times i.e., time resolutions which can be obtained without an intensive specialized effort are usually between 100 and 1000 times larger and hence time resolutions have proceeded from microsecond to the nanosecond range during the last four years. Since the characteristic velocity in fluid mechanics is the speed of sound, such an increase in time resolution seems hardly important, but there do exist problems for which such resolution is essential, e.g., shock or ionization front speeds in electromagnetically driven shock tubes have reached a few percent of the velocity of light and in more conventional facilities an exploration of the topography of a shock or detonation wave front requires nanosecond resolution times as do measurements of reaction rates.

2. The use of correlation techniques to extract signals buried in noise, developed largely in the Radar field, has become more or less a standard technique in instrumentation. The sensor amplitude modulates a harmonic wave, i.e., gives an output of the form of, say, $a \cos \omega t + n(t)$ where $n(t)$ is the random noise and a the signal. The output is obtained in the form $\overline{\cos \omega t (a \cos \omega t + n(t))}^T$ where the bar indicates a time average over a time T, i.e., the output is the covariance of the unmodulated harmonic wave and the sensor output. The term $\overline{n(t) \cos \omega t}$ is expected to vanish for a sufficiently long time average. Applied to electronic circuits this approach is usually called phase locking but the same principle is applied in beam methods as well, e.g., chopping a light beam, electron beam, or even molecular beam for absorption measurements leads to an amplitude modulated output which can be similarly handled. Unfortunately, another and unpleasant discovery in recent years reduces the effectiveness of these noise reduction techniques: The power spectral density of the noise of many semiconducting elements, lasers, etc., is not uniform as expected but at low frequencies *inversely* proportional to a power of the frequency. This fact influences all measurements which rely on long time averages to decreases the noise effect, including phase locking.

3. Optical methods in fluid mechanics experiments are, of course, not new. Shadow and Schlieren photography and interferometry have been used for a long time. The obvious advantage of the absence of unwanted flow disturbances was however, often largely offset by the lack of localized measurements which restricted the quantitative applicability of the methods to two-dimensional or axisymmetric flows. The same restrictions used to apply to optical absorption measurements. In recent times the use of correlation measurements with multiple beams, focussing systems, and fiber optics has largely overcome this restriction. However, the major new device in flow measurements is evidently the laser. The two most important uses of the laser in fluid dynamics seems to me the use of the Doppler effect for true velocity measurements and Raman scattering for chemical species detection.

Since speeding cars have been spotted by Doppler effect measurements for some time, the use of the same principle for fluid flow measurement seems an obvious extension, and microwaves and microwave Doppler measurements of ionizing shock fronts have been made. However, the combination of an exact phase and frequency definition with a short enough wave length to permit localized measurements in fluid flow was simply impossible to achieve before the advent of the laser. The relative frequency shift $\Delta f/f$ due to Doppler effect is equal to the ratio of a velocity component u of the scatterer to the velocity of light, i.e., for slow fluid flow of the order $\Delta f/f \sim 10^{-8}$. For visible light this amounts to wave length shifts of the order of 10^{-4} Å, too small for resolution by the usual spectrometric methods. However, the corresponding frequency shift of $\Delta f \sim 10^{+6}$ Hz is very convenient for electronic measurements, e.g., by heterodyning.

The velocity component u measured is the one in the plane of and along the line bisecting the angle made by the incoming and scattered beams (the problem is equivalent to the classical one of reflection from a moving mirror). Hence, the direction of u is within limits at the discretion of the experimenter; the sensitive volume can be made quite small, of the order of $(1\ \mathrm{mm})^3$ or less, i.e., comparable to the hot-wire anemometer. Fluctuations in the velocity component u appear as an additional frequency modulation about the mean beat frequency. Figure 15 illustrates a typical laser velocimeter for use in turbulent flow. The output signal can be processed in the usual fashion. However, the method is capable of distinguishing the *sign* of the velocity component and hence, in contrast to the hot wire, capable of turbulence measurements at zero mean flow velocity, a fact which is very important indeed since it allows measurements in separated flows, "turbulence in a box" and similar set ups.

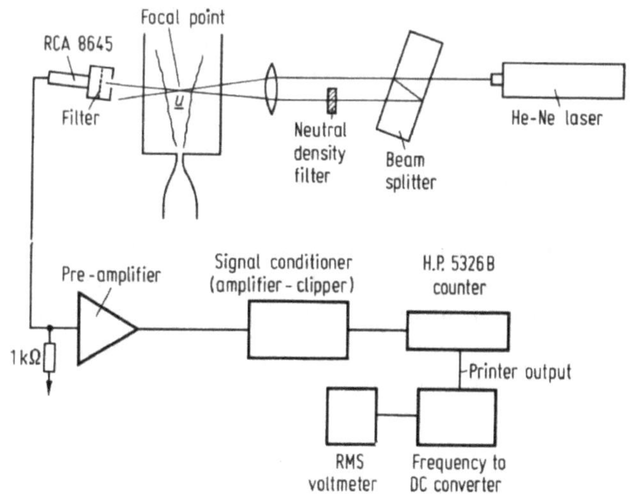

Fig. 15. Laser Doppler velocimeter (*Barker*).

Combustion, some chemical processes, gas dynamic and chemical lasers depend critically on molecular reactions in turbulent flow. To follow a specific reaction in time and space requires very discriminating, fast sensors. Raman scattering provides one approach. The effect, quantum mechanically, amounts to inelastic photon scattering with the energy difference being transferred to or from one of the internal molecular modes which are characteristic for a specific molecule. Classically, the effect is interpreted as the result of the interaction of a harmonically oscillating electrical field of frequency ω with the molecule characterized by its polarizability, which in turn depends on the internal modes with frequencies ω_k. The interaction thus gives rise to terms of the form $\cos \omega t \cos \omega_k t$ i.e., to a set of lines with frequencies $\omega - \omega_k$ and $\omega + \omega_k$ from which the ω_k can be determined. In gases the Raman lines are quite weak and hence require for their observation an intense source of monochromatic light. The possibility of producing intense, short wave laser radiation has now brought the Raman effect in gases from a research subject down to a diagnostic method.

Finally, a more or less new way of looking at optical methods, sometimes called "modern optics" is rapidly taking over the field. In essence one can look at the system to be investigated as a black box characterized by a generalized susceptibility or refractive index tensor. This box is "interrogated" by a wave passing through it with well defined wave fronts. The reply information is contained in the emerging wave fronts and stored in amplitude, frequency and phase modulation. It is recoverable by various correlation techniques, e.g., holography or holographic interferometry is one way to extract part of the information contained. It is not difficult to produce light of almost any frequency with a well controlled amplitude. It is quite a different matter to produce optical waves with closely controlled phases and frequencies. Prior to the laser, good phase and frequency control could only be achieved for vanishing intensity and only the coarsest features of the system susceptibility studied. At present, frequency modulation of one part in 10^8 and better represents little problem. Evidently any waves or beams can be looked upon in the same way. Thus interrogation by acoustic waves using similar techniques is quite feasible and extends the use of wave optical methods to optically opaque systems. Indeed, the study of the propagation of sonic booms through the turbulent atmosphere may well be viewed as a method to study atmospheric turbulence.

Given the appropriate signal in analog form, usually in terms of an electric voltage, what then? The signal has to be stored, the useful information extracted and transformed into a table or graph. The process involves writing, reading and computing. Several choices can now be made: One can rely on a central computing facility and terminate the experiment with a signal written in analog form or transmitted by human hands into digital form. The other extreme is the coupling of a small special-purpose computer to the experiment. Between these two extremes a large number of other possible solutions exist. The continuing improvement in digital components, "chips" and the astonishing decrease in the cost of such components has made the second approach quite feasible. The decision where to put the interface between computing and experimenting and indeed what the difference really is, depends largely on the problem at hand, the availability of general computing facilities, etc., but even more so on a choice of the desired degree of active interaction between the computer and sensor parts of a given experiment. This active interaction can be primitive, e.g., it may amount to nothing more than the motion of a traversing mechanism once a data point has been recorded, or much more sophisticated, e.g., amounting to a selection of measured data to be recorded and processed according to a programmed criterion. It sems to me very likely that a large number of experiments will require active interaction and hence, I believe, the trend will be in the direction of small local memory devices and computers, with the very large and central facilities reserved for numerical analysis rather than data handling.

References

Shock Structure

Bird, G. A.: JFM 30 (1967) 479.

Liepmann, H. W., Narasimha, R., Chahine, M. T.: Phys. Fluids 5 (1962) 1313.

Muntz, E. P.: Phys. Fluids 5, 1 (1962) 80.

Robben, F., Talbot, L.: Phys. Fluids 9 (1966) 633.

Schmidt, B.: JFM 39 (1969) 361.

Coherent Structures in Turbulent Flow

Boundary Layers and Turbulence, Phys. Fluids Supplement 1967, ed. K. F. Bowden, F. N. Frenkiel, and I. Tani, New York.

Blackwelder, R. F., Kaplan, R. E.: XIII Int. Congr. of Theor. and Appl. Mech., Moscow 1972.

Coles, D., Van Atta, C.: AIAA J. 4 (1966) 11.

Kim, H. T., Kline, S. J., Reynolds, W. C.: JFM 50 (1971) 133.

Kline, S. J., Reynolds, W. C., Schraub, F. A., Runstadler, P. W.: JFM 30 (1967) 741.

Kovasznay, L. S. G., Kibens, V., Blackwelder, R. F. JFM 41 (1970) 283.

Willmarth, W. W., Lu, S. S.: AGARD, Fluid Dynamics Panel Specialists Meeting, London, Sept. 1971.

Mixing Zone

Brown, G. L., Rebollo, M. R.: AIAA J. 10, 5 (1972) 649.

Brown, G. L., Roshko, A.: XIII Int. Congr. Theor. and Appl. Mech., Moscow 1972.

Chemical Reactions in Turbulent Mixing

Batt, R. G., Kubota, T., Laufer, J.: AIAA Conf. on Reacting Flows, San Diego, June 1970.

Wegener, P. P.: J. Chem. Phys. 41, 5 (1964) 1512.

Polymer Additives and Lasers

Barker, S. J.: submitted to JFM.

Greated, C. A.: Nature (GV) 224 (1969) 2297.

Goldstein, R. J., Hagen, W. F.: Phys. Fluids 10 (1967) 1349.

Logan, S. E.: AIAA J. 10, 7 (1972) 962.

Rudd, M. J.: Nature (GV) 224 (1969) 587.

Cryogenic Fluid Mechanics

Liepmann, H. W., Cummings, J. C., Rupert, V. C:. Phys. Fluids, 16, 2 (1972) 332.

The mechanics of the surf zone

M. S. Longuet-Higgins

Cambridge, U.K., and Wormley, Godalming, U.K.

1. Introduction and summary

In recent years many previously baffling phenomena to do with the action of waves on sloping beaches have been sufficiently understood physically that they are now amenable to mathematical analysis. This in turn has suggested new problems, which it will be the purpose of this paper to point out.

The chief phenomena to be discussed are the following:

When waves or swell approach a shoreline they produce a change in the mean level of the water, at first a depression, and then—inside the surf zone—a sharp rise (see Fig. 1). This so-called *wave set-up* may amount to several feet in magnitude and is of practical importance in constructing coastal defences against hurricanes (*Harris*, 1963).

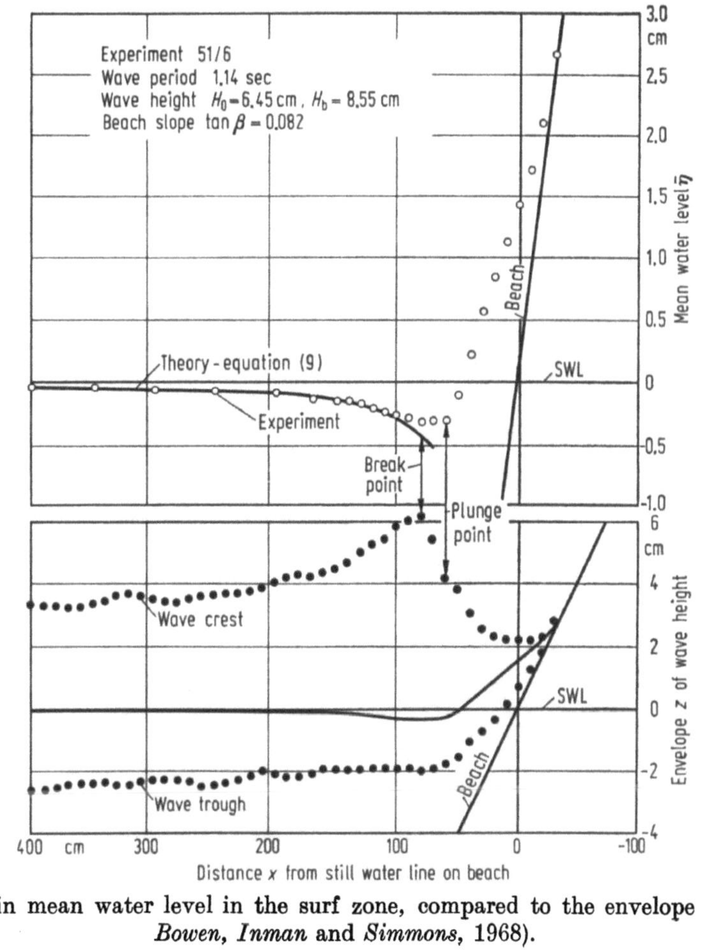

Fig. 1. The changes in mean water level in the surf zone, compared to the envelope of the waves (from *Bowen, Inman* and *Simmons*, 1968).

Secondly, waves approaching a coastline obliquely are found to produce a strong current running parallel to the shore (Fig. 3) and confined mainly to the surf zone. Associated with this *longshore current* there is a longshore *transport of sand* or sediment amounting typically to 10^6 m^3 per year on an exposed coastline. This is certainly of importance both to coastal engineers and to marine geologists.

It is also found that under the action of the waves narrow *rip-currents* flow out to sea at right-angles to the coastline. These are now known to be associated with *edge waves* (see Section 7) and with the more-or-less regular indentations in the shoreline called *beach cusps* (Section 9).

A basic idea underlying the explanation of these phenomena is that of the *momentum flux due to the waves*, which is expressed mathematically by the *radiation stress tensor* defined in Section 2 below.

After discussing the special cases mentioned above we shall present, in Section 8, a unified analysis of these effects, based on the linearised theory of small-amplitude waves. For any gradual variation of the bottom topography it is shown that the dynamical effect of the radiation stress can be expressed as the sum of two terms. One term can always be balanced by a change in mean water level. The other, which is proportional to the local dissipation of wave energy, can generally be balanced only by frictional forces.

In Section 9 these results are applied to the discussion of a non-uniform coastline, in which there are variations both of wave amplitude and angle of incidence. The possibility of an equilibrium form for the beach profile is also discussed.

In Section 10 we discuss the longshore *transport of sand and sediment* and give an explanation for the observed relation between this and the incident waves. As a result we may now envisage a theory for the *interaction* between waves and beaches. Some suggestions for further advances are made in Section 11.

2. The radiation stresses

Consider a progressive wave of amplitude a, wavenumber k and radian frequency σ, in water of uniform depth h. Let (x, y, z) denote rectangular co-ordinates with the origin in the mean water level, the x-axis in the direction of wave propagation and the z-axis vertically upwards. If (u, v, w) denote the components of the velocity, then the flux of x-momentum across the plane $x =$ constant is given by

$$\int_{-h}^{\zeta} (p + \varrho u^2)\, dz, \tag{2.1}$$

where p denotes the pressure, ϱ the density and $z = \zeta(x, y, t)$ the equation of the free surface. In the absence of any waves we have $u = 0$, $\zeta = 0$ and $p = -\varrho g z$ where g denotes gravity. So the *excess flux of momentum* due to the presence of the waves is given by

$$\overline{\int_{-h}^{\zeta} (p + \varrho u^2)\, dz} - \int_{-h}^{0} (-\varrho g z)\, dz = S_{xx}, \tag{2.2}$$

say. An overbar denotes the average with respect to the time t. It can easily be shown (*Longuet-Higgins* and *Stewart*, 1960, 1964) that S_{xx} is always positive. Moreover for waves of small amplitude, in which the orbital velocity is given by

$$\left.\begin{aligned} u &= \frac{a\sigma}{\sinh kh} \cosh k(z + h) \cos (kx - \sigma t), \\ v &= 0, \\ w &= \frac{a\sigma}{\sinh kh} \sinh k(z + h) \sin (kx - \sigma t) \end{aligned}\right\} \tag{2.3}$$

and

$$\sigma^2 = gk \tanh kh \tag{2.4}$$

(see *Lamb*, 1932 C.9) it is easily found that, to order $(ak)^2$,

$$S_{xx} = E\left(\frac{2kh}{\sinh 2kh} + \frac{1}{2}\right), \tag{2.5}$$

where E denotes the total energy density (kinetic plus potential), namely

$$E = \frac{1}{2} \varrho g a^2. \tag{2.6}$$

In a similar way we may evaluate the other components of the radiation stress tensor:

$$S_{yy} = \overline{\int_{-h}^{\zeta} (p + \varrho v^2)\, dz} - \overline{\int_{-h}^{0} (-\varrho g z)\, dz} \tag{2.7}$$

and

$$S_{xy} = S_{yx} = \overline{\int_{-h}^{\zeta} \varrho u v\, dz} \tag{2.8}$$

with the result

$$\begin{bmatrix} S_{xx} & S_{xy} \\ S_{yx} & S_{yy} \end{bmatrix} = E \begin{bmatrix} \dfrac{2kh}{\sinh 2kh} + \dfrac{1}{2}, & 0 \\ 0 & \dfrac{kh}{\sinh 2kh} \end{bmatrix}. \tag{2.9}$$

We note that in deep water ($kh \gg 1$)

$$\begin{bmatrix} S_{xx} & S_{xy} \\ S_{yx} & S_{yy} \end{bmatrix} = E \begin{bmatrix} \dfrac{1}{2} & 0 \\ 0 & 0 \end{bmatrix} \tag{2.10}$$

and in shallow water ($kh \ll 1$)

$$\begin{bmatrix} S_{xx} & S_{xy} \\ S_{yx} & S_{yy} \end{bmatrix} = E \begin{bmatrix} \dfrac{3}{2} & 0 \\ 0 & \dfrac{1}{2} \end{bmatrix}. \tag{2.11}$$

The flux of *energy* on the other hand is given by

$$F = (F_x, F_y) = E(c_g, 0), \tag{2.11}$$

where c_g denotes the group-velocity:

$$c_g = \partial\sigma/\partial k \tag{2.12}$$

calculated from the dispersion relation (2.4).

3. Waves in water of slowly varying depth

Consider now waves approaching normally to a coastline. The frequency σ may be assumed constant, but the wave amplitude a and wavenumber k will be functions of the coordinate x normal to the shoreline. Let us suppose that the depth $h(x)$ varies sufficiently slowly that the dispersion relation (2.4) is locally valid. Writing for short

$$kh = \xi, \quad \sigma^2 h/g = \eta \tag{3.1}$$

in (2.4) we have

$$\eta = \xi \tanh \xi \tag{3.2}$$

and since η (the non-dimensional depth) is a function of x, so is ξ. Moreover the phase-velocity $c = \sigma/k$, relative to its value c_∞ in deep water is given by

$$\frac{c}{c_\infty} = \frac{\sigma}{k}\Big/\frac{g}{\sigma} = \frac{\sigma^2}{gk} = \frac{\eta}{\xi}, \tag{3.3}$$

and the group-velocity c_g is given by

$$\frac{c_g}{c_{g\infty}} = \frac{\partial\sigma}{\partial k}\Big/\frac{g}{2\sigma} = \frac{1}{g}\frac{\partial(\sigma^2)}{\partial k} = \frac{d\eta}{d\xi}. \tag{3.4}$$

The functions η and $d\eta/d\xi$ are shown in Fig. 2. Clearly η is a monotonic function of ξ, but must have a point of inflexion, at which $d\eta/d\xi$ has a maximum slightly exceeding 1. This occurs where

$$\xi = \coth\xi = d\eta/d\xi = 1.200; \quad \eta = 1.440. \tag{3.5}$$

As the waves enter shallow water h and η gradually diminish, as also do $\xi = kh$ and $\eta/\xi = c/c_\infty$. However the group velocity, proportional to $d\eta/d\xi$, first *increases* to a maximum and then falls to zero.

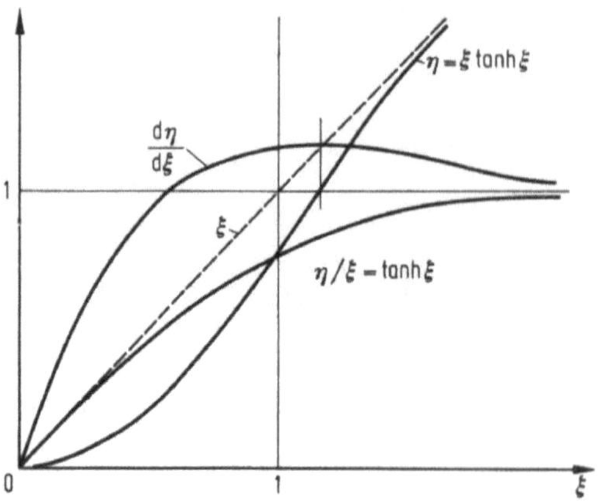

Fig. 2. The functions η, η/ξ and $d\eta/d\xi$ given by Eqs. (3.2), (3.3) and (3.4).

We may distinguish two regions:

(a) *Outside the surf zone* where the dissipation of energy is relatively slight, we may assume that energy is conserved and hence the flux is constant:

$$Ec_g = E_\infty c_{g\infty}. \tag{3.6}$$

From (2.6) and (3.4) it follows that

$$a/a_\infty = (c_g/c_{g\infty})^{-1/2} = (d\xi/d\eta)^{1/2}. \tag{3.7}$$

Hence the wave amplitude first diminishes to a minimum at $\eta = 1.440$ and then increases to the breaking point.

(b) *Inside the surf zone* the wave amplitude steadily diminishes. Empirically (*Munk*, 1949; *Bowen, Inman* and *Simmons*, 1968) it is found that the wave height is roughly proportional to the local mean depth, so that we can assume

$$a = \alpha(h + \bar{\zeta}) \doteqdot \alpha h, \tag{3.8}$$

where α is constant lying between about 0.3 and 0.5 and $\bar{\zeta}$ denotes the mean level of the free surface.

4. Wave set-up

The wave set-up is now readily calculated from the equation of mean momentum flux, namely

$$\varrho g(h + \bar{\zeta})\,\partial\bar{\zeta}/\partial x + \partial S_{xx}/\partial x = 0. \tag{4.1}$$

If $\bar{\zeta} \ll h$ then we have simply

$$\frac{\partial\bar{\zeta}}{\partial x} = -\frac{1}{\varrho g h}\frac{\partial}{\partial x}S_{xx}. \tag{4.2}$$

Now *outside* the surf zone it is found that S_{xx} always increases with x. Hence $\partial\bar{\zeta}/\partial x < 0$, and we obtain a "set-down". In fact Equation (4.2) may be integrated (*Longuet-Higgins* and *Stewart*

1962) to give

$$\bar{\zeta} = -\frac{a_\infty^2 k_\infty}{4}\frac{d}{d\eta}\left(\frac{\xi}{\eta}\right) = -\frac{a_\infty^2 k_\infty \coth^2 kh}{4kh + 2\sinh 2kh} \tag{4.3}$$

in good agreement with observation (*Longuet-Higgins* and *Stewart* 1963). Equation (4.3) can also be derived from the general theory given below in Section 8.

Inside the surf zone, on the other hand, S_{xx} constantly diminishes with x, so $\partial\bar{\zeta}/\partial x$ is positive and the mean water level tends to *rise*. In fact from Eqs. (2.6) and (2.11)

$$S_{xx} = \frac{3}{4}\varrho g \alpha^2 (h + \bar{\zeta})^2 \tag{4.4}$$

and so from (4.1) it is easily found that

$$\partial\bar{\zeta}/\partial x = -K\, dh/dx, \qquad K = 3\alpha^2/(3\alpha^2 + 2); \tag{4.5}$$

in other words the tilt of the mean surface level is simply proportional to the bottom gradient. This prediction can be seen to be verified in Fig. 1.

It will also be seen that the effect of a higher deepwater wave amplitude is to make the waves break *sooner* (i.e., in deeper water) and hence to produce a higher "set-up" at the shoreline.

5. Longshore currents: the driving forces

Consider waves approaching a straight coastline at an oblique angle, as in Fig. 3a. It is found both in field observations and laboratory experiments (*Galvin* 1967) that a strong current is generated parallel to the coast. This "longshore current" is confined mainly to the surf zone, that is to say the region between the "breaker line", where the waves first break, and the shoreline. We shall now outline the theory of these currents as given in recent papers by *Bowen* (1969) and *Longuet-Higgins* (1970a, b).

We take axes with the x and y directions normal and parallel to the coastline, so that $h = h(x)$ as before. k denotes the local wavenumber, $c = \sigma/k$ the local phase velocity and Θ

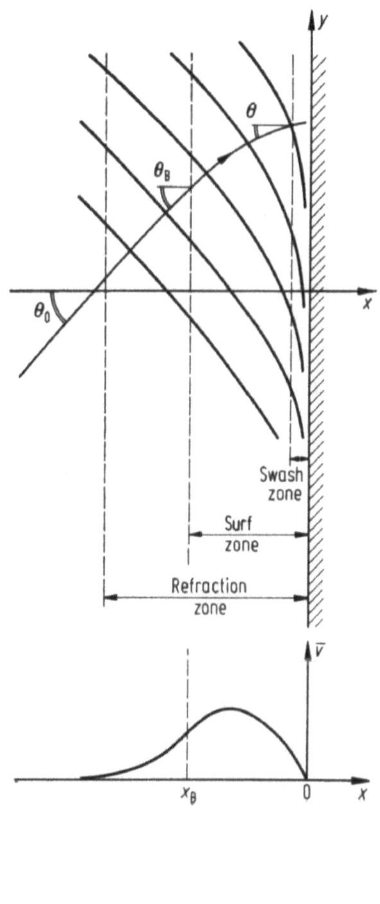

Fig. 3. A schematic view of waves approaching a coastline obliquely, and the profile of the longshore current.

the angle of incidence. We assume that the depth varies sufficiently slowly that locally the dispersion relation (2.4) is satisfied.

We have the following conservation laws. First, assuming that there is no wave splitting, the wavenumber in the y direction is constant, that is

$$k \sin \Theta = \text{constant} = k_\infty \sin \Theta_\infty \tag{5.1}$$

a suffix ∞ denoting the values in deep water. Next, if the effect of the current itself on the waves is at first ignored,

$$\sigma = \text{const} = \sigma_\infty . \tag{5.2}$$

combining these two relations we have Snell's law:

$$\frac{\sin \Theta}{c} = \text{const} = \frac{\sin \Theta_\infty}{c_\infty} . \tag{5.3}$$

Outside the surf zone we assume that the dissipation of energy is negligible, so that the flux of energy towards the shoreline is a constant:

$$F_x = E c_g \cos \Theta = \text{const} = E_\infty c_{g\infty} \cos \Theta_\infty . \tag{5.4}$$

Inside the surf zone we have more generally

$$\partial F_x / \partial x = -D , \tag{5.5}$$

where D denotes the rate of energy dissipation per unit time and horizontal area. At the shoreline itself

$$F_x = 0 \tag{5.6}$$

and we assume that the proportion of energy reflected is negligible.

Consider now the flux of y-momentum across a line $x =$ constant, parallel to the shoreline. This flux is given by the radiation stress component

$$S_{xy} = \overline{\int_{-h}^{\zeta} \varrho u v \, dz} . \tag{5.7}$$

This can be evaluated directly, either by substituting for u und v the known orbital velocities, or else by using Eq. (2.9) and the usual tensor law (*Longuet-Higgins* 1970a). In either case it is found that

$$S_{xy} = E (c_g / c) \cos \Theta \sin \Theta \tag{5.8}$$

or more simply

$$S_{xy} = F_x \left(\frac{\sin \Theta}{c} \right) . \tag{5.9}$$

The last factor is the "phase-velocity parallel to the coast" and is a constant, by Snell's law (5.3). An alternative proof of the very simple result (5.9) has been given by *Longuet-Higgins* (1972), showing that it is valid also for waves of finite amplitude, and is independent of the theory of small-amplitude waves.

Consider now the consequences of Eq. (5.9). *Outside* the breaker zone, since F_x is constant by (5.4) we have

$$S_{xy} = \text{const} = H \tag{5.10}$$

say. On the other hand at the shoreline S_{xy} vanishes, by (5.6). Thus H represents the *total longshore thrust* exerted by the waves on the water in the surf zone. This is also the total force available for moving sand and sediment parallel to the shore, over the rough bottom (see Section 10). The thrust may be evaluated very simply by substituting in (5.9) the values of the quantities in deep water:

$$H = (E c_g \cos \Theta)_\infty \left(\frac{\sin \Theta_\infty}{c_\infty} \right) = \left(\frac{c_g}{c} \right)_\infty E_\infty \cos \Theta_\infty \sin \Theta_\infty = 1/4 \, E_\infty \sin 2\Theta_\infty \tag{5.11}$$

(since in deep water $c_g / c = 1/2$). Clearly for a given energy density E_∞ in deep water, the total longshore thrust is a maximum when $\Theta_\infty = 45°$.

Consider now the *distribution* of the thrust H. From the balance of momentum flux we see that the waves exert a local stress τ_y parallel to the shoreline given by

$$\tau_y = -\partial S_{xy}/\partial x\,. \tag{5.12}$$

Substituting from (5.9) and remembering that $\sin\Theta/c$ is independent of x we have

$$\tau_y = -\frac{\partial F_x}{\partial x}\left(\frac{\sin\Theta}{c}\right) \tag{5.13}$$

or from (5.5)

$$\tau_y = D\left(\frac{\sin\Theta}{c}\right). \tag{5.14}$$

In other words, the local shear stress is simply proportional to the rate of energy dissipation. Outside the surf zone, where D vanishes, so also does τ_y; no dissipation, no driving force.

This conclusion is very similar to the analogous result in acoustical streaming found by *Rayleigh* (1884) and by *Medwin* and *Rudnick* (1953). The acoustical driving force is confined to zones of *dissipation* of acoustical energy, particularly the boundary-layers near solid objects.

6. Longshore currents: the velocity profile

In a steady state, and on a straight coastline, one would expect a simple momentum balance of the form

$$\text{driving force } \tau_y + \text{bottom friction} + \text{lateral friction} = O. \tag{6.1}$$

A theory on this basis, assuming a beach of uniform gradient:

$$dh/dx = -s, \qquad h = -sx, \tag{6.2}$$

has been given by *Bowen* (1969) and *Longuet-Higgins* (1970a, b).

Outside the surf zone, where D is negligible, we have $\tau_y = 0$. *Inside* the surf zone it is assumed that the linear theory of waves in shallow water is applicable, as in Sections 2 to 4. We have then

$$F_x = Ec_g = \frac{1}{2}\varrho g a^2\sqrt{gh} \doteq \frac{1}{2}\alpha^2\varrho g^{3/2}h^{5/2} \tag{6.3}$$

and so from (5.13)

$$\tau_y = \frac{5}{4}\alpha^2 s\varrho(gh)^{3/2}\left(\frac{\sin\Theta}{c}\right). \tag{6.4}$$

In other words the driving force is proportional to $x^{3/2}$.

For the bottom friction $\boldsymbol{B}$, *Bowen* (1969) assumed that $\overline{B}_y \propto \bar{v}$, where $\bar{v}$ is the time-averaged longshore velocity. On the other hand *Longuet-Higgins* (1970a, b) assumed the Czezy law:

$$\boldsymbol{B} = C\varrho\boldsymbol{u}\,|\boldsymbol{u}|, \tag{6.5}$$

where C is a constant and $\boldsymbol{u}$ is the horizontal velocity, consisting of both mean and oscillatory components. On taking time-averages, this leads to the result

$$\overline{B}_y = \frac{2}{\pi}C\alpha\varrho(gh)^{1/2}\bar{v}, \tag{6.6}$$

indicating that $\overline{B}_y$ is proportional to $x^{1/2}\bar{v}$.

If the lateral friction is altogether neglected the balance of the two remaining terms in Eq. (6.1) then loads to a linear variation of $\bar{v}$ within the surf zone:

$$\bar{v} = \begin{cases} v_0(x/x_B), & 0 < |x| < |x_B|, \\ 0, & |x_B| < x| < \infty, \end{cases} \tag{6.7}$$

where

$$v_0 = \frac{5\pi}{8}\frac{\alpha s}{C}(gh_B)^{1/2}\sin\Theta_B, \tag{6.8}$$

and the suffix B denotes the value at the breaker line. There is necessarily a discontinuity in $\bar{v}$ at the breaker line itself.

The introduction of lateral friction in Eq. (6.1) smooths out the discontinuity in $\bar{v}$ and reduces the value of $\bar{v}$ at the breaker line to somewhat less than $v_0/2$. For the lateral friction one may assume a term of the form $\partial/\partial x(\mu_e h\, \partial\bar{v}/\partial y)$ where μ_e is an eddy-viscosity. On physical grounds *Longuet-Higgins* (1970b) took

$$\mu_e = N\varrho\,|x|\,(gh)^{1/2}, \tag{6.9}$$

where N is a numerical constant which may be expected to be less than about 0.16, and probably nearer one-tenth of this value. Incorporating these assumptions into Eq. (6.1) and choosing the non-dimensional variables

$$X = x/x_B, \quad V = \bar{v}/v_0, \quad P = \frac{\pi}{2}\frac{Ns}{\alpha C}, \tag{6.10}$$

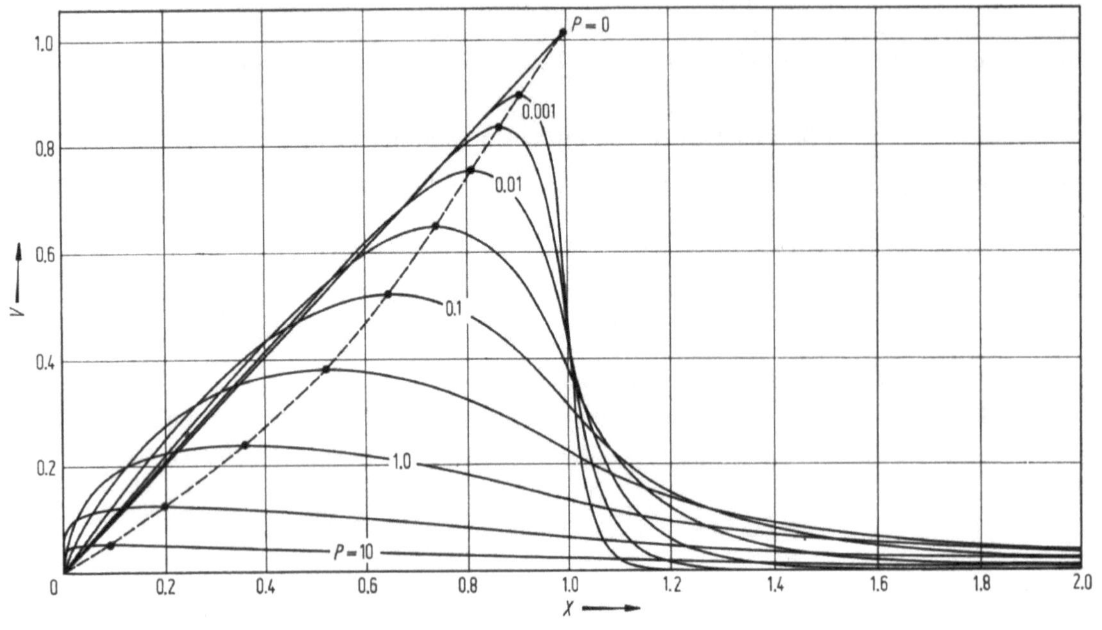

Fig. 4. The solutions to Eq. (6.11), giving the velocity profile of the longshore current, for different values of the parameter P (from *Longuet-Higgins* 1970 b).

we obtain the differential equation

$$P\frac{d}{dx}\left(X^{5/2}\frac{dV}{dX}\right) - X^{1/2}V = \begin{cases} -X^{3/2}, & 0 < X < 1, \\ 0, & 1 < X < \infty \end{cases} \tag{6.11}$$

with the boundary conditions that V shall be finite as $X \to 0$ or ∞ and that V and dV/dX shall be continuous at $X = 1$ (corresponding to the breaker line). The solution to (6.11) is expressible in terms of elementary functions (see *Longuet-Higgins* 1970b) and is shown in Fig. 4 for different values of P. Zero lateral friction ($P = 0$) gives the triangular distribution, described earlier. Generally, it is found that the distributions are not critically dependent on the value of P, but are much more dependent on the coefficient of bottom friction C, which determines the scale velocity v_0.

The case $P = 2/5$ is of some interest, as being the limiting value, below which the velocity gradient dV/dX remains finite at the shoreline. In this case we have simply

$$V = \begin{cases} \frac{10}{49}X - \frac{5}{7}X\ln X, & 0 < X < 1, \\ \frac{10}{49}X^{-5/2}, & 1 < X < \infty. \end{cases} \tag{6.12}$$

In Fig. 5 is seen a comparison between this profile (crosses) and the idealised velocity profiles observed by *Galvin* and *Eagleson* (1965, Fig. 37) for unaccelerated flow (Fig. 5a) and accelerated flow (Fig. 5b). The agreement in Fig. 5b is remarkable.

The theoretical curves of Fig. 4 have also been compared with both field observations and with the laboratory measurements of *Galvin* and *Eagleson* (1965) with very reasonable agreement, on the basis of a friction coefficient C of order 0.01 (see *Longuet-Higgins*, 1970a, b). Assuming a value of P equal to 0.4, and with a beach slope $s = 0.11$, this corresponds to a value of N also of order 0.01. Values of N of this order have also been found in field observations by *Inman, Tait* and *Nordstrom* (1971).

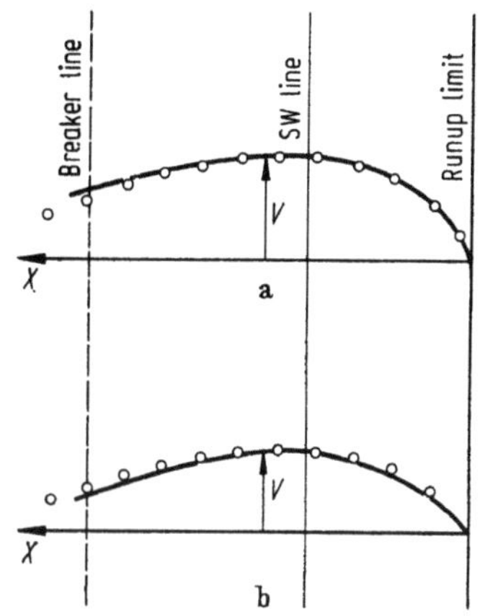

Fig. 5. A comparison of the "idealised" longshore velocity profile as observed by *Galvin* and *Eagleson* (1965) (full curve) with the profile given by Eq. (6.12) (dots). (a) unaccelerated flow; (b) accelerated flow.

7. Rip currents

The two-dimensional pattern of longshore currents is often broken up by a sequence of narrow currents flowing at right-angles to the shoreline. These are known as "rip currents". *Bowen* and *Inman* (1969) pointed out that these could be associated with a periodic longshore variation in the wave amplitude, higher waves producing a larger set-up, and hence a longshore pressure gradient. Based on this idea, *Bowen* (1969b) evaluated the driving forces due to the radiation stress gradients, and balanced these against either (a) bottom friction or (b) horizontal stresses due to lateral mixing. In either case he demonstrated that a pattern of circulating cells would tend to occur. In case (b), he showed that the higher the Reynolds number the more unsymmetrical the flow, with stronger and narrower currents flowing seawards (Fig. 6).

What causes the variations in amplitude? *Bowen* (1969a, b) presents evidence to show that rip currents are often associated with standing "edge-waves" of the form,

$$\zeta = b\Phi_n(x) \cos my \cos \sigma t. \tag{7.1}$$

In the simplest form of these ($n = 0$), discovered by *Stokes* (1847), the surface elevation decreases exponentially with distance $|x|$ off shore. In higher modes (*Ursell* 1952) the elevation has one or more modal lines. There is a simple relation between the frequency σ, the beach angle ψ, the longshore wavenumber m and the modal number n, namely

$$\sigma^2 = gm \sin (2n + 1) \psi. \tag{7.2}$$

By model experiments *Bowen* and *Inman* (1969) showed that when long-crested waves of frequency σ approach a beach normally, then edge-waves of the same frequency may occur simultaneously. Where the edge-wave is in-phase with the incident wave the effective wave amplitude is locally enhanced, and where the waves are out-of-phase the effective amplitude is reduced. So one obtains a pattern of rip-currents with the same longshore spacing as the longshore wavelengths of the edge wave.

Field observations of rip currents on the coast of California (*Bowen* 1969b) tend to confirm these ideas quantitatively. It appears, however, that the most favoured edge-wave mode is not necessarily the lowest, but more usually corresponds to the mode with $n = 1$ or 2. Further field investigations have been conducted by *Tait* (1971).

For the association of rip currents with "beach-cusps" see Section 10.

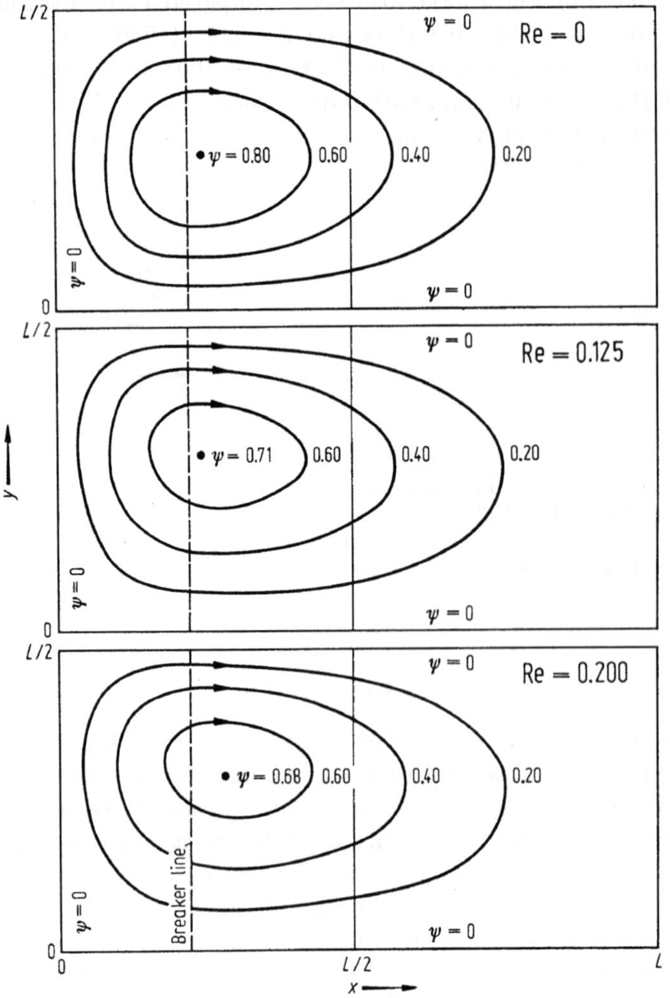

Fig. 6. The theoretical circulation pattern due to waves of nonuniform amplitude striking a straight coastline, when the driving forces are balanced by lateral friction: Re = 0.0, 0.125 and 0.2 (from *Bowen* and *Imnan* 1969b).

8. General theory

We shall now give a general theory for the calculation of wave set-up and mean currents generated by waves in water of non-uniform depth, it being assumed that the variation in the depth is sufficiently gradual that the wave field may be sufficiently well evaluated by the elementary laws of refraction. In this way we shall obtain some simple general results which will include as special cases the basic theory for a straight coastline given in earlier sections.

The proofs will be given in the context of the linear theory of wave motion already found so successful in Sections 2—6.

The general law of refraction. Let us denote by Θ_1 and Θ_2 the angles (measured in the anti-clockwise sense) which the direction of wave propagation makes with the axes of x_1 and x_2, so that $\Theta_2 = \Theta_1 - \pi/2$. We note that $\cos \Theta_i$ and $\sin \Theta_i$ are vectors, and that $\cos \Theta_i \cos \Theta_j$ and $\sin \Theta_i \sin \Theta_j$ are tensors such that

$$\cos \Theta_i \cos \Theta_j + \sin \Theta_i \sin \Theta_j = \delta_{ij}. \tag{8.1}$$

Now the rate of change of Θ_i in a direction *normal* to the wave crests is the curvature of an orthogonal trajectory drawn in that direction. By Huyghen's principle this is equal to the proportional rate of change of the phase velocity in the direction at right angles. Thus

$$\cos \Theta_j \frac{\partial \Theta_i}{\partial x_j} = \sin \Theta_j \left(\frac{1}{c} \frac{\partial c}{\partial x_j} \right). \tag{8.2}$$

Now

$$\frac{1}{c} \frac{\partial c}{\partial x_j} = \frac{\partial}{\partial x_j} \log \left(\frac{c}{c_\infty} \right) = \frac{\partial}{\partial x_j} \log \left(\frac{\eta}{\xi} \right) \tag{8.3}$$

by Eq. (3.3), So (8.2) may be written

$$\cos \Theta_j \frac{\partial \Theta_i}{\partial x_j} = -\sin \Theta_j \frac{\eta}{\xi} \frac{\partial}{\partial x_j} \left(\frac{\xi}{\eta} \right). \tag{8.4}$$

Let

$$k_i = k \cos \Theta_i \tag{8.5}$$

denote the vector wavenumber. Then, since $k/k_\infty = \xi/\eta$ we have

$$\cos \Theta_j \frac{\partial}{\partial x_j} \left(\frac{k_i}{k_\infty} \right) = \cos \Theta_j \frac{\partial}{\partial x_j} \left(\frac{\xi}{\eta} \cos \Theta_j \right)$$

$$= \cos \Theta_i \cos \Theta_j \frac{\partial}{\partial x_j} \left(\frac{\xi}{\eta} \right) - \sin \Theta_i \cos \Theta_j \frac{\xi}{\eta} \frac{\partial \Theta_i}{\partial x_j}. \tag{8.6}$$

Using (8.4) we now have

$$\begin{aligned} \cos \Theta_j \frac{\partial}{\partial x_j} \left(\frac{k_i}{k_\infty} \right) &= (\cos \Theta_i \cos \Theta_j + \sin \Theta_i \sin \Theta_j) \frac{\partial}{\partial x_j} \left(\frac{\xi}{\eta} \right) \\ &= \delta_{ij} \frac{\partial}{\partial x_j} \left(\frac{\xi}{\eta} \right) \\ &= \frac{\partial}{\partial x_i} \left(\frac{\xi}{\eta} \right) \\ &= \frac{\partial \eta}{\partial x_i} \frac{d}{d\eta} \left(\frac{\xi}{\eta} \right). \end{aligned} \tag{8.7}$$

Hence, the normal derivative of the wavenumber k_i is in the direction of the bottom gradient $\partial h/\partial x_i$. If we consider the component of the wavenumber k_i parallel to the bottom contours, we see that its derivative in the direction of wave propagation vanishes. Thus (8.7) is the appropriate generalisation of the law of refraction (Snell's law) for a straight coastline.

Equation (8.7) can also be derived from the Hamiltonian equations

$$\dot{x}_i = -\partial \mathscr{H}/\partial p_i, \quad p_i = \partial \mathscr{H}/\partial x_i, \tag{8.8}$$

where x_i and p_i denote the co-ordinates and the momenta of a wave packet and $\mathscr{H}$ is the Hamiltonian:

$$\mathscr{H} = -i\sigma = -i(gk \tanh kh)^{1/2}. \tag{8.9}$$

For, identifying $\dot{x}_i$ with the group velocity c_{gi} and p_i with ik_i, Eqs. (8.8) become

$$c_{gi} = \partial \sigma/\partial k_i, \quad c_{gi}\, \partial k_j/\partial x_i = -\partial \sigma/\partial x_i. \tag{8.10}$$

In gravity waves, the group velocity is normal to the wave crests, so that $c_{gi} = c_g \cos \Theta_i$ and the second equation becomes

$$c_g \cos \Theta_i \frac{\partial k_i}{\partial x_j} = -\frac{\partial \sigma}{\partial x_i} \tag{8.11}$$

and hence

$$\cos \Theta_j \frac{\partial k_i}{\partial x_j} = -\frac{1}{c_g} \frac{\partial \sigma}{\partial x_i} \tag{8.12}$$

a result equivalent to (8.7).

The wave amplitude. The energy flux vector being given by

$$F_i = F \cos \Theta_i, \tag{8.13}$$

(where $F = Ec_g$), by the conservation of wave energy we have (in the absence of large currents)

$$\partial F_i/\partial x_i + D = 0, \tag{8.14}$$

where D denotes the rate of energy dissipation. When D vanishes then

$$\partial F_i/\partial x_i = 0\,. \tag{8.15}$$

Once the ray trajectories have been determined then this equation, together with the value of F_i in deep water, is sufficient to determine F everywhere, and hence E and a.

The driving force. The radiation stresses are equivalent to a body force P_i per unit volume, where

$$P_i = -\frac{1}{h}\frac{\partial S_{ij}}{\partial x_j} \tag{8.16}$$

and S_{ij} denotes the radiation stress tensor. Now S_{ij}, when referred to principle axes (x, y), is given by Eq. (2.9). By Eqs. (3.3) and (3.4) this can also be written

$$\begin{bmatrix} S_{xx} & S_{xy} \\ S_{yx} & S_{yy} \end{bmatrix} = \frac{\sigma F}{g}\begin{bmatrix} \left(\frac{2\xi}{\eta} - \frac{d\xi}{d\eta}\right) & 0 \\ 0 & \left(\frac{\xi}{\eta} - \frac{d\xi}{d\eta}\right) \end{bmatrix}. \tag{8.17}$$

So in general, since Θ_i is the angle made by the principal axis (i.e., the direction of wave propagation) with the axis of x_i we have

$$\begin{aligned} S_{ij} &= \frac{\sigma F}{g}\left[\left(\frac{2\xi}{\eta} - \frac{d\xi}{d\eta}\right)\cos\Theta_i\cos\Theta_j + \left(\frac{\xi}{\eta} - \frac{d\xi}{d\eta}\right)\sin\Theta_i\sin\Theta_j\right] \\ &= \frac{\sigma F}{g}\left[\frac{\xi}{\eta}\cos\Theta_i\cos\Theta_j - \eta\frac{d}{d\eta}\left(\frac{\xi}{\eta}\right)\delta_{ij}\right], \end{aligned} \tag{8.18}$$

using (8.1). Substitution in (8.16) then gives

$$P_i = \frac{G}{\eta}\left[\frac{\partial}{\partial x_i}\left\{F\frac{d}{d\eta}\left(\frac{\xi}{\eta}\right)\right\} - \frac{\partial}{\partial x_j}\left\{\frac{F\xi}{\eta}\cos\Theta_i\cos\Theta_j\right\}\right], \tag{8.19}$$

where

$$G = \sigma^3/g^2\,, \tag{8.20}$$

a constant. On performing the differentiation and using the law of refraction (8.7) we obtain

$$P_i = G\left[\frac{\partial}{\partial x_i}\left\{F\frac{d}{d\eta}\left(\frac{\xi}{\eta}\right)\right\} - \frac{\xi}{\eta^2}\cos\Theta_i\frac{\partial}{\partial x_j}(F_j\cos\Theta_j)\right]. \tag{8.21}$$

Now using Eqs. (8.13) and (8.14) we find simply

$$P_i = \frac{\partial}{\partial x_i}\left\{GF\frac{d}{d\eta}\left(\frac{\xi}{\eta}\right)\right\} + \frac{GD\xi}{\eta^2}\cos\Theta_i\,. \tag{8.22}$$

First, suppose there is no loss of energy, so that $D = 0$. Then from (8.22) we see that P_i is the gradient of a scalar function and can be balanced precisely by a rise in the mean level ("set-up"), given by

$$\bar{\zeta} = \frac{GF}{\varrho g}\frac{d}{d\eta}\left(\frac{\xi}{\eta}\right). \tag{8.23}$$

There is then no current.

Equation (8.23) is clearly the generalisation of equation (4.3).

Where on the other hand the dissipation of energy is not negligible we obtain an additional stress per unit area given by h times the second term on the right of (8.22), that is to say

$$\tau_i = (hGD\xi/\eta^2)\cos\Theta_i \tag{8.24}$$

or from Eqs. (3.1) to (3.3)

$$\tau_i = (D/c)\cos\Theta_i\,. \tag{8.25}$$

Hence, in addition to set-up given by (8.23), the *dissipation* of wave energy produces a stress τ_i equal in magnitude to D/c and in the same direction as the phase velocity. This produces currents, in general, which must be balanced by frictional forces, either bottom friction or lateral mixing.

It will be seen that Eq. (8.25) is the generalisation of (5.14).

Equation (8.23) may be obtained by a more direct argument as follows (cf. *Longuet-Higgins* 1965).

If the motion is irrotational, then we may apply Bernoulli's equation throughout the fluid. After averaging with respect to the time t this gives

$$\bar{p} + \frac{1}{2}\,\overline{\varrho(u^2 + v^2 + w^2)} = \text{const}. \tag{8.26}$$

On the other hand, from the vertical flux of momentum applied to a column of water bounded by $z = 0$ and $z = \zeta$ we have

$$(\bar{p} + \varrho\overline{w^2})_{z=0} = \varrho g\bar{\zeta}, \tag{8.27}$$

terms of order $(ak)^3$ being neglected. From these two equations it follows that

$$g\bar{\zeta} = \frac{1}{2}\,\overline{\varrho(w^2 - u^2 - v^2)}, \tag{8.28}$$

the constant vanishing since $\bar{\zeta}$ vanishes in deep water. Substitution from the expressions (2.3) for the orbital motion now gives

$$g\bar{\zeta} = -a^2\sigma^2/(4\sinh^2 kh) \tag{8.29}$$

which is equivalent to (8.23).

The wave set-up (or wave set-down) given by Eqs. (8.23) or (8.29) is usually quite small. In most applications the important term is (8.25), representing the effect of breaking or bottom friction.

9. Applications

Equations (8.23) and (8.25), together with suitable assumptions as to bottom friction and lateral mixing, provide us with a complete framework for determining the wave set-up and the mean currents generated by any given wave field in water where the mean depth is also given as a function of the horizontal co-ordinates.

One example of the application of such a theory (though in less simple form) is the calculation by *O'Rourke* and *LeBlonde* (1972) of the currents generated by waves entering a semicircular bay. These authors adopt a form of boundary-layer theory in which the surf zone is assumed narrow compared to the radius of the bay. The angle of incidence at the breaker line is assumed to be given by $\Theta_B = \Theta_0 \sin\phi$ where ϕ is the polar co-ordinate for the bay ($\phi = 0$ corresponding to the line of symmetry); and the depth is assumed to be a linear function of the radical co-ordinate, i.e., $h = h_0(r_0 - r)\,(1 - m\cos 2\phi)$, where h_0 and m are constants and r_0 is the radius of the bay. When $m \ll 1$ the wave amplitude is then nearly uniform. The driving forces are balanced by bottom friction in the surf zone only, and lateral mixing is altogether neglected.

With these assumptions, *O'Rourke* and *LeBlonde* show that the effects of the oblique angle of incidence may be separated from the effects of a non-uniform wave amplitude. The two tend to produce circulations in roughly opposite senses.

A second example, related to the first, is the calculation of the currents associated with "beach-cusps". These are the regularly spaced projections of the coastline often seen on beaches exposed to incoming waves. It is found that such cusps are frequently associated with rip currents. However, sometimes the rip currents make their appearance opposite the projecting cusps and sometimes opposite the bays. Model experiments (*Komar* 1972) have confirmed this observation. It appears that rip currents may first develop, which transport material in the lee of the rip, so building up the cusps. But at a certain stage of development the currents suddenly cease or are even reversed.

This phenomenon may well be due to the refraction of the incoming waves by the projecting cusps, which by convergence of the ray orthogonals increases the wave amplitude on the cusp. In addition, the angle of incidence on each side of the cusp is such as to tend to drive the current in the opposite direction.

Komar (1972) has suggested that there may be an equilibrium form of the cusp in which the driving forces are exactly balanced. It is perhaps worthwhile to consider under what special conditions this would occur. From Eq. (8.25), the corresponding part of the surface elevation would be given by

$$\begin{aligned} \varrho g(h + \bar{\zeta})\, \partial\bar{\zeta}/\partial x &= (D/c) \cos \Theta\,, \\ \varrho g(h + \bar{\zeta})\, \partial\bar{\zeta}/\partial y &= (D/c) \sin \Theta\,, \end{aligned} \tag{9.1}$$

where x and y are now curvilinear coordinates normal and parallel to the contours of $(h + \bar{\zeta})$, and Θ is the local angle of incidence. We neglect terms arising from the curvature of the coordinates, which is assumed sufficiently small. Equations (9.1) imply that

$$\frac{\partial}{\partial y}\left(\frac{D}{h + \bar{\zeta}} \cdot \frac{\cos \Theta}{c}\right) = \frac{\partial}{\partial x}\left(\frac{D}{h + \bar{\zeta}} \cdot \frac{\sin \Theta}{c}\right). \tag{9.2}$$

Now from Section 6 we see that for small angles Θ the dissipation D is proportional to $(h + \bar{\zeta})^{3/2}\, \partial(h + \bar{\zeta})/\partial x$, while c is proportional to $(h + \bar{\zeta})^{1/2}$. So writing

$$\frac{\sin \Theta}{c} = Q(y) \tag{9.3}$$

and assuming that

$$c = \sqrt{g(h + \bar{\zeta})} = c(x) \tag{9.4}$$

we have

$$-\frac{\partial(h + \bar{\zeta})}{\partial x} \sin \Theta \frac{\partial \Theta}{\partial y} = Q \frac{\partial}{\partial x}\left[c \frac{\partial(h + \bar{\zeta})}{\partial x}\right]. \tag{9.5}$$

Since from (9.3)

$$\sin \Theta = Qc, \qquad \partial\Theta/\partial y = c\, dQ/dy \tag{9.6}$$

it follows that if $Q \neq 0$,

$$\frac{dQ}{dy} = -\frac{(c\, \partial/\partial x)^2\, (h + \bar{\zeta})}{c^3\, \partial(h + \bar{\zeta})/\partial x}. \tag{9.7}$$

In this separated equation each side must be a constant. This implies that there is no solution in which Q is an oscillatory function of y.

How then could an equilibrium be maintained? We suggest that the explanation lies in the existence of an edge-wave occurring simultaneously with the incoming wave. For edge-waves are *trapped* motions, their energy being propagated along rays which strike the shoreline at angles of incidence *greater* than the critical angle (see for example *Shen*, *Meyer* and *Keller* 1968). Such a wave will tend to drive the currents in a direction opposite to the incoming wave and so it may then be possible to establish an equilibrium. This, however, must depend critically on both the amplitude and the relative phase of the edge-wave. A precise solution remains to be found.

10. Sand movement

So far we have dealt exclusively with the motion of the water. Of greater practical interest is the sand movement caused by the waves.

It is found experimentally (see *Komar* and *Inman* 1970) that on a straight coastline there is a linear relationship between the total rate of transport of the sand past a given point, and a quantity related to the wave energy. This relationship can be expressed as

$$I = KHc_B \tag{10.1}$$

(see *Longuet-Higgins* 1972), where I is the "immersed weight transport rate"[1], H the longshore wave thrust (Eq. (5.11)), c_B the phase velocity at the breaker line and K a numerical constant, equal to 0.77.

---

[1] I is defined as $\alpha(\varrho' - \varrho)\, gS$ where ϱ' and ϱ are the densities of stand and water, S is the sand volume transport rate and α the volume concentration, $\alpha \doteq 0.6$ for closely packed sand.

Equation (10.1) can be very simply explained on the hypothesis that there is a constant coefficient of friction γ between the mobile layer of sand and the relatively immobile beach beneath.

For, the sand moves up and down the beach in a zig-zag fashion, the direction of motion making on the average a small angle δ with the mean orbital motion. The orbital velocity in the surf zone being of order αc we may expect the mobile sand velocity to be of this order. Let it be βc where β is about 0.4. Then the mean longshore sand velocity is $\delta\beta c$ and the local immersed weight transport per unit distance offshore is

$$i = (\delta\beta c)\,W, \tag{10.2}$$

where W is the local immersed weight per unit area.

But the horizontal stress exerted by the moving sand on the layer below is γW, and the mean longshore component is therefore $\delta\gamma W$. Neglecting lateral mixing, this force must be balanced by the longshore stress τ_y. Hence

$$\tau_y = \delta\gamma W \tag{10.3}$$

and from (10.2) and (10.3) it follows that

$$i = K'\tau_y c, \qquad K' = \beta/\gamma. \tag{10.4}$$

Now on a beach of constant slope, c varies as $x^{1/2}$ in the surf zone, while τ_y, as we have seen varies as $x^{3/2}$. So

$$\left.\begin{aligned} H &= \int\limits_{-x_B}^{0} \tau_y\,dx = \frac{2}{5}(\tau_y)_B\,|x_B|\,, \\ I &= \int\limits_{-x_B}^{0} i\,dx = \frac{K'}{3}(\tau_y c)_B\,|x_B| \end{aligned}\right\} \tag{10.5}$$

from which it follows that

$$I = \frac{5K'}{6}\,H c_B. \tag{10.6}$$

This is equivalent to (10.1) provided $K' = 1.2K = 0.92$. Then taking $\beta = 0.4$ we have

$$\gamma = \beta/K' \doteqdot 0.43. \tag{10.7}$$

The above value of γ is not inconsistent with the independent measurements by *Bagnold* (1966) who found that for dry sand $0.37 \leq \gamma \leq 0.75$.

The law of longshore transport (10.4) can, by (8.25), also be expressed in the form

$$i = K'D\sin\Theta, \qquad K' = 0.92, \tag{10.8}$$

where D is the local rate of wave energy dissipation and Θ is the local angle of incidence.

We have taken into account only the "bed-load" transport of sand, which usually is dominant within the surf zone. A certain proportion of the transported sand may also be carried in suspension (see *Bagnold*, 1966). *Thornton* (1972) has given experimental evidence in support of the relation

$$i = C_1 D\sin\Theta + C_2 D\bar{v}/w_s, \tag{10.9}$$

where w_s denotes the settling velocity of the sand (which depends on particle size) and where C_1 and C_2 are constants. *Thornton* finds that in the surf zone $C_1 = 1.0$, in good agreement with (10.8).

The transport of sand across the contours is less simple to predict. It might be assumed that, when the sand is in motion, a certain fraction, say ε, of the radiation stress gradient goes to maintain the wave set-up and the remainder to maintaining the slope of the beach against the down-slope component of gravity. The experiments of *Bowen*, *Inman* and *Simmons* (1968) mentioned in Section 4, which were performed with a solid beach, would then correspond to $\varepsilon = 1$. In general the appropriate value of ε must depend on a number of circumstances, including the permeability of the beach material, and the effective depth of the mobile layer of sand. This is still a matter for experimental investigation.

11. Discussion and conclusions

Already the consideration of momentum flux, together with certain assumptions regarding the frictional forces, has gained us a new understanding of the factors which determine wave set-up and longshore currents. The linearised theory described above gives us surprisingly good agreement with observation. Indeed, for a qualitative understanding of situations where the bottom topography is complicated, the linear theory may serve us very well.

Further, the known laws of longshore sand transport now make it possible to study rationally certain aspects of the *interaction* between waves and beaches. With further knowledge of the mean onshore movement of sand it should eventually be possible to construct a complete theory.

The surprising success of the linear theory can to some extent be accounted for by considering non-linear theories which take full account of the asymmetry of the orbital motion in waves of finite amplitude and the interaction of the waves with the induced mean current. Such studies by *James* (1972), which make use of *cnoidal* wave theory in place of Stoke's small-amplitude theory, have suggested that these factors make little quantitative difference to the predicted currents and set-up, the differences involved being much less than an order of magnitude.

The detailed consideration of these effects does, however, bring to light a generally unsuspected gap in our knowledge, namely the dynamical description of breaking waves and surf. We know surprisingly little about the pattern of flow in a breaking wave either immediately before or in the mixed turbulent and laminar regime just following the instant of breaking. There is not space here to say more than that this is likely to be a fruitful field for mathematicians, guided by experiment.

References

Bagnold, R. A.: An approach to the sediment transport problem from general physics. Geological Survey Professional Pap. 422-1, U.S. Govt. Printing Office, Washington, D.C. (1966).

Bowen, A. J.: J. Mar. Res. 27 (1969a) 206—215.

Bowen, A. J.: J. Geophys. Res. 74 (1969b) 5467—5478.

Bowen, A. J., Inman, D. L.: J. Geophys. Res. 74 (1969b) 5479—5490.

Bowen, A. J., Inman, D. L., Simmons, V. P.: J. Geophys. Res. 73 (1968) 2569—2577.

Eckart, C. E.: Phys. Rev. 73 (1948) 68—76.

Galvin, C. J., Eagleson, P. S.: Experimental study of longshore currents on a plane beach. U.S. Army Coastal Eng. Res. Center, Tech. Mem. 10 (1965).

Harris, D. L.: Characteristics of the hurricane storm surge. U.S. Weather Bureau, Tech. Pap. No. 48, Govt. Printing Off., Washington, D.C. (1963).

Inman, D. L., Tait, R. J., Nordstrom, C. E.: J. Geophys. Res. 76 (1971) 3493—3514.

James, I. D.: Some nearshore effects of ocean waves. Ph. D. Thesis, University of Cambridge, England (1972).

Komar, P. D.: Bull. Geolog. Soc. Amer. 82 (1971b) 2643—2650.

Komar, P. D., Inman, D. L.: J. Geophys. Res. 75 (1970) 5914—5927.

Longuet-Higgins, M. S.: J. Geophys. Res. 75 (1970a) 6778—6789.

Longuet-Higgins, M. S.: J. Geophys. Res. 75 (1970b) 6790—6801.

Longuet-Higgins, M. S., Stewart, R. W.: J. Fluid Mech. 8 (1960) 565—583.

Longuet-Higgins, M. S., Stewart, R. W.: J. Fluid Mech. 13 (1962) 481—504.

Longuet-Higgins, M. S., Stewart, R. W.: J. Mar. Res. 21 (1963) 4—10.

Longuet-Higgins, M. S., Stewart, R. W.: Deep-Sea Res. 11 (1964) 529—562.

Longuet-Higgins, M. S.: Recent progress in the study of longshore currents. Proc. Advanced Seminar on Waves and Beaches, Univ. of Winconsin, Madison, Oct. 11—14, 1971 (1972).

Medwin, H., Rudnick, I.: J. Acoust. Soc. Amer. 25 (1953) 538—540.

Munk, W. H.: Ann. N.Y. Acad. Sci. 51 (1949) 376—424.

Rayleigh, Lord: Phil. Trans. Roy. Soc. A 175 (1884) 1—21.

O'Rourke, J. C., LeBlond, P. H.: J. Geophys. Res. 77 (1972) 444—452.

Saville, T.: Proc. 2nd Conf. on Hurricances, U.S. Dept. of Commerce, Nat. Hurricance Res. Proj., Rep. 50 (1961) 242—252

Shen, M C., Meyer, R. E., Keller, J. B.: Phys. Fluids 11 (1968) 2289—2304.

Shepard, F. P., Inman, D. L.: Proc. 1st Conf. Coastal Eng. 50—59. Council on Wave Res., Univ. of California, Berkeley (1951).

Tait, R. J.: Rip currents. Ph. D. Thesis, Univ. of California. San Diego (1971).

Thornton, E. B.: Proc. 13th Int. Conf. on Coastal Engineering, Vancouver, B.C. July 10—14, 1972.

Ursell, F.: Proc. Roy. Soc. Lond. A 214 (1952) 79—97.

Inverse methods in two-dimensional elasticity

E. H. Mansfield, F. R. S.

Farnborough, U.K.

1. Introduction

In this talk I propose to discuss briefly some analytical techniques which stem from the idea of designing certain plane stress structures (flat sheets with compact line reinforcement) to exhibit various prescribed stress features. I will first discuss some simple load diffusion problems in which the varying taper of the reinforcement is the design variable and then go on to consider the neutral hole, in which the shape of the hole as well as the stiffness of the reinforcement is determined by the prescribed stress field. Next, consideration is given to the optimum variation of the reinforcement around a circular hole in a sheet subjected to uniaxial tension. Finally some examples are given of sheets with optimum thickness variation.

2. Taper required to give constant stress edge members to long rectangular panel

This particularly simple example is shown in Fig. 1. The transverse edge is unloaded while the conditions along the longitudinal edges are that

$$\begin{aligned} \sigma_x &= \sigma_e = \text{const}, \\ \sigma_y &= 0, \end{aligned} \tag{1}$$

the latter condition implying that the edge members are assumed to be flexible.

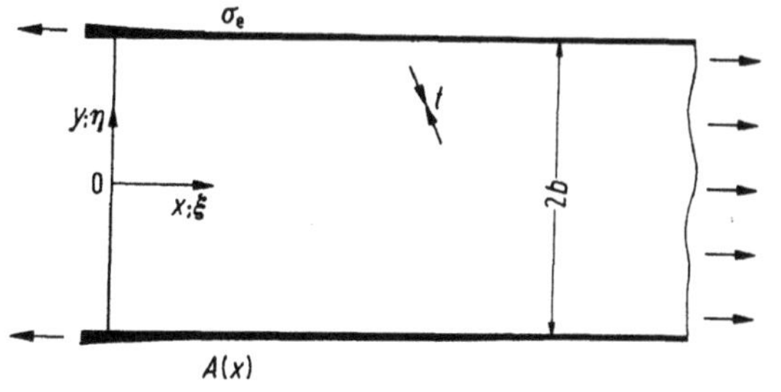

Fig. 1. Long rectangular panel with constant stress edge members.

These stress boundary conditions determine the stress field throughout the sheet and it may be shown that

$$\begin{aligned} \frac{\sigma_x}{\sigma_e} &= 1 - \frac{2}{\pi}\left\{\frac{\xi \cosh \xi \cos \eta}{\cosh^2 \xi - \sin^2 \eta} + \tan^{-1}\left(\frac{\cos \eta}{\sinh \xi}\right)\right\}, \\ \frac{\sigma_y}{\sigma_e} &= \frac{2}{\pi}\left\{\frac{\xi \cosh \xi \cos \eta}{\cosh^2 \xi - \sin^2 \eta} - \tan^{-1}\left(\frac{\cos \eta}{\sinh \xi}\right)\right\}, \\ \frac{\tau_{xy}}{\sigma_e} &= -\frac{2}{\pi}\left\{\frac{\xi \sinh \xi \sin \eta}{\cosh^2 \xi - \sin^2 \eta}\right\}, \end{aligned} \tag{2}$$

where

$$\xi = \pi x/2b, \qquad \eta = \pi y/2b.$$

Of particular interest is the variation of τ_{xy} along the longitudinal edges:

$$\frac{(\tau_{xy})_e}{\sigma_e} = \frac{2\xi}{\pi \sinh \xi} \tag{3}$$

and this may be integrated to give the required variation of the section area A of the constant stress edge members:

$$A = A_0 - \frac{4bt}{\pi^2} \int_0^{\xi} \frac{\xi \, d\xi}{\sinh \xi}. \tag{4}$$

3. Load diffusion into a semi-infinite sheet to give prescribed shear input

The problem considered is shown in Fig. 2 and we will determine a relationship between the prescribed shear input $q(x)$ and the section area of the stiffener $A(x)$. We first require to solve the subsidiary problem of a point load δP applied at a distance X from the free edge. This subsidiary problem is readily solved by combining the known solution of a point load applied in an infinite sheet with solutions for normal and shear stresses applied to the edge of a semi-infinite sheet. The direct stresses along the x axis are given by

$$\begin{aligned} \sigma_x &= -\frac{2\delta P x^2\{(2+\nu)X + x\}}{\pi t(X-x)(X+x)^3}, \\ \sigma_y &= \frac{2\delta P X^2(X - \nu x)}{\pi t(X-x)(X+x)^3} \end{aligned} \tag{5}$$

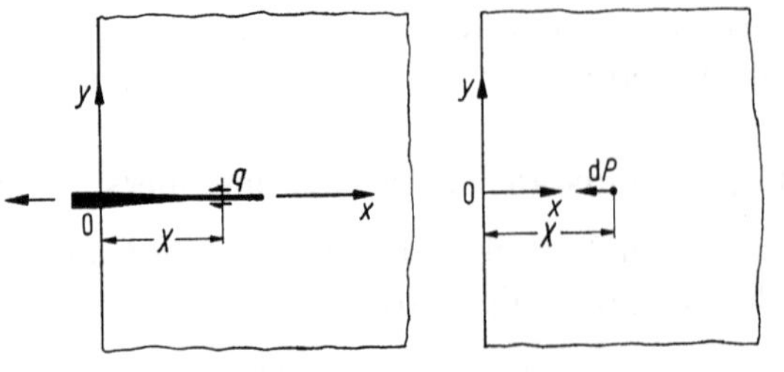

Fig. 2. Load diffusion into semi-infinite sheet.

and hence the direct strain along the x axis is given by

$$\varepsilon_x = (\sigma_x - \nu\sigma_y)/E. \tag{6}$$

Now the shear input applied by the stiffener to a length of sheet δX at a distance X from the free edge can be represented by a point load of

$$\delta P = 2q(X)\, t\, \delta X \tag{7}$$

which may be integrated to give the load in the stiffener at a section distance x from the root

$$P(x) = 2t \int_x^{\infty} q\, dX, \tag{8}$$

it being assumed that all the load is eventually absorbed by the sheet.

Now from Eqs. (5), (6) we know the strain in the sheet due to each elemental point load δP and hence, by integration, the strain due to a distributed point load system—and therefore, to the shear stress distribution, $q(x)$. If it is assumed that there is no relative slip between stiffener and sheet, this direct strain in the sheet along the line of action of the applied shear stresses is also the strain in the stiffener $P(x)/EA$, and this condition of equality of strain in sheet and stiffener leads to the following equation

$$\frac{t}{A} \int_x^{\infty} q\, dX + \frac{2}{\pi} \int_0^{\infty} \frac{\{\nu X^3 - \nu^2 X^2 x + (2+\nu) X x^2 + x^3\}\, q\, dX}{(X-x)(X+x)^3} = 0 \tag{9}$$

from which $A(x)$ is readily determined in terms of $q(x)$.

4. The neutral hole

In the previous examples we saw how it was possible to vary the stiffness of a line-stiffener to give a prescribed stress variation along the line of a stiffener. If we wish to prescribe a complete stress field we find that the boundary shape of the stiffeners as well as the variation of their stiffness is determined by the stress field. Such considerations lead us naturally to a study of the *neutral hole* defined as a reinforced hole in a flat sheet which does not affect the stress distribution by its presence, i.e., the stresses in a sheet containing a neutral hole are identical to the stresses in the uncut sheet.

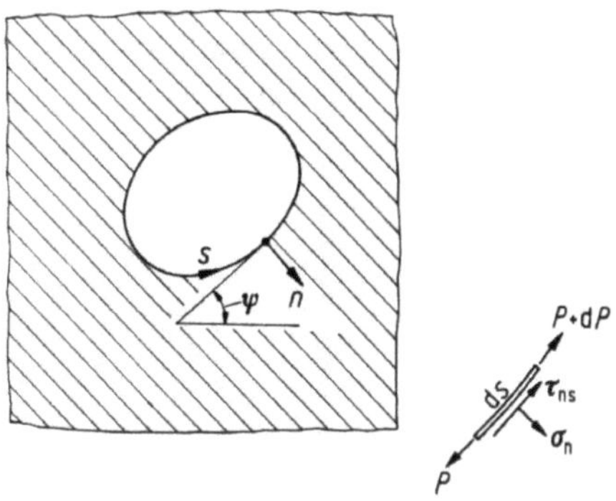

Fig. 3. Sheet with neutral hole.

Let the prescribed stress distribution be defined by the stress function Φ and let Fig. 3 refer to a sheet with a neutral hole and the forces acting on an element of the boundary reinforcement. Now it can be shown that in practice the bending stiffness of a compact reinforcing member is negligible in comparison with its tensile stiffness so that the reinforcing member has the properties of a chain in that the line of action of the load P in the member is directed along the length of the member. Thus for the element shown in Fig. 3 the conditions of equilibrium normal (n) and tangential (s) to the boundary are

$$\begin{aligned} P\frac{\partial\psi}{\partial s} - t\sigma_n &= 0\,, \\ \frac{\partial P}{\partial s} + t\tau_{ns} &= 0\,. \end{aligned} \tag{10}$$

When the stresses are expressed in terms of Φ Eq. (10) may be integrated to give the following equation for determining the shape of the neutral hole

$$\Phi = 0 \tag{11}$$

and we note that terms of the form $(ax + by + c)$ may be added to Φ as they do not affect the stresses. The load in the reinforcing member is given by

$$P = t\frac{\partial\Phi}{\partial n} \tag{12}$$

and the section area of the reinforcing member, obtained from a knowledge of P and the strain ε_s, is given by

$$A = \frac{t\dfrac{\partial\Phi}{\partial n}}{\dfrac{\partial^2\Phi}{\partial n^2} - \nu\dfrac{\partial\Phi}{\partial n}\dfrac{\partial\psi}{\partial s}}\,, \tag{13}$$

where $\partial\psi/\partial s$ is the local curvature. Finally we note that Eqs. (11), (12) are derived purely from considerations of statics and are independent of the elastic properties of the sheet.

4.1 Neutral hole in 2:1 stress field

A well-known example of a neutral hole is the $\sqrt{2}:1$ ellipse in a 2:1 stress field, such as occurs in a pressurised cylinder. The appropriate section area A is nearly constant around the ellipse but with a localised thickening towards the ends of the major axis. Figure 4 shows photoelastic fringe patterns for a reinforced circular hole and for a "nearly" neutral hole ("nearly" because, for ease of manufacture, the reinforcement was constant) in a 2:1 compressive stress field. The fringe patterns around the boundary stem from a multiple screw method of loading and, indeed, provide a check on the equality of the two stress fields. The superiority of the (nearly) neutral hole is evidenced by the relative absence of fringes around the hole.

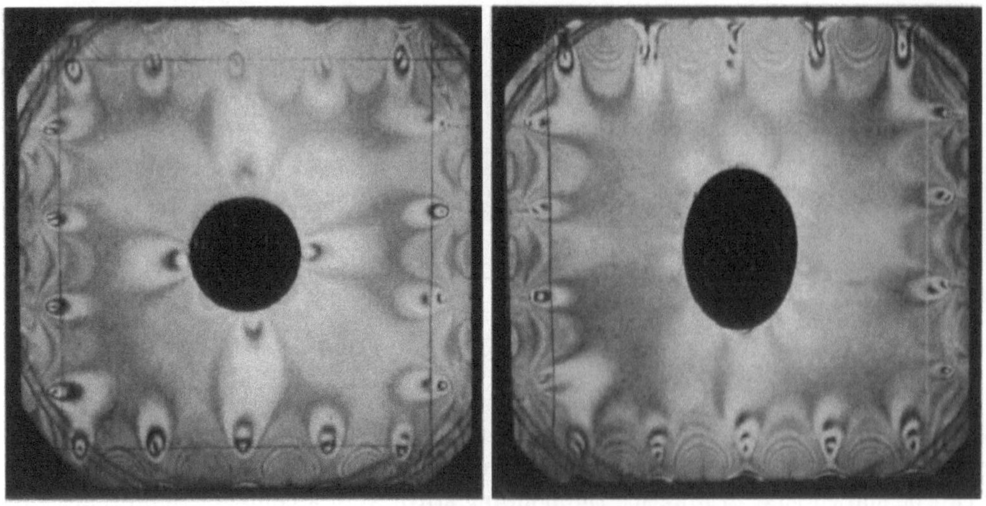

Fig 4. Reinforced holes in sheet with 2:1 stress field

4.2 Neutral hole with body forces

Equation (10) may also be used to determine the cross-section of a flexible tube subjected to internal hydrostatic gravity stresses—or a moment-less tube subjected to external gravity stresses. For this problem we find

$$P = \text{constant}$$

so that

$$\frac{\partial \psi}{\partial s} \sim y, \tag{14}$$

where y, measured vertically downwards, is proportional to the hydrostatic pressure. Note that Eq. (14) is the same as that governing the shape of an elastica.

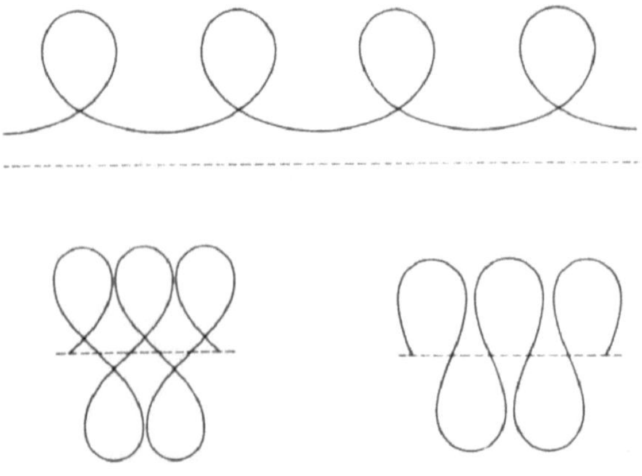

Fig 5. The elastica and "neutral holes" for hydrostatic gravity stress field.

Figure 5 shows typical curves which satisfy Eq. (14); the broken line is the line of zero hydrostatic stress which corresponds to the line of thrust (or tension) in the elastica. The reason why the curves are not closed stems from the fact that the gravity stress field is not self-equilibrating. The shaded area A in Fig. 6 is appropriate to a flexible (or moment-less) trough filled with fluid. The shaded area B is appropriate to a tube resting on a horizontal foundation. The shaded area C is appropriate to the tube supported by a vertical line load; the shape is that of a ladies handbag containing a typical homogeneous assortment of oddments!

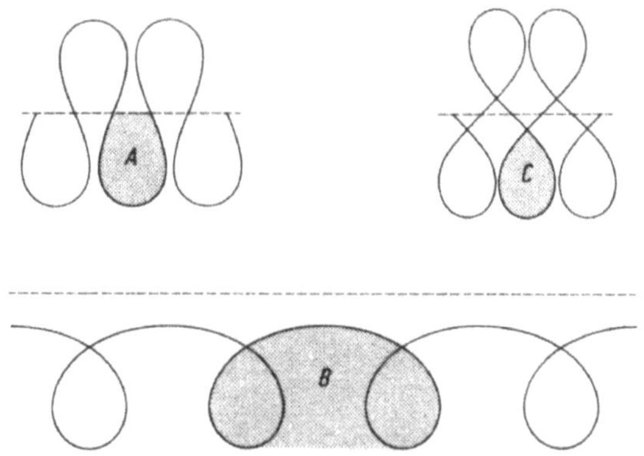

Fig. 6. Cross-sections of neutral or moment-less throughs and tubes.

5. Optimum variation of reinforcement around a circular hole in a sheet under uniaxial tension

There are many cases where for aesthetic or practical considerations a hole of circular shape is specified and the problem now is to determine what distribution of reinforcing material gives the least stress concentration, and if this optimum distribution is too heavy, what then are the optimum distributions for reinforcements of given smaller total weights. This problem may be treated by an inverse technique based on Eqs. (11) and (13) for the neutral hole, because these equations are necessarily valid whether the stress function is known *a priori* or not.

The most general form of the stress function Φ, which satisfies Eq. (11) and the requirements of symmetry together with the conditions of unit uniaxial stress in the direction $\theta = \pi/2$ remote from the hole, is given by

$$\Phi = \frac{1}{4}\left\{2\alpha R^2 \ln(r/R) + (r^2 - R^2)\left(1 + \cos 2\theta + \frac{\beta R^2}{r^2}\cos 2\theta + \frac{\gamma R^4}{r^4}\cos 4\theta + \frac{\delta R^6}{r^6}\cos 6\theta + \cdots\right)\right\}, \tag{15}$$

where R is the radius of the hole and the parameters $\alpha, \beta, \gamma, \delta, \ldots$ give rise to progressively more localised perturbations of the general stress field. In what follows, a truncated form of the stress function is chosen which contains only these specified terms.

Now in evaluating A/Rt it is convenient to express $1/R\ \partial\Phi/\partial n$ and $\partial^2\Phi/\partial n^2$ in powers of $\cos 2\theta\ (= C)$:

$$\begin{aligned}\frac{1}{R}\frac{\partial\Phi}{\partial n} &= [\sigma_r]_R = \frac{1}{2}(1 + \alpha - \gamma) + \frac{1}{2}(1 + \beta - 3\delta)\,C + \gamma C^2 + 2\delta C^3,\\ \frac{\partial^2\Phi}{\partial n^2} &= [\sigma_\theta]_R = \frac{1}{2}(1 - \alpha + 7\gamma) + \frac{1}{2}(1 - 3\beta + 33\delta)\,C - 7\gamma C^2 - 22\delta C^3.\end{aligned} \tag{16}$$

The reason for expressing Eq. (16) in this form stems from the fact that in any optimum variation of A/Rt the hoop stress in the reinforcement vanishes at certain points and hence, because reinforcements with infinite section area are not possible, the stresses $[\sigma_r]_R$ and

$[\sigma_\theta]_R$ vanish simultaneously at such points specified by $C = \phi_1$, say, and we can write

$$[\sigma_r]_R = \frac{1}{2}(\phi_1 - C)(\phi_4 + \phi_5 C - 4\delta C^2),$$
$$[\sigma_\theta]_R = \frac{1}{2}(\phi_1 - C)(\phi_2 + \phi_3 C + 44\delta C^2), \quad \text{say.} \tag{17}$$

By comparing coefficients of different powers of C in Eqs. (16), (17) it is possible to express $\phi_1 \cdots \phi_5$ and δ in terms of α, β, γ. Thus ϕ_1 is determined by the equation

$$4\gamma\phi_1^2 + (6 + 4\beta)\phi_1 + (6 + 5\alpha - 2\gamma) = 0,$$

and

$$\begin{aligned} \phi_2 &= (1 - \alpha + 7\gamma)/\phi_1, \\ \phi_3 &= 44\delta\phi_1 + 14\gamma, \\ \phi_4 &= (1 + \alpha - \gamma)/\phi_1, \\ \phi_5 &= -4\delta\phi_1 - 2\gamma, \end{aligned} \tag{18}$$

where

$$\delta = \frac{1 + \beta + 2\gamma\phi_1 + \phi_4}{3 - 4\phi_1^2}.$$

Equations (11), (17), (18) thus enable A/Rt to be expressed in terms of θ and the parameters α, β, γ:

$$\frac{A}{Rt} = \frac{\phi_4 + \phi_5 C - 4\delta C^2}{(\phi_2 - \nu\phi_4) + (\phi_3 - \nu\phi_5)C + (44 + 4\nu)\delta C^2}, \tag{19}$$

and this may be integrated to yield an expression for the total weight of reinforcement

$$W = \frac{1}{\pi R t}\int_0^{2\pi} A\, d\theta = W(\alpha, \beta, \gamma). \tag{20}$$

We now choose a given numerical value for W, represented by $\overline{W}$, and use the equation

$$W(\alpha, \beta, \gamma) = \overline{W}$$

to determine γ for given numerical values of α, β. Next we determine the stress concentration factor S and choose those values of α, β which minimise S. Figure 7 shows the optimum W, S relationship where S is based on the Mises-Hencky equivalent stress. Figure 8 shows some typical variations of A/Rt around a quadrant of the hole boundary.

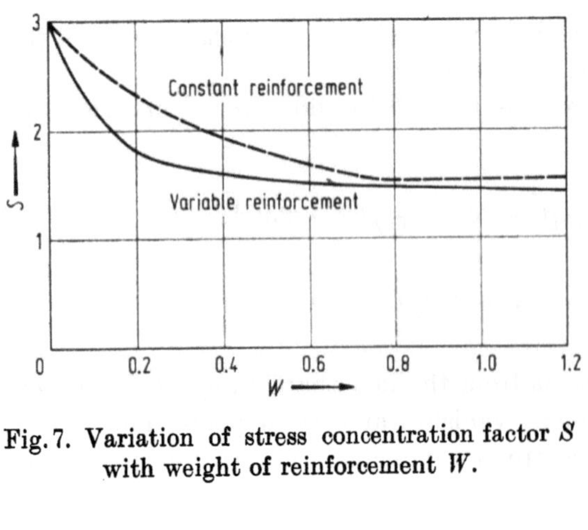

Fig. 7. Variation of stress concentration factor S with weight of reinforcement W.

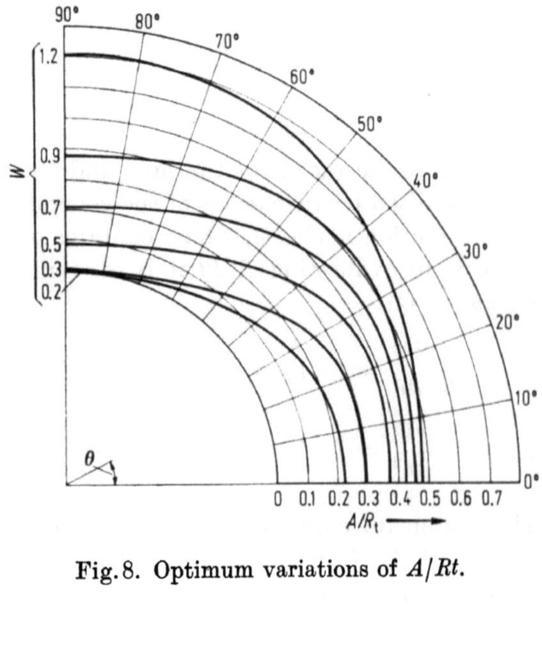

Fig. 8. Optimum variations of A/Rt.

The technique outlined in this section is, of course, applicable to any stress field; indeed, for those cases—such as the 2:1 stress field—where the hoop stress in the reinforcement does not change sign, the analysis is markedly simpler.

6. Plates with optimum variations of thickness

Here we consider the optimum variation of thickness t in the simplest of cases, namely near a circular hole of radius r_0 in a sheet subjected to uniform radial stress at infinity. We assume that the midplane of the sheet is flat and that conditions of plane stress exist. The case in which the sheet material satisfies the Tresca yield criterion is considered in Section 6.1, while the corresponding case of the Mises-Hencky yield criterion is outlined in Section 6.2. It is found that the stress distribution is determined by the yield criterion and the equation of compatibility, and the thickness variation is then determined by the equation of equilibrium.

6.1 Sheet material satisfying the Tresca yield criterion

In the problem considered the principal stresses are of the same sign so that the "equivalent" stress is the greater of the two principal stresses and is therefore equal to the hoop stress, so that

$$\sigma_\theta = \sigma, \tag{21}$$

where σ is the uniform radial stress applied remote from the hole.

Now because the sheet and the applied loading are axisymmetric the tangential displacement v and the shear stress $\tau_{r\theta}$ are zero. Accordingly the stress-strain relations are simply

$$\left.\begin{aligned} \varepsilon_r &= \frac{du}{dr} = \frac{1}{E}(\sigma_r - \nu\sigma_\theta), \\ \varepsilon_\theta &= \frac{u}{r} = \frac{1}{E}(\sigma_\theta - \nu\sigma_r), \end{aligned}\right\} \tag{22}$$

from which u may be eliminated to yield the equation of compatibility

$$\sigma_r - \nu\sigma_\theta - \frac{d}{dr}\{r(\sigma_\theta - \nu\sigma_r)\} = 0. \tag{23}$$

Substitution of Eq. (21) in Eq. (23) yields an equation for σ_r which may be integrated to give

$$\sigma_r = \sigma(1 - \varrho^{-1/\omega}),$$

where

$$\varrho = r/r_0, \tag{24}$$

and

$$\omega = \nu/(1 + \nu).$$

Finally, equilibrium of an element in the radial direction is given by

$$t\sigma_\theta - \frac{d}{dr}(rt\sigma_r) = 0, \tag{25}$$

which may be combined with Eqs. (21), (24) and integrated to yield

$$t = t_\infty(1 - \varrho^{-1/\omega})^{\omega - 1}. \tag{26}$$

This optimum thickness variation has the practical disadvantage that $t \to \infty$ at the edge of the hole, and it would be necessary to introduce a hybrid reinforcement in which the thickness is constant over an annular region adjacent to the hole and Eq. (26) is valid beyond this annulus. Nevertheless, in discussing the weight penalty associated with this type of reinforcement, we will consider this optimum thickness variation as it represents the lightest possible structure. A convenient and nondimensional measure of the weight penalty is afforded by the expression

$$W = \frac{\text{weight of additional sheet}}{\text{weight of sheet removed}}, \tag{27}$$

where the additions and removals are related to a continuous sheet of thickness t_∞. In this connection it is worth noting that for a constant thickness sheet with a "neutral" reinforcement —whether compact or annular—the nondimensional measure of the weight penalty is given by

$$W^*, \text{ say, } = 2/(1-\nu),$$

and it is in comparison with W^* that the present variable thickness reinforcement should be judged. It is found that

$$[W]_{\nu=1/3} = 2.311, \quad \text{cf.} \quad W^* = 3,$$
$$[W]_{\nu=1/4} = 2.174, \quad \text{cf.} \quad W^* = 2.667.$$

6.2 Sheet material satisfying the Mises-Hencky yield criterion

If the sheet material satisfies the Mises-Hencky yield criterion the hoop stress is no longer constant but is related to the radial stress by the equation

$$\sigma_r^2 - \sigma_r\sigma_\theta + \sigma_\theta^2 = \sigma^2, \tag{28}$$

which can be conveniently expressed in the following parametric form

$$\begin{aligned} \sigma_r &= \sigma\left(\sin\lambda - \frac{1}{\sqrt{3}}\cos\lambda\right), \\ \sigma_\theta &= \sigma\left(\sin\lambda + \frac{1}{\sqrt{3}}\cos\lambda\right). \end{aligned} \tag{29}$$

Equation (29) traces out the boundary of the Mises-Hencky ellipse as λ varies from zero to 2π. In the present problem σ_r and σ_θ are of the same sign and $\sigma_\theta \geq \sigma_r$ so that the parameter λ is confined to the range $\pi/6 \leq \lambda \leq \pi/2$, where the upper limit corresponds to conditions remote from the hole.

Substitution of Eq. (29) into the compatibility Eq. (23) yields

$$\frac{2}{r}\frac{dr}{d\lambda} + \sqrt{3}\left(\frac{1-\nu}{1+\nu}\right) - \tan\lambda = 0 \tag{30}$$

which may be integrated to give

$$\frac{r}{r_0} = \frac{f(\lambda)}{f(\lambda_0)},$$

where

$$f(\lambda) = \sec^{1/2}\lambda e^{-\alpha\lambda/2}, \tag{31}$$

$$\alpha = \sqrt{3}\left(\frac{1-\nu}{1+\nu}\right)$$

and λ_0, at present unspecified, is the value of λ at $r = r_0$. Now $f(\lambda)$ attains a minimum value when

$$\lambda = \tan^{-1}\alpha,$$

$$\left(= \frac{\pi}{4} \text{ if } \nu = 0.268\ldots\right)$$

and hence, because solutions for $r < r_0$ are not meaningful, it is necessary that $\lambda_0 \geq \tan^{-1}\alpha$. This in turn means that σ_r is non-zero at the edge of the hole and it is necessary to equilibrate that radial stress with a reinforcing ring. If the hoop stress in this ring is limited to σ an additional restriction is imposed on the value of λ_0.

We will confine attention here to the limiting case when the reinforcing ring can be regarded as a compact member. The hoop strain in the ring can be equated to the hoop strain in the adjacent skin to yield the following expression for λ_0

$$\lambda_0 = \sin^{-1}\left(1 - \frac{3\nu^2}{2(1-\nu+\nu^2)}\right). \tag{32}$$

The stress distribution is now given by Eqs. (31), (32) and the equilibrium Eq. (25) yields the following relation between t and λ

$$\frac{2t}{1+\nu} + \left(\tan\lambda - \frac{1}{\sqrt{3}}\right)\frac{dt}{d\lambda} = 0. \tag{33}$$

Now, away from the hole $\lambda \to \pi/2$ and $t \to t_\infty$ and accordingly the integral of Eq. (33) is given by

$$\frac{t}{t_\infty} = \exp\{\beta(\lambda - \pi/2)\} \cdot \left(\sin\lambda - \frac{1}{\sqrt{3}}\cos\lambda\right)^{-\beta\sqrt{3}} \tag{34}$$

where

$$\beta = \frac{\sqrt{3}}{2(1+\nu)}.$$

The section area A of the reinforcing ring is determined by radial equilibrium with the adjacent sheet:

$$A\sigma = [rt\sigma_r]_{r=r_0},$$

whence

$$\frac{A}{r_0 t_\infty} = \frac{\exp\{\beta(\lambda_0 - \pi/2)\}}{\left(\sin\lambda_0 - \frac{1}{\sqrt{3}}\cos\lambda_0\right)^{\gamma}}, \tag{35}$$

where

$$\gamma = \frac{1-2\nu}{2(1+\nu)}.$$

The nondimensional measure of the weight penalty associated with the varying thickness reinforcement plus ring is given by

$$[W]_{\nu=1/3} = 1.633 \quad \text{and} \quad [W]_{\nu=1/4} = 1.692.$$

The values are markedly lower than those for a Tresca material.

7. Plates with optimum buckling characteristics

Finally, we consider a problem in which the thickness variation is chosen to maximise the onset of buckling. Figure 9 shows a simply-supported strip whose thickness varies smoothly across the width. The strip is subjected to a uniform compressive stress, but in the subsequent analysis it is initially convenient to consider an arbitrary variation of stress across the width. We intend to search for a thickness variation such that buckling occurs simultaneously in infinitely many modes. This does not guarantee a minimum weight design, but it probably does because any small variation of thickness designed to raise the buckling stress in certain modes is likely to lower it in other modes. Anyway, the proof of the pudding is in the eating, and we will see later that our solution yields a buckling stress some 80% higher than that of the corresponding constant thickness plate.

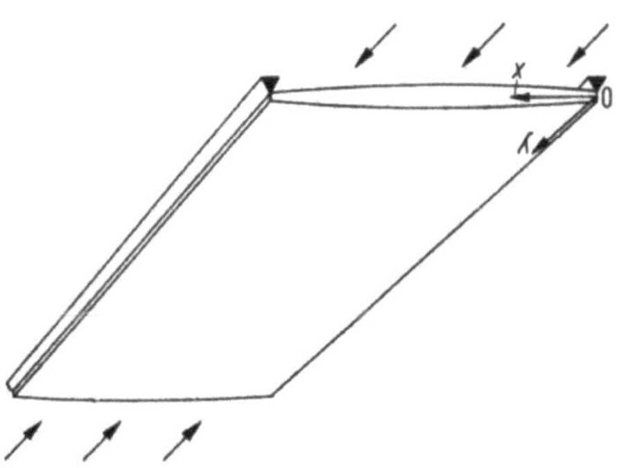

Fig. 9. Simply-supported strip under compression.

In the analysis we find it convenient to derive the (compressive) longitudinal stresses from a force function Φ according to the relation

$$t\sigma_y = \frac{d^2\Phi}{dx^2}, \tag{36}$$

and in determining the magnitude of Φ which causes buckling it is convenient to write

$$\Phi = \gamma\Phi_0, \tag{37}$$

where γ is a constant of proportionality while the magnitude of Φ_0 is fixed. A proportional increase in the compressive stresses throughout the plate is now represented by an increase in γ.

At the onset of buckling the release of middle-surface strain energy is equal to the strain energy of bending and this gives rise to the following expression for γ:

$$\gamma = \frac{\iint D\left[(\nabla^2 w)^2 + 2(1-\nu)\left\{\left(\frac{\partial^2 w}{\partial x\,\partial y}\right)^2 - \frac{\partial^2 w}{\partial x^2}\frac{\partial^2 w}{\partial y^2}\right\}\right] dx\,dy}{\iint \frac{d^2\Phi_0}{dx^2}\left(\frac{\partial w}{\partial y}\right)^2 dx\,dy} \tag{38}$$

in which D is the flexural rigidity and the deflexion w is to be chosen to minimise γ.

Now by integration by parts it can be shown that, subject to certain boundary conditions satisfied by our solution,

$$\iint \frac{d^2\Phi_0}{dx^2}\left(\frac{\partial w}{\partial y}\right)^2 dx\,dy = 2\iint \Phi_0\left\{\left(\frac{\partial^2 w}{\partial x\,\partial y}\right)^2 - \frac{\partial^2 w}{\partial x^2}\frac{\partial^2 w}{\partial y^2}\right\} dx\,dy. \tag{39}$$

Now substitution into Eq. (38) shows immediately that the particular force function given by

$$\Phi_0 = D \tag{40}$$

has just the properties that we are looking for because

$$\gamma = 1 - \nu + \frac{\iint D(\nabla^2 w)^2\, dx\,dy}{2\iint D\left\{\left(\frac{\partial^2 w}{\partial x\,\partial y}\right)^2 - \frac{\partial^2 w}{\partial x^2}\frac{\partial^2 y}{\partial y^2}\right\} dx\,dy} \tag{41}$$

whose minimum value $(1-\nu)$ occurs when

$$\nabla^2 w = 0,$$

which yields infinitely many modes of the form

$$w = \sum w_n \sin ny \sinh nx.$$

In terms of the thickness it may be shown that the critical force-function $(1-\nu)D$ yields stresses given by

$$\sigma_y = G\left\{\left(\frac{dt}{dx}\right)^2 + \frac{1}{2}t\frac{d^2t}{dx^2}\right\}, \tag{42}$$

where G is the shear modulus. Referring back to our original problem in which σ_y is constant we see that the optimum variation of thickness is a linear variation subtending an angle

$$\frac{dt}{dx} = \left(\frac{\sigma_y}{G}\right)^{1/2}. \tag{43}$$

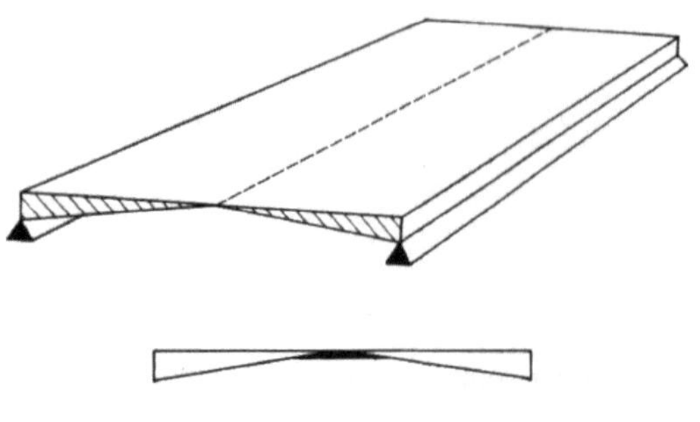

Fig 10. Optimum thickness variation.

This solution is shown in Fig. 10a. Of course, the vanishing thickness along the centre-line makes the solution impractical as it stands and it would be necessary to increase the thickness locally in this region as shown in Fig. 10b.

8. Conclusions

The problems that I have outlined in this talk have all yielded simple solutions, but my main aim has been to suggest that there is still a wide variety of unsolved "inverse" problems in the general field of two-dimensional elasticity.

Electrohydrodynamics

J. R. Melcher

Cambridge, Mass., USA

I. Introduction

It is as a nuisance that electromechanical coupling between electric fields and fluids is most commonly encountered by those working in the areas of fluid mechanics. Examples are the "static electrification" problems that plague aerodynamic systems and fuel transfer operations, and the "electrochemical" difficulties with hydrodynamic instrumentation. Yet, electrohydrodynamic phenomena are basic to highly significant technologies. Examples are found in electrostatic precipitators for air pollution control [1], electrostatic coalescers for crude oil processing [2], electrostatic sprayers for efficiently painting conducting objects [3] and electrostatic powder coating processes. That these applications have developed largely through invention rather than basic research is testimony to the need for a cohesive effort toward sorting out an inordinate number of interrelated phenomena. The 1969 IUTAM-IUPAP Symposium on Electrohydrodynamics demonstrated an international recognition of this need [4]. Incentives for basic research now come from applications to static [5] and dynamic imaging [6], new techniques for air pollution control [7], problems in atmospheric electrification [8] and insulation performance [9].

As in any attempt to give some unity to a subject, there is an inherent conflict betwen giving a fair presentation of existing knowledge and emphasizing how much there is to be learned. In the interest of giving an overview, but at the risk of oversimplification, the few macroscopic models introduced in Section II are used for reviewing a wide range of phenomena. Interactions, represented by Sections III—VI, are classified in Section II after terminology has been established.

The sole concern here is with electro-quasi-static phenomena, where processes connected with the propagation time of electromagnetic waves are complete in times short compared to those of interest, and electric dominates magnetic energy storage [10]. Thus, the effects of magnetic induction are negligible and the electric field intensity $\boldsymbol{E}$ and potential Φ satisfy

$$\nabla \times \boldsymbol{E} = 0; \quad \boldsymbol{E} = -\nabla\Phi. \tag{1}$$

Even so, rate processes are a central concern. Truly "electrostatic" phenomena seldom exist. As an example intended to clarify the difference between magnetohydrodynamic and electrohydrodynamic phenomena, suppose that a material has a uniform electrical conductivity σ, permittivity ε and permeability μ, and the system is characterized by a single dimension l. Then, two of three time constants are generally involved: the electrical relaxation time $\tau_e = \varepsilon/\sigma$, the magnetic diffusion time $\tau_m = \mu\sigma l^2$ and the transit time for an electromagnetic wave $\tau_{em} = \sqrt{\mu\varepsilon l^2}$. Because $\tau_{em} = \sqrt{\tau_e\tau_m}$, it is clear that so long as τ_e and τ_m are not on the same order, one of these time constants must be long compared to the wave transit time. If $\tau_m \gg \tau_e$, then magnetohydrodynamic phenomena are of interest, displacement current and the associated net charge are ignorable, but magnetic induction is essential. By contrast, if $\tau_e \gg \tau_m$, the interaction is electrohydrodynamic and characterized by τ_e, magnetic induction is negligible, but effects of net charge on charge conservation are essential.

In all sections to follow, bulk mechanics are represented by an incompressible Newtonian fluid having mass density ϱ and dynamic viscosity η. By "incompressible", it is meant that the pressure enters only by contributing a force density $-\nabla p$ to the momentum equation. For example, mass density has a negligible dependence on p.

II. Macroscopic models

A. Polarization

In classical electromagnetic texts [11], the effect of the material on the macroscopic field is accounted for by means of macroscopically defined field sources. These are the free charge density ϱ_f and the polarization charge density $\varrho_P = -\nabla \cdot \boldsymbol{P}$. For convenience, the polarization density $\boldsymbol{P}$ is absorbed in the electric displacement $\boldsymbol{D}$ by defining $\boldsymbol{D} \equiv \varepsilon_0 \boldsymbol{E} + \boldsymbol{P}$, so that Gauss' law is

$$\nabla \cdot \boldsymbol{D} = \varrho_f. \tag{2}$$

Most fluids are well represented by linear polarization constitutive laws

$$\boldsymbol{D} = \varepsilon \boldsymbol{E}; \quad \boldsymbol{P} = (\varepsilon - \varepsilon_0)\, \boldsymbol{E}, \tag{3}$$

where ε is the permittivity (1.89 for hexane, 81 for water) with $\varepsilon_0 = 8.854 \times 10^{-12}$ Farads/m.

B. Conduction

Unlike polarization charge, ϱ_f is governed by conduction through the agent of the free current density $\boldsymbol{J}_f$. Conservation of net free charge is an independent physical law.

$$\nabla \cdot \boldsymbol{J}_f + \partial \varrho_f / \partial t = 0 \tag{4}$$

which usually underlies electrical rate processes of interest.

i) *Unipolar Conduction*: Based on the simple thought-model of massless charged particles pulled by the field through a viscous fluid, itself moving with the velocity $\boldsymbol{v}$, the simplest conduction law attributes to charge carriers of a single species the current density

$$\boldsymbol{J}_f = \varrho_f(\pm b\boldsymbol{E} + \boldsymbol{v}) - K_D \nabla \varrho_f, \tag{5}$$

where upper and lower signs respectively refer to positive and negative carriers having the positive mobility b. Included in Eq. (5) is a self-diffusion current with diffusion coefficient K_D. Given that $\nabla \cdot \boldsymbol{v} = 0$ and K_D is constant, Eqs. (2)—(5) combine to give

$$\frac{d\varrho_f}{dt} = \mp \frac{\varrho_f^2 b}{\varepsilon} \pm \varrho_f b \boldsymbol{E} \cdot \frac{\nabla \varepsilon}{\varepsilon} \mp \varrho_f \boldsymbol{E} \cdot \nabla b + K_D \nabla^2 \varrho_f \quad \text{on} \quad \frac{d\boldsymbol{r}}{dt} = \pm b\boldsymbol{E} + \boldsymbol{v}. \tag{6}$$

That is, the time rate of change of ϱ_f in the frame of reference of the charge carrier can be attributed to one of four effects: a self-field repulsion always tending to dissipate the charge density, non-uniformities in permittivity and mobility which can either increase or decrease the local density and diffusion of charge from the local region.

Much of what is known about ion dynamics in homogeneous fluids derives from the "imposed flow and field" approximation where $\boldsymbol{v}$ is known at the outset (usually meaning that there is little reaction of the field on the flow) and $\boldsymbol{E}$ is induced by charge exterior to the region of interest. Because the terms on the right in Eq. (6) are then neglected, it follows that the charge density is constant along the particle lines found by integrating the characteristic equation. If particle transit times are short compared to times of interest (for example the charging time of a water drop in an ambient flow, field and ion flux) [12] the solenoidal nature of $\boldsymbol{v}$ and $\boldsymbol{E}$ makes it possible to explicitly solve the characteristic equation using stream functions.

ii) *Ohmic (ionic) Conduction*: By an Ohmic conductor, it is meant that $\boldsymbol{J}_f' = \sigma \boldsymbol{E}'$, where primes denote variables measured in the frame of reference moving with the macroscopic fluid velocity $\boldsymbol{v}$. In the electro-quasi-static approximation [13] $\boldsymbol{E}' = \boldsymbol{E}$ and $\boldsymbol{J}_f' = \boldsymbol{J}_f - \varrho_f \boldsymbol{v}$ so that the conduction law expressed in terms of laboratory frame variables is

$$\boldsymbol{J}_f = \sigma \boldsymbol{E} + \varrho_f \boldsymbol{v}. \tag{7}$$

The consequent distribution of free charge density is evidenced by substituting Eq. (7), along with Eqs. (2) and (3), into Eq. (4).

$$\frac{D\varrho_f}{Dt} = -\frac{\varrho_f}{(\varepsilon/\sigma)} - \frac{\sigma}{\varepsilon}\boldsymbol{E}\cdot\nabla\varepsilon - \boldsymbol{E}\cdot\nabla\sigma; \quad \frac{D}{Dt} \equiv \frac{\partial}{\partial t} + \boldsymbol{v}\cdot\nabla. \tag{8}$$

The time-rate of change of ϱ_f local to a given element of the macroscopic fluid is given by the terms on the right. In uniform fluids charge simply dissipates with the relaxation time ε/σ. In non-uniform media permittivity and especially conductivity gradients imply local charge accumulation or dissipation.

Underlying the empirical relation, Eq. (7), is conduction involving at least two species of charge carriers. As an example, suppose that two oppositely charged species are involved, each giving rise to its own current density

$$\boldsymbol{J}_\pm = \pm\varrho_\pm b_\pm \boldsymbol{E} - K_\pm \nabla\varrho_\pm \tag{9}$$

and satisfying a conservation equation in the form of Eq. (4). The net charge is then $\varrho_+ - \varrho_-$ and the conductivity model implies a large self-neutralizing background density ϱ_0 of each species such that the net charge is comparatively small. To specialize matters even more, suppose that Einstein's relation between mobility and diffusion coefficient holds, as in electrolytes. Then, $b_\pm = z_\pm K_\pm F/RT$ where z is the number of elementary charges per carrier and $F/RT = 25.3\times10^{-3}$ volts at room temperature. For a fluid in which $b_\pm$, $K_\pm$ and ε remain constant and in a range of electric field such that

$$\frac{RT/F}{l} \ll |\boldsymbol{E}| \ll \left(\frac{RT/F}{l}\right)\left(\frac{\varrho_0}{\varrho_f}\right), \tag{10}$$

where l is a characteristic length, Eqs. (2)—(4) (with the latter written using Eq. (9) for each of the species) combine to show that [14]

$$D\sigma/Dt = K_{\text{eff}}\nabla^2\sigma; \quad \sigma \equiv \varrho_0(b_+ + b_-); \quad K_{\text{eff}} \equiv (K_+b_- + K_-b_+)/(b_+ + b_-) \tag{11}$$

and that ϱ_f is given by Eq. (8) with $\nabla\varepsilon \to 0$. That is, the conductivity diffuses at a rate typical of molecular diffusion while the charge relaxes with the time constant ε/σ. Often, molecular diffusion times l^2/K_{eff} are long compared to dynamical times of interest, and hence Eq. (11) reduces to the statement that the conductivity local to a given fluid element remains constant. This promotes a model in which σ depends on the local state, for example on the temperature.

Many phenomena can be successfully modelled by Eq. (8) based upon empirical arguments. However the conditions for electrolytic behavior as represented by Eq. (10) are often not met. Clearly, the usefulness of the Ohmic model is sometimes in describing phenomena that are not sensitive to the details of the conduction phenomena. In a sense, the model represents what may be a number of processes, each with its own characteristic time-constant, by a dominant relaxation time. *Felici* gives a useful review of conduction phenomena in liquids at high field strengths [15] and emphasizes the need for understanding the combined influences of "microphysical", electrochemical and electrohydrodynamic processes [16]. As experimental controls are refined so as to justify detailed representations of conduction, it is likely that the Ohmic model will remain useful for summarizing behavior in common to broad classes of semi-insulators.

C. Stress-tensor, force density and surface force density

The microscopic electrical force, acting on individual charge carriers, is transmitted to the macroscopic media through collisions. As suggested by Fig. 1, this is accomplished either by individual charges or by pairs of charges which transmit the net electrical force to the fluid. Associated with the individual charges is the force density $\varrho_f\boldsymbol{E}$, while the force density on a dilute distribution of dipoles is $\boldsymbol{P}\cdot\nabla\boldsymbol{E}$ [17]. The latter does not account for interactions between dipoles which, as a practical matter, are best included by using thermodynamic arguments to derive a force density based on parameters that are themselves macroscopically defined. With the assumption that $\boldsymbol{D}$ and $\boldsymbol{E}$ are co-linear, the force density $\boldsymbol{F}$ is written in terms of the Maxwell-stress-tensor T_{ij} as

$$F_i = \partial T_{ij}/\partial x_j; \qquad T_{ij} = E_i D_j - \delta_{ij} w'; \qquad w' \equiv \int \boldsymbol{D} \cdot \delta \boldsymbol{E}. \tag{12}$$

In the most important case of an incompressible linear dielectric, the co-energy density $w' = (1/2D^2)/\varepsilon$. In the much less important more general case where the fluid is compressible so that for a given fluid element $\varepsilon = \varepsilon(\varrho)$, there is an additional "electrostriction" term in w' involving $\partial\varepsilon/\partial\varrho$.

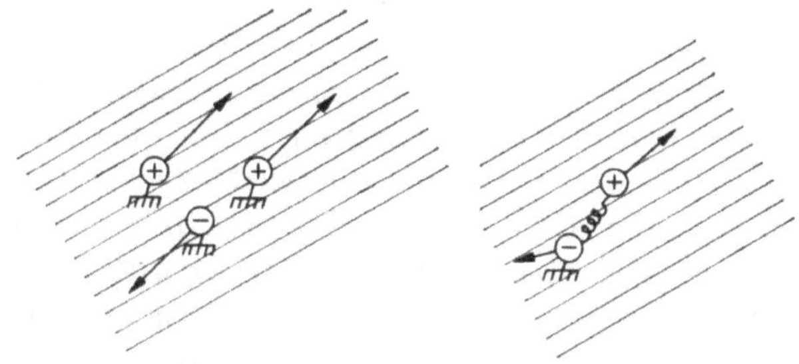

Fig. 1. Free charges individually transfer their momentum to the neutral fluid, while charges forming dipoles do so in pairs.

Carrying out the tensor divergence of Eq. (12) and using Eqs. (1) and (2) gives the force density [18, 19, 20, 21]

$$\boldsymbol{F} = \varrho_f \boldsymbol{E} - \frac{1}{2} E^2 \nabla \varepsilon. \tag{13}$$

For compressible media, there is an added pressure-like "electrostriction" term $\nabla(1/2\varrho(\partial\varepsilon/\partial\varrho)E^2)$.

In thinking of the macroscopic force density as the resultant of forces on microscopic particles, it is best to think of the polarization part not as $-1/2E^2 \nabla\varepsilon$, but rather as $\boldsymbol{P} \cdot \nabla \boldsymbol{E}$. These differ by the gradient of a pressure, and hence, though they may even act over different regions of the fluid, are completely consistent so long as the fluid is incompressible in the sense defined in Section I.

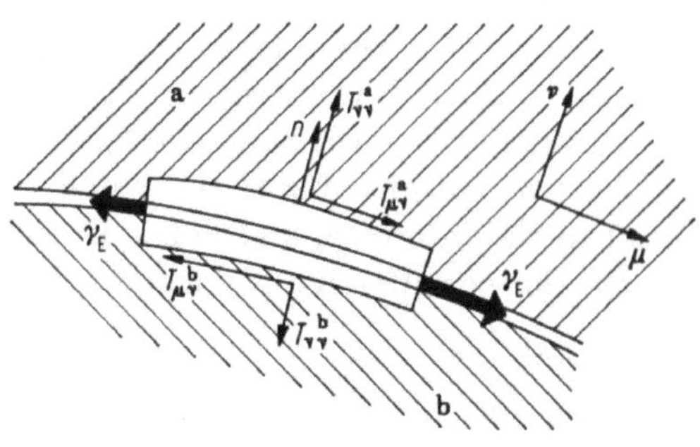

Fig. 2. Surface force density at an interface between regions (a) and (b) is decomposed into stress contributions due to the "external" fields, represented by T^a_{ij} and T^b_{ij}, and due to "internal" (double-layer) fields, represented by γ_E.

At an interface between regions (a) and (b), depicted by Fig. 2, there is a surface force density, consistent with Eqs. (12) and (13), given by

$$\mathsf{T} = \boldsymbol{n} \cdot [\![\mathsf{T}]\!] - \boldsymbol{n}\gamma_E \nabla \cdot \boldsymbol{n} + \nabla_\Sigma \gamma_E; \quad \gamma_E \equiv -\int_{0^-}^{0^+} w'\, dv. \tag{14}$$

This is found by integrating the stress tensor over the surface of an incremental volume enclosing a section of the interface having unit normal $\boldsymbol{n}$. The sum of the reciprocal radii of curvature is $\nabla \cdot \boldsymbol{n}$, ∇_Σ is the surface gradient and $[\![A]\!] \equiv A_a - A_b$. Included is the possibility that the interfacial region contains a double-layer with sufficiently large $\boldsymbol{n}$ directed internal fields that the integral of the electric stress over the edges cutting through the interface is significant. Thus, the "internal" contributions giving rise to γ_E are not significant unless the "internal" electric field is large compared to the "external" field. The double-layer surface force density takes the same form as that attributed to surface tension.

D. Classification of interactions

Electrohydrodynamic phenomena are usually in one of four categories, respectively taken up in the following sections.

Section III is concerned with coupling that is entirely due to *polarization*. Effects of free charge, and hence of conduction, are ignorable. Sometimes termed dielectrophoretic phenomena [22], these interactions are analogous to ferrohydrodynamic phenomena constituting the area of ferrofluid mechanics [23]. Whereas ferrofluids are synthesized stable suspensions of magnetic dipoles, discussion here is confined to ordinary liquids. In the remaining three categories polarization effects are in the background acting in consort with forces on free charges.

Section IV illustrates the most common electrohydrodynamic coupling involving conduction phenomena. The free charge is largely confined to interfaces, so that the charge can be characterized as a *monolayer*. In the interfacial region, the free charge distribution takes the spatial form of an impulse.

Section V considers interactions with the free charge distribution throughout a significant part of the *bulk* region of interest.

Section VI by contrast considers *double-layer* coupling. Not only is the free charge density confined to an interface, but so also is the dominant electric field. The charge distribution in the interfacial neighborhood is a doublet with the dominant electric field sandwiched between regions of positive and negative charge density.

III. Polarization interactions

An adaptation of Pellat's classic experiment [24] for demonstrating the nature of the polarization force density is shown in Fig. 3. The diverging plates, made from transparent tin oxide coated glass [25, 26] are equipotentials. The electric field between is approximated by $\boldsymbol{E} = (V_0(t)/\theta_0 r)\, \boldsymbol{i}_\theta$, with $V_0(t)$ alternating at a frequency high enough relative to the reciprocal relaxation time of the liquid to insure that there is no free charge either on the interface or in the bulk. Only the second term in Eq. (13) is operative. The frequency is also high enough that the mechanical response is essentially to the time average E^2.

The experiment is classic for two reasons. First, the liquid rises in such a way that the electric field remains tangential to the interface, and hence is undisturbed by the liquid geometry. Second, in the liquid bulk the permittivity is constant, from Eqs. (2) and (3) $\boldsymbol{E}$ and $\boldsymbol{P}$

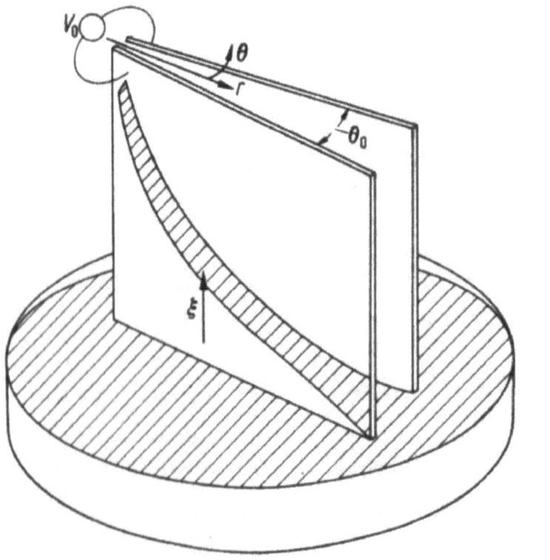

Fig. 3. Adaptation of Pellat's experiment for demonstrating polarization force on dielectric liquid.

are solenoidal and there is therefore no polarization charge. Because $\boldsymbol{E}$ is tangential to the interface, there is also no polarization surface charge at the interface. Clearly, the experiment shows that the polarization force density is not $\varrho_P \boldsymbol{E}$, and emphasizes the necessity for distinguishing between microscopic and macroscopic concepts.

Surface Coupling: The diverging plate experiment is an example of systems now considered composed of homogeneous liquids separated by interfaces. The force density represented as in Eq. (13) has the advantage of being confined to interfaces, where ε is discontinuous. With the field tangential to an interface, as it is in the experiment of Fig. 3, the surface force density follows from Eqs. (12) and (14) as $1/2(\varepsilon - \varepsilon_0) E^2$ acting outward from the region of high permittivity. Thus, the liquid rises between the plates to the altitude

$$\xi = \frac{\varepsilon - \varepsilon_0}{2\varrho g}\left(\frac{V_0}{\theta_0 r}\right)^2. \tag{15}$$

The static equilibrium exemplifies what must be the stationary state (if it exists) resulting from a polarization interaction with externally constrained effectively constant potentials and charges. A closed stationary system of perfect insulators can accept no current, and hence no electric power. Since motion of a real fluid implies dissipation, the only stationary state is a static equilibrium. Consistent with this observation, aimed mainly at those that would like to make a $d-c$ pump using the polarization interaction, is the general fact that the polarization surface force density given by the first term in Eq. (14) can not have a shearing component, as is best seen directly from Eq. (13).

Albeit typically weak, the polarization interaction has a nature well suited to orientation, levitation and ducting of liquids. Incentive for developing this class of fluid mechanics has come from the need for management of liquids in zero gravity environments [27]. Two types of innovations have made it possible to make significant advances in quantitatively relating observations and theories. First, the control of conduction and electrode processes afforded by the use of high frequency fields, as already mentioned. Theories based on perfectly insulating models with static excitations are found to be directly applicable, provided the rms values of the fields are used. Second, fluids and configurations have been found that make dominant the relatively small polarization forces involved. (To achieve a rise of 10 cm in the experiment of Fig. 3, the gas environment is compressed so as to permit an $|\boldsymbol{E}|$ approaching 10^7 v/m [25, 26].) Since gravity is a main competitor for the field, use of liquids having nearly the same densities is one way to enhance electrical effects. But generally, those few fluid pairs that are not miscible, while being both highly insulating and of disparate permittivities, are viscous and hence not capable of demonstrating the inertial effects that are essential in applications of interest.

The fluid mechanician who appreciates the significance of gravity-capillary waves for the dynamics of free surface systems will recognize why the basic interest here is in the interfacial waves and instabilities associated with the polarization. To make matters more interesting, $\boldsymbol{E}$, unlike the gravitational acceleration $\boldsymbol{g}$ of terrestrial applications, is itself influenced by the fluid deformation and moreover can be imposed in a non-uniform way. Coupling phenomena, especially for small amplitude motions, are profitably divided into "self-field" effects associated with the change in surface force caused by the field distortion due to the fluid deformation itself, and "imposed-field" effects, where changes in electrical surface force density result from motion of an interface through a non-uniform imposed field [28].

"Self-field" effects alone determine the perturbation dynamics of a planar interface between fluids (a) and (b) stressed by a uniform electric field having tangential field component E and perpendicular components E_a and E_b such that $\varepsilon_a E_a = \varepsilon_b E_b$. The field components have superimposable effects. With increasing perpendicular field, gravity-capillary waves are slowed by polarization until instability is incipient at the Taylor-wavelength $2\pi \sqrt{\gamma/g(\varrho_a - \varrho_b)}$ as the applied field reaches a magnitude such that [29]

$$\frac{(\varepsilon_a - \varepsilon_b)^2 E_a E_b}{(\varepsilon_a + \varepsilon_b)\, 2\sqrt{g(\varrho_a - \varrho_b)\,\gamma}} = 1, \tag{16}$$

where γ is the surface tension.

An experimental study of the destabilizing effects of the perpendicular field support such theoretical models [30]. Magnetization interactions, which should be analogous, have also been studied, not only in regard to small amplitude consequences of instability [31, 21], but for large amplitude static deformations of the interface as well [32].

That the tangential field has no "self-field" effects on surface deformations with peaks and valleys parallel to the field is illustrated by the experiment of Fig. 3. Thus, even though perturbations propagating along the lines of electric field are stabilized by the field, giving rise to a self-field polarization wave having square phase velocity

$$V_s^2 = \frac{(\varepsilon_a - \varepsilon_b)^2 E^2}{(\varrho_a + \varrho_b)(\varepsilon_a + \varepsilon_b)} \tag{17}$$

adding to that of the gravity-capillary waves, the field does not stabilize an interface that is unstable to begin with.

Self-field polarization [29, 30] and magnetization [21] wave propagation along force lines have been studied experimentally by measuring field effects on surface wave resonance frequencies.

In terrestrial gas-liquid systems, self-field effects are usually dominated by gravity and capillarity. The orientation experiment of Fig. 4 is intended to illustrate how "imposed-field" effects, through the mechanism of strong field gradients, can greatly enhance the role of the field. The electrode structure, which was studied as a means of preventing disorientation of liquid fuel in the S-IVB at engine cut-off [33], consists of concentric rings at alternate potentials. Thus, the electric field is concentrated between the electrodes, falling off in the fringing regions at a rate determined by the interelectrode spacing s.

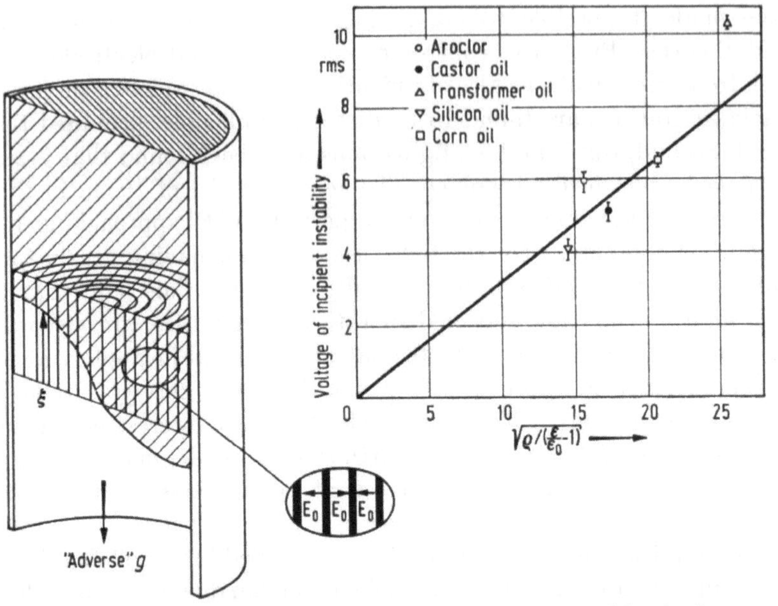

Fig. 4. Polarization force used to stabilize orientation of liquid supported by hydrostatic pressure of gas through action of field gradient concentrated in neighborhood of interface. Experimental data indicate success of simple model in predicting minimum voltage consistent with stable equilibrium.

In the laboratory experiment the cylinder is filled with liquid and stoppered at the top. With the field applied, gas bubbles introduced from the bottom stop at the electrodes. Hence, liquid is expelled from the region below the electrodes, leaving a static equilibrium with the liquid supported by the gas. The electrodes are wetted, and hence surface tension is not responsible for the stable equilibrium, as is easily seen by removing the voltage (gas rises and liquid falls) and re-applying it (stable interface re-established). To appreciate the experiment, it must be recognized that the polarization surface force density would not support 1/10 of the liquid above. (In fact, the surface force density consistent with Eq. (13) is downward not upward.) It is the static pressure in the gas below that holds the liquid up. Mainly, the electric field stabilizes the equilibrium.

Any slight tipping of the interface results in a restoring surface force density. For a given electric field between the electrodes, the stiffness of this surface force density is increased by decreasing s. A simple model incorporating this "imposed-field" picture of the dynamics shows that instability should be incipient as the electric field between the electrodes is reduced below the value

$$E_0 = \sqrt{(\varepsilon - \varepsilon_0)/2\varrho g s}\,. \tag{18}$$

Experimental observations are compared to this prediction in Fig. 4.

Orientation with field gradients distributed [34] and oscillations with gradients concentrated [33] (as in Fig. 4) have also been studied.

Work on the stability of cylindrical equilibria can be understood physically in terms of these same "self" and "imposed" field effects [35, 36, 28].

To understand "imposed-field" dynamics, it is also convenient to think in terms of polarization electromechanical waves, but with velocity proportional to $(\varepsilon - \varepsilon_0)\,E^2$ and not $(\varepsilon - \varepsilon_0)^2\,E^2/(\varepsilon + \varepsilon_0)$. These waves play a key role in flow systems using polarization as the mechanism for ducting a liquid, in a way analogous to gravity in channel flows. It was recognized at an early stage in work on drainage of zero gravity storage tanks [37] that of the various classes of "imposed-field" waves that could exist on flow structures, a principle-mode is most essential to understanding the flow characteristics. *Jones* has a current interest [38] in this class of electromechanical flow structures for applications to heat pipes, where the polarization ducting is used to replace that of capillarity.

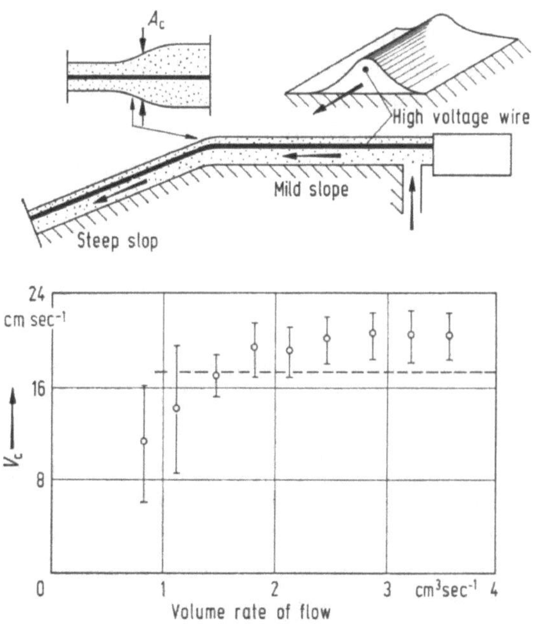

Fig. 5. Example of "wall-less pipe" flow structure consisting of wire at high voltage parallel to conducting plane. Change in slope forces transition from super-critical to subcritical flow, so that flow velocity is equal to velocity of "imposed-field" type of polarization wave at transition point. As a function of flow rate, observed critical velocities are compared to independently measured wave velocity indicated by broken line.

In a typical flow structure [25], or "wall-less pipe", shown in Fig. 5, a wire at high voltage runs parallel to a conducting wall. The liquid is contained between the wire and the wall by the field, and forced to flow longitudinally (the z direction) by external means. Note that with the electrical force density written as Eq. (13) this means that the internal pressure of the "pipe" is less than that of the exterior.

In the principle mode, the interior pressure is uniform over a given cross-section. For a wide range of geometries, a coenergy per unit length, W', can be defined such that

$$p = -\partial W'/\partial A; \quad W' \equiv \int_s w' \, da, \tag{19}$$

where the cross-sectional area $A(z, t)$ of the liquid is a variable generalizing the liquid depth in a gravity channel flow, and S is the total flow-structure cross-section. It follows that the long-wave equations of motion, written in terms of the longitudinal velocity v, take the form [25]

$$\varrho\left(\frac{\partial v}{\partial t} + v\frac{\partial v}{\partial z}\right) - \frac{\partial^2 W'}{\partial A^2}\frac{\partial A}{\partial z} = 0; \quad \frac{\partial A}{\partial t} + \frac{\partial}{\partial z}(vA) = 0. \tag{20}$$

The particular dependence of W' on A is determined by the specific structure.

The analogy to gravity-wave channel flows helps to make it clear that embodied in these equations is a representation of such subcritical and supercritical phenomena as conservative transitions, shocks, etc, but here there is a greater variety of effects stemming from the wide range of possible dependences of W' on A. Shown in Fig. 5 is a sample experiment in which the polarization "imposed-field" wave velocity is determined by measuring (as a function of flow-rate) the critical area A_c at a subcritical to supercritical transition in the flow.

That the field ducts the flow, but provides no pumping action, is illustrated by a flow structure used as a siphon [39].

A form of Freon that is liquid under room conditions is particularly convenient for these flow experiments because it is not easily ignited, it is highly insulating, less viscous than water and has a vapor that resists electrical breakdown. With such a liquid the pulsating component of the imposed $a-c$ field has an observable effect even at 400 Hz, and a sometimes catastrophic effect at 60 Hz. Instability resulting from the periodic component of applied surface force density (sometimes termed parametric or dynamic instability) is analogous to that found for fluid interfaces in an oscillatory acceleration field [40]. Although a number of theoretical and experimental studies have been made on this class of instabilities, observations are at best explained in an after-the-fact fashion [25, 36, 41, 42]. Useful overall criteria, quantitatively substantiated by observations, have yet to be obtained. Dynamic stabilization, also caused by the oscillatory component of force density, has seen considerable recent interest in a purely fluid-mechanical context [43] and for stabilizing plasmas [44] and deserves attention in the context of electrohydrodynamics.

Considerable interest in polarization surface instabilities comes from those attempting to augment two-phase heat transfer. Particularly in zero-gravity the film boiling and condensation heat transfer are significantly influenced [45, 46]. In much of this work insufficient care has been taken to distinguish between polarization coupling, and that dominated by free charge monolayers to be taken up in the next section.

Bulk Coupling: Although generally weaker than the coupling at interfaces, bulk interactions in liquids are nevertheless demonstrable. Most studies involve bulk permittivity gradients achieved by thermal stress. Because the permittivity is a function of mass density, and hence of temperature, thermal gradients are tantamount to gradients in permittivity. Note that so long as ϱ is not an appreciable function of pressure, the electrostriction term left out of Eq. (13) has no effect.

Between plane parallel horizontal electrodes supporting a vertical thermal flux such that vertical equilibrium density and permittivity gradients are $D\varrho$ and $D\varepsilon$, and with parameter variations small compared to their mean equilibrium values, the bulk is unstable [47] as the applied field E is raised beyond a value such that

$$E^2(D\varepsilon)^2/(-g\varepsilon\, D\varrho) = 1. \tag{21}$$

Here, the competition is between a "self-field" destabilization and the stabilizing effect of gravity. Proof of exchange of stabilities under more general conditions and experimental verification have been reported [47]. The connection between conventional gravity surface waves and internal waves helps to suggest the relation here between the polarization surface waves (underlying the instability predicted by Eq. (16)) and internal electromechanical waves (underlying the bulk instability predicted by Eq. (21)).

Thermal diffusion times are often long compared to those of interest, thus justifying the view (inherent to the previous remarks) that ε is convected by the fluid. Particularly as gravity stabilization becomes negligible, thermal diffusion effects become important. In a manner suggested by the Bénard problem, thermal diffusion effects on instability of a static fluid have been investigated [48]. In a configuration where field and permittivity gradients are horizontal, careful experiments involving vertical convection between parallel plates have been related to a stability theory for the onset of cellular convection [49]. As in the work on initially static equilibria, the major incentive for this investigation is the field induced increase in heat transfer between the plates.

IV. Charge monolayer interactions

It is clear from Eq. (8) that in a system of fluids, each having uniform permittivity and conductivity, the volume free charge density local to a given element of fluid is zero unless it has evolved from a source of net charge. Charges, which tend to be absent from the bulk, accumulate at the interfaces. There, properties are discontinuous and hence gradient terms on the right in Eq. (8) are not zero but rather impulsive singularities. In many common poorly insulating liquids (relaxation times of a second or less) this model, with its underlying assumption of an Ohmic conductivity, is remarkably useful.

The salient difference between the surface force density on an interface supporting a monolayer of free charge and the polarization surface force density highlighted in the previous section is its shearing component. Unless the interface is an equipotential, it is subject to electrical shear stresses. This is obvious from the simple picture of a surface charge combined with the tangential field implied by a potential that varies in the interfacial plane. Formally, it follows from Eq. (14).

It is difficult to demonstrate simultaneous monolayer and double-layer effects. Hence, γ_E is taken as zero in this section.

Coupling at Equipotential Surfaces: If a homogeneous liquid has a short relaxation time, portions of its interface abuting relatively less conducting fluid are essentially equipotentials. Because the resulting electromechanical coupling is through a surface force density that has no shearing component, the resulting effects are often similar to those described in Section III. The celebrated Rayleigh limit, $Q = 8\pi \sqrt{\varepsilon_0 \gamma R^3}$, on the amount of charge Q that can be retained by a water drop without causing instability [50] is an example deriving from this class of interactions. The tendency of surface charge to accumulate on any saliency of an equipotential interface is a "self-field" effect that causes the buckling of even a planar interface stressed by a sufficiently large uniform perpendicular field [51, 52] E_0.

$$E_0 = \{4g\gamma(\varrho_a - \varrho_b)/\varepsilon^2\}^{1/4}. \tag{22}$$

This critical field and the general tendency of the coupling to reduce the frequency of surface waves are well verified by experiments [51, 53, 54, 55] as is also the Rayleigh limit [56]. Of particular fascination are the motions of relatively conducting jets in electric fields [29, 57, 58], where instability can take the form of an amplifying wave which can be given a sinusoidal steady state excitation [59].

Jet disintegration into drops can be, in some respects, viewed in terms of the equipotential model, but as will be discussed shortly, additional effects can easily play an important role. Because of meteorological applications, drops and their interactions in electric fields modelled by equipotentials are a subject of continual research [60, 61, 62].

In gravitational forms of *Rayleigh-Taylor* instability, a sinusoidal perturbation grows into bulbous protrusions of the heavy fluid into the lighter one below. In the polarization version, instability leads to a new static equilibrium in the form of "fingers" of the fluid of higher permittivity reaching into the other fluid region. The tips of the fingers remain smoothly rounded [31]. By contrast with both of these, the equipotential surface tends to sharpen into a point [29, 52]. Because of the associated field concentration, new physical processes are almost inevitably brought into play and are at the heart of understanding the sometimes bewildering variety of phenomena.

Evaluation of Eq. (22) for an air-water interface shows E_0 to be close to the breakdown strength of air under nominal conditions. Thus it is to be expected that in many cases electrical discharge phenomena are an important factor [56]. As current is discharged through the interface, electrical shear stresses can come into play.

Shear-Stress Coupling: A review, devoted to the role of monolayer shearing surface force densities, gives background on research in this area [63]. Further discussion here is confined to three recent research projects.

Current driven jets [64, 65, 66] are a remarkable demonstration of the combined action of the polarization forces of Section III and monolayer shearing forces. A controlled experiment has the arrangement of Fig. 6. In the absence of an electric field, a liquid such as glycerin drips from a pipette onto a metallic rod. With the application of voltage between pipette and rod the dripping is abruptly replaced by a steady stable stream, typically accelerating until it reaches the electrode in the form of a fine thread. Obvious questions are: why is the stream stabilized against the capillary instability accounting for the formation of drops in the absence of the field, and why is the stream accelerated?

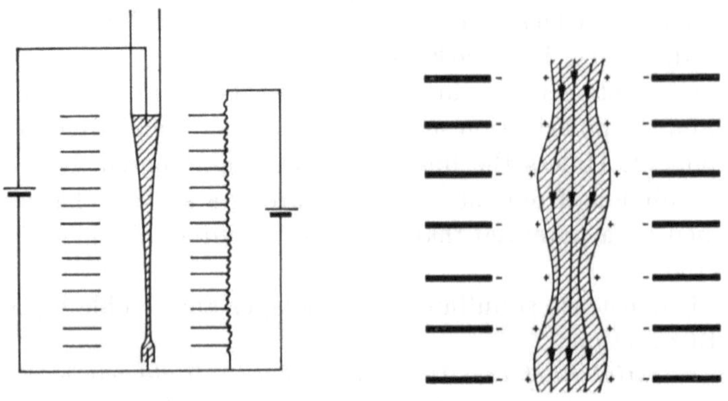

Fig. 6. Cross-sectional view of current driven jet with potential controlling washer shaped electrodes. Detail of jet and electrodes suggests how current is concentrated by pinching and hence restoring radial polarization surface force density is generated. Also shown are surface charges induced by means of surrounding electrodes.

The polarization surface force density on a dielectric interface in a tangential electric field (Fig. 1) provides the first answer. Because the total current through any cross-section of the stream must be the same, the field intensity is increased where the stream "necks" in and decreased where it bulges. Hence, the surface force density, acting as it does from liquid outward, tends to restore the interface to its initial radius, in effect counteracting the surface tension [66, 67]. The tangential field associated with the longitudinal current also conspires with surface charges to create a shearing surface force density that can accelerate the stream. Just which direction it acts depends on the sign of the surface charge, and if the stream geometry is taken as fixed, that depends on the electrodes in the surrounding region. In Fig. 6 washer shaped electrodes surround the jet. These are controlled in relative potential so as to provide some control over the distribution and sign of induced charge. The use of the potential controlling structure has made it possible to make a long-wave model for the self-consistent dynamics of sausage deformations on the stream [66]. At least for relatively viscous liquids, such as glycerin, the model successfully predicts the shape of the jet, which can be made to decelerate as well as accelerate. The fact that transistions between accelerating and decelerating flows are abrupt suggests that there is more to predicting flow regimes than simply knowing the direction of the surface force density. It is found that the stream dynamics are intimately related to whether or not the flow is supercritical or subcritical relative to the propagation velocity of electromechanical waves.

A practical example of a current driven jet is found when high voltage transmission cables are subjected to foul weather conditions. The film of liquid on the conductor surface is ruptured by the electric stress. The sharp tips formed by the perturbances become the sites for corona discharge. Of considerable engineering interest is the associated undesirable audible noise [68].

Here, what is of interest is the interplay of the corona discharge and the advanced stages of deformation of the liquid surface. Figure 7 shows an instantaneous picture of a jet directed from a cylindrical conductor (9/8 inch diameter) toward a surrounding coaxial cage [69], and the same jet viewed by its own corona discharge through a time exposure. In the absence of a field, the water simply drips from the conductor without forming a stream. Under ordinary lighting, the stream appears stable down to a well defined point at which it abruptly breaks into a "brush" of fine droplets. Audible noise also indicates the corona discharge.

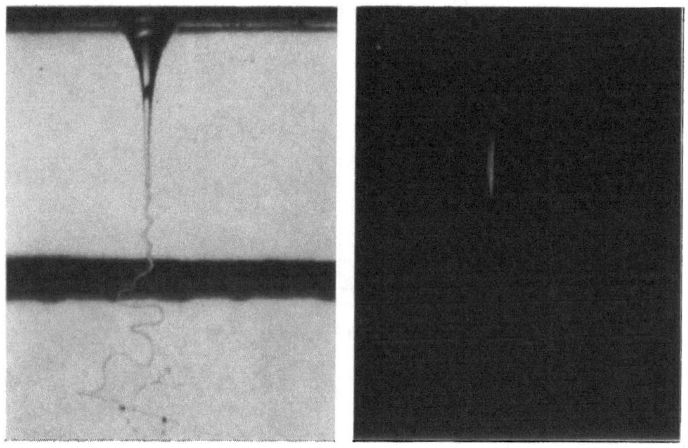

Fig. 7. Instantaneous view of current driven water jet on high voltage cable under practical conditions, together with time exposure of corona discharge on the same jet. Note that corona light appears just above the section of violent instability.

The crucial point is that the corona discharge occurs on a smooth section of the jet before the violent breakup. Measurements on such corona terminated streams in an apparatus such as shown in Fig. 6 (with the lower electrode replaced by the discharge) substantiate the following picture of the dynamics. Much of the stream is stabilized by the pinch mechanism described previously. As the jet narrows, the normal electric field rises, as shown in Fig. 8 where observed stream radius and E_r are shown as a function of axial position. Also plotted is the electric stress

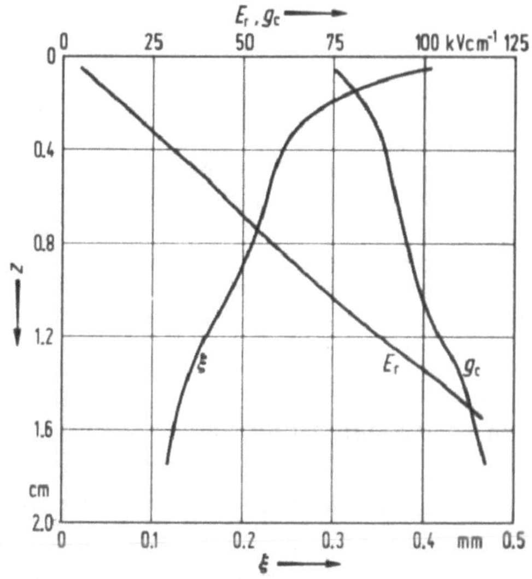

Fig. 8. Measurements on current driven water jet in configuration of Fig. 6 but with corona serving as lower electrode. Observed jet radius ξ, electric field intensity normal to jet surface, E_r, and Peek's critical gradient for initiation of corona discharge, g_c, as functions of distance along axis of jet, z. Position where $g_c = E_r$ is essentially where corona discharge is observed.

for incipience of corona discharge from a metallic cylindrical conductor, as determined empirically by *Peek* [70]. The observed discharge is found at just about the position at which E_r and g_c cross. Thus, the water surface behaves much as a metallic electrode in causing electrical breakdown. (This is itself a point of some interest, since studies of electrical breakdown from water surfaces are plagued by electrohydrodynamic instabilities [71]). Once the current has left the stream, it has the nature of an equipotential jet in a radial field [29], and is violently unstable.

Monomolecular film dynamics can be studied by means of the shearing component of electrical surface force density [72]. A configuration chosen to develop this application is shown in Fig. 9, where the film A is at the upper interface of a layer of distilled water B contained in an insulating trough C. The object here is to provide a standing-wave of shearing force density, instantaneously acting as shown by the arrows, so as to study the compressional attributes of the film.

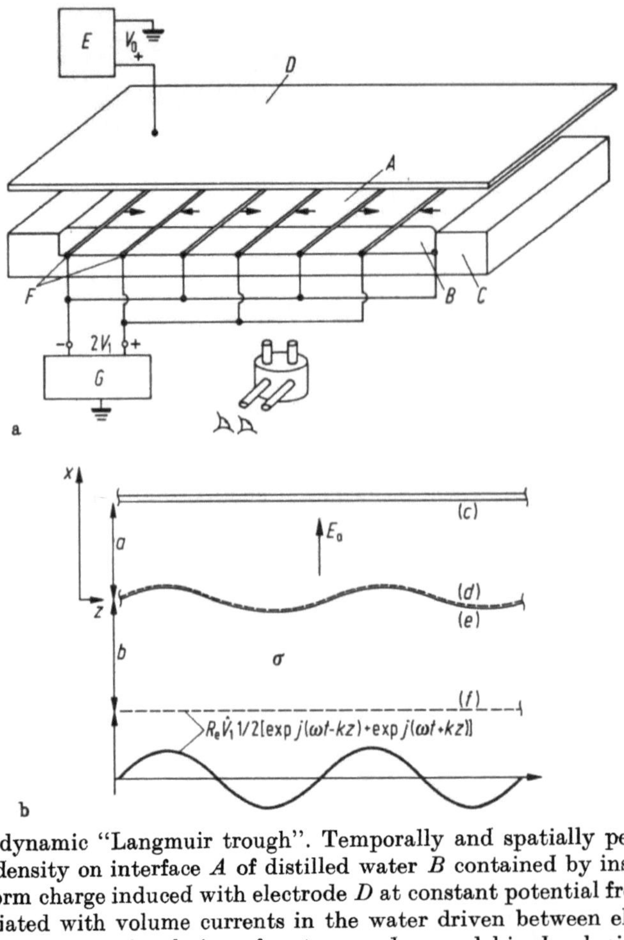

Fig. 9. (a) Electrohydrodynamic "Langmuir trough". Temporally and spatially periodic standing wave of electrical surface force density on interface A of distilled water B contained by insulating trough C is produced by means of uniform charge induced with electrode D at constant potential from source E and tangential electric fields associated with volume currents in the water driven between electrodes F by means of $a-c$ voltage source G. (b) Cross-sectional view of system used as model in developing frequency response of horizontal motions of film placed on interface.

Surface charge and a tangential field at the interface are required. To this end, the electrodes F on the insulating bottom of the trough are driven by an oscillating voltage from the source G, to produce a tangential component of $\mathbf{E}$ at the interface. A uniform monolayer of charge is induced on the interface by a parallel electrode D having the constant potential V_0.

The cross-section of the liquid layer and electrodes, also shown in Fig. 9, helps to characterize the type of analytical model involved. The configuration is one of several examples to be described in this review where the system is forced to respond at a specified frequency ω and with a specified wavenumber k. This "imposed $\omega - k$" approach to the study of linear dynamics of continua has the virtue of allowing the straightforward analytical representation of complex phenomena which, if stable, can be critically studied by varying the frequency and wavenumber.

As an example, Fig. 10 shows the measured and predicted horizontal component of displacement response at the interface, as a function of driving frequency. The antiresonance is due to coupling to the gravity-capillary wave system by the essentially shearing modes under study. The resonance is the result of the film elasticity $\mathscr{E}_0$ (newt/m) coupled to the inertia of the water substrate through the bulk viscous stresses. Well removed from the gravity antiresonance, this film resonance is at the frequency

$$\omega = \left(\mathscr{E}_0 k^2/\sqrt{2\varrho\eta}\right)^{2/3}. \tag{23}$$

Thus, the experiment provides a dynamic measurement of $\mathscr{E}_0$ which has been checked against measurements using a conventional Langmuir balance. Because the frequency and wavelength can be independently controlled, it is possible to study rate processes such as connected with surface and bulk diffusion and surface dilitational viscosity.

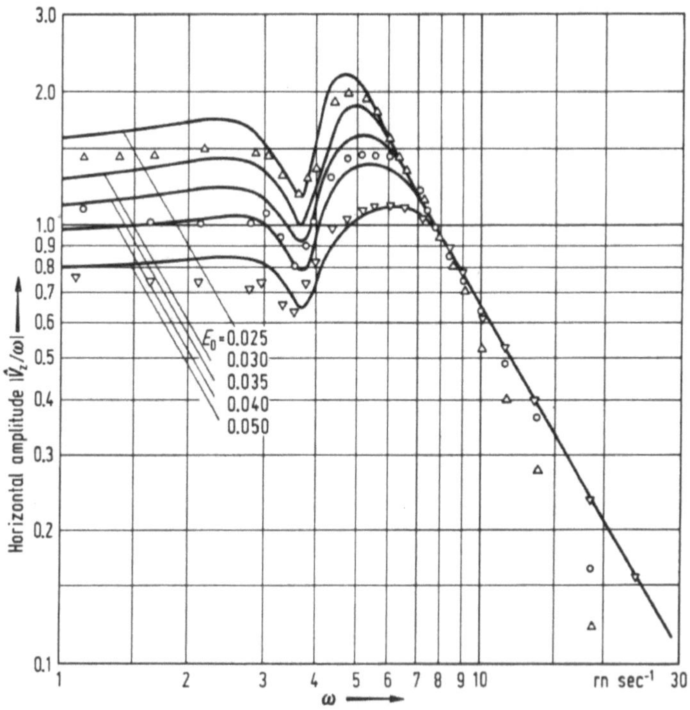

Fig. 10. Frequency response of peak horizontal film displacement with film elasticity E_0 as a parameter. Experimental results are for oleic acid film with independently measured elasticity coefficients $E_0 = 0.026$, △△△; $E_0 = 0.035-0.04$, ○○○ and $E_0 = 0.041-0.052$, × × ×.

By revising the upper electrode so that it consists of the "chess-board" of positively and negatively constrained sections shown in Fig. 11, it is possible to study the film shear viscosity η_{ss} (newt.sec./m) with an apparatus otherwise as in Fig. 9. Because the rows of upper electrodes now induce surface charges that reverse sign as a function of y in spatial phase with the tangential electric field, the electric shearing surface force density has the distribution of strips, essentially uniform in the y direction and a standing-wave in the z direction. The resulting motions of the interface are shown in the figure. To avoid end effects, the system can be closed on itself in the y direction.

The response is determined not only by the film shear stresses (the film shear viscosity) but also by the bulk viscous stresses tending to carry along the substrate. The response is sensitive to the film shear viscosity provided that the frequency is low compared to the critical frequency

$$\omega = 2\eta_{ss}^2 k^4/\varrho\eta. \tag{24}$$

Although motions of this type have been qualitatively observed, the technique has yet to be perfected.

Charge monolayers on interfaces of highly insulating liquids have mechanical attributes similar to those of a monomolecular film. With the water substrate, the fields are for all practical purposes not influenced by the motion so that electromechanical effects can be separated from those due to the film. But, if the liquid is highly insulating, convection of the surface charge plays an essential role in the compressional dynamics of the interface.

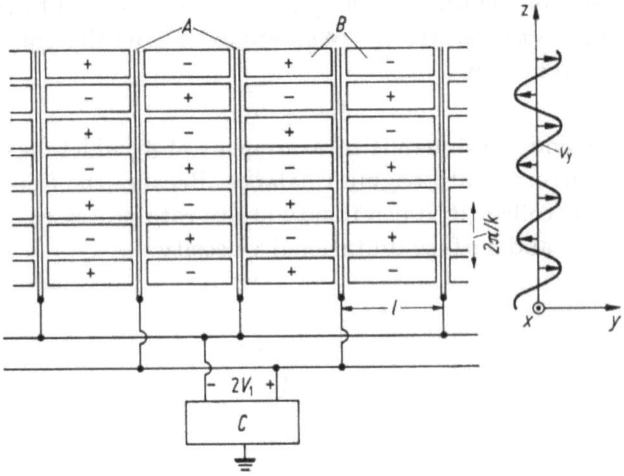

Fig. 11. Top view of electrode structure for adapting apparatus of Fig. 9 to study of shearing motions within plane of interface. Wires (A) are in bottom of trough, as in Fig. 9. Electrodes (B) replace electrode (D) of Fig. 9, and are driven by potential having polarity indicated. Voltage source C is alternating. Interfacial response has the instantaneous form shown, independent of y.

The strong tendency of a monolayer to make an insulating interface behave as an equipotential has been recognized in regard to incipience of normal field instability. With a monolayer even highly insulating liquids display incipient instability predicted by Eq. (22) rather than Eq. (16) [73]. Large effects are also found for tangential "self-field" interactions [74]. But stability studies give relatively little insight concerning the detailed nature of what is new about the monolayer coupling: the existence of electromechanical compressional modes on the interface.

In the simplest possible models the equilibrium monolayer is composed of a single species of charged particles held at the interface by attraction toward images on an electrode above as sketched in Fig. 12. The layer has a finite thickness because of self-diffusion into the liquid volume. If the current at the interface vanishes, and the gap field E_0 is completely terminated

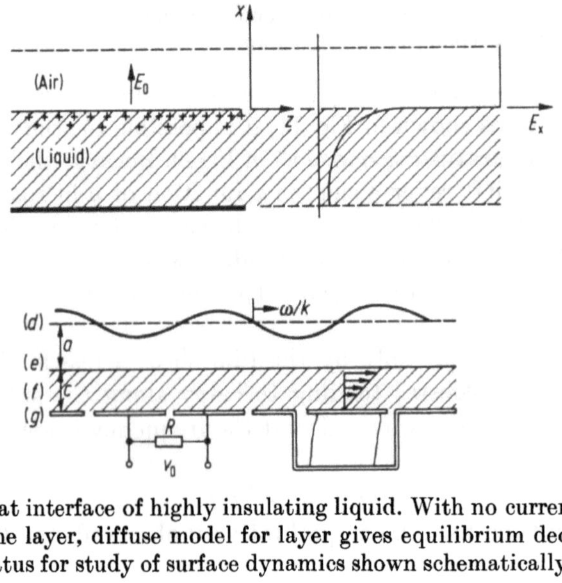

Fig. 12. Charge monolayer at interface of highly insulating liquid. With no current through interface and all of field E_0 terminated in the layer, diffuse model for layer gives equilibrium decay of E_x over distances on the order of 10^{-7} m. Apparatus for study of surface dynamics shown schematically in cross-section at bottom.

by the monolayer, then Eqs. (2), (4) and (5) show that the equilibrium layer thickness is typified by the length $2(RT/zF)/E_0(\varepsilon_0/\varepsilon)$, where the diffusion coefficient has been related to the mobility by the Einstein relation. Thus, at room temperature where $RT/F \simeq 25$ mv and with a field $E_0 = 10^6$ v/m, the thickness is on the order of 10^{-7} m. It is easy to criticize this model. For example, the possibility exists that portions of the charge layer are compact.

The cross-sectional schematic of an experiment for studying the dynamical nature of monolayers is shown at the bottom in Fig. 12. Instead of a standing-wave, a travelling-wave is imposed by means of electrodes just above the interface. These are also used to terminate E_0. Because it is extremely difficult to unambiguously measure motions of insulating liquids in electric fields, they are inferred from measurements designed to leave the fields and flow undisturbed. The oscillatory component of the response is measured by means of the electrodes shown in Fig. 12 embedded in the bottom. For hexane a typical voltage per unit frequency, measured as a function of frequency, is shown in Fig. 13. Surface charge density is the parameter in the measurements and in the three theoretical curves. The latter are based on the model of surface charge confined to the interface and capable of moving only by virtue of liquid convection.

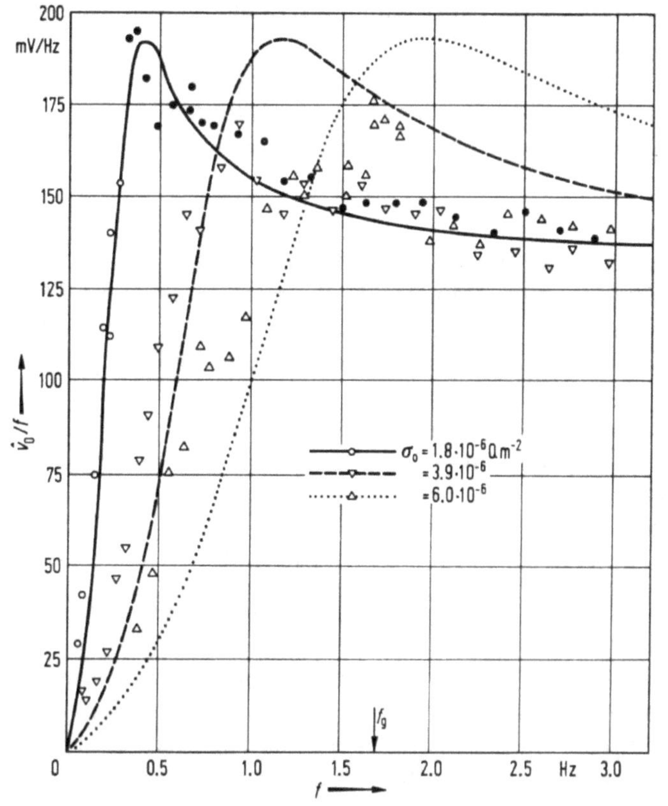

Fig. 13. Frequency response of voltage per unit frequency induced on electrodes at bottom of apparatus in Fig. 12. Resonance is caused by stiffness of interface due to space charge and inertia of bulk coupled to interface through viscous shears.

The salient characteristic of the layer, clearly displayed for the low charge density, is the resonance and low frequency attenuation. At low frequencies, the electromechanically induced redistribution of surface charge attenuates fields that would otherwise pass through the interface. The surface charge also contributes a stiffness to the interface that resonates with the inertia of the liquid below through the viscous coupling, in much the same manner as for the molecular films. The resonance frequency, given by

$$\omega = 1.43 \left\{ \frac{k\sigma_0^2}{\varepsilon_0 \sqrt{\varrho\eta} \left(\coth ka + \dfrac{\varepsilon}{\varepsilon_0} \coth kc \right)} \right\}^{2/3}, \tag{25}$$

is tuned by varying the surface charge density. Although the resonance shift is observed as expected, there is a decrease in the "Q" of the resonance that requires investigation.

A second type of measurement also reflects the interfacial electromechanics without disturbing the flow. There is a time average component of the electrical surface force density acting to produce plane Couette flow. This results because of interactions between the travelling-wave of potential and the travelling-wave of induced surface charge. As would be expected on the basis of similar interactions resulting from charge relaxation to the interface from the volume [75], the time average force density peaks in the neighbourhood of the frequency given by Eq. (25).

Using the "imposed $\omega - k$" configuration, it appears possible to study the properties of the diffuse layer using models that fall within the domain of the next section.

V. Bulk interactions

The diffuse monolayer is atypical of bulk interactions that have received attention. Typically, dimensions are large enough that diffusion currents can be ignored. But, other mechanisms for the accumulation and dissipation of volume free charge are also represented by the right hand sides of Eqs. (6) and (8).

Constant Charge Continuum: Particularly for those prone to think of $\varrho_f \boldsymbol{E}$ as being analogous to $\varrho \boldsymbol{g}$, it is tempting to obtain physical insight by considering the limit of zero mobility, or zero conductivity, as the case may be. Then, if diffusion is also ignorable, Eqs. (6) and (8) both reduce to

$$D\varrho_f/Dt = 0. \tag{26}$$

For unipolar conduction, this limit is justified if times of interest are short compared to the mean $\varepsilon/\varrho_f b$ and velocities are high compared to the mean ion velocity $b\boldsymbol{E}$. In the Ohmic conductor, the mean relaxation time ε/σ must be long compared to times of interest. Prediction of sufficiently high instability growth rates and responses at high frequencies are justifiably made using Eq. (26).

A configuration of fundamental importance is again a liquid layer in static equilibrium between parallel electrodes, this time with a space charge gradient and electric field perpendicular to the plates [47, 76, 77]. The principle of exchange of stabilities holds, and a sufficient condition for stability is

$$gD\varrho - ED\varrho_f < 0 \tag{27}$$

which is a condition expected because of the analogy to the bulk Rayleigh-Taylor stability problem.

Stability of even linear bulk coupled systems is usually complicated by the need for dealing with differential equations having space varying coefficients. Regions of uniform charge density separated by "step" discontinuities is one model that approximately represents a distribution of charge density. Because the electric force density within the homogeneous regions is then $\varrho_f \boldsymbol{E} \to -\nabla(\varrho_f \Phi)$, a redefinition of the pressure "lumps" the coupling at the interfaces. The result is a surface coupled representation of the dynamics, much as in Sections III and IV. Early work applied this model in nuclear physics [78] and meteorology [79]. A methodical approach to handling the coupling through the boundary conditions is useful in obtaining high frequency and growth rate responses [80].

The basic mechanism for instability is evident from the electric shear-stress that must accompany the motion of a charged sheet of liquid in the direction of the equilibrium field. Suppose $\boldsymbol{E} = \boldsymbol{i}_z E_z > 0$, $\varrho_f > 0$ and $D\varrho_f < 0$. Then, by Gauss' law, translation of a liquid layer (in the plane $x = 0$) in the z direction must result in x components of electric field directed outward from the layer. Hence, a translation in the positive z direction results in a shear-stress $\varepsilon E_x E_z$ tending to carry the sheet in the same direction, tending to promote instability.

Because a spatially varying equilibrium charge distribution is required to cause instability, it is not easy to demonstrate the basic interactions in terms of a simple "back of the envelope" calculation. But the same electric shear-stress mechanism for inducing instability is inherent in the constant potential continuum.

Constant Potential Continuum: An Instantaneous Relaxation Model: Especially in systems having sufficiently large electric stresses and charge densities to be of electromechanical interest, the free charge distribution is often determined in a very short time compared to times characterizing the motions and excitations. Then, the charge distribution is given by Eqs. (6) or (8) with $\partial(\ )/\partial t = 0$ and $\boldsymbol{v} = 0$ in what is termed the "instantaneous relaxation limit".

An appealing model that has considerable value in understanding how volume instabilities are created, and in identifying essential parameters, represents the potential of each fluid element as constant:

$$D\Phi/Dt = 0. \tag{28}$$

This statement does not follow in any simple way from the conduction laws, because it pertains to a rather special sort of continuum. As suggested by Fig. 14a, suppose that a static equilibrium exists in which a fluid which is homogeneous in all mechanical regards has regions of relatively high conductivity that are "sheet-like" and separated by insulating layers. Then, an initially uniform electric field E_0, applied co-linear with the periodic conductivity gradient, results in little current flow. Continuum deformations of the fluid carry along the highly conducting layers in such a way that on a macroscale, each element tends to retain its initial potential.

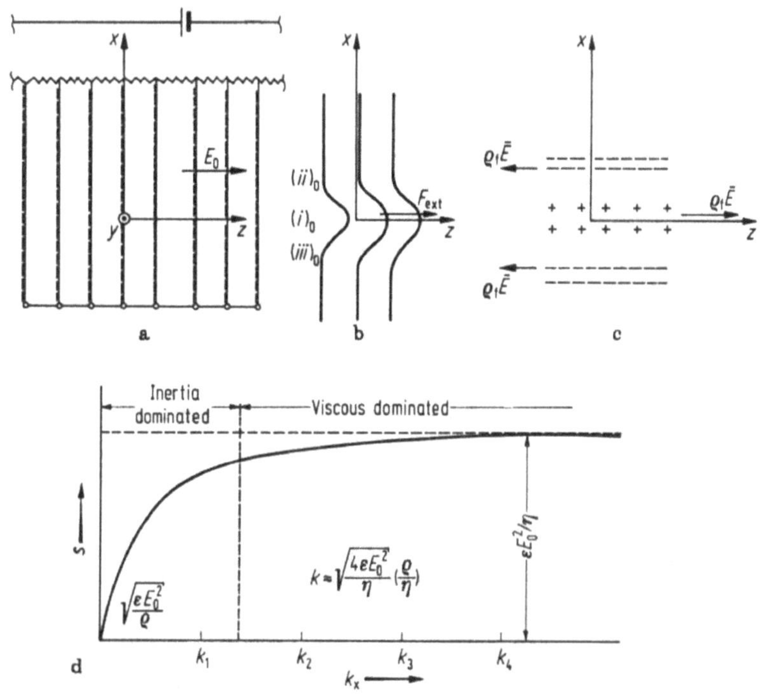

Fig. 14. Potential conserving fluid consisting of numerous highly conducting layers buffered by relatively insulating layers. On a macroscale, a given fluid region retains its potential as it deforms. (b) Shearing displacement causing elevation of potential in plane (i) relative to that in planes (ii) and (iii). (c) Charge density implied by potential conservation, showing electrical force induced in adjacent layers by the motion. (d) Dependence of shear stress instability rate of growth s on wavenumber k_x.

It follows from the solenoidal character of $\boldsymbol{v}$ and the irrotational nature of $\boldsymbol{E}$ that for z directed motions depending only on x, v_x remains zero and E_z remains E_0, even as the fluid deforms. The z component of the fluid force equation can then be written as

$$\partial^2\Phi/\partial t^2 = -(\varepsilon E_0^2/\varrho)\,\partial^2\Phi/\partial x^2 + (\eta/\varrho)\,\partial^2\Phi/\partial x\,\partial t \tag{29}$$

by making use of Eq. (28) to relate v_z and Φ, and expressing the electrical force density in terms of the stress tensor, Eq. (12).

The electrohydrodynamics represented by Eq. (29) is the antithesis of the magnetohydrodynamics found when a uniform magnetic field H_0 permeates a perfectly conducting fluid.

Instead of a constant magnetic flux linked by contours of fixed identity, it is the local potential that is constant, and instead of Alfvén waves caused by magnetic shear-stress and characterized by the squared velocity uH_0^2/ϱ, instabilities are characterized by $-\varepsilon E_0^2/\varrho$. Essential to the instabilities more than the waves is the viscous damping represented by the last term in Eq. (29).

To see why the instability occurs, diagrams similar to Alfvén's are helpful [81]. Suppose that a midplane sheet is displaced to the right, as shown in Fig. 14b. The equipotential lines then appear as shown, and currents exist such that charges accumulate to retain the net field normal to the equipotentials. Thus, volume charges are induced as in Fig. 14c. The midplane layer is thus encouraged to translate even more, while the adjacent layers are set into motion in the opposite directions. As in the constant charge continuum, the electric shear stress is the essence of the instability. In the constant charge continuum this shear stress may or may not encourage further motion, depending on the sign of the equilibrium charge gradient. Here, E_x is induced by E_0 itself, and so the shear stress always tends to produce instability.

For solutions of the form $R_e \hat{\Phi} \exp(st - jk_x x)$, the growth rate s has the dependence on wavenumber sketched in Fig. 14d. The highest growth rate is at the shortest wavelengths, where inertial effects are dominated by viscosity.

To appreciate bulk interactions, it is important to recognize that the growth rate $\varepsilon E^2/\eta$ can be significant even at relatively modest field strengths. For example, distilled water supporting the postulated periodic conductivity striations (perhaps by dint of ions concentrated in layers) and subjected to 100 volts between electrodes having a 1 cm spacing would have a growth rate of $s \sim 70$ secs^{-1}. As should be expected, instabilities on interfaces between insulating miscible liquids are also characterized by the growth rate $\varepsilon E^2/\eta$. Instability in this context implies an easily demonstrated mechanism for field induced mixing.

Space-Charge Unipolar Coupling: This is the most popular category of electrohydrodynamic interactions. For example, most work in "EGD power generation", where fluid convection of charge is used to generate high voltage and some electrical power, is based on a unipolar space-charge model. The carriers can be ions, but for low mobility and hence high efficiency charged liquid or solid particles are preferable. The distribution of free charge in this category of coupling is determined by the first term on the right of Eq. (6), to the exclusion of the others. For a description of work on generators using gases as the working fluid recent reviews should be consulted [4, 82]. With the objective of making both pumps and generators, liquids have also received attention [83, 84, 85]. In all of these applications it is clear that progress depends on recognizing basic mechanisms by which fields tend to induce instability and often turbulence. In liquids incentive also comes from the effect of instability on conduction in highly insulating [9] as well as relatively conducting [15, 16] fluids, for permanent imaging in thermoplastics [5] and for dynamic imaging in liquid crystals [6, 86].

That a rational view could be developed for dealing with bulk instabilities induced in a highly insulating liquid has been demonstrated by *Watson* and *Schneider* [87, 88]. Charge injected at an upper "free" surface gives rise to space-charge limited conduction and an associated charge density that decays monotonically through the layer.

In Eq. (27), $ED\varrho_f < 0$, so that the layer tends to be unstable. Because the mass density is uniform, there is only the electrical relaxation to provide a stabilization mechanism. That is, the convection of the equilibrium charge is in part cancelled by conduction, which tends to retain the original distribution. With the implicit understanding that conduction is space-charge limited so that voltage V_0 across a layer having thickness L is related to the volume charge distribution, the growth rate takes the form

$$s = C\,(\varepsilon E^2 - \varepsilon E_c^2)/\eta; \quad E \equiv V_0/L; \quad E_c \equiv V_c/L, \tag{30}$$

where the constant $C \sim 0.06$. The key role of the same growth rate $\varepsilon E^2/\eta$ found with the potential continuum is further emphasized by the finding that instability is incipient as

$$\left\{\frac{\varepsilon (V_c/L)^2}{\eta}\right\} \Big/ \left(\frac{b V_c}{L^2}\right) = \frac{\varepsilon V_c}{\eta b} = 99 \tag{31}$$

which is the ratio of growth rate $\varepsilon E^2/\eta$ to the effective reciprocal transit-time for an ion.

Control of conduction in polar liquids and rationalization of electrically induced turbulence are two of the most challenging problems in bulk electrohydrodynamics. Steady progress in both of these areas is evident from work reported by *Felici* and his colleagues [15, 16]. Advances have come through control of both the "ionic purity" of polar liquids and injection of charge by means of ion exchange membranes [15]. *Atten* and *Moreau* [89] use a model similar to that just described to study critical conditions for instability, not only under space-charge limited conditions but with weak injection as well. They also identify the condition for incipient instability in terms of the parameter given by Eq. (31) and connect the prediction of the remarkably low voltage of 16 volts for instability in nitrobenzene with experiments [90].

In more recently reported experiments [91], *Hopfinger* and *Gosse* describe the transient by which a state of uniform turbulence is created between parallel electrodes subsequent to application of a field and charge injection from one plate. A front, behind which the liquid is turbulent, propagates from one plate to the other at a velocity shown to be controlled by electrohydrodynamic processes rather than mobility. The time required for the front to establish itself is the reciprocal of the familiar growth rate $\varepsilon E^2/\eta$ [91, 92].

Electrothermally Induced Conductivity Gradient Coupling: For common commercial liquids, like corn-oil and chocolate, a significant and reasonably reproducible temperature dependence on an Ohmic conductivity is observed. A number of studies support the use of a thermally induced conductivity gradient model. Space-charge generation is thus due to the third term of the right in Eq. (8). Steady pumping, using $d-c$ fields imposed by means of electrodes contacting the thermally stressed liquids, is an example [93]. Using the phase-shifts caused by finite relaxation time effects, it has also been shown that a travelling-wave of potential can induce charge in the bulk, and hence a time average force density reasonably explained by the conductivity gradient model [94]. But, organized bulk motions can only be created if there is some additional mechanism for stabilizing the fluid. Typically, as suggested by Eq. (27), thermally induced "stabilizing" density gradients and gravity serve this purpose. Applications of too large a field result in instability [94].

Turnbull has studied electrothermally induced instability of a static equilibrium of semi-insulating oils between a pair of horizontal electrodes [95, 96]. These both impose a vertical thermal flux (hot over cold) and electric field. Thus gravity stabilizes, and (on the basis of an instantaneous relaxation model) the field is found to induce overstability at a threshold given by

$$s/\omega_g \approx 8. \tag{32}$$

Here, $s = (\varepsilon E^2/\eta)\,(\Delta\sigma/\sigma)$, or the typical growth rate adjusted by the fractional change in conductivity between the plates, and ω_g is the frequency of the short-wavelength internal gravity waves that would exist without the field. More recent work considers similar instabilities in boundary layers [97].

VI. Double-layer interactions

Eletrocapillary phenomena fall in this category. Typically, transduction is confined to interfaces between liquids that by electrohydrodynamic standards are highly conducting, for example, an electrolyte and mercury. In the bulk, conduction is well represented by a uniform Ohmic conductivity. To appreciate why the surface coupling is the result of the last two "internal-stress" terms of Eq. (14) rather than the first "external-stress" term, consider a typical situation. Because even the electrolyte is highly conducting, external (bulk) fields are seldom more than 10^2 v/m. Hence, the "external-stress" is typically $1/2\varepsilon E^2 \sim 10^{-5}$ newt./m^2. The shearing component of the double-layer surface force density, the last term in Eq. (14), is typified by recognizing that variations in γ_E are on the order of 10% of the surface tension, and if that variation occurs over a distance 1 cm then $\nabla_\Sigma \gamma_E \sim (0.1)\,(0.4\text{newt./m})/10^{-2} = 4$. Thus, the contribution of the "internal-stresses" far exceeds that of the "external-stresses".

Because the electric field is irrotational, the discontinuity in tangential electric field is related to the jump in potential across the double-layer by

$$[\![E_t]\!] = -\nabla_\Sigma [\![\Phi]\!]. \tag{33}$$

Mercury and an electrolyte are an example where, because one fluid is much more highly conducting than the other, the assumption is justified that the potential on one side is zero, and hence so also is E_t on that side.

There are three consistent ways of viewing the shearing component of the double-layer surface force density: first as an extension of surface thermo-dynamics including the fact that γ is a function of double-layer potential difference; second, in terms of electric stresses, as already discussed; and third, in terms of fields and charges. The first is the pragmatic approach ultimately underlying most macroscopic formulations, and the second gives considerable insight about scaling the electromechanical transduction. To pursuse the last, consider a clean interface at which $\gamma_E = \gamma_E(\vartheta_d)$, where $\vartheta_d \equiv [\![\Phi]\!]$. Then, the surface force density, Eq. (14), can be written as

$$\mathbf{T} = -\boldsymbol{n}\gamma_E \nabla \cdot \boldsymbol{n} - \sigma_d \nabla_\Sigma \vartheta_d; \quad \sigma_d \equiv - \partial\gamma_E/\partial\vartheta_d, \tag{34}$$

where σ_d is the double-layer surface charge density. Writing the shearing component as in Eq. (34) suggests that the positive part of the double-layer charge is subject to a tangential electric field that exceeds that for the negative half. Thus there is a net surface force density. That the tangential field is "external" gives the misleading impression that the last term in Eq. (34) is associated with external stresses. In fact, because $\boldsymbol{E}$ must remain irrotational, the tangential field is implicitly related to the internal field.

The double-layer charge σ_d is by definition a free charge, since it is subject to external control. Conservation of charge for the double-layer then requires that charge can accumulate in a local region either because of conduction normal to the layer through the bulk or because of convection within the plane of the interface. Thus, in a self-consistent picture of the electromechanics, conservation of charge requires that

$$\partial\sigma_d/\partial t = -\sigma(\boldsymbol{E} \cdot \boldsymbol{n}) - \nabla_\Sigma \cdot (\sigma_d \boldsymbol{v}). \tag{35}$$

To appreciate the significance of this boundary condition, note that a typical volume charge relaxation time for the electrolyte is usually less than $\varepsilon/\sigma \sim 10^{-9}/10^{-2} = 10^{-7}$ secs. Yet, rate processes connected with charge storage and convection can play important parts in the dynamics. This is because electric energy storage is primarily within the layer while the dissipation is in the surrounding bulk. For example, according to Eq. (35), a spatially periodic perturbation of σ_d from a uniform equilibrium on a static planar layer would result in a transient characterized by the time constant $C_D/\sigma k$, where C_D is the differential double-layer capacitance per unit area and k is the wavenumber of the initial perturbation. Typically, $C_D \sim 0.25 \mathrm{F/m^2}$, and therefore a typical relaxation time based on $k = 10^3\ \mathrm{m^{-1}}$ is 0.025 secs.

Levich [98] gives a summary of basic research that is particularly useful to those approaching the area from the viewpoint of continuum electromechanics. Migration of a mercury drop through an electrolyte supporting an imposed electric field is an example that aptly illustrates transduction propensities of the double-layer. Integrals of Eq. (34) over a closed surface, such as that of a drop, can give no net force. Yet, a mercury drop falling through an electrolyte is readily deflected by an imposed electric field on the order of only 10^2 v/m.

The swimming mechanism that makes it possible for the double-layer to cause a net translation but no net electric force is also illustrated by the configuration of Fig. 15. In what might be termed Levich's drop "turned inside out" the electrical excitation takes the form of a dipole current source I at the center of a semi-conducting (telectrolytic) region (b). The surrounding region (a) is filled by a second liquid separated from the drop by an ideally polarized double-layer. Using a low-Reynolds number approximation, it can be shown that the drop propels itself through the surrounding fluid with the velocity

$$U = \frac{Id}{4\pi R^2 \sqrt{\sigma_b\left(\eta_a + \dfrac{3}{2}\eta_b\right)}} \frac{H_e}{(1 + H_e^2)}; \quad H_e \equiv \frac{\sigma_0}{\sqrt{\sigma_b\left(\eta_a + \dfrac{3}{2}\eta_b\right)}}. \tag{36}$$

Here, σ_0 = average σ_d, Id is the moment of the dipole current source, R is the drop radius and to make a connection with the shear-stress electrohydrodynamics of monolayers [63], the parameter H_e might be termed an electric Hartmann number. The ratio of that part of the electric surface force density caused by the motion itself to the viscous surface force density is H_e. Too much double-layer surface charge and fields induced by the convection counteract those induced by the driving current to limit the motion.

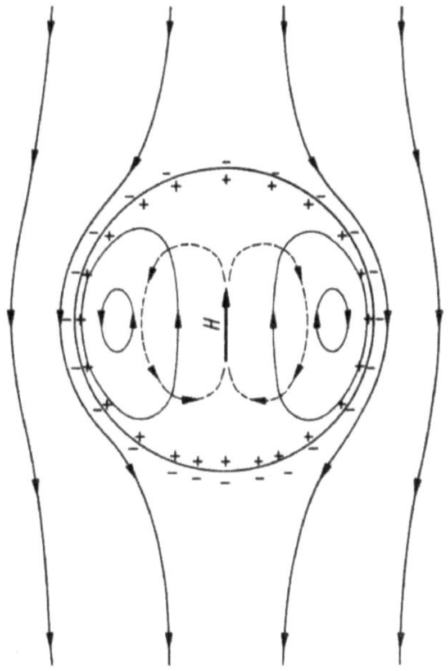

Fig. 15. Cross-sectional view of semi-conducting drop separated from surrounding conducting liquid by ideally-polarized double-layer. Doublet current source I at center of drop drives current within drop in such a way as to produce swimming in whichever direction the source is pointed. Flow lines are shown as viewed from frame fixed with respect to drop. There are no external electric fields far from the drop.

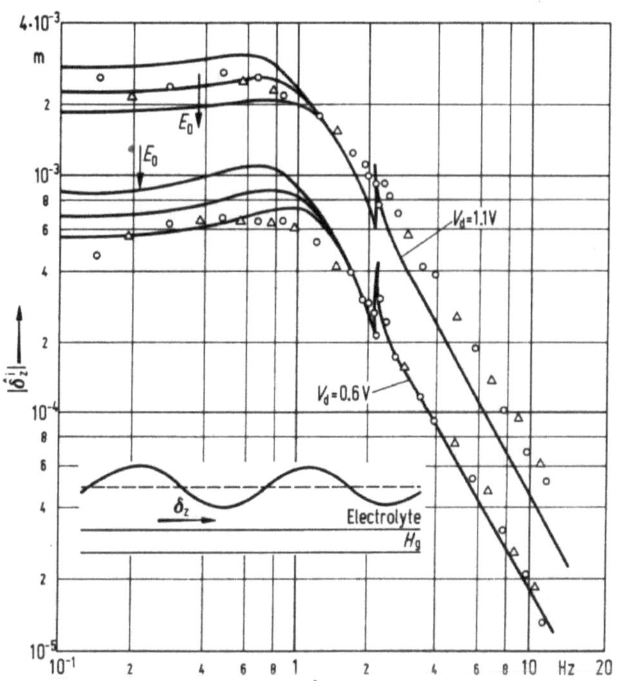

Fig. 16. Horizontal displacement frequency response for double-layer system shown schematically in inset. The frequency and wavenumber are imposed by means of the standing wave of current injected at the top into the electrolyte (.01 Na_2Co_3). Experimental and theoretical results are shown for two different bias voltages. In each case, the three solid lines are based on film elasticities of 0.02, 0.025 and 0.03 newt./m.

That the entity represented by the drop can propel itself in whatever direction the dipole points (the source might be of an electrochemical nature), suggests why double-layer electromechanics is likely to be of increasing interest in the biological sciences.

Finally [99], consider the double-layer configuration shown by the insert to Fig. 16. This "imposed $\omega - k$" configuration is reminiscent of that portrayed in Figs. 9 and 12. But by contrast with those studies, typical voltages are now volts rather than kilovolts. A planar mercury layer is covered by electrolyte. Platinum electrodes are used to impose a standing-wave of current in a plane of the electrolyte parallel to the interface. The result is an induced standing-wave of shear force density. A typical horizontal displacement response to the electrical excitation is shown as a function of frequency in Fig. 16. The solid curves are theoretical. The response illustrates the now familiar film resonance associated with an interfacial film. The niche is caused by coupling to the gravity-capillary wave system.

The objective of such experiments is to develop a technique for studying complicated rate dependent electromechanical processes under conditions where theoretical models can be derived that take full account of inertia, viscosity, conduction and diffusion in the bulk and of the complex electrochemical properties of the interface.

Acknowledgment: This paper was written while the author was a guest of the Department of Applied Mathematics and Theoretical Physics at Cambridge University. Experimental work on monolayer dynamics at highly insulating interfaces was carried out at the Cavandish Laboratory with the encouragement of *G. I. Taylor*.

References

1. White, H. J.: Industrial Electrostatic Precipitation, Reading, Mass.: Addison-Wesley 1963.
2. Waterman, L. C.: Chemical Engineering Progress 61 (1965) 51.
3. Ransburg, E. M.: U.S. Patent No. 2,658,009 (1953).
4. Melcher, J. R.: J. Fluid Mech. 40 (1970) 641.
5. Cressman, P. J.: J. Appl. Phys. 34 (1963) 2327.
6. Hass, W., Adams, J.: Appl. Optics 7 (1968) 1203.
7. Melcher, J. R., Sachar, K. S.: Charged Droplet Technology for Removal of Particulates from Industrial Gases. Final report under Task No. 8, Contract No. 68-002-0018, Air Pollution Control Office, Durham, N.C.
8. Gunn, R.: Proc. of I.R.E. (Oct. 1957) 1331.
9. Watson, P. K.: National Academy of Sciences-National Research Council Publication No. 1356 (1966) 53.
10. Woodson, H. H., Melcher, J. R.: Electromechanical Dynamics, Part 1, New York: Wiley 1968.
11. Panofsky, W. K. H., Phillips, M.: Classical Electricity and Magnetism, Reading, Mass.: Addison-Wesley 1956).
12. Whipple, F. J. W., Chalmers, J. A.: Quart. J. of Royal Met. Soc. 70 (1944) 103.
13. ibid (10), part II, 370.
14. Levich, V. G.: Physicochemical Hydrodynamics, Englewood Cliffs, N.J.: Prentice-Hall 1962, 280.
15. Felici, N. J.: Direct Current 2 (1972) 90.
16. Felici, N. J.: Direct Current 2 (1972) 147.
17. ibid. (11), 16.
18. ibid (11), 92—104.
19. Stratton, J. A.: Electromagnetic Theory, Chap. II, New York: McGraw-Hill 1941.
20. Penfield, P., Haus, H. A.: Electrodynamics of Moving Media, Chap. 4, Cambridge, Mass.: MIT Press 1967.
21. Zelazo, R. E., Melcher, J. R.: J. Fluid Mech. 39 (1969) 1.
22. Pohl, H. A.: Scientific American 203 (1960) 107.
23. Rosensweig, R. E.: Internat. Sci. Tech. 55 (1966) 48.
24. Pellat, M. H.: Comptes Rendus 123 (1896) 691.
25. Jones, T. B.: Dynamics of Electromechanical Flow Structures, Ph. D. Thesis, Dept. Elect. Eng., MIT Cambridge, Mass.: 1970.
26. Melcher, J. R.: Electric Fields and Moving Media, Film produced for National Committee on Electrical Engineering Films by Education Development Center 55 Chapel St. Newton, Mass 02160.
27. Blackmon, Jr. J. B.: Spacecraft Rockets 2 (1965) 391.
28. Calvert, R. T., Melcher, J. R.: J. Fluid Mech. 38 (1969) 721.
29. Melcher, J. R.: Field-Coupled Surface Waves, Cambridge, Mass.: MIT Press 1963.
30. Devitt, E. B., Melcher, J. R.: Phys. Fluids 8 (1965) 1193.
31. Cowley, M. D., Rosensweig, R. E.: J. Fluid Mech. 30 (1967) 671.
32. Gailitis, A.: Magnetohydrodynamics 5 (1969) 44.
33. Melcher, J. R., Guttman, D. S., Hurwitz, M.: J. Spacecraft and Rockets 6 (1969) 25.
34. Melcher, J. R., Hurwitz, M.: J. Spacecraft and Rockets 4 (1967) 864.
35. Nayyer, N. K., Murty, G. S.: Proc. Phys. Soc., London 75 (1960) 369.
36. Reynolds, J. M.: Phys. Fluids 8 (1965) 161.
37. Melcher, J. R., Hurwitz, M., Fax, R. G.: J. Spacecraft and Rockets 6 (1969) 961.

38. Jones, T. B.: The Feasibility of Electrohydrodynamic Heat Pipes, Rep. No. 1 on NASA Grant NGR-06-002-127, Dept. Elect. Eng. Colorado State Univ., Fort Collins, Colorado.
39. Jones, T. B., Perry, M. P., Melcher, J. R.: Science 174 (17 Dec. 1971) 1232.
40. Chang, C. T.: Phys. Fluids 2 (1959) 656.
41. Nesterov, S. V.,: J. Moscow University 1 (1964) 56.
42. Magd, A. A.: Some Problems in Electrohydrodynamics (Ph. D. Thesis, Univ. Delhi, Dept. Math., Delhi, India 1971.
43. Wolf, G. H.: Z. Physik 227 (1969) 291.
44. See many papers in Proc. of the American Institute of Physics Conference, Feedback and Dynamic Control of Plasmas, Pub. by A.I.P., New York 1970.
45. Johnson, R. L.: AIAA J. 6 (1968) 1456.
46. Velkoff, H. R., Miller, J. H.: J. of Heat Transfer 87, Series C (1965).
47. Turnbull, R. J., Melcher, J. R.: Phys. Fluids 12 (1969) 1160.
48. Turnbull, R. J.: Phys. Fluids 12 (1969) 1809.
49. Savkar, S. D.:Phys. Fluids 14 (1971) 2670.
50. Lord Rayleigh: Phil. Magazine 14 (1882) 184.
51. ibid. (29), 64.
52. Taylor, G. I., McEwan, A. D.: J. Fluid Mech. 22 (1965) 1.
53. Larmor, J.: Proc. Cambridge Phil. Soc. 7 (1890) 69.
54. Melcher, J. R., Warren, E. P.: Phys. Fluids 9 (1966) 2085.
55. Curzon, F. L., Ionides, G. N.: Canadian J. Phys. 49 (1971) 458.
56. Abbas, M. A., Latham, J.: J. Fluid Mech. 30 (1967) 663.
57. Basset, A. B.: American J. Math. 16 (1894) 93.
58. Michael, D. H.: Canadian J. Physics 45 (1967) 3179.
59. Melcher, J. R.: Complex Waves II, Film, produced for the National Committee on Electrical Engineering Films by the Education Development Center, 55 Chapel St., Newton, Mass. 02160.
60. Taylor, G. I.: Proc. Roy. Soc. A 280 (1964) 383.
61. Latham, J., Rosburgh, I. W.: Proc. Royal Soc. A 295 (1966) 84.
62. Brazier-Smith, P. R., Latham, J.: Nature 220 (1968) 689.
63. Melcher, J. R., Taylor, G. I.: Electrohydrodynamics: A Review of the Role of Interfacial Shear Stresses, Chap. in: Annual Review of Fluid Mechanics, 1, Palo Alto, Calif.: Annual Reviews Inc. 1969.
64. Taylor, G. I.: Proc. Royal Soc., London A 313 (1969) 453.
65. Hendricks, C. D., Carson, R. S., Hogan, J. J., Schneider, J. M.: AIAA J. 2 (1964) 733.
66. Melcher, J. R., Warren, E. P.: J. Fluid Mech. 47 (1971) 127.
67. Saville, D. A.: Phys. Fluids 13 (1970) 2987.
68. Taylor, E. R., Chartier, V. L., Rice, D. N.: IEEE Trans. on Power Apparatus and Systems, PAS-88 (1969) 666.
69. Hoburg, J. F.: Modes of Behavior of Water on High Voltage Transmission Lines, M. S. Thesis, Dept. Elect. Eng., M.I.T., Cambridge, Mass., 1971.
70. Peek Jr., F. W.: Dielectric Phenomena in High-Voltage Engineering, 3rd. ed., New York: McGraw-Hill 1929, p. 77.
71. Loeb, L. B.: Electrical Coronas, Their Basic Physical Mechanisms. Berkeley, Calif.: Univer-ofCalifornia Press 1965, pp. 248–266.
72. Zelazo, R. E.: Dynamic Interactions of Monomolecular Films with Imposed Electric Fields Ph. D. Thesis, Dept. of Elect. Eng., M.I.T., Cambridge, Mass., 1971.
73. Melcher, J. R., Smith Jr., C. V.: Phys. Fluids 12 (1969) 778.
74. Melcher, J. R., Schwartz Jr., W. J.: Phys. Fluids 11 (1968) 2604.
75. Melcher, J. R.: Phys. Fluids 9 (1966) 1548.
76. Barston, E. M.: Phys. Fluids 13 (1971) 2876.
77. Turnbull, R. J., Melcher, J. R.: Phys. Fluids 13 (1971) 2879.
78. Bohr, N., Wheeler, J. A.: Phys. Rev. 56 (1939) 426.
79. Colgate, S. A.: J. Geophys. Res. 72 (1967) 479.
80. Zahn, M., Melcher, J. R.: Phys. Fluids 15 (1972) 1197.
81. Alfvén, H., Fälthammar, C. G.: Cosmical Electrodynamics. Oxford, England: Oxford and the Clarendon Press 1963, 72.
82. Lawson, M., von Ohain, H.: J. of Eng. for Industry, Paper No. 70-Ener-A, 1, 1970.
83. Stuetzer, O. M.: J. Appl. Phys. 31 (1960) 136.
84. Pickard, W. F.: The Electrical Pumping of Dielectric Liquids, Cruft Laboratory Report, No. 350, 1961.
85. Secker, P. E.: Electrostatic Generation of Electricity, Spectrum No. 47, 2–4, U.K. Gov. Central Office of Information, 1968.
86. Heilmeier, G. H., Zanoni, L. A., Barton, L. A.: IEEE Trans. 17 (1970) 22.
87. Schneider, J. M.,Watson, P. K.: Phys. Fluids 13 (1970) 1948.
88. Watson, P. K., Schneider, J. M., Till, J. R.: Phys. Fluids 13 (1970) 1955.
89. Atten, P., Moreau, R.: Compt. Rend. Acad. Sci. Paris 270 A (1970) 415.
90. Filippini, J. C., Gosse, J. P., Lacroix, J. C., Tobazeon, R.: Compt. Rend. Acad. Sci. Paris 269 B (1969) 167.
91. Hopfinger, E. J., Gosse, J. P.: Phys. Fluids 14 (1971) 1671.
92. Felici, N.: Rev. Gen. Elect. 78 (1969) 717.
93. Wong, J., Melcher, J. R.: Phys. Fluids 12 (1969) 2264.
94. Melcher, J. R., Firebaugh, M. S.: Phys. Fluids 10 (1967) 1178.
95. Turnbull, R. J.,: Phys. Fluids 11 (1968) 2588.
96. Turnbull, R. J.: Phys. Fluids 11 (1968) 2597.
97. Turnbull, R. J.: J. Fluid Mech. 47 (1971) 231.
98. ibid. (14), 472–590.
99. Zuercher, J. C.: Electromechanics of Fluid Double-Layer Systems, Ph. D. Thesis, Dept. of Elect. Eng., M.I.T., Cambridge, Mass., 1971,

A review of optimal structural design

F. I. Niordson P. Pedersen

Lyngby, Denmark

Introduction

Structural design problems are concerned with the transmission of forces and moments in space. The engineer usually selects his design from a subset of the total class of possible solutions according to the value of some *objective functional* defined on the subset (for instance mass or, more generally, cost).

This subset is usually arbitrarily confined to a type of structure (for instance an I-beam or a truss). The subset is further restricted to a *feasible subspace* by certain conditions regarding the behaviour of the structure, usually in response to applied loads. The term "optimal design" will have a meaning when the objective functional on the feasible subspace has a minimum.

We intend to present here an overall view of the results achieved, on the one hand, with numerical methods of mathematical programming, and on the other, with analytical methods, which attempt to establish necessary and sufficient conditions for the optimal structure. These are the two fields in which work on optimal structural design has been and is still being concentrated.

There are already about thousand papers on optimal structural design and new ones are appearing at the rate of fifty to a hundred a year. Several reviews are available, some of which contain extensive reference lists (*Wasiutyński* & *Brandt* [1], *Sheu* & *Prager* [2], *Pope* & *Schmit* [3]). Further surveys are given by *Barnett* [4], *Gerard* [5], *Save* [6], *Prager* [7, 8] and others.

The first two sections of our paper deal with the delimitation of the feasible subspace and list the most important factors. The third and fourth sections are devoted to the mathematical formulation and numerical solution of optimal design problems, and the following sections are devoted to solutions to problems. The review is mainly based on recent papers and no attention is paid to the historical development. The last section discusses new trends and indicates areas for future work.

1. Design space and subspace

It is convenient to conceive a structure as a point in an abstract *design space*. In this space the coordinates of a point define a structure by specifying all geometrical dimensions and all material constants. These specifications, which will be called the *parameters* of the structure, may be given as real numbers, functions, or multi-variables (sets of real numbers).

For a fuller understanding of the design space and its characterization, we shall list the parameters that would actually be used by a designer to specify a structure. First, there is a group of geometrical parameters:

Cross-sectional shape of rod, column, or beam.

Tapering function of rod, column, or beam.

Thickness function of disc, plate, or shell.

Boundary shape of disc, plate, or shell,

Spatial position of joints of truss or frame.
Configuration of elements of composite structures.
Shape of middle surface of shell.

Secondly, the material of the structure is characterized by:

Elastic moduli.
Mass density.
Coefficients of thermal expansion and conductivity.
Coefficients of constitutive laws governing plastic deformation, visco-elasticity, and creep.
Yield stress and ultimate stress.
Material constants of fatigue life.

Finally, the state of prestressing of a structure may also be regarded as a design parameter.

This list is certainly not complete, but it does include all commonly used parameters.

Now, a properly posed optimization problem is always restricted to a *design subspace*. Thus, the designer selects an arbitrary type of structure; he decides beforehand if he wants, say, an I-beam or a truss. This selection will generally involve bounds on the design parameters, for instance, the maximum height of the I-beam. Although not strictly necessary, the designer usually also decides in advance what material to use and thus restricts his design space still further. Material constants may well be left among the *design variables* to be determined in the optimization process, and though very little work has been done in this field, it may well furnish a rewarding area for research in the future provided material data is available in a suitable form.

2. Feasible subspace

The design subspace contains structures that do not fulfil a number of requirements necessary for functional acceptability of the structure under loading. Conditions of sufficient strength, lifetime, stiffness etc., in general, *bounds on response to the loads*, may be conceived as constraints that further divide the design subspace into a *feasible subspace* and a *non-feasible subspace*.

Among conceivable constraints, the following are commonly encountered:

Maximum stress.
Maximum displacement.
Minimum factor of safety against buckling or plastic collapse.
Maximum imperfection sensitivity.
Minimum fundamental frequency.
Maximum strain rate in stationary creep.
Minimum fatigue life.

It may be noted that different modes of failure are taken into account through the appropriate constraints on the design subspace.

Unfortunately, the constraints are usually expressed as bounds on functionals defined on the design subspace and thus delimit this subspace only *implicitly*. This is, perhaps, the complication that causes most difficulties in optimal design problems. A further complication arises if the loads are determined by the design, for instance, in the case of inertia forces. Still further complications follow when random rather than deterministic loads are to be considered, or when the loads are of a dynamic rather than a static nature. Finally, in the case of multiple loading, it may well happen that different constraints apply to different load systems. This is usually the case when testing or overload conditions differ from service conditions.

It is worth bearing in mind that constraints on response bounds are frequently not physical necessities but rather, code regulations. When this is the case, an interesting problem relating to optimal design arises, viz., the sensitivity of the optimal solution to small changes in the design code.

In the mathematical formulation of the problem, the constraints usually appear as inequalities and, of course, we have no a priori knowledge of which are "active" (critical) for the optimal structure.

3. Mathematical formulation

When the design variables have been selected, the problem in optimum design may be formulated as follows:

Find a design S such that

$$f_k(S) = 0 \quad \text{for} \quad k = 1, 2, \ldots, K,$$
$$h_j(S) \leq 0 \quad \text{for} \quad j = 1, 2, \ldots, J, \tag{3.1}$$
$$\Phi(S) \text{ is a minimum}.$$

Here, S is a point in the design space, characterized by the selected variables. In most problems, the conditions on the functionals f_k and h_j are due to the constraints on the response to loads, but some of them may well express given delimitations of the design subspace. The objective functional is denoted by Φ [(1)].

Whether a problem, defined in general terms by (3.1), has a solution or not is an open question that can only in rare instances be answered on the basis of physical intuition, and this also applies to questions concerning the uniqueness of solutions.

From (3.1) it follows that if S is an optimal design, then for any small variation δS within the design subspace

$$\delta f_k(S) = 0 \quad \text{for} \quad k = 1, 2, \ldots, K,$$
$$\delta h_j(S) \leq 0 \quad \text{for all } j \text{ for which} \quad h_j(S) = 0, \tag{3.2}$$
$$\delta\Phi(S) \geq 0.$$

This classical variational formulation is hence a necessary condition for an optimal solution.

The condition may be stated in another, often more useful, form. For the sake of simplicity, let us assume that the design variables are N real numbers [(2)] so that the design space can be interpreted as an N-dimensional Euclidean.

Let S be a feasible design and δS an arbitrary variation within the feasible design subspace. As $f_k(S) = 0$, the variation δS is perpendicular to all vectors $\nabla f_k(S)$ $(k = 1, 2, \ldots, K)$ [(3)]. Similarly, the active inequality constraints $h_j(S) = 0$ imply that δS has no component in the positive direction of $\nabla h_j(S)$.

Hence we may infer that for any real numbers λ_k and any $\gamma_j \geq 0$, the projection of δS on the vector

$$\sum_{k=1}^{K} \lambda_k \nabla f_k(S) + \sum_j{}' \gamma_j \nabla h_j(S) \tag{3.3}$$

is non-positive. ($\sum'$ denotes the fact that the summation is restricted to values of j for which $h_j(S) = 0$). In other words, any direction that has a positive component in any of the directions (3.3) leads into the non-feasible subspace.

In order to decrease the objective functional Φ, it is necessary to move in a direction that has a positive component in the negative direction of $\nabla\Phi$, but if this direction $(-\nabla\Phi)$ is any of the directions (3.3), no move within the feasible subspace will decrease the object functional. Therefore, at the point of optimality, $-\nabla\Phi$ is one of the directions (3.3). Utilizing this fact, we see that

if S is an optimal solution, there exists a set of real numbers λ_k and non-negative numbers $\gamma_j \geq 0$, such that

$$-\nabla\Phi(S) = \sum_{k=1}^{K} \lambda_k \nabla f_k(S) + \sum_j{}' \gamma_j \nabla h_j(S). \tag{3.4}$$

This is called the Kuhn-Tucker condition [9]. We observe that when there are no active inequality constraints, the λ_k's may be interpreted as Lagrangian multipliers. For the fully unconstrained problem, the condition reduces to $\nabla\Phi = 0$.

As all *stationary* solutions fulfil the conditions (3.2) and (3.4), these alone cannot ensure global optimality. Nevertheless, they form the basis for most work on optimal design problems.

[(1)] Problems of maximizing Φ and conditions $h_j \geq 0$ may easily be transformed into the standard form (3.1) by proper changes of sign.

[(2)] A generalization to functions is rather obvious.

[(3)] ∇ is the Nabla-operator.

The use of additional tests will help to ensure the globality of any minimum attained. In particular, if the feasible design space is convex and if the objective functional is either convex or concave everywhere, then some theorems of non-linear programming (cf. Chapter 3 of *Hadley* [10]) can give important information about the globality and/or positions of possible solutions.

It should be realized that problems that are mathematically totally different when formulated in terms of (3.1) may express what is essentially one and the same physical problem. Thus, for instance, to find the tallest possible column of given material and volume (taking buckling under its own weight into account) is essentially the same as finding the minimum volume of a column of given material and height. However, their formulation in terms of (3.1) would be quite different. This fact is not trivial, since here one formulation leads to a more easily soluble problem.

For some very special structural models (as for instance an elastic beam, the stiffness of which is proportional to the mass) with a single or a few design constraints that are characterized by extremum principles of structural theory (minimum potential energy and Rayleigh's principle), the necessary conditions obtained by variational methods can be supplemented by sufficient conditions. Depending on whether the structural minimum principle is of a global or local character, the resulting condition is also sufficient for global or local optimality. Most of the work on establishing sufficient criteria has been carried out by Prager and his associates at Brown University. Section 5 below lists the results obtained.

The application of Pontryagin's maximum principle (cf. ref. [11]) and results from control theory have so far been of limited use in formulating optimal design in mechanics, but this state of affairs is very likely to change soon.

4. Methods of solution

Having discussed the formulation of the mathematical problem, let us devote this section to methods used for solving it i.e. to solving the problem stated in terms of (3.1) or in a variational form. The fact that the formulations are essentially different affects the selection of methods employed for solving the problem. Problems formulated in terms of (3.1) are solved mainly by the use of iteration procedures in which each step is designed to provide a "better" solution[1]. Variational problems on the other hand, lead to sets of differential equations and boundary conditions. Only in very exceptional cases is a solution in closed form known or to be found and we are generally forced to devise schemes for obtaining numerical solutions. Due to the fact that the differential equations are often highly non-linear and have no regular solution (singularities very often appear at the boundaries) such problems may be rather cumbersome and tricky to solve and the whole field seems to be rather unclarified.

At this stage we should like to emphasize an important difference between the variational formulation and the more general formulation (3.1). The variational formulation deals with *the optimal* (or rather, stationary) solution (assuming its existence) while the functionals of (3.1) are readily applied to *any* feasible design. This difference becomes especially noticeable when the optimal solution is singular or non-existent. Roughly speaking, this means that a solution procedure based on (3.1) may yield a "better" design even when there is no "best" design.

The iterative procedures mentioned above are methods of *mathematical programming* (cf. *Fox* [12]). Their common feature is the generation of a sequence of points in the design subspace

$$S_1, S_2, \ldots, S_q, \ldots \tag{4.1}$$

from an arbitrary starting point S_1. The step ΔS_q in $S_{q+1} = S_q + \Delta S_q$ is determined from the gradients of the constraint functionals ∇f_k, ∇h_j, and the objective functional $\nabla \Phi$ at the point S_q. The difference between the various methods lies in the way the gradients and the step ΔS_q are determined.

[1] The procedure may for instance be interrupted when the improvement in objective value becomes sufficiently small.

The simplest case, of linear constraints and linear objective functional, can be solved by the well-known methods of linear programming (cf. *Dantzig* [13], *Hadley* [14] & *Gass* [15]). For linear constraints and certain types of non-linear objective functionals there are the methods of quadratic programming etc. (cf. ref. [10]).

More general cases of non-linear problems may be solved either by *direct* or *indirect* methods. The direct methods are a) the gradient projection method (*Rosen* [16]), b) the method of usable feasible directions (*Zoutendijk* [17, 18]), and c) the SLP method, which employs a *sequence of linear programs* with move limits (*Griffith* & *Stewart* [19]).

In the gradient projection method, the starting point S_1 is selected on the boundary of the feasible subspace[1]. The path followed is then determined by the projection of $\nabla\Phi$ on the boundary. (Of course, the optimal design cannot generally be reached if it is an interior point.) With a non-linear objective functional but linear constraints this method is reported to be very efficient (cf. [12], p. 205).

In the method b) of usable feasible directions, S_1 may be taken either inside or on the boundary of the feasible subspace. The step ΔS_q is determined by

$$\Delta S_q = \alpha_q D_q, \tag{4.2}$$

where D_q is an "optimal" direction that is feasible over a finite distance and α_q is an "optimal" step length. There is a variety of different practices for determing α_q and D_q.

Similarly, the SLP method c) allows stepping *into* the feasible subspace. However, the sequence of linear programs determines "optimal" ΔS_q's directly, i.e., D_q and α_q simultaneously.

Problems formulated in terms of (3.1) may sometimes be advantageously solved *indirectly* using a *sequence of unconstrained minimization techniques.* (The so-called SUMT method, cf. *Fiacco* & *McCormick* [20]). The idea is to replace the inequality constraints by penalty terms, added to the objective functional. The sum is the so-called *penalty function* ψ. If we are to consider a sequence of feasible designs, a common form is the so-called *interior* penalty function

$$\psi_p(S) = \Phi(S) - r_p \sum_{j=1}^{J} (h_j(S))^{-1} \tag{4.3}$$

with an arbitrary positive number $r_p > 0$. For a given p, the problem is reduced to minimizing ψ_p over the design subspace. Next, r_p is replaced by a smaller positive number $0 < r_{p+1} < r_p$, and the procedure is repeated starting with the design S, that minimized the preceding penalty function. The equality constraints $f_k = 0$ however, are not so easily taken into account, and as stated by *Fox* (ref. [12], p. 127), "the problem of equality constraints is considered an underdeveloped area".

The *external* penalty function gives progress of convergences from the non-feasible domain. A common formulation is

$$\psi_p(S) = \Phi(S) + r_p \bar{\sum_{j}} (h_j(S))^2 \tag{4.4}$$

with increasing constants $r_{p+1} > r_p > 0$, where $\bar{\Sigma}$ denotes that the summation is restricted to values of j for which $h_j(S) > 0$.

When the penalty function is choosen, a procedure for minimizing this function may be selected disregarding the constraints (cf. ref. [12]). It has been pointed out (cf. ref. [3], p. 21) that formulations using penalty functions can give rise to certain difficulties. Firstly, the functions generated may be difficult to minimize. Secondly, local minima present in the basic problem do not vanish and additional minima may be created. In general, strict equivalence between the original problem and the transformed one cannot be guaranteed.

In addition to the methods treated above some procedures which are of a different nature have been applied to certain special cases. These procedures include *geometric programming* by *Duffin*, *Peterson* & *Zener* [21] and *dynamic programming* by *Bellman* & *Dreyfus* [22].

This concludes the section on methods in optimal design. The rest of our work shall be devoted to a review of solved problems, both elements and compound systems.

---

[1] Or else, one follows the direction of steepest descent directly to the boundary.

5. Structural elements

The work on structural optimization may be classified according to the method employed, the constraint (s) or the type of structure. Although we have chosen the structural type classification, we will also to some extent follow the methods employed, since problems concerning structural elements are mostly treated by variational methods, while mathematical programming is preferred for systems.

For the sake of convenience, we may as well take the total mass Φ of the structural element as our objective functional, since the problems we want to discuss are either formulated in this way or can be readily reformulated.

Restricting our attention first to one-dimensional elements, and letting $\phi(x)$ denote the mass per unit length, we may state the condition of optimality in terms of finding the shape S, for which

$$\Phi = \int \phi(x)\, dx \quad \text{is a minimum}. \tag{5.1}$$

Most of the problems studied concern a relation of the form

$$\phi(x) = b^2 (s(x))^{1/p} \tag{5.2}$$

between the specific mass $\phi(x)$ and the specific stiffness $s(x)$. Here, b and p are given real numbers, and the shape of the structure is described by either $\phi(x)$ or $s(x)$.

The structural elements with which we shall deal are beams, columns, arches, rods, plates, discs, and shells. Great attention has been paid to the formulation of optimality conditions. We shall see that these conditions generally lead to non-linear differential equations.

Optimal tapered beams

It is hardly surprising that a great deal of work has been devoted to the optimization of such comparatively simple structural elements as tapered beams.

Dynamic as well as *static* constraints have been treated. Constraints on the *fundamental natural frequency* ω_0 have been considered by *Niordson* [26], and *Karihaloo* & *Niordson* [27, 28] for $p = 2$, and $1, 2, 3$, respectively. It may be observed that $p = 2$ corresponds to the case of geometrically similar cross sections. For $p > 1$ a necessary condition for an optimal beam can be given in terms of the *amplitude function* $u(x)$[(1)], which has to satisfy the following non-linear differential equation

$$\left\{ \left(\frac{\lambda^2 + u^2}{p(u'')^2} \right)^{p/(p-1)} u'' \right\}'' - \left(\frac{\lambda^2 + u^2}{p(u'')^2} \right)^{1/(p-1)} u = 0, \tag{5.3}$$

where λ^2 is an eigenvalue, determined by the frequency ω_0.

Prager & *Taylor* [23] have shown that for the *linear* case $p = 1$, the necessary and sufficient conditions for global optimality are given by the equation

$$e[u] - b^2 \omega_0^2 u^2 = 1, \tag{5.4}$$

where $e[u]$ is the *specific elastic strain energy density* of the beam. For lateral vibrations of a linearly elastic beam, $e[u] = (u'')^2$, and the condition reduces to the form

$$(u'')^2 - b^2 \omega_0^2 u^2 = 1. \tag{5.5}$$

This condition was also obtained by *Brach* [24], while *Sheu* [25] extended Eq. (5.4) to beams having piecewise uniform stiffness such that the condition for any segment i ($i = 1, 2, 3, \ldots$) is

$$\frac{1}{b_i^2 l_i} \int_0^{l_i} (e[u] - b_i^2 \omega_0^2 u^2)\, dx = 1. \tag{5.6}$$

Optimal tapering of beams under forced lateral vibrations with a given frequency $\omega < \omega_0$ have been studied by *Icerman* [29] and *Mroz* [30] for the linear case $p = 1$. They have found that in the simplest case of specified *dynamic compliance*[1], a necessary and sufficient condition

---

[(1)] The stiffness $s(x)$ (i.e. the design), is easily derived from $u(x)$ in all cases below.

[1] The compliance is defined as the scalar product of the given generalized external forces and the resulting displacements.

for global optimality is that $u(x)$ satisfies the differential equation

$$\mu^2[(u'')^2 - \omega^2 b^2 u^2] = 1 \quad (\omega < \omega_0), \tag{5.7}$$

where the constant μ^2 is determined by the magnitude of the given response constraint. The analogy between the conditions for free and forced vibrations is apparent when (5.5) and (5.7) are compared.

For the case in which the *displacement* is prescribed at a *specified* point (not necessarily the point of loading) *Plaut* [31] has found a necessary condition in the form

$$\mu^2(u''\tilde{u}'' - \omega^2 b^2 u\tilde{u}) = 1 \quad (\omega < \omega_0), \tag{5.8}$$

where $\tilde{u}(x)$ is the displacement function due to a unit load at the specified point.

The more general case of restrictions on the displacement at an *unspecified* point in dynamic response to forced vibrations has been discussed by *Brach* [32] and *Plaut* [33] but the problem does not seem to be clarified yet.

The above conditions for dynamic constraints also seem to hold for $\omega = 0$, corresponding to *static* cases. Conditions for the linear $p = 1$ parallel to (5.7) and (5.8) are given by *Sheu* & *Prager* [34] and *Shield* & *Prager* [35], respectively. By introducing the principle of stationary *mutual potential energy* Shield and Prager obtain a necessary condition for the optimum.

An extension to cover thermo-elastic design is given by *Prager* [36], who obtains formally identical conditions, although u'' is identified as the difference between the resulting curvature and the given thermal curvature. The effect of shear deformation is taken into account by *Huang* [37]. If the *displacement* at a *specified* point is prescribed, the necessary and sufficient conditions for global optimality are found by *Huang* & *Tang* [38] to be

$$\mu^2 u''\tilde{u}'' = 1 \quad (u''\tilde{u}'' \geq 0), \tag{5.9}$$

where the side condition ensures a significant solution. We find it worthwhile to stress that the analysis used, which is related to mathematical programming, ensures here a *global* optimum.

Problems of *static compliance* have also been treated for $p \neq 1$. Thus, *Huang* [39] has found the necessary and sufficient conditions for a *local* optimum to be

$$\mu^2 e[u]\, s^{(p-1)/p} = 1 \tag{5.10}$$

including the special case $e[u] = (u'')^2$. When there is a restriction on the displacement everywhere, the necessary conditions for an optimal solution has been found by *Huang* [40] to be contained in the equation

$$\mu^2 u''\tilde{u}'' s^{(p-1)/p} = 1. \tag{5.11}$$

Chern [41] has treated the related case with the presence of body forces.

A *multiplicity of constraints* may arise either from several constraints related to the same load system and/or from essentially the same constraint related to several load systems. The resulting conditions are obtained as linear combinations of the form

$$\frac{d\phi}{ds} = \sum_k \mu_k^2(u_k''\tilde{u}_k'' - \omega_k^2 b^2 u_k\tilde{u}_k) \tag{5.12}$$

which includes the conditions for single constraints given above as special cases[1]. This has been demonstrated by *Prager* [42], *Sheu* & *Prager* [43], *Martin* [44], and *Chern* & *Prager* [45]. When there is a prescribed lower bound on the stiffness, the equality sign of (5.12) is supplemented by an inequality sign, cf. *Chern* & *Martin* [46].

One major difficulty encountered in solving problems with a multiplicity of constraints is that we cannot know in advance which constraints are *active* at the point. Therefore, in general, all possible combinations must be investigated separately.

Optimal design of beams when the mode of failure is *viscous* (*creep*) *collapse* has been studied in a recent paper by *Martin* [47] using a viscous power law:

$$\dot{u}'' = \dot{K}\left(\frac{M}{Ks}\right)^n, \tag{5.13}$$

[1] The condition becomes more complicated when $p \neq 1$ and the forces are determined by the design, cf. (5.1).

where K is a function of time only, having the dimension of curvature, M is the bending moment, and n is a positive odd integer. Thus, the case $n = 1$ corresponds to linear visco-elastic behaviour, and $n = \infty$ to perfertly plastic behaviour. If the objective functional $\Phi(s)$ is *convex* towards the s-axis, then a necessary and sufficient condition for global optimality is

$$\frac{d\phi}{ds} = \mu^2(\dot{u}'')^{(n+1)/n}. \tag{5.14}$$

When the function ϕ is determined by (5.2), it is convex only if $p \leq 1$. The case $p = 1$ yields

$$\mu^2(\dot{u}'')^{(n+1)/n} = 1. \tag{5.15}$$

This condition cannot be used in the limit $n = \infty$. However, for this case necessary and sufficient conditions for global optimality (also given in ref. [47]) are

$$s \cdot \dot{u}_p'' = [M], \tag{5.16}$$

where $\dot{u}_p''$ is the plastic rate of change of curvature. This result is in agreement with earlier result by *Marcal* & *Prager* [48] and *Prager* & *Shield* [49], who used the cocept of an „associated" non-linear elastic beam.

The optimal plastic design of rings has been investigated by *Prager* [50] and by *Batterman* & *Felton* [51].

The importance of the assumption of convexity of ϕ for problems in optimal plastic design has been discussed by *Megarefs* & *Hodge* [52].

It has been pointed out by *Chan* [53] and in a note by *Spillers* [54] that methods of mathematical programming are very useful for formulating and solving problems in optimal plastic design.

Inequality *constraints on stresses* in optimal beam design have not received much attention. However, important results for elastic beams have been achieved by *Haug* & *Kirmser* [55] and *Masur* [56].

Optimal tapered columns

In most work on the optimal design of columns under compressive load, tapering is treated as the design variable[1].

Defining the function $\psi(x)$ by the equation

$$s = \psi^{2p/(1+p)}. \tag{5.17}$$

Tadjbakhsh & *Keller* [58] have shown that a necessary condition for optimality is given by

$$\psi'' + \mu^2\psi^{(1-p)/(1+p)} = 0. \tag{5.18}$$

The linear case $p = 1$ has been treated by *Prager* & *Taylor* [23], who have obtained necessary and sufficient conditions for global optimality in the form

$$(u'')^2 = 1. \tag{5.19}$$

The same result has been obtained by *McIntosh* & *Eastep* [59]. *Taylor* & *Liu* [60] include minimum bounds on $s(x)$, and *Huang* & *Sheu* [61] include the column weight.

The non-linear case of geometrically similar cross sections ($p = 2$) has been treated by *Keller* [57] and *Taylor* [62]. *Trahair* & *Booker* [63] include minimum bounds on $s(x)$ and *Keller* & *Niordson* [64] include the weight of the column. This last problem gives rise to a more complex condition than (5.18) above.

The case $p = 3$ has been considered by *Simitses, Kamat* & *Smith* [65] who also take elastic supports into account.

Optimal arches have been investigated by *Budiansky, Frauenthal* & *Hutchinson* [66], who have found a necessary condition in the form

$$\mu^2 s^{(p-1)/p}(u'' + \alpha^2 u)^2 = 1, \tag{5.20}$$

where 2α is the opening angle of the arch.

---

[1] A discussion of optimal shapes of cross sections is, however, available in a paper by *Keller* [57].

Recently, *non-conservative* problems of stability in optimal design have received attention. Reference is made to papers by *Zyczkowski & Gajewski* [67], *Gajewski & Zyczkowski* [68], *Plaut* [69], and *Vepa* [70].

Optimal tapered rods

The deformation of a rod is given by the longitudinal displacement and/or the angle of twist.

Problems of optimal design with prescribed *fundamental frequency* for longitudinal vibrations have been treated by *Turner* [71] and *Taylor* [72]. A necessary and sufficient condition for global optimality is found to be

$$(u')^2 - b^2\omega_0^2 u^2 = 1 \tag{5.21}$$

in analogy with (5.5) for transverse vibrations of beams. In fact, most results for optimal rods can be obtained from the corresponding results for beams by substituting u' for u''. The case of segmentwise uniform stiffness is given by (5.6).

The result (5.21) also holds good for *torsional vibrations* when $u(x)$ is interpreted as the angle of twist. This was found by *Ashley & McIntosh* [73]. The result is based upon the assumption that $p = 1$, which generally does not hold for torsional vibrations.

Prescribed *dynamic compliance* gives results analogous to the corresponding beam problems. Thus, for $\omega < \omega_0$ the necessary and sufficient condition for global optimality (cf. ref. [29]) for a rod having piecewise uniform stiffness is

$$\mu^2 \frac{1}{l_i} \int_0^{l_i} ((u')^2 - \omega^2 b^2 u^2)\, dx_i = 1 \tag{5.22}$$

for any segment i.

In problems with constraints on *static* displacements Eq. (5.9) is valid when u'' is replaced by u'. Torsional compliance is considered in ref. [59] and [73]. Optimal design of rods with external loads and temperature fields determined by the design have been studied by *Chern* [74] using the principle of mutual potential energy. A necessary condition for optimality is shown to be

$$\mu^2(\tilde{u}'(u' - \alpha\theta) - r\tilde{u} - \alpha I) = 1, \tag{5.23}$$

where α is the coefficient of linear thermal expansion, $\theta(x)$ the temperature, $r(x)$ the load per unit of stiffness, and $I(x, s)$ is a functional that is due to the heat conduction. *Ashley* and *McIntosh* also discuss non-conservative problems, cf. ref. [73].

Optimal plates

Although plates generally involve much more complicated problems than beams, a considerable number of optimal design problems for plates have been studied. This section deals only with *unstiffened* plates.

The design problems of plates suffering *plastic collapse* have been studied extensively using either the Tresca or the von Mises yield criterion. As might perhaps be expected, the choice of yield criterion does not significantly influence the result. Most of the work has been concerned with cost as objective functional which is generally assumed to depend on plastic resistance according to Eq. (5.2). For $p = 1$, the linear case of sandwich plates, there are the solutions of *Megarefs* [75, 76, 77][1] and *Reiss & Megarefs* [78], who apply a statical method of stress variations. Other solutions are given by *Sheu & Prager* [79], *Eason* [80], *Shield* [81], *Prager & Shield* [82], *Onat, Schumann & Shield* [83], and *Drucker & Shield* [84], using a theorem of constant rate of dissipation due to *Drucker & Shield* [85].

Multiple loading has been considered by *Shield* [86], who obtains a condition containing a linear combination of the terms involved in the individual loading systems.

For the case $p < 1$, *Marcal* [88] has made further use of the concept of "associated structure" (cf. ref. [48]). A discussion of the case $p > 1$ is given by *Kozlowski & Mroz* [89], who demonstrate the necessity of introducing ribs.

---

[1] In these papers special attention is paid to the singular behaviour of the solution for plates at the boundary. (See also *Brotchie* [87] with regard to ribs.)

In addition to the work on plastic collapse, some progress has been made with respect to other constraints. Thus, *elastic strength* is discussed by *Taylor* [90], and *static compliance* by *Huang* [39]. *Huang* finds for $p = 1$, the condition of global optimality is equivalent to constant strain energy density (cf. also *Mroz* [91]). *Dynamic compliance* has been considered by *Mroz* [30], who points out that only in special cases can global optimality be concluded.

The optimal design of plates with a constraint on the *fundamental natural frequency* has been studied by *Olhoff* [92, 93], and *Armand* [94]. *Olhoff* treats the case of solid plates ($p = 3$), and in his papers special attention is paid to the solution procedures of the resulting fourth order, non-linear and singular differential equations. *Frauenthal* [95, 96] has considered plates with prescribed *buckling* load, and his analysis includes the post-buckling behaviour of optimal plates.

Optimal discs

Plastic collapse and *static compliance* have been studied in the design of optimal discs by *Kozlowski* & *Mroz* [97] with the boundary shape and thickness function as design variables. The analysis is related to that of "Michell" structures (see below). *Chern* & *Prager* [98] have applied the principle of mutual potential energy to optimal discs with prescribed *displacements*, giving the conditions for a stationary solution.

In addition, a considerable amount of work has been done on *discs of uniform strength*, but it is felt that this falls outside the scope of our review.

Optimal shells

In an extension of ref. [85], *Shield* [99] has treated optimal circular cylindrical sandwich shells under *plastic collapse*. Again, a great deal of work has been done on membrane shells of *uniform* strength, which we shall not review here.

6. Structural systems

Problems in the optimal design of systems, concern the *layout* and *dimensions* of the individual elements which form the structural system. Although work in this field has been done on such really complicated structures as, for instance, aircraft structures, we shall confine our review to simple systems like trusses, frames, stiffened panels and stiffened cylinders, for which more basic solutions may be obtained.

The conditions for the optimality of systems cannot be expressed in a form analogous to that of the last section, on elements. When dealing with systems, we are usually confined to the field of mathematical programming, which is still under rapid development.

Optimal trusses

If the layout of a truss is given (i.e., the location of all joints and all bars is known) the optimization problem is restricted to determining the cross-sectional area of each bar, these areas being the design parameters.

For *statically determinate* trusses subject to elastic *stress* constraints only, the problem of optimal design is trivial, since the internal forces are independent of the design. Hence it is evident that all bars should be fully stressed for at least one of the load systems. *Displacement* constraints on statically determinate trusses have been treated by *Toakley* [100].

The optimal design of *statically indeterminate* trusses leads to non-linear problems that can be solved by iteration as described in Section 4. Solutions have been obtained by *Reinschmidt, Cornell* & *Brotchie* [101], *Brown* & *Ang* [102], *Pope* [103], *Romstad* & *Wang* [104], *Johnson* & *Brotton* [105], *Melosh* & *Luik* [106], and *Moses* & *Kinser* [107]. *Displacement* constraints have been taken into account by some of the authors (ref. [102, 103, 104]). The value of fully stressed designs is discussed by *Razani* [108] and *Kicher* [109]. Constraints on *stability* and *natural frequency* have been treated by *Zarghamee* [110, 111, 112].

The problem of *optimal configuration*, i.e., the selection of bars when the location of all joints are prescribed, has only been solved for statically determinate trusses. *Fleron* [113] has proved the *existence* of a statically determinate structure (usually not unique) when the maximum stresses are assumed to be independent of forces. A solution can be determined by linear programming, cf. *Dorn, Gormery & Greenberg* [114]. This solution has been extended by *Pedersen* [115, 116] to permissible stresses (local buckling) that are dependent on forces with proof of the existence of an optimal design (cf. ref. [116]). When there is a multiplicity of loading systems and/or constraints on the displacement, the optimal design usually becomes statically indeterminate.

Problems of optimal *layout* include the problem of finding the optimal location of some or all joints. A very early paper by *Michell* [117] treats this problem, assuming a single load system and stress constraints independent of force. *Michell* proves that under these conditions, all bars are fully stressed and are arranged along the orthogonal lines of principal strain of a field having principal strains of constant magnitude. Several later papers are based on this principle, cf., for instance, *Hemp* [118], *Richards & Chan* [119], and *Hegemier & Prager* [120]. In general, there is no proof of the existence of a Michell structure and no general methods are available for the construction of such a field. In *two* dimensions, the field is identical with the slip line field in plane flow of a perfectly plastic solid, and the standard procedures for constructing the slip lines may be used. It has also been shown (cf. ref. [120]) that the Michell structure is optimal with respect to constraints on elastic *static compliance* and *compliance in stationary creep*. Furthermore, the *layout* is the same in the case of constraints on the *fundamental frequency*, but then the cross-sectional areas generally vary. Michell structures, however, do not provide solutions for optimal design problems when there is a multiplicity of load systems, general displacement constraints and permissible stresses dependent on force. For such cases, iterative procedures have been applied and solutions obtained by *Pedersen* [115, 121] for both statically determinate and indeterminate trusses (cf. also *Vanderplaats & Moses* [122]).

Optimal frames

To our knowledge, the problem of optimal *layout* of frames has not been studied yet. Publications on optimal design of frames are restricted to problems in which the dimensions of the beams are the only design variables. *Plastic collapse* has been dealt with in early papers by *Foulkes* [123] and *Heymann* [124, 125]. While later results have been obtained by *Megarefs & Sidhu* [126], *Toakley* [127], *Ridha & Wright* [128] and *Yokoo & Nakamura* [129].

Elastic *stress constraints* for statically indeterminate frames appear in ref. [102] and [104] and for grillages, in a paper by *Kavlie & Moe* [130]. Dynamic response is treated by *Fox & Kapoor* [131].

The description of *concrete* frames requires further design variables. Some progress has been made in this field by *Grierson & Cohn* [132] and *Rozvany & Cohn* [133].

It may safely be stated that the field of optimal design of frames still offers a wide variety of unsolved problems.

Stiffened plates

It is well known that the load-carrying capacity of a plate in compression may be increased considerably by stiffeners of ribs. The optimal design of stiffened plates has therefore attracted much attention, and there is a large number of papers on the subject. The most relevant works for this review are, perhaps, those by *Schmit, Kicher & Morrow* [134] on waffle plates, *Mroz* [135, 136], *Mroz & Shamiev* [137] on reinforced slabs, and *Taylor* [138] on optimal prestressing.

In general, an optimal design is characterized by a point located in a *corner* of the feasible design space. Thus, when for instance two modes of failure are considered, we may generally expect both modes to be active at the optimal point. It is therefore reasonable to expect that the optimally designed stiffened panel is such that local and overall buckling occurs simultaneously when the load is increased. However, *Tvergaard* [139] has shown that such a panel is highly sensitive to geometrical imperfections and therefore, in practice, hardly optimal. This had already been disclosed earlier in the case of thin-walled compression members and built-up columns, by *van der Neut* [140], *Koiter & Kuiken* [141], and *Thompson & Lewis* [142].

Stiffened cylinders

Singer & *Baruch* [143] have studied a variation of stiffener cross-sections as an approach to optimization of cylindrical and conical shells for elastic stability. Cylinders, stiffened externally with rings and stringers of rectangular cross section, have been treated by *Burns* & *Almroth* [144] in a study containing six design variables. Similar cases have been treated by *Kicher* [145], *Stroud* & *Sykes* [146], *Melchers* & *Rozvany* [147], *Rozvany* & *Adidam* [148], and many others.

7. Summary and future problems

Structural optimization is concerned with the important problem of designing lighter, better and cheaper structures. With the advent of large computers research in this field has accelerated along two different lines: analytical methods based upon the calculus of variations for simple structural elements, and mathematical programming for compound systems. Structural *analysis* is the foundation and the most important ingredient of all methods of optimization. Thus, the designer has to select a definitive type of structure before he can attempt an optimization. The type he selects determines the method of analysis to be used (although not necessarily uniquely), and only with a thorough understanding of this method can he attempt a systematic synthesis or an optimization.

We have seen that structural elements are usually handled by analytical methods based on the calculus of variations, while structural systems can only be dealt with efficiently by mathematical programming. Except for the simplest cases, both methods lead to extensive numerical problems.

There is a number of open questions and new fields for research in structural optimization. We have seen that problems in which the stability or the dynamic response of the structure has to be accounted for lead to highly non-linear eigenvalue problems, for which the theory is still mainly unexplored. The numerical solution of the differential equations encountered presents difficult problems, only a few of which have been solved.

One of the more important problems is to span the gap between the two main fields of structural optimization. This could to some extent be done if we could solve the problem of composite structures by variational methods—for instance, by obtaining an "integrally" stiffened panel by using variational methods on plates of variable thickness.

An important field for the designer and engineer is the optimal design of fillets and grooves, which has as yet hardly been touched upon. For instance, no-one seems to know how to minimize the maximum elastic stress in a bar in tension, for which the cross-sectional areas (different in size) at two given sections are known. The classical calculus of variations does not seem to be immediately applicable for this case.

From the practical point of view, material selection is very important. In optimal design, however, the question of material selection is hardly ever raised. This is doubtless due to the lack of data on material properties in a suitable parametric form. In the absence of better data material parameters must be treated as *discrete* design variables, and techniques for dealing with such variables (integer programming) must be improved before any real progress can be achieved in this field.

References

1. Wasiutyński, Z., Brandt, A.: Mech. Rev. 16 (1963) 341–350.
2. Sheu, C. Y., Prager, W.: Appl. Mech. Rev. 21 (1968) 985–992.
3. Pope, G. G., Schmit, L. A.: Structural design applications of mathematical programming techniques. Agardograph No. 149, AGARD, 1971.
4. Barnett, R. L.: Survey of optimum structural design. Experimental Mechanics 19A–26A (1966).
5. Gerard, G.: J. Spacecraft 3 (1966) 5–18.
6. Save, M. A.: Some aspects of minimum-weight design. Engineering Plasticity, Cambridge University Press (1968) 611–626.
7. Prager, W.: J. of Optimization, Theory and Appl., 6 (1970) 1–21.

8. Prager, W.: Optimality criteria in structural design. AGARD Report No. 589, 1971.
9. Kuhn, H. W., Tucker, A. W.: Nonlinear programming, in: Proc. 2nd Berkeley Symp. on Math. Statist. and Probl., Univ. of California Press, 481—492, 1951.
10. Hadley, G.: Nonolinear and dynamic programming. Reading, Mass.: Addison-Wesley,1964, student ed. 1970, p. 484.
11. Pontryagin, L. S., Bol'tanskii, V. G., Gamkrelidze, R. S., Mischenko, E. F.: The math. theory of optimal proc., Pergamon Press 1964, 340.
12. Fox, R. L.: Optimization methods for engineering design. Reading, Mass.: Addison-Wesley 1971, 270.
13. Dantzig, G. B.: Linear programming and extensions, Princeton, N.J.: Princeton University Press 1963.
14. Hadley, G.: Linear programming, Reading, Mass.: Addison-Wesley 1962.
15. Gass, S. I.: Linear programming, 3rd ed., Tokyo: Kógakusha 1969.
16. Rosen, J. B.: J. SIAM Control 8 (1960) 181—217.
17. Zoutendijk, G.: Methods of feasible directions, Amsterdam: Elsevier 1960.
18. Zoutendijk, G.: J. SIAM Control 4 (1966) 194—210.
19. Griffith, R. E., Stewart, R. A.: Man. Sci. 7 (1961) 379—392.
20. Fiacco, A., McCormick, G. P.: Man. Sci. 10 (1964) 601—617.
21. Duffin, R. J., Peterson, E. L., Zener, C. M.: Geometric programming. New York: Wiley 1967.
22. Bellman, R. E., Dreyfus, S. E.: Applied dynamic programming. Princeton University Press, New York: 1962, 363.
23. Prager, W., Taylor, J. E.: Problems of optimal structural design. J. Appl. Mech. 35 (1968) 102—106.
24. Brach, R. M.: Int. J. Solids Structures 4 (1968) 667—674.
25. Sheu, C. Y.: Elastic minimum-weight design for specified fundamental frequency. Int. J. Solids Structures 4 (1968) 953—958.
26. Niordson, F. I.: Quart. Appl. Math. 23 (1964) 47—53.
27. Karihaloo, B. L., Niordson, F. I.: Optimum design of vibrating cantilevers (to appear in J. Optimization, Theory & Appl.).
28. Karihaloo, B. L., Niordson, F. I.: Optimum design of vibrating beams under axial compression (to appear in Arch. Mechanics, Warszawa).
29. Icerman, L. J.: Int. J. Solids Structures 5 (1969) 473—490.
30. Mroz, Z.: ZAMM 50 (1970) 303—309.
31. Plaut, R. H.: Quart. Appl. Math. (1971) 315—318.
32. Brach, R. M.: Int. J. Mech. Sci. 10 (1968) 429—439.
33. Plaut, R. H.: ZAMP 21 (1970) 1004—1010.
34. Sheu, C. Y., Prager, W.: J. Optimization, Theory & Appl. 2 (1968) 179—186.
35. Shield, R. T., Prager, W.: ZAMP 21 (1970) 513—523.
36. Prager, W.: Int. J. Mech. Sci. 12 (1970) 705—709.
37. Huang, N. C.: Int. J. Solids Structures 7 (1971) 321—326.
38. Huang, N. C., Tang, H. T.: J. Optimization, Theory & Appl. 4 (1969) 277—298.
39. Huang, N. C.: Int. J. Solids Structures 4 (1968) 689—700.
40. Huang, N. C.: J. Appl. Mech. (1971) 1078—1081.
41. Chern, J. M.: Int. J. Solids Structures 7 (1971) 373—382.
42. Prager, W.: Int. J. Solids Structures 4 (1968) 469—475.
43. Sheu, C. Y., Prager, W.: ZAMP 20 (1969) 289—297.
44. Martin, J. B.: J. Optimization, Theory & Appl. 6 (1970) 22—40.
45. Chern, J. M., Prager, W.: J. Optimization, Theory & Appl. 5 (1970) 424—431.
46. Chern, J. M., Martin, J. B.: ZAMP 22 (1971) 834—855.
47. Martin, J. B.: J. Solids Structures 7 (1971) 63—81.
48. Marcal, P. V., Prager, W.: J. de Mécanique 3 (1964) 509—530.
49. Prager, W., Shield, R. T.: J. Appl. Mech. (1967) 184—186.
50. Prager, W.: Optimal plastic design of rings. Contributions to Mech., Pergamon Press (1969) 163—169.
51. Batterman, S. C., Felton, L. P.: Optimal plastic design of doubly symmetric closed structures. T.A.E. Report No. 126, Technion, Israel, 1971.
52. Megarefs, G. J., Hodge, P. G.: Quart. Appl. Math. 21 (1963) 91—103.
53. Chan, H. S. Y.: Int. J. Solids Structures 4 (1968) 885—895.
54. Spillers, W. R.: J. Appl. Mech. 38 (1971) 1073—1074.
55. Haug, B. J., Kirmres, P. G.: J. Appl. Mech. (1967) 999—1004.
56. Masur, E. F.: ASCE, EM 5 (1970) 621—639.
57. Keller, J. B.: Arch. Rational Mech. Anal. 5 (1960) 275—285.
58. Tadjbakhsh, I., Keller, J. B.: J. Appl. Mech. (1962) 159—164.
59. McIntosh, S. C., Eastep, F. E.: AIAA J. 6 (1968) 962—964.
60. Taylor, J. E., Liu, C. Y.: AIAA J. 6 (1968) 1497—1502.
61. Huang, N. C., Sheu, C. Y.: J. Appl. Mech. 35 (1968) 285—288.
62. Taylor, J. E.: J. Appl. Mech. 34 (1967) 486—487.
63. Trahair, N. S., Booker, J. R.: Int. J. Mech. Sci. 12 (1970) 973—983.
64. Keller, J. B., Niordson, F. I.: The tallest column. J. Math. Mech. 16 (1966) 433—446.

65. Simitses, G. J., Kamat, M. P., Smith, C. V.: The strongest column by the finite element displacement method. AIAA Paper No. 72—141, 1972.
66. Budiansky, B., Frauenthal, J. C., Hutchinson, J. W.: J. Appl. Mech. 36 (1969).
67. Zyczkowski, M., Gajewski, A.: Proc. IUTAM Symp., Karlsruhe 1969, Springer 1971, 295—301.
68. Gajewski, A., Zyczkowski, M.: ZAMP 21 (1970) 806—818.
69. Plaut, R. H.: J. Optimization, Theory & Appl., 7 (1971) 52—59.
70. Vepa, K.: Optimally stable structural forms. Ph. D. Thesis, Solid Mechanics Division, University of Waterloo, 1972.
71. Turner, M. J.: AIAA J. 5 (1966) 406—412.
72. Taylor, J. E.: AIAA J. 5 (1967) 1911—1913.
73. Ashley, H., McIntosh, S. C.: Application of aeroelastic constraints in structural optimization. Proc. IUTAM Congress, Stanford 1968, Springer 1969, 100—113.
74. Chern, J. M.: J. Appl. Mech. 38 (1971) 538—540.
75. Megarefs, G. J.: ASCE 92, EM 6 (1966) 79—99.
76. Megarefs, G. J.: ASCE, EM 6 (1967) 245—269.
77. Megarefs, G. J.: ASCE, EM 1 (1968) 177—198.
78. Reiss, R., Megarefs, G. J.: Int. J. Solids Structures 7 (1971) 603—623.
79. Sheu, C. Y., Prager, W.: J. Mech. Phys. Solids (1968) 11—16.
80. Eason, G.: ZAMP 11 (1960) 368—375.
81. Shield, R. T.: Quart. Appl. Math. 18 (1960) 131—144.
82. Prager, W., Shield, R. T.: ZAMP 10 (1959) 421—425.
83. Onat, E. T., Schumann, W., Shield, R. T.: ZAMP 8 (1957) 485—499.
84. Drucker, D. C., Shield, R. T.: Quart. Appl. Math. 15 (1957) 269—281.
85. Drucker, D. C., Shield, R. T.: Proc. 9th Int. Congr. Appl. Mech., Brussels 1956, 212—222.
86. Shield, R. T.: ZAMP 14 (1963) 38—45.
87. Brotchie, J. F.: ASCE, EM 5 (1967) 173—175.
88. Marcal, P. V.: Int. J. Solids Structures 3 (1967) 427—443.
89. Kozlowski, W., Mroz, Z.: Int. J. Solids Structures 5 (1969) 781—794.
90. Taylor, J. E.: ASCE, EM 3 (1969) 653—663.
91. Mroz, Z.: Archiwum Mechaniki Stosowanej 1 (1963) 63—76.
92. Olhoff, N.: Int. J. Solids Structures 6 (1970) 139—156.
93. Olhoff, N.: Optimal design of vibrating rectangular plates (to be published).
94. Armand, J. L.: AIAA J. 9 (1971) 1739—1745.
95. Frauenthal, J. C.: Constrained optimal design of circular plates against buckling. Harvard University, Cambridge, Report No. SM-50, 1971.
96. Frauenthal, J. C.: Initial postbuckling behavior of optimally designed columns and plates. Tufts University Mechanical Engineering Rep. 72-4.
97. Kozlowski, W., Mroz, Z.: Int. J. Mech. Sci. 12 (1970) 1007—1021.
98. Chern, J. M., Prager, W.: J. Optimization, Theory & Appl., 6 (1970) 161—170.
99. Shield, R. T.: J. Appl. Mech. (1960) 316—322.
100. Toakley, A. R.: Int. J. Mech. Sci. 10 (1968) 115—127.
101. Reinschmidt, K. F., Cornell, C. A., Brotchie, J. F.: ASCE, ST 6 (1966) 281—318.
102. Brown, D. M., Ang, H. S.: ASCE, ST 6 (1966) 319—339.
103. Pope, G. G.: Int. J. Mech. Sci. 10 (1968) 251—263.
104. Romstad, K. M., Wang, C. K.: ASCE, ST 12 (1968) 2817—2845.
105. Johnson, D., Brotton, D. M.: ASCE, ST 12 (1969) 2589—2610.
106. Melosh, R. J., Luik, R.: Approximate multiple configuration analysis and allocation for least weight structural design. USAF, AFFDL-TR-67-59, 1967.
107. Moses, F., Kinser, D. E.: AIAA J. 5 (1967) 1152—1158.
108. Razani, R.: AIAA J. 3 (1965) 2262—2268.
109. Kicher, P.: Optimum design. ASCE, ST 6 (1966) 265—279.
110. Zarghamee, M. S.: ASCE, ST 8 (1970) 1697—1710.
111. Zarghamee, M. S.: ASCE, ST 6 (1969) 1139—1152.
112. Zarghamee, M. S.: AIAA J. 6 (1968) 749—750.
113. Fleron, P.: Bygningsstatiske Meddelser 35 (1964) 81—96.
114. Dorn, W. S., Gomory, R. E., Greenberg, H. J.: J. de Mecanique 3 (1964) 25—51.
115. Pedersen, P.: AGARD Conf. Proc., No. 36, 11.0—11.17, 1970.
116. Pedersen, P.: Optimal layout af gitterkonstruktioner. DCAMM, Report No. S1, (in Danish), 1970.
117. Michell, A. G. M.: Phil. Mag. S. 8 (1904) 589—597.
118. Hemp, W. S.: Proc. 11th Int. Congr. Appl. Mech., Munich 1964, Springer 1966, 621—628.
119. Richards, D. M., Chan, H. S. Y.: Developments in the theory of Michell optimum structures. AGARD Rep. No. 543, 1966.
120. Hegemier, G. A., Prager, W.: Int. J. Mech. Sci. 11 (1969) 209—215.
121. Pedersen, P.: Int. J. Comp. & Structures 2 No. 5/6, 1972.
122. Vanderplaats, G. N., Moses, F.: ASCE, ST 3 (1972) 671—689.
123. Foulkes, J.: Proc. R. Soc. A 223 (1954) 482.
124. Heyman, J.: Quart. J. Mech. and Appl. Math. 12 (1959) 314.
125. Heyman, J.: Int. J. Mech. Sci. 1 (1959) 121.
126. Megarefs, G. J., Sidhu, H. S.: ASCE, ST 12 (1968) 2985—2998.
127. Toakley, A. R.: ASCE, ST 5 (1968) 1219—1241.
128. Ridha, R. A., Wright, R. N.: ASCE, ST 4 (1967) 165—183.

129. Yokoo, Y., Nakamura, T.: Int. J. Solids Structures 6 (1970) 353—369.
130. Kavlie, D., Moe, J.: Int. J. Numerical Methods in Engng. 1 (1969) 351—378.
131. Fox, R. L., Kapoor, M. P.: AIAA J. 8 (1970) 1798—1804.
132. Grierson, D. E., Cohn, M. Z.: J. Appl. Mech. (1970) 356—360.
133. Rozvany, G. I. N., Cohn, M. Z.: ASCE, EM 6 (1970) 1013—1029.
134. Schmit, L. A., Kicher, T. P., Morrow, W. M.: Structural synthesis capability for integrally stiffened waffle plates. AIAA J. 1 (1963) 2820—2836.
135. Mroz, Z.: Acta Mech. 3 (1965) 34—55.
136. Mroz, Z.: Int. J. Solids Structures 6 (1970) 859—870.
137. Mroz, Z., Shamiev, F. G.: Archiwum Inzynierii Ladowej 16 (1970) 575—584.
138. Taylor, J. E.: Int. J. Solids Strcutures 7 (1971) 213—223.
139. Tvergaard, V.: Imperfection-sensitivity of a wide integrally stiffened panel under compression (to appear in Int. J. Solids Structures).
140. Neut, A. van der: Proc. 12th Int. Congr. Appl. Mech., Stanford 1968, Springer 1969, 389—399.
141. Koiter, W. T., Kuiken, G. D. C.: The interaction between local buckling and overall buckling on the behaviour of built-up columns. Technische Hogeschool Delft University 1971.
142. Thompson, J. M. T., Lewis, G. M.: J. Mech. Phys. Solids 20 (1972) 101—109.
143. Singer, J., Baruch, M.: Recent studies on optimisation for elastic stability of cylindrical and conical shells. Technion, Israel Inst., Haifa.
144. Burns, A. B., Almroth, B. O.: J. Spacecraft 3 (1966) 19—25.
145. Kicher, T. P.: J. Spacecraft 5 (1968) 62—67.
146. Stroud, W. J., Sykes, N. P.: AIAA J. 7 (1969) 1599—1601.
147. Melchers, R. E., Rozvany, G. I. N.: ASCE, EM 6 (1970) 1093—1105.
148. Rozvany, G. I. N., Adidam, S. R.: Int. J. Solids Structures 8 (1972) 661—677.

Flow in porous media

J. R. Philip

Canberra City, Australia

1. Introduction

The problems of flow and transport in porous media present many fascinating challenges to students of mechanics: and it is surprising that, over the years, these questions have received relatively little attention at the Congresses and Symposia of IUTAM and of its predecessors. The 1964 Grenoble Symposium (*Kravtchenko* and *Sirieys* 1966) touched on certain special aspects and the 1966 Vienna Symposia (*Parkus* and *Sedov* 1968) had some relevance; but the first specific involvement of the Union with flow in porous media was a useful but limited meeting, sponsored jointly by IUTAM and IUGG, at Calgary in 1971.

I am gratified that "flow in porous media" is the title of a sectional lecture at this Congress. I may read too much into this, but, as a porous-medium enthusiast, I take it as a sign that IUTAM is beginning to suspect that the mechanics of flow and transport in porous media may not be a wholly trivial matter.

My pleasure is, however, offset by my alarm at the choice of lecturer. I should like to be able to offer what the occasion demands: a systematic interpretation of flow in porous media in the context of the various great established sectors of mechanics. But such an exposition is beyond my powers; and you must resign yourselves to a personal (and perhaps idiosyncratic) account of some selected aspects of the topic of flow in porous media.

As I have remarked elsewhere (*Philip* 1970b), a peculiar difficulty of the topic of flow in porous media is that it arises, often in subtly different forms, in several separate fields of natural science and in a large number of branches of technology. The categories are far from clear-cut, but an enumeration must include fluid mechanics, surface and colloid physical chemistry, physiology, rheology, geophysics, soil physics, hydrology, soil mechanics, agricultural engineering, sanitary engineering, petroleum engineering, powder metallurgy, metal casting, ceramic engineering, and the technologies of wood, paper, textiles, leather, cement, food, and solid fuels.

We must remark on a consequence of this fragmentation. Certain basic concepts of flow in porous media have relevance in many distinct fields. Yet it is clear from the literature that the date of their recognition and use in one application may lag behind that in another by many decades. Sometimes the concepts have been imported into the new field from a more mature one, but in some cases progress has depended on what appears to be *de novo* rediscovery. The problem of the exchange of information and ideas between various fields involving flow in porous media is a real one: a mitigating circumstance is the the growing body of reviews, monographs, symposia, and scientific groups which recognize flow in porous media as a topic in its own right.

Almost inevitably, fragmentation has produced wide variations of terminology. Once it is recognized, this presents no insuperable difficulties, but it does mean that some care must be exercised in relating work from different fields. There is some scope for standardization, but it seems unrealistic to expect a wholly uniform terminology to be acceptable in all fields.

Within this wide and diverse topic, I limit myself here to the following: after some remarks (Section 2) on the *scales of discourse* in porous-medium flow studies, I shall review in broadest outline the theoretical mechanics of flow in unsaturated nonswelling media (Section 3) and its extension to saturated and unsaturated swelling media (Section 4)[1]. This major theme is

[1] The configuration of the solid component of "nonswelling" media is constant, whereas that of "swelling" media is free to change in the course of an unsteady flow process.

supplemented by remarks on momentum exchange between immiscible fluids in porous media (Section 5) and on some mathematical developments which have arisen in the porous-medium context (Section 6).

2. The scales of discourse

A well-based development of the theoretical mechanics of flow in porous media must begin with recognition of the appropriate scales of discourse: and it is worth-while for us to spend a little time on this question.

Writers on continuum mechanics (e.g. *Truesdell* and *Noll* 1965, *Sedov* 1971) are careful to distinguish between two scales of discourse: on the one hand, what has been called the "mass-point", or "microscopic" or "statistical" approach; and, on the other, the "continuum", or "macroscopic" or "phenomenological" approach. They have, very properly, stressed the greater relevance and power of macroscopic theory *vis à vis* many problems of mechanics.

In porous-medium mechanics, however, we must discriminate between not two, but *three*, scales of discourse (e.g. *Raats* 1965): (i) the *molecular* scale; (ii) the scale sometimes called "microscopic" in porous medium studies, which will be called here, less ambiguously, the *Navier-Stokes* scale; and (iii) the scale sometimes termed "macroscopic", which we prefer to call the *Darcy* scale.

The molecular scale and the Navier-Stokes scale are, of course, precisely the scales called "microscopic" and "macroscopic" in continuum mechanics. On the other hand, discourse on the Darcy scale deals with physical quantities related to the *averages* of analogous quantities on the Navier-Stokes scale, the averages being taken over a fluid volume (or a cross-section) which is large compared with that of the individual pore (*Philip* 1969a). A formal description of a suitable averaging process has been given, for example, by *Zaslavsky* (1968).

I need hardly remark that Darcy-scale theory is the ultimate goal of porous-medium studies: the physical observations are made on this scale; and, in any case, a full theory on the Navier-Stokes scale would be needlessly elaborate even if it were feasible. But studies on the Navier-Stokes scale remain important in their own right: well-founded theory on this scale is at hand to guide us as to the form Darcy theory should take and as to the limits of its applicability. Darcy theory must be at least consistent with what we know on the Navier-Stokes scale (*Philip* 1970b). In my opinion the optimal strategy lies in constant interplay between the effort to construct useful theory on the Darcy scale and critical studies on the Navier-Stokes scale. As it happens, the succeeding Sections 3 and 4 review Darcy-scale theory, whilst Section 5 is a discussion on the Navier-Stokes scale.

Failure to distinguish clearly between the scales of discourse has led to confusion and to misplaced effort in many studies. General understanding of the mechanics of flow in porous media has tended to be impeded by the superficial resemblance of Poiseuille's law (which refers to an *exactly linear* flow in a special and precise geometry) and Darcy's law (which applies to *conditionally linear* flows in a generally irregular and unknown geometry). Once the scales of discourse are recognized, the fundamental and wide-reaching differences (both kinematic and dynamic) between the two flows become obvious. Uncritical use of capillary-bundle models of porous media suggests, however, that many workers neglect to make this discrimination (*Philip* 1957i, 1958c, 1969e, 1970b). The confusion carries over into the challenging field of solute transport during flow in porous media: current attempts at Darcy-scale theory (*Fried* and *Combarnous* 1971) take little account of the high degree of order and of kinematic reversibility on the Navier-Stokes scale. Flow in porous media is (approximately) an ordered flow in a disordered geometry: all too often and without explanation, it is transmogrified by the models into a disordered flow in an ordered geometry (*Philip* 1969d, 1970b, 1972a). A peculiar but instructive example of the confusion of scales concerns *vorticity*. This concept is immensely useful on the Navier-Stokes scale, but leads only to physical unreality on the Darcy scale (*Philip* 1968a).

3. Flow in unsaturated nonswelling media

The theory of this Section and the next has been developed largely in the context of water movement in soil and is central to many phenomena arising in agronomy, hydrology, and soil mechanics. It is closely related also to processes occurring in fields such as chemical engineering, petroleum engineering, and physiology. Here we shall use the terminology of soil physics, with a few attempts at translation into the general vocabulary of mechanics. Extensions to other media and to other incompressible Newtonian fluids will be fairly obvious.

Unsaturated soil is a porous medium made up of three components: the soil material, water, and air. We outline the theory in the simplification (often justified in the soil application) that any effect of air pressure differences may be neglected[1]. For the nonswelling media of this Section, the theory then deals explicitly with the flow of only one component, water: details of the air-flow follow from continuity.

3.1 Darcy's law for unsaturated nonswelling media

We may write Darcy's law for *saturated* media, specialized to water flow, as

$$V = -K\nabla\Phi. \tag{1}$$

V is the vector flow velocity, Φ is the total potential, and K is the *hydraulic conductivity*[2]. The engineering device of expressing potentials per *unit weight* simplifies our equations and units: K then has the dimension [length] [time]$^{-1}$. We note that

$$\Phi = p/\varrho g + \Xi,$$

where p is the pressure, ϱ the density of the water, g the acceleration due to gravity, and Ξ the potential of the external forces.

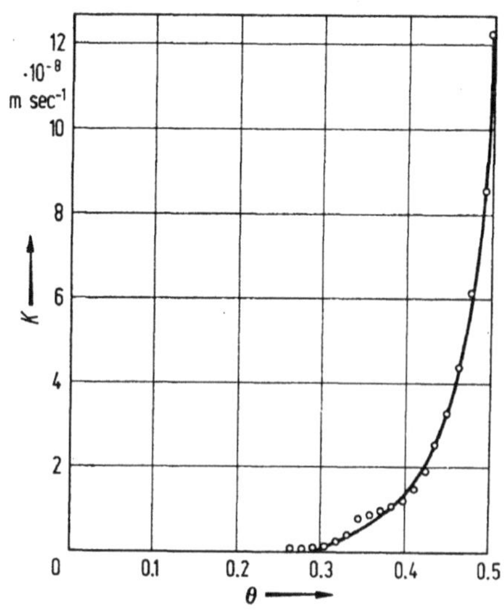

Fig. 1. Relation between hydraulic conductivity, K, and volumetric moisture content, θ, for Yolo light clay (*Moore* 1939).

---

[1] The slightly more general formalism for the case where air-pressure differences are not neglected is identical to that for two-fluid flow (with capillarity taken into account) occurring, for example, in petroleum engineering.

[2] As we state above, V and Φ are averages over regions with dimensions large compared with that of the individual pore. Much effort has gone into alleged "proofs" of Darcy's law: but it suffices to observe that, on the Navier-Stokes scale, there are both hydrodynamic and statistical elements to the law. The *hydrodynamic* element is that the Navier-Stokes equation (combined with the incompressibility and no-slip conditions) is linear in the limit as the Reynolds number approaches zero; the *statistical* element is that, although it is not feasible to know details of the internal geometry or of the distribution of velocity and potential on the Navier-Stokes scale, the medium is sufficiently homogeneous that local mean quantities such as V and Φ can be defined and exist (*Philip* 1969a, 1970b).

K is a scalar for isotropic media and a symmetrical second-order tensor for anisotropic media. The subsequent discussion refers specifically to isotropic media, but extension to anisotropic media is straightforward.

The basic concepts of flow in unsaturated nonswelling media are due primarily to *Buckingham* (1907), who is better known in general mechanics for the Π-theorem of dimensional analysis (*Buckingham* 1914). He suggested that Darcy's law should hold for unsaturated media in a modified form with K a function of θ, the volumetric moisture content (the fraction of the bulk volume occupied by water). *Richards* (1931), *Moore* (1939), *Childs* and *Collis-George* (1950), and others provided experimental confirmation and established the general character of $K(\theta)$. For obvious physical reasons (*Philip* 1954, 1957a) K decreases rapidly, through as much as six or more decades (*Gardner* 1960), as θ decreases from its saturation value through the range of interest. Figure 1 shows a typical $K(\theta)$ relationship. We therefore rewrite (1) in the form appropriate to unsaturated nonswelling media:

$$V = -K(\theta)\,\nabla\Phi. \tag{2}$$

3.2 Total potential and moisture potential of water in nonswelling media

In (water-wet) unsaturated media the water is not free in the thermodynamic sense because of capillarity, adsorption, and electrical double layers (*Edlefsen* and *Anderson* 1943, *Schofield* 1935a). Capillarity is dominant in wet, coarse-textured media, and adsorption assumes its greatest importance in dry media. Double-layer effects may be significant in fine-textured media exhibiting colloidal properties. *Buckingham* (1907) was the first to appreciate that the conservative forces governing the equilibrium and movement of soil-water are amenable to treatment through their associated scalar potentials.

We define such potentials relative to the reference state of water (of composition identical to the soil solution) at atmospheric pressure and datum elevation $z = 0$. Here z is the vertical ordinate, conveniently taken positive downward. We then have

$$\Phi = \Psi - z. \tag{3}$$

Ψ, the *moisture potential*[1], is the potential of the forces arising from local interactions between solid and water (*Philip* 1970a). It is not essential either to know or to specify these forces in detail: it suffices that Ψ can be measured by well-established techniques (*Croney, Coleman* and *Bridge* 1952; *Richards* 1965; *Holmes, Taylor* and *Richards* 1967). In water-wet *nonswelling media* $\Psi = 0$ at saturation and decreases with θ to very large negative values (typically -10^4 m) at the dry end of the moisture range of interest. Figure 2 depicts the $\Psi(\theta)$ relation for the soil for which $K(\theta)$ is given in Fig. 1.

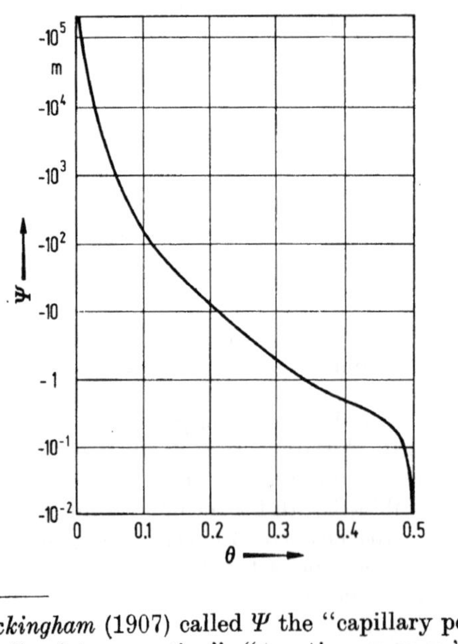

Fig. 2. Relation between moisture potential, Ψ, and volumetric moisture content, θ, for the soil of Fig. 1 (*Moore* 1939).

[1] *Buckingham* (1907) called Ψ the "capillary potential". Other names for Ψ (or $-\Psi$) include "moisture tension", "moisture suction", "negative pressure", "pressure head", "matric potential" (*Richards* 1949; *Childs* and *Collis-George* 1950; *Miller* and *Klute* 1967; *Rose* 1966) and, in swelling media "swelling pressure" (*Bolt* 1956). Petroleum engineers use "capillary pressure".

The partial volumetric Gibbs free energy associated with the local solid-water interaction is $\varrho g\Psi$ and it follows that (in the absence of solutes) the liquid and vapour systems are connected at equilibrium by the relation

$$H = \exp g\Psi/RT\,, \tag{4}$$

where H is the relative humidity, R is the gas constant for water vapour, and T is the absolute temperature.

3.3 General partial differential equation of flow in unsaturated nonswelling media

Combining (2) and (3) with the continuity requirement yields

$$\partial\theta/\partial t = \nabla\cdot(K\nabla\Psi) - \partial K/\partial z\,, \tag{5}$$

where t denotes time. When the relations between K, Ψ, and θ are single-valued, (5) may be rewritten in terms of a single dependent variable. In terms of θ, the equation is

$$\frac{\partial\theta}{\partial t} = \nabla\cdot(D\nabla\theta) - \frac{dK}{d\theta}\frac{\partial\theta}{\partial z}. \tag{6}$$

Both the *moisture diffusivity* D, defined by

$$D = K\, d\Psi/d\theta\,,$$

and the coefficient $dK/d\theta$ are, in general, strongly-varying functions of θ. $D(\theta)$ for the soil of Figs. 1 and 2 is shown in Fig. 3. In terms of Ψ, (5) becomes

$$\frac{d\theta}{d\Psi}\frac{\partial\Psi}{\partial t} = \nabla\cdot(K\nabla\Psi) - \frac{dK}{d\Psi}\frac{\partial\Psi}{\partial z}, \tag{7}$$

with the various coefficients functions of Ψ.

Richards (1931) developed (5) and (7). *Childs* and *George* (1948) recognized the diffusion character of (6) for a horizontal one-dimensional system. *Klute* (1951) explicitly derived (6). *Philip* (1954, 1955a, 1957c) extended the approach to include water transfer in vapour and adsorbed phases in the same formalism.

The strong nonlinearity of Fokker-Planck equations (6) and (7) cannot be ignored, and progress in unsaturated flow studies depends centrally on their solution. We return to a discussion of nonlinear Fokker-Planck and diffusion equations in Section 6.

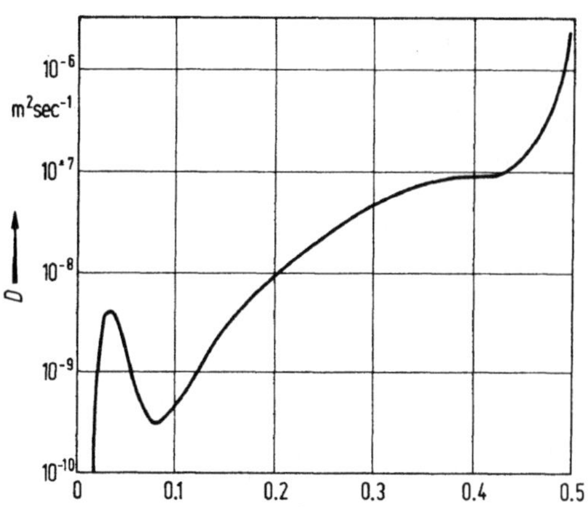

Fig. 3. Relation between moisture diffusivity, D, and volumetric moisture content, θ, for the soil of Fig. 1 (*Philip* 1955a). Note that D for $\theta \leq 0.06$ includes dominant contribution in vapour phase.

3.4 Applications and extensions

Appropriate solutions of (6) and (7) accord with experiment and offer insight into the mechanics of such diverse phenomena as infiltration (*Philip* 1954, 1957c—g, 1958a, b, 1966a), capillary rise (*Philip* 1966b), evaporation from porous bodies (*Philip* 1954, 1957a), drainage and retention of soil-water (*Philip* 1957a, *Staple* and *Lehane* 1954, *Day* and *Luthin* 1956,

Whisler and *Watson* 1968), and extraction of soil-water by plant roots (*Philip* 1957a, *Gardner* 1960, *Cowan* 1965).

Limitations to this approach are reviewed elsewhere (*Philip* 1969a). Extensions are required, for example, to deal with complications due to capillary hysteresis (*Childs* 1964) and to aggregation (or cracking) (*Philip* 1968b, c). Generalization to nonisothermal systems was made by *Philip* and *de Vries* (1957, *de Vries* 1958, *de Vries* and *Philip* 1959). This involves simultaneous moisture and heat transfer and provides, *inter alia*, the quantitative theory of drying of porous bodies (*Philip* 1957h). Stability during infiltration in heterogeneous media (*Hill* and *Parlange* 1972) and simultaneous moisture and solute transfer (*Scotter* and *Raats* 1970) are topics of recent research (*Philip* 1972a).

4. Flow in saturated and unsaturated swelling media

The developments of Section 3 have provided a fruitful theoretical framework for the study of the hydrology of nonswelling soils, but the need for generalization to swelling soils has long been recognized (*Philip* 1958c). In this Section we review some modest steps towards the required extension. They are also modest steps towards integration of two generally divergent viewpoints: on the one hand that of soil physics (preoccupied with soil-water and its flow); and, on the other, that of soil mechanics (concerned with soil volume-change and the associated stresses).

I must warn at once that these extensions are quite limited in character: they are, for the most part, restricted to *one-dimensional* systems; and they do not purport to treat irreversible structural changes which may occur in the presence of large enough stresses. Nevertheless, they have important consequences: they reveal, for example, that many classical concepts of soil- and ground-water hydrology, based on the behaviour of nonswelling media, fail completely for swelling soils.

4.1 The extension to swelling media

Flow in both saturated and unsaturated swelling media is amenable to the same general formalism, essentially because K, Ψ, and ϑ[1] are all free to vary, whatever the saturation. The void ratio e is defined as the ratio of void volume[2] to particle volume. Saturated or two-component (solid, water) swelling systems necessarily exhibit "normal" volume change (*Keen* 1931, *Marshall* 1959) and $e \equiv \vartheta$. Unsaturated or three-component (solid, water, air) systems exhibit "residual" volume change, with e dependent both on ϑ and on the normal stress P. The particles are taken to be inelastic.

Three new basic elements enter the extension of the flow theory to swelling media:

A. In unsteady swelling systems, the soil particles are, in general, in motion, so that it must be recognized (*Gersevanov* 1937) that Darcy's law applies to flow relative to the soil particles. We therefore replace (2) by

$$V_r = -K(\vartheta)\, \nabla\Phi, \tag{8}$$

where V_r is the vector flow velocity in the local rest frame of the particles.

B. For one-dimensional systems involving self-weight and/or surface loading, Eq. (3) must be generalized to include the *overburden potential* Ω (*Philip* 1969b), so that it becomes

$$\Phi = \Psi + \Omega - z. \tag{9}$$

It is convenient to take Ψ as the "unloaded" moisture potential, and Ω is then the contribution to Φ due to the normal stress, P. See Subsection 4.3 below.

C. Whereas $K(\theta)$ and $\Psi(\theta)$ provide a sufficient hydrodynamic characterization of nonswelling media (for nonhysteretic processes), we now require $K(\vartheta)$, $\Psi(\vartheta)$, and $e(\vartheta, P)$, as well as the particle specific gravity γ_s. Figures 4 and 5 show typical $\Psi(\vartheta)$ and $e(\vartheta, P)$ relations for a swelling soil. For mineral soils $\gamma_s \approx 2.7$.

[1] For swelling media, ϑ, the *moisture ratio*, is more convenient than θ. ϑ, the ratio of the volume occupied by water to that occupied by particles, is equal to $\theta(1 + e)$.

[2] Void volume is based on the medium as a whole: any cracks, for example, count as voids.

Unsteady flow problems may be subjected to either Eulerian analysis (*Prager* 1953, *Philip* 1968d) in the physical space coordinate, or to Lagrangian analysis (*Hartley* and *Crank* 1949, *McNabb* 1960; *Smiles* and *Rosenthal* 1968) in material coordinate m such that

$$\nabla m = (1 + e)^{-1}. \tag{10}$$

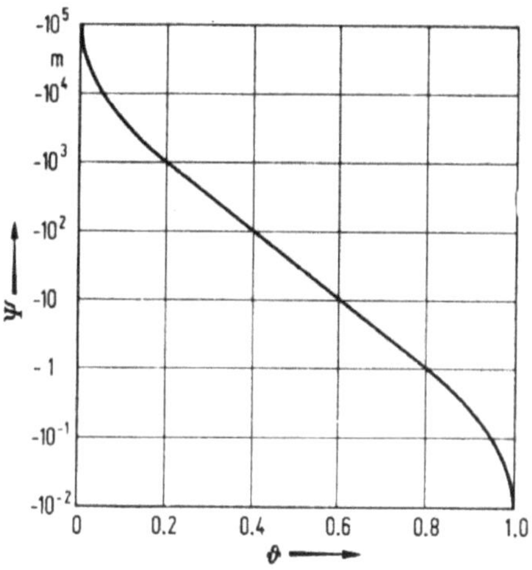

Fig. 4. Typical relation between moisture potential, Ψ, and moisture ratio, ϑ, for a swelling soil (*Philip* 1969b), adopted here for illustrative purposes.

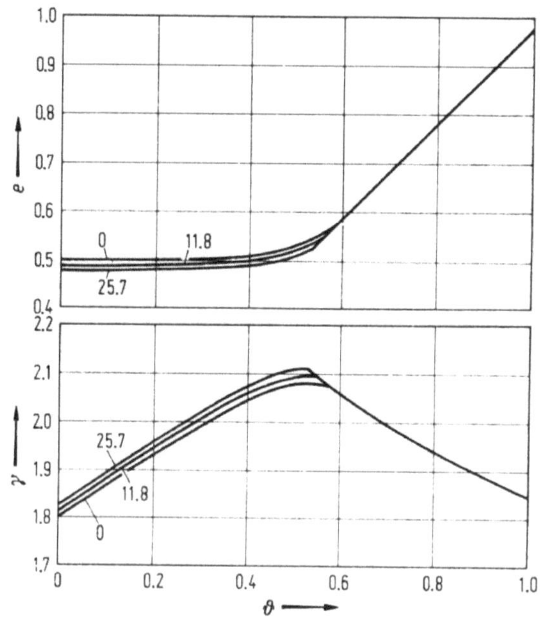

Fig. 5. The functions $e(\vartheta, P)$ and $\gamma(\vartheta, P)$ for the illustrative swelling soil (*Philip* 1971a). e is the void ratio, γ the apparent wet specific gravity, and P the normal stress. The numerals on the curves denote values of P(m).

4.2 Unsteady two-component and (unloaded) three-component horizontal systems

For two-component and (unloaded) three-component horizontal systems, application of the continuity requirement to (8) yields

$$\left(\frac{\partial \vartheta}{\partial t}\right)_m = \frac{\partial}{\partial m}\left[K \frac{\partial \Psi}{\partial x}\right],$$

where x is the horizontal coordinate. Combination with (10) then yields the nonlinear diffusion equation in Lagrangian coordinates

$$\frac{\partial \vartheta}{\partial t} = \frac{\partial}{\partial m}\left[D^* \frac{\partial \vartheta}{\partial m}\right], \tag{11}$$

where

$$D^* = (1 + e)^{-1} K \, d\Psi/d\vartheta.$$

(*Smiles* and *Rosenthal* 1968, *Philip* 1968d, *Philip* and *Smiles* 1969). The classical theory of consolidation and swelling (*Terzaghi* 1923) takes K and $d\Psi/d\vartheta$ constant and neglects mass flow due to particle movement. Relevant solutions of (11) (which is free of both limitations) indicate that the mass flow may be of the same order as the Darcy flow relative to the particles (*Philip* 1968d). The experiments of *Smiles* and *Rosenthal* (1968) agree with (11). Equation (16) below is the generalization of (11) to loaded three-component systems.

4.3 Hydrostatics in swelling media

The overburden potential Ω enters the analysis of vertical and loaded systems. It has been shown by *Bolt* (*Philip* 1970c, *Groenevelt* and *Bolt* 1972) that

$$\Omega = \alpha P, \quad \text{with} \quad \alpha(\vartheta, P) = P^{-1} \int_0^P (\partial e/\partial \vartheta)_P \, dP_{\vartheta=\vartheta}. \tag{12}$$

For a vertical column at equilibrium, then, the use of (12) in (9) yields

$$\Phi = \Psi - z + \alpha\left[P(0) + \int_0^z \gamma \, dz\right] = \text{constant} = -Z. \tag{13}$$

Here $P(0)$ is the vertical stress due to loading at the upper surface $z = 0$ and γ is the *apparent wet specific gravity*, defined by

$$\gamma = (\vartheta + \gamma_s)/(1 + e).$$

Figure 5 shows γ for our illustrative swelling soil. In the hydrological context Z is the "water table depth". With γ_s and characterizing functions $\Psi(\vartheta)$ and $e(\vartheta, P)$ known, (13) may be solved to yield the equilibrium moisture profile corresponding to a given Z.

Soil-mechanical data and exploratory calculations indicate that the P-dependence of α is weak and also that γ is a slowly-varying function of P; and it appears that the variation of these functions with P may be neglected without serious error in a P-range of, say, 20 m (corresponding to the top 10 m of the soil) (*Philip* 1971a).

In this approximation Eq. (13) reduces to a linear first-order differential equation, for which the solution $z(\vartheta)$ is available in closed form. It is found (*Philip* 1969b) that there are three classes[1] of solution of (13) representing three distinct types of equilibrium profile. These are defined through ϑ_P, the moisture ratio at which γ assumes its maximum value. ϑ_P is designated the *pycnotatic point* (Greek πυκνότατος = "densest"). We have: (A) *Hydric* profiles for which the surface moisture ratio $\vartheta_0 > \vartheta_P$ and $\partial\vartheta/\partial z < 0$; (B) *Pycnotatic* profiles with $\vartheta_0 = \text{constant} = \vartheta_P$; (C) *Xeric* profiles for which $\vartheta_0 < \vartheta_P$ and $\partial\vartheta/\partial z > 0$. See Fig. 6.

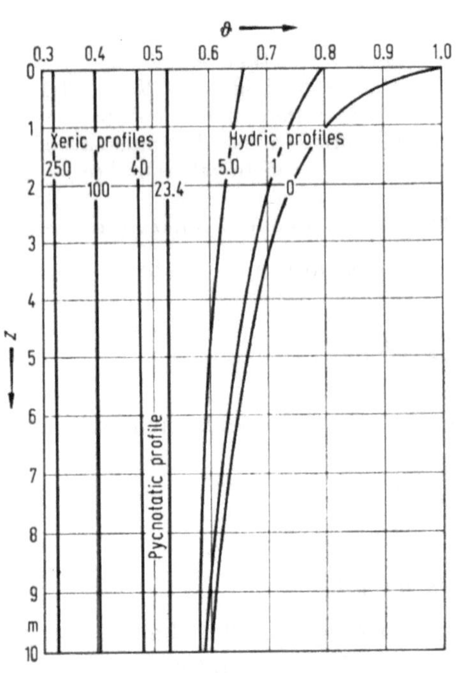

Fig. 6. Equilibrium moisture profiles computed for the soil characterized by Figs. 4 and 5 (*Philip* 1969b). z is depth beneath the surface. The numerals on the profiles denote values of Z, the depth to the water table in m.

4.4 Steady vertical flows

Combining (8), (9), and (12) gives the equation for vertical flow. With the P-dependence of α and γ neglected, this is

$$V_r = -K\left[\left\{\frac{d\Psi}{d\vartheta} + \frac{d\alpha}{d\vartheta}\left[P(0) + \int_0^z \gamma \, dz\right]\right\}\frac{\partial\vartheta}{\partial z} + \alpha\gamma - 1\right]. \tag{14}$$

For steady flows this reduces to a linear differential equation of the first order in $d\vartheta/dz$ and ϑ, which may be integrated to yield $z(\vartheta)$ in closed form. The solution is singular under certain

[1] This result requires $\gamma_s > 1$.

conditions: steady vertical flows are possible only for certain combinations of values of ϑ_0 and ϑ_∞ (the moisture ratio at large z). The details are complicated (*Philip* 1969d).

4.5 Unsteady vertical flows

Applying continuity to (14) and using (10), we obtain the equation for unsteady flow and volume-change in vertical swelling systems,

$$\frac{\partial\vartheta}{\partial t} = \frac{\partial}{\partial m}\left[\frac{K}{1+e}\left\{\frac{d\Psi}{d\vartheta} + P(0)\frac{d\alpha}{d\vartheta}\right\}\frac{\partial\vartheta}{\partial m}\right] - \frac{\partial}{\partial m}\left[K\left\{1 - \gamma\alpha - \frac{d\alpha/d\vartheta}{1+e}\int_0^m \gamma(1+e)\,dm \cdot \frac{\partial\vartheta}{\partial m}\right\}\right]. \tag{15}$$

The theory of various unsteady processes in vertical systems may be developed through solution of (15) subject to appropriate conditions: we have in mind infiltration, capillary rise, drainage, and evaporation; and also the soil-mechanical processes of consolidation and swelling of layers so thick that self-weight cannot be neglected. Equation (15) is a generalization of the nonlinear Fokker-Planck equation: in certain important special cases it reduces to a Fokker-Planck equation. Several methods developed for studying the nonlinear Fokker-Planck equation (Section 6 below) promise to be useful in connexion with (15) (*Philip* 1969d, 1970b). For example, the second term on the right of (15) is negligibly small at small t for many unsteady phenomena of interest, so that solutions of the nonlinear diffusion equation

$$\frac{\partial\vartheta}{\partial t} = \frac{\partial}{\partial m}\left[D^{**}\frac{\partial\vartheta}{\partial m}\right] \tag{16}$$

with

$$D^{**} = (1+e)^{-1}\,K\{d\Psi/d\vartheta + P(0)\,d\alpha/d\vartheta\}$$

form the basis of perturbation methods of solving the full equation. Equation (16) is the counterpart of (15) for horizontal systems.

4.6 Applications

Although much further work is required, it is clear that hydrologic theory based on Eqs. (13), (14), and (15) differs profoundly from the classical theory which takes no account

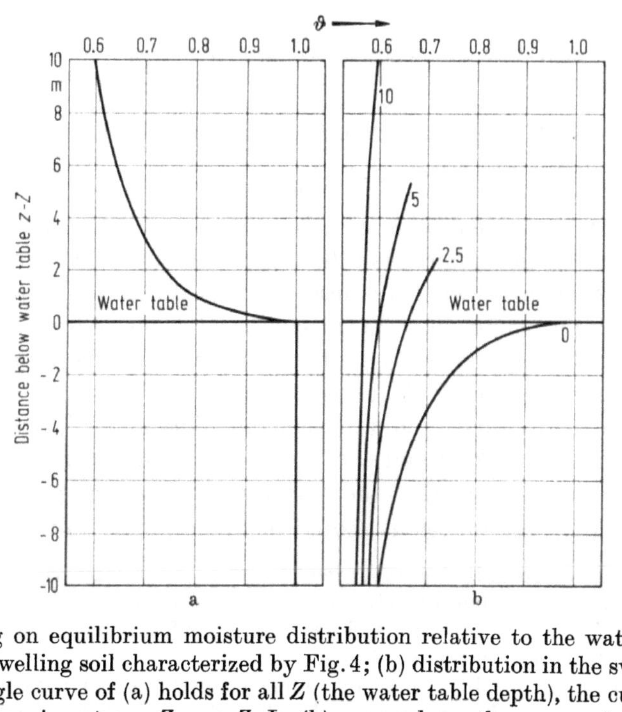

Fig. 7. Effect of swelling on equilibrium moisture distribution relative to the water table (*Philip* 1969c): (a) distribution in a nonswelling soil characterized by Fig. 4; (b) distribution in the swelling soil characterized by Figs. 4 and 5. The single curve of (a) holds for all Z (the water table depth), the curve terminating in each case at the soil surface, i.e. at $z - Z = -Z$. In (b) numerals on the curves signify values of Z(m).

of swelling and neglects the contribution to Φ of the overburden potential Ω[1]. Perhaps the simplest general statement one can offer on the influence of swelling is this: the net effect of gravity on the equilibrium and flow of water in swelling media is approximately $(1 - \gamma\alpha)$ times that in nonswelling ones. For a mineral soil this factor is about -1 in the normal range, decreasing to 0 at $\vartheta = \vartheta_P$, and approaching $+1$ as $\alpha \to 0$ at small values of ϑ. Various "intuitions" of the hydrologist are therefore invalidated.

Equilibrium moisture distributions for a swelling soil are thus totally different in character from those for a nonswelling soil (with $d\vartheta/dz$ necessarily positive and the moisture distribution invariant with respect to the water table) (see Figs. 7 and 8). Any attempt to interpret equilibria in a swelling soil through classical concepts is therefore doomed to failure (*Philip* 1969c)

This *bouleversement* carries over to vertical flow processes. For example, the course of infiltration in a swelling soil evidently has analogies with that of capillary rise in a nonswelling one: and evaporation from an initially wet swelling soil will not exhibit the sharp transition between constant-rate and falling-rate phases which is characteristic of nonswelling media (*Philip* 1954, 1957h). These developments have evident practical consequences for groundwater hydrology (*Philip* 1971a) and irrigation technology (*Philip* 1972e).

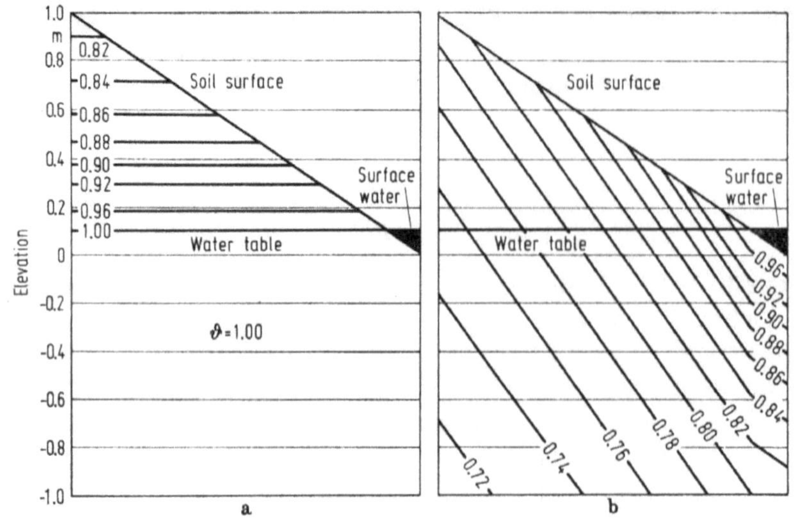

Fig. 8. Effect of swelling on equilibrium moisture distribution in a landscape (*Philip* 1969c): (a) distribution in a nonswelling soil characterized by Fig. 4; (b) distribution in the swelling soil characterized by Figs. 4 and 5. Surface water ponded to maximum depth 0.1 m. Datum of elevation taken at lowest point of soil surface. Horizontal scale is arbitrary. Calculation for (b) assumes soil surface is not too steep.

4.7 Comment

Before leaving the Darcy-scale discourse of Sections 3 and 4 I wish to offer a perspective on these Sections. To put these developments into a form recognizable to members of at least one school of continuum mechanics, we may describe Eqs. (2) and (3) (plus continuity) as the constitutive equations governing flow in unsaturated nonswelling media: and, similarly, Eqs. (8) and (9) (plus continuity) as the constitutive equations governing one-dimensional flow and volume-change in swelling media. These equations are of fair generality in that the functions characterizing the individual medium, such as $K(\theta)$ and $\Psi(\theta)$ for the (nonhysteretic) nonswelling medium and $K(\vartheta)$, $\Psi(\vartheta)$ and $e(\vartheta, P)$ for the (nonhysteretic) swelling medium, are arbitrary —beyond the requirement that they satisfy a few general physical constraints. There is no good reason why any such functions should be of simple or universal form[2]; and, accord-

[1] But some authors have recognized overburden effects in hydrology. *Schofield* (1935b) saw that Φ should include an overburden contribution. *Coleman* and *Croney* (1952) introduced the "compressibility factor" α, though their definition, evaluation, and use of α differ from those reviewed here. See also *Collis-George* (1961) and *Rose*, *Stern* and *Drummond* (1965).

[2] Functions of limited "universality" may of course be useful in connexion with media of internal geometries which are similar but differ in length scale (cf. *Miller* and *Miller* 1956).

ingly, we do not insist that nature obligingly fit particular functions, but aim, rather, to develop the methodology at a level where the various characterizing functions are treated as empirical—experimentally determined for each individual medium.

This Darcy-scale attack on problems of flow in porous media evidently has some affinities with continuum mechanics, but there are profound differences: primarily differences associated with the scale of discourse, but also differences of philosophy. One is tempted, nevertheless, to remark that many of us workers in porous medium physics are rather like *M. Jourdain* (*Molière* 1671) who spoke prose for forty years without knowing it: we have practised something akin to continuum mechanics for about as long and never knew it.

5. Momentum exchange between immiscible fluids in porous media

Buckingham (1907) considered the boundary conditions governing water flow on the Navier-Stokes scale in a soil containing both water and air. He concluded "in view of the general behavior of dirty surface films on water" that the no-slip condition is probably appropriate to the air-water interface as well as to the solid-water interface. Buckingham's suggestion is supported by the behaviour of small rising bubbles and small falling droplets, when observed terminal velocities are consistent with skin rigidity (*Miller* and *Miller* 1956, *Lane* and *Green* 1956). It seems probable that, in many circumstances, the interfaces between immiscible fluids in porous media do behave rigidly. In this case there can be *no momentum exchange* between immiscible fluids flowing in a medium: each fluid would move in a self-contained way under its own potential gradient. And that would be an end to the matter.

If, on the other hand, we admit the possibility of nonrigid interfaces, the appropriate condition is continuity across the interface of both velocity and shearing stress parallel to it. *Momentum exchange between the fluids is then possible.* In a medium containing n immiscible fluids[1], the flow velocity of each fluid would depend on the potential gradient in all n fluids. The $n \times n$ matrix of coefficients entering the extension of Darcy's law required in this case would necessarily be a function of $(n - 1)$ saturations: a prospect to delight the generalizers and the manipulators of symbols.

Yuster (1951) assumed that the interfaces were nonrigid. His model consists of a circular tube with the first fluid occupying an outer annular region and the second fluid the central circular core. The central fluid is at no point constrained by a *direct* no-slip condition. *Philip* (1957j) made calculations for a slightly more realistic model: the space between two parallel walls is occupied by two fluids; each fluid is in contact with one wall and the interface between the fluids is a plane parallel to the walls.

Both models grossly overestimate the possibilities of momentum exchange in real media, even if the interfaces were nonrigid. In real media the interfaces tend to occur in small isolated regions, each surrounded by a solid-fluid-fluid line of contact and by solid-fluid interfaces at which the no-slip condition holds (Fig. 9). Each small patch of fluid-fluid interface is thus strongly constrained by the adjoining no-slip surfaces, so that it seems unlikely that conditions at the interface can have much influence on Darcy-scale flow velocities.

This conjecture has prompted calculations on viscous flows subject to mixed no-slip and no-shear conditions. *Philip* (1972b, c), using complex variable methods, has studied unidirectional flows in various systems with longitudinal no-shear slots in no-slip walls and plane Stokes flows over no-slip planes containing transverse no-shear slots. *Mahony* (1972), using the vector stream function of *Benjamin* and *Mahony* (1971), has obtained the solution for Stokes flows over a no-slip plane containing a circular no-shear patch. These various calculations yield bounds on the magnitude of momentum exchange across nonrigid interfaces and its influence on flows on the Darcy scale. The fractional increase in flow due to nonrigidity varies as ε^2 when ε, the fraction of the flow boundary area occupied by fluid-fluid interface, is small. For ε as large as 0.1, it is estimated that the flow (of, for example, water in an air-water system) may increase by about 1%.

[1] The case $n = 3$ (brine, oil, gas) is common in petroleum engineering.

The indications are that the question of momentum exchange between fluids is, for most practical purposes, a non-problem. Although this will come as a relief to practising porous-medium physicists, it is unlikely to discourage the searchers after generality.

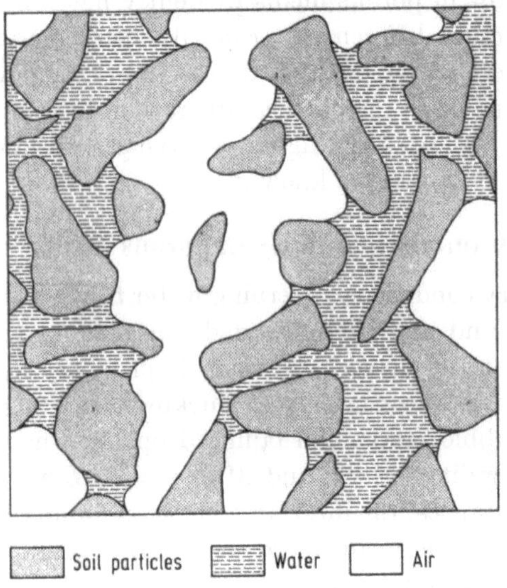

Fig. 9. Schematic cross-section of unsaturated soil, showing distribution of fluid-fluid (air-water) interfaces.

6. Nonlinear diffusion and Fokker-Planck equations

Many mathematical problems of flow in porous media are of some significance for mechanics in general. This penultimate section is a brief review of work on the nonlinear diffusion and Fokker-Planck equations occurring in the context of Sections 3 and 4.

Cole (1970) identified four fields of what he called the "New Math of mechanics": perturbation theory; similarity methods; dispersive wave theory; and numerical analysis of partial differential equations. He saw these fields as deeply rooted in mechanics and emphasized "that problems of mechanics provide an indispensible inspiration for mathematical research and that mathematical abstraction can illuminate the fields of mechanics". Without bending definitions too far, we can say that all four of Cole's fields have their place in the following account.

The equations which concern us may, of course, be solved by the brute-force use of high-speed computers, so long as the problem is well-posed and the computation scheme stable. But effective progress depends on knowledge of the fundamental structure of the solutions, and we are therefore concerned here mainly with quasi-analytical and analytical methods. "Quasi-analytical" solutions depend on the methods of mathematical analysis for the establishment of their basic form, even though some "coefficients" require to be determined numerically. When the solution may be found completely by mathematical analysis, we call the solution "analytical". Both analytical and quasi-analytical solutions are "exact" in the (somewhat optimistic) usage of fluid mechanics (*Van Dyke* 1964).

Philip (1969a) reviewed in detail many of the principal exact solutions and relevant techniques, so the present treatment is very condensed. See also *Braester, Dagan, Neuman* and *Zaslavsky* (1971). We give some attention, however, to connexions with work in other sectors of mechanics and to recent developments. We discuss primarily the transient problems which arise when water is suddenly supplied or removed in consequence of a step-function change of θ or Ψ on (at least part of) the boundary of a mass of a homogeneous medium at some initial uniform moisture content.

6.1 Fundamental similarity solution

A fundamental solution is that for horizontal one-dimensional systems. This is a similarity solution of the one-dimensional nonlinear diffusion equation (*Boltzmann* 1894), namely

$$x(\theta, t) = \phi_1(\theta)\, t^{1/2}, \tag{17}$$

where x is the horizontal coordinate. The function ϕ_1 is the solution of a nonlinear ordinary equation which may be found by a simple and accurate numerical technique (*Philip* 1955b, 1957c). ϕ_1 may be obtained analytically for an indefinitely large class of functional forms of $D(\theta)$ (*Philip* 1960). The importance of this solution is that it provides the leading term of the perturbation solution of related problems involving the complications of (arbitrary) two- and three-dimensional geometry and/or gravity. The latter is embodied in the first-order term on the right of (6).

6.2 Perturbation solutions: diffusion in two and three dimensions, infiltration

The solutions for unsteady flow from (or to) a cylindrical or spherical cavity (no gravity) take the form (*Philip* 1966a)

$$r(\theta, t) - r_0 = \phi_1(\theta)\, t^{1/2} + \phi_{*2}(\theta)\, r_0^{-1} t + \phi_{*3}(\theta)\, r_0^{-2} t^{3/2} + \cdots \tag{18}$$

Here r is the radial coordinate, with the cavity surface at $r = r_0$. The functions ϕ_{*2}, ϕ_{*3}, etc. are solutions of linear ordinary equations which are readily solved numerically (*Philip* 1957b). For three dimensions the function ϕ_{*2} is just twice that for two dimensions. The radius of practical convergence of expansion (18) is limited. The three-dimensional solution is supplemented by a steady exact solution asymptotically valid for large t.

The solution for one-dimensional infiltration (vertically downward flow) is of the form (*Philip* 1954, 1957b, c)

$$z(\theta, t) = \phi_1(\theta)\, t^{1/2} + \phi_2(\theta)\, t + \phi_3(\theta)\, t^{3/2} + \cdots. \tag{19}$$

Here again ϕ_2, ϕ_3, etc. are the easily-obtained solutions of linear ordinary equations. For the example depicted in Fig. 10, the first four terms of (19) suffice to yield an accurate solution (error $\leq 0.5\%$) for t as large as 10^6 sec. The same solution applies for capillary rise (vertical upward flow) (*Philip* 1966b) with the sign of the odd terms on the right of (19) reversed. The radius of practical convergence is, however, less; and the solution is supplemented by the exact solution in the limit as $t \to \infty$.

A double perturbation of (17) has been used in a study of the early stages of unsteady flow from cylindrical and spherical cavities (and semi-circular furrows and hemispherical basins) with gravity included (*Philip* 1969f).

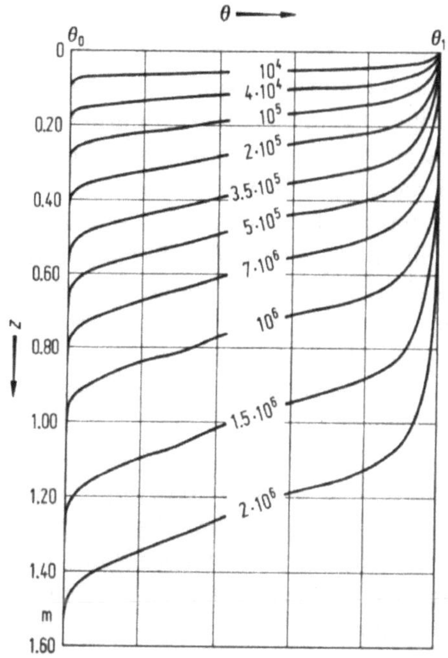

Fig. 10. Computed moisture profiles for one-dimensional infiltration in Yolo light clay (*Philip* 1954, 1957c, d). Numerals on each profile represent value of t (sec) at which profile is realized. Profiles for $t \leq 10^6$ sec calculated from the first four terms of series (19); those for $t > 10^6$ sec based on (20).

These various solutions based on perturbation of a similarity solution have close analogies in more conventional areas of fluid mechanics—for example the perturbation by *Howarth* (1938) of the *Blasius* (1908) flat plate boundary layer. As *Cole* (1970) observes, linearity of the equations for all "coefficients" except the first is characteristic of the perturbation of similarity solutions.

6.3 Asymptotic (large t) shock-wave-like solution

Solution (19) is supplemented by the asymptotic solution, valid for large t,

$$z(\theta, t) = (t - t_0)\, v + \zeta(\theta). \tag{20}$$

The velocity of the moisture profile, v, is given by

$$v = [K(\theta_1) - K(\theta_0)]/(\theta_1 - \theta_0),$$

where θ_0, θ_1 are the initial and final values of θ. ζ follows from a simple quadrature and t_0 from a matching procedure (*Philip* 1957d). The solution for large t thus consists of the uniform downward movement of a wave of constant shape. For the example of Fig. 10, this solution is used for $t > 10^6$ sec.

The one-dimensional vertical form of (6) is recognizable as a "more nonlinear" variant of the equation of *Burgers* (1948). As *Lighthill* (1956) showed, the quasi-equilibrium between convection and diffusion implicit in appropriate asymptotic solutions of Burgers' equation yields a plane shock-wave of constant shape and uniform velocity—of the same general character as (20). Lighthill relates this result to the perceptive analysis of *Taylor* (1910). *Taylor* had resolved the paradoxes of the previous purely convective theory of shock-waves by including *diffusive* processes.

6.4 Other exact solutions

Steady two- and three-dimensional solutions may be found through quasilinearization (*Philip* 1966a, 1968e, 1969a, 1971b, *Wooding* 1968, *Raats* 1970, 1971, 1972) based on a transformation due to *Kirchhoff* (1894). Limited ranges of exact solutions are available for other types of boundary condition (*Philip* 1969a) and for heterogeneous media (*Philip* 1967, 1972d).

6.5 Integral methods

Finally, some brief comments on integral methods of solving nonlinear diffusion and Fokker-Planck equations. Where similarity methods apply they have the great virtue of reducing the number of independent variables by one. Integral methods offer the prospect of effectively achieving this same simplification for a vast constellation of problems of the real world which stubbornly refuse to fit the Procrustean bed of similarity[1].

In essence, some guessed "shape" of the solution is put into an integral of the equation with respect to the variable to be eliminated. (For Eq. (6) the eliminated variable is θ, so that the technique implicitly involves changing the dependent variable.) The "shape" may be a once-and-for-all guess, it may contain parameters fixed by the boundary conditions (*Pohlhausen* 1921), or it may be subject to iterative improvement.

Green and *Ampt* (1911), the first Australian soil physicists, unconsciously applied a primitive integral method to one-dimensional (horizontal and vertical) forms of (6). They simply supposed the advancing moisture profiles to be step-functions and solved the resulting ordinary equations. *Philip* (1957f, 1972e) showed that the Green-Ampt assumption was exact for D proportional to a Dirac delta-function at the wet end of the moisture range and designated the class of such solutions "delta-function" solutions (*Philip* 1966a). Delta-function solutions and linearized solutions (*Philip* 1966a, b, c, 1969a) have been used to estimate, and to establish bounds on, the integral properties of nonlinear solutions.

---

[1] Another means of carrying over into nonsimilar problems some of the economies of similarity solutions has been explored by *Drake* and *Ellingson* (1970). This work on two-dimensional nonlinear diffusion arose in the porous-medium context.

Parlange (1971a, b, c) proposed an integral method involving iteration. Parlange's first approximation is that of *Macey* (1959): his higher approximations abandon integral continuity, with the consequence that his procedure is nonconvergent and oscillatory even in favourable cases (*Knight* and *Philip* 1974). *Philip* and *Knight* (1974) have developed a related technique which enables greatly improved initial estimates of "shape"[2] and preserves integral continuity in all approximations. The iterative procedure is consequently stable and is rapidly convergent in favourable cases.

7. Conclusion

This lecture has fallen far short of a comprehensive treatise. I am especially conscious that I have said nothing about the excellent studies of flow in porous media carried out here in the Soviet Union—from the pioneering work of Academician *Polubarinova-Kochina* (1952) to the diverse researches of the present day. Perhaps this deficiency will be remedied in the sessions of contributed papers at this Congress.

Mr. Chairman, I hope that this talk, for all its shortcomings, will give to those who are interested some intimation of the challenges offered by this messy, but non-trivial, sector of mechanics.

[2] The characteristic used is the "flux-concentration relation", the dependence of flux density on θ in appropriate reduced form (*Philip* 1973).

References

Benjamin, T. B., Mahony, J. J.: J. Fluid Mech. 49 (1971) 385—390.
Blasius, H.: Z. Math. u. Phys. 56 (1908) 1—37.
Bolt, G. H.: Géotechnique 6 (1956) 86—93.
Boltzmann, L.: Ann. Phys. (Leipzig) 53 (1894) 959—964.
Braester, C., Dagan, G., Neuman, S., Zaslavsky, D.: A survey of the equations and solutions of unsaturated flow in porous media. Rep. Hydrodynamics Hydr. Eng. Lab., Technion, Israel (1971).
Buckingham, E.: U.S. Dept. Agr. Bur. Soils Bull. 38 (1907).
—: Phys. Rev. Ser. 2, 4 (1914) 345—376.
Burgers, J. M.: Adv. Appl. Mech. 1 (1948) 171—199.
Childs, E. C.: Soil Sci. 97 (1964) 173—178.
—, George, N. C.: Disc. Faraday Soc. 3 (1948) 78—85.
—, Collis-George, N.: Advan. Agron. 2 (1950) 233—272.
—: Proc. Roy. Soc. A 201 (1951) 392—405.
Cole, J. D.: Proc. 6th U.S. Nat. Congr. Appl. Mechs. (1970). 3—10
Coleman, J. D., Croney, D.: Road Res. Lab. Note RN/1709/JDC.DC (unpublished) (1952).
Collis-George, N.: Soil Sci. 91 (1961) 306—311.
Cowan, I. R.: J. Appl. Ecol. 2 (1965) 221—239.
Croney, D., Coleman, J. D., Bridge, P. M.: Road Res. Tech. Paper 24 (1952).
Day, P. R., Luthin, J. N.: Soil Sci. Soc. Amer. Proc. 20 (1956) 443—447.
Drake, R. L., Ellingson, N. B.: J. Comp. Phys. 6 (1970) 200—218.
Edlefsen, N. E., Anderson, A. B. C.: Hilgardia 16 (1943) 31—299.
Fried, J. J., Combarnous, M. A.: Adv. Hydroscience 7 (1971) 169—182.
Gardner, W. R.: Soil Sci. 89 (1960) 63—73.
Gersevanov, N.M.: The foundations of dynamics of soils. Stroiizdat: Moscow-Leningrad, 3rd Ed. (1937).
Green, W. H., Ampt, G. A.: J. Agr. Sci. 4 (1911) 1—24.
Groenevelt, P. H., Bolt, G. H.: Soil Sci. 113 (1972) 238—245.
Hartley, G. S., Crank, J.: Trans. Faraday Soc. 45 (1949) 801—818.
Hill, D. E., Parlange, J.-Y.: Soil Sci. Soc. Amer. Proc. 36 (1972) 697—702.
Holmes, J. W., Taylor, S. A., Richards, S. J.: pp. 275—303 of Irrigation of Agricultural Lands (eds. Hagen, R. M., Haise, H. R., Edminster, T. W.): Amer. Soc. Agronomy, Madison (1967).
Howarth, L.: Proc. Roy. Soc. A 164 (1938) 547—579.
Keen, N. A.: The physical properties of the soil. London: Longmans (1931).
Kirchhoff, G.: Vorlesungen über die Theorie der Wärme. Leipzig: Barth (1894).
Klute, A.: Soil Sci. 73 (1951) 105—116.
Knight, J. H., Philip, J. R.: On solving the unsaturated flow equation: 2. Critique of Parlange's method. Soil. Sci. (in press) (1974).
Kravtchenko, J., Sirieys, P. M.: (eds). IUTAM Symposium Grenoble. Berlin: Springer (1966).
Lane, W. R., Green, H. L.: pp. 162—215 of Surveys in Mechanics (eds.: Batchelor, G. K., Davies, R. M.). Cambridge: University Press (1956).
Lighthill, M. J.: pp. 250—251 of Surveys in Mechanics (eds. Batchelor, G. K., Davies, R. M.). University Press: Cambridge (1956).
Macey, R. I.: Bull. Math. Biophys. 21 (1959) 19—32.
McNabb, A.: Q. Appl. Math. 17 (1960) 337—347.
Mahony, J. J.: Bull. Australian Math. Soc. 7 (1972) 77—89.
Marshall, T. J.: Relations between water and soil. Commonwealth Agricult. Bureau: Farnham Royal (1959).
Miller, E. E., Klute, A.: pp. 209—244 of Irrigation of Agricultural Lands (eds.: Hagen, R. M.,

Haise, H. R., Edminster, T. W.). Amer. Soc. Agronomy: Madison (1967).
Miller, E. E., Miller, R. D.: J. Appl. Phys. 27 (1956) 324—332.
Molière (Poquelin, J. B.): Le Bourgeois gentilhomme. Paris: Le Monnier 1671.
Moore, R. E.: Hilgardia 12 (1939) 383—426.
Parkus, H., Sedov, L. I.: (eds.) IUTAM Symposia Vienna. Wien: Springer (1968).
Parlange, J.-Y.: Soil Sci. 111 (1971a) 134—137.
—: Soil Sci. 111 (1971b) 170—174.
—: Soil Sci. 112 (1971c) 313—317.
Philip, J. R.: J. Inst. Engrs. Australia 26 (1954) 255—259.
—: Proc. Natn. Acad. Sci. India (Allahabad) 24A (1955a) 93—104.
—: Trans. Faraday Soc. 51 (1955b) 885—892.
—: Proc. 3rd. Int. Congr. Irrig. Drainage, San Francisco 8.125—8.154 (1957a).
—: Australian J. Phys. 10 (1957b) 43—53.
—: Soil Sci. 83 (1957c) 345—357.
—: Soil Sci. 83 (1957d) 435—448.
—: Soil Sci. 84 (1957e) 163—178.
—: Soil Sci. 84 (1957f) 257—264.
—: Soil Sci. 84 (1957g) 329—339.
—: J. Meteorology 14 (1957h) 354—366.
—: Proc. 2nd Australian Conf. Soil Sci. 1 (1957i) 63.1—63.9.
—: Trans. Amer. Geophys. Un. 38 (1957j) 782—784.
—: Soil Sci. 85 (1958a) 278—286.
—: Soil Sci. 85 (1958b) 333—337.
—: Spec. Rep. Highway Res. Board, Washington 40 (1958c) 147—163.
—: Australian J. Phys. 13 (1960) 1—12.
—: Symposium Wageningen, UNESCO 1 (1966a) 503—525.
—: Symposium Wageningen, UNESCO 2 (1966b) 559—564.
—: Symposium Wageningen, UNESCO 1 (1966c) 471—478.
—: Review No. 1457. Appl. Mech. Rev. 21 (1968a) 205.
—: Australian J. Soil Res. 6 (1968b) 1—19.
—: Australian J. Soil Res. 6 (1968c) 21—30.
—: Australian J. Soil Res. 6 (1968d) 249—267.
—: Water Resources Res. 4 (1968e) 1039—1047.
—: Adv. Hydroscience 5 (1969a) 214—296.
—: Australian J. Soil Res. 7 (1969b) 99—120.
—: Australian J. Soil Res. 7 (1969c) 121—141.
—: Water Resources Res. 5 (1969d) 1070—1077.
—: pp. 25—48 of Circulatory and Respiratory Mass Transport. Symposium Ciba Fdn. (eds. Wolstenholme, G.E.W., Knight, J.) London: Churchill (1969e).
—: Australian J. Soil Res. 7 (1969f) 213—221.
—: Soil Sci. 109 (1970a) 294—298.
—: Ann. Rev. Fluid Mech. 2 (1970b) 177—204.
—: Water Resources Res. 6 (1970c) 1248—1251.
—: pp. 95—107 of Salinity and Water Use (eds. Talsma, T, Philip, J. R.). London: Macmillan (1971a).
—: Soil Sci. Soc. Amer. Proc. 35 (1971b) 867—871.
—: Soil Sci. 113 (1972a) 294—300.
—: ZAMP 23 (1972b) 353—372.
Philip, J.R.: ZAMP 23 (1972c) 960—968.
—: Soil Sci. Soc. Amer. Proc. 36 (1972d) 268—273.
—: Proc. 8th Int. Congr. Irrig. Drainage, Varna. C.13—C.28 (1972e).
—: On solving the unsaturated flow equation: 1. The flux-concentration relation. Soil Sci. (in press) (1973).
—, Knight, J. H.: On solving the unsaturated flow equation: 3. New quasi-analytical technique. Soil Sci. (in press) (1974).
—, de Vries, D. A.: Trans. Amer. Geophys. Un. 38 (1957) 222—232.
—, Smiles, D. E.: Australian J. Soil Res. 7 (1969) 1—19.
Pohlhausen, K.: ZAMM 1 (1921) 252—268.
Polubarinova-Kochina, P. Ya.: Theory of Ground Water Movement. Gosizdat. Tekhn.-teor. Liter.: Moscow 1952.
Prager, S.: J. Chem. Phys. 21 (1953) 1344—1347.
Raats, P. A. C.: Development of equations describing transport of mass and momentum in porous media, with special reference to soils. Doctoral thesis, Univ. of Illinois 1965.
—: Soil Sci. Soc. Amer. Proc. 34 (1970) 709—714.
—: Soil Sci. Soc. Amer. Proc. 35 (1971) 689—694.
—: Soil Sci. Soc. Amer. Proc. 36 (1972) 399—401.
Richards, L. A.: Physics 1 (1931) 318—333.
—: Soil Sci. 68 (1949) 95—112.
—: Desert Research. Res. Council Israel Spec. Rep. 2 (1965) 523—546.
Rose, C. W.: Agricultural Physics. London: Pergamon Press 1966.
— Stern, W. R., Drummond, J. E.: Australian J. Soil Res. 3 (1965) 1—9.
Schofield, R. K.: Trans. 3rd Congr. Soil Sci. 1 (1935a) 30—33.
—: Trans. 3rd Int. Congr. Soil Sci. 2 (1935b) 37—48.
Scotter, D. R., Raats, P. A. C.: Soil Sci. 109 (1970) 170—178.
Sedov, L. I.: A Course in Continuum Mechanics, Vol. 1. Groningen: Wolters-Noordhof (1971).
Smiles, D. E., Rosenthal, M. J.: Australian J. Soil Res. 6 (1968) 237—248.
Staple, W. J., Lehane J. J.: Canadian J. Agr. Sci. 34 (1954) 329—342.
Taylor, G. I.: Proc. Roy. Soc. A 84 (1910) 371—377.
Terzaghi, K.: Sitzb. Akad. Wiss. Wien. Abt 2a, 132 (1923) 125—138.
Truesdell, C., Noll, W.: The non-linear field theories of mechanics. Handbuch der Physik III/3. Berlin: Springer (1965).
Van Dyke, M.: Perturbation Methods in Fluid Mechanics. New York: Academic Press (1964).
Vries, D. A. de: Trans. Amer. Geophys. Un. 39 (1958) 909—916.
— Philip, J. R.: J. Geophys. Res. 64 (1959) 386—388.
Whisler, F. D., Watson, K. K.: J. Hydrol. 6 (1968) 277—296.
Wooding, R. A.: Water Resources Res. 4 (1968) 1259—1273.
Yuster, S. T.: Proc. 3rd World Petroleum Congr. The Hague II (1951) 437—445.
Zaslavsky, D.: Soil Sci. 106 (1968) 358—362.

Turbulent shear layer prediction on the basis of the transport equations for the Reynolds stresses

J. C. Rotta

Göttingen, F. R. Germany

1. Introduction

Turbulence is the most complicated type of flow. Its theoretical solution is, in the case of an incompressible Newtonian fluid, the general integral of the Navier-Stokes equations. The exact treatment of turbulence fails because the general solution of the Navier-Stokes equations can not be determined. Engineering problems are treated with approximative calculation methods which rely, to a large extent, on hypotheses and suppositions and which are based on experimental results. Such calculations, the basis of which is a set of equations for some relevant statistical mean quantities, are usually labelled as semiempirical methods. There are a few suggestions to develop calculation methods on the basis of stochastic turbulence models. However, none of these ideas has yet led to results which are applicable to engineering problems.

2. The problem of turbulent flow calculation

The usual way to describe the velocity field of turbulent flow by the sum of the local mean velocity, U_i, and fluctuating velocity, u_i, leads to the Reynolds equations for the mean velocity components. For steady mean flow in a Cartesian coordinate system, these equations read

$$U_j \frac{\partial U_i}{\partial x_j} = -\frac{1}{\varrho}\frac{\partial P}{\partial x_i} - \frac{\partial \overline{u_i u_j}}{\partial x_j} + \nu \frac{\partial^2 U_i}{\partial x_j\,\partial x_j}, \tag{1}$$ [1]

where P is the mean value of pressure, ϱ the density, ν kinematic viscosity, and the overbar denotes time averaged quantities. In these equations the effect of turbulence occurs in the form of the Reynolds stresses, or apparent stresses, $-\varrho\overline{u_i u_j}$. The mean velocity has to satisfy the continuity equation

$$\partial U_i/\partial x_i = 0\,. \tag{2}$$

The magnitudes of the Reynolds stresses are in general much larger than those of the Stokes stresses

$$\tau_{ij} = \varrho\nu(\partial U_i/\partial x_j + \partial U_j/\partial x_i)\,. \tag{3}$$

The Reynolds stresses appear as unknown quantities in the equations of mean motion. Thus, the first problem of turbulent flow calculation is to provide functional relations between the Reynolds stresses and the mean flow field. In the simplest way this is achieved by algebraic equations. The oldest and best known relations of this kind are *Boussinesq*'s eddy viscosity relation and *Prandtl*'s mixing length formula. It has been pointed out many times that these relationships are too simple to describe properly the true behavior of turbulent flows.

Bradshaw, recently mentioned in his brilliant Reynolds-Prandtl Lecture [1] that the eddy viscosity concept is nearly 100 years old, Prandtl published his first paper on the mixing length formula in 1926. Thus, this relation has been existing for almost 50 years. However,

[1] The summation convention for repeated indices applies.

little use was made of both relationships for a relatively long period of time. The reason was that numerical integration of partial differential equations with the aid of mechanical computing machines and slide rule is difficult and time consuming. Therefore, the integral methods have been devised, beginning with the *v. Kármán-Pohlhausen* method for laminar boundary layers. Originally, the integral methods were developed as a particular technique for solving partial differential equations. The attraction of the integral methods for turbulent boundary layers lay in the fact that detailed assumptions about the mechanism of turbulence can be replaced by global assumptions. The first methods of this kind have been given by *Buri* [2] and *Gruschwitz* [3] (about 1930). Over a period of more than three decades such integral methods have been the prevalent procedures for turbulent boundary layer computation. The latest review of these integral methods is found in the Proceedings of the 1968 Stanford Conference [4]. After the electronic computer had been developed to a stage that partial differential equations could quickly be solved on the basis of the finite-difference technique, the eddy viscosity and mixing length concept was used with some degree of success. The inherent disadvantage of these methods, which are called "mean field methods" after the 1968 Stanford Conference, is that their application to a particular type of flow requires ad hoc assumptions with respect to the distributions of eddy viscosity or mixing length. This fact, together with the physical insufficiencies, impedes the formulation of a unified computational method for turbulent flow. There is, however, need for calculation methods which are applicable to a wide class of turbulent flows.

Today, powerful computers are available which enable us to solve complicated systems of partial differential equations. Thus, the calculation of turbulent flow on the basis of transport equations for the relevant mean quantities is now practicable for engineering problems. The purpose of this paper is to give a review of the basic concepts and the progress that has been achieved. The numerical treatment will not be discussed.

The idea to replace the older algebraic equations by transport equations is not new. *Prandtl* [5], using the transport equation for kinetic fluctuation energy, proposed a formalism which inspired many other investigators. My own work of 1951 [6] was stimulated by *Prandtl*'s paper. Actually, *Kolmogorov* [7] published a paper already in 1942, in which he gave a complete system of equations for turbulent shear flow calculation. The local properties of turbulence are described by the kinetic fluctuation energy and a "mean frequency". In *Prandtl*'s work a length scale appeared as a second quantity for which additional assumptions had to be made. Thus *Prandtl*'s formalism was not a complete one. Except for the fact that *Kolmogorov* gave a transport equation for the mean frequency, there is a remarkable similarity in the concepts of the two authors. Unfortunately, *Kolmogorov*'s pioneering work remained generally unknown outside the Soviet Union for a very long time.

3. Transport equations for the Reynolds stresses

There are two ways of deriving transport equations in closed form. One is to derive the equations in a fully empirical manner as *Kolmogorov* and *Prandtl* did. The other way is to derive exact transport equations from the Navier-Stokes equations and to close them by introducing approximative assumptions for relevant terms of the equations. The latter way has the advantage that the exact transport equations sometimes reveal effects which may be overlooked otherwise.

The following set of equations for the tensor of the Reynolds stresses can be derived from the time dependent Navier-Stokes equations (*Rotta* [8]):

$$\underbrace{U_k \frac{\partial \overline{u_i u_j}}{\partial x_k}}_{\text{convection}} + \underbrace{\overline{u_k u_j} \frac{\partial U_i}{\partial x_k} + \overline{u_k u_i} \frac{\partial U_j}{\partial x_k}}_{\text{production}} - \underbrace{\frac{1}{\varrho} \overline{p \left(\frac{\partial u_i}{\partial x_j} + \frac{\partial u_j}{\partial x_i} \right)}}_{\text{redistribution}} + \underbrace{2\nu \overline{\frac{\partial u_i}{\partial x_k} \frac{\partial u_j}{\partial x_k}}}_{\text{dissipation}}$$

$$+ \underbrace{\frac{\partial}{\partial x_k} \left[\overline{u_k u_i u_j} + \frac{1}{\varrho} \overline{(\delta_{jk} u_i + \delta_{ik} u_j) \, p} - \nu \frac{\partial \overline{u_i u_j}}{\partial x_k} \right]}_{\text{diffusion}} = 0, \tag{4}$$

where p denotes the pressure fluctuations and δ_{ik} is the unit tensor of second rank. These equations hold for mean steady flow of an incompressible Newtonian fluid and are called transport equations for the Reynolds stresses. The sum of the diagonal terms of the Reynolds stress matrix, divided by two, yields the value of kinetic fluctuation energy. If the notation $\overline{q^2} = \overline{u_i u_i}$ is introduced, we obtain the transport equation for the kinetic fluctuation energy, $\overline{q^2}/2$, in the following form:

$$\underbrace{U_k \frac{1}{2}\frac{\partial \overline{q^2}}{\partial x_k}}_{\text{convection}} + \underbrace{\overline{u_k u_i}\frac{\partial U_i}{\partial x_k}}_{\text{production}} + \underbrace{\nu\overline{\left(\frac{\partial u_i}{\partial x_k}\frac{\partial u_i}{\partial x_k} + \frac{\partial u_i}{\partial x_k}\frac{\partial u_k}{\partial x_i}\right)}}_{\text{dissipation}}$$
$$+ \frac{\partial}{\partial x_k}\underbrace{\left[\overline{u_k(q^2/2 + p/\varrho)} - \frac{\nu}{2}\frac{\partial \overline{q^2}}{\partial x_k} - \nu\frac{\partial \overline{u_i u_k}}{\partial x_i}\right]}_{\text{diffusion}} = 0. \tag{5}$$

It is seen that the Reynolds stresses appear in different combinations (the first three terms of Eq. (4)). If there were only terms of this kind, the transport equations in combination with the equations for the mean motion would give a closed set of equations. However, there are additional terms, which are related to the Reynolds stresses in no simply recognizable way. There are terms involving velocity fluctuation gradients, correlations with pressure fluctuations, and mean products of three velocity fluctuation components. The transport equations contain, indeed, even more supernumerary unknowns than the equations for the mean motion. It is known that for each term new equations can be derived from the Navier-Stokes equations, which again contain new unknowns. Proceeding in this way, one ends up with a system of an endless number of equations.

4. Closure of the transport equations

For practical purposes, the difficulties are circumvented by introducing empirical closure assumptions. This means that the terms are approximated by algebraic relations. These relations must, of course, be in agreement with the physical content of the original terms, and must be dimensionally correct. The spatial structure of turbulence does not appear explicitly in Eqs. (4) and (5), but its effects are implied in the velocity gradient fluctuations and the pressure fluctuations. In order to approximate these effects, certain length scales are introduced which require additional transport equations. A correct interpretation of the terms of the transport equations is very important. Therefore the meaning of the terms is indicated in Eqs. (4) and (5). It may be distinguished between source/sink terms and transport terms. The main source terms are the production terms. Production means generation by interaction of turbulent motion with mean flow. The sink terms are viscous dissipation (destruction by viscous forces). The correlation of pressure and rate of strain fluctuations effects redistribution of fluctuation energy and destruction of correlation between different velocity fluctuations. The transport consists of convection (transport by mean flow), turbulent diffusion (transport by turbulent mixing) and diffusion by viscous forces.

The transport equations are substantially simplified if the considerations are restricted to high Reynolds number flow. The viscous diffusion terms can then be neglected, and the remaining viscous terms, which are of type $\nu\overline{(\partial u_i/\partial x_k)(\partial u_j/\partial x_k)}$, can be expressed by a single scalar quantity, ε, if the concept of local isotropy is applied. The idea behind this assumption is that, for high Reynolds numbers, the turbulent spectra extend over a very large range of wave numbers, and that the contribution to $\overline{(\partial u_i/\partial x_k)(\partial u_j/\partial x_k)}$ comes mainly from the large wave number part of the spectra. The motion of the large wave number components is statistically independent of the energy containing part of the spectra and assumes a universal, isotropic form. The isotropic tensor has the form

$$\nu\overline{(\partial u_i/\partial x_k)(\partial u_j/\partial x_k)} = \delta_{ij}\varepsilon/3, \tag{6}$$

where ε equals the rate of energy dissipation per unit mass and time

$$\varepsilon = \nu\overline{(\partial u_i/\partial x_k)(\partial u_i/\partial x_k + \partial u_j/\partial x_k)}. \tag{7}$$

At high Reynolds numbers, the process of viscous dissipation is governed by a continued energy transfer from big to small eddies, such that the rate of dissipated energy is independent of viscosity. The magnitude of ε may then be expressed by the equation

$$\varepsilon = c(\overline{q^2}/2)^{3/2}/L, \tag{8}$$

where L is a length scale of the big eddies and c is a dimensionless coefficient. Many authors have made use of this simplified approach, which is referred to as high Reynolds number approximation. However, an extension of the validity of the relations to lower Reynolds numbers is desirable with regard to the flow near solid walls and laminar turbulent transition.

The correlations of pressure fluctuations with rate of strain fluctuations, $\overline{p(\partial u_i/\partial x_j + \partial u_j/\partial x_i)}$, have the same importance as the production and generation terms, but their estimation is more difficult. The pressure fluctuations can be expressed as a space integral of the velocity field and are presented as a sum of two parts,

$$p = p_T + p_M, \tag{9}$$

the one, p_T, being caused by the interaction of velocity fluctuations; the other, p_M, by the interaction of the mean velocity gradients with velocity fluctuations. Thus the pressure-strain term is approximated by two parts. The relation for the first part,

$$\frac{1}{\varrho}\overline{p_T\left(\frac{\partial u_i}{\partial x_j} + \frac{\partial u_j}{\partial x_i}\right)} = -k_p c\sqrt{\overline{q^2}/2}\,(\overline{u_i u_j} - \overline{q^2}\,\delta_{ij}/3)/L \tag{10}$$

with coefficient k_p is now accepted by many authors. With regard to the second part, one obtains from the Poisson equation

$$\frac{1}{\varrho}\overline{p_M\frac{\partial u_i}{\partial x_j}} = -\frac{1}{2\pi}\int_{V(r)}\frac{\partial}{\partial r_l}U_k(\boldsymbol{x}+\boldsymbol{r})\left[\frac{\partial^2}{\partial r_j\,\partial r_k}R_{il}(\boldsymbol{x},\boldsymbol{r}) - \frac{\partial^2}{\partial x_j\,\partial r_k}R_{il}(\boldsymbol{x},\boldsymbol{r})\right]\frac{d\boldsymbol{r}}{r}. \tag{11}$$

Here

$$R_{jl}(\boldsymbol{x},\boldsymbol{r}) = \overline{u_j(\boldsymbol{x})\,u_l(\boldsymbol{x}+\boldsymbol{r})} \tag{12}$$

is the two-point correlation tensor, defined in Fig. 1. In general, the available information about the correlation tensor is not sufficient to exactly evaluate the integrals, Eq. (11), which extend over the whole volume of $\boldsymbol{r}$. But the equation may serve as a guide for making estimates. For example, it can be evaluated by assuming the isotropic form for the correlation functions.

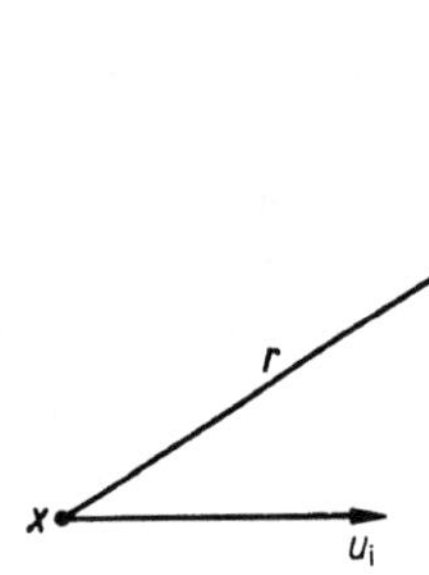

Fig. 1. Two-point velocity correlation function $R_{ij}(\boldsymbol{x},\boldsymbol{r}) = \overline{u_i(\boldsymbol{x})\,u_j(\boldsymbol{x}+\boldsymbol{r})}$.

5. Review of investigated models

A calculation method comprises a turbulence model, i.e. the mathematical formalism, and the numerical method to solve this formalism. It is not surprising that the transport equations have practically been used in a variety of combinations. Clearly, the tensor of the Reynolds stresses has 6 independent coordinates in the general case, thus the complete set of transport equations contains 6 equations. In the case of turbulent flow with axisymmetric or plane mean velocity fields, usually only one coordinate of the Reynolds stress tensor is important with respect to the mean velocity field, namely the shear stress $-\varrho\overline{uv}$. However, in the transport equation for this shear stress, there appears the normal stress $\overline{v^2}$. In the transport equation for $\overline{v^2}$, there occur the normal stresses $\overline{u^2}$ apd $\overline{w^2}$. Thus, strictly speaking, at least four transport equations are required from the point of view of the exact Reynolds stress transport equations.

With the empirical relationships, which are introduced for the various terms, eventually new flow quantities, for example, length scales, enter the equations. For these quantities, additional transport equations are needed in order to establish a complete system of equations. A great amount of empirical input is necessary to close the equations, and numerical solutions of such a system of partial differential equations, in conjunction with equations for the mean motion, require considerable expenditure of computer and programming time. On the other hand, computational methods, which are to be used by engineers, should be as simple as possible. Most of the practical methods are, therefore, based on a smaller number of transport equations. We know models which use only one transport equation, but there are also models which comprise as many as 5 transport equations even in simple cases of incompressible fluid flow.

In a review of what has been achieved up to the present time, it appears worthwhile to make a classification of the various schemes that have been used. One criterion of such a classification can be the number of transport equations used. Another criterion seems to be more significant in this context. It is the question: "Does the turbulence model form a closed set of partial differential equations which can be applied to different types of flow characterized by different boundary conditions, or is the empirical input tailored to a particular flow geometry?" For example, the length scale is often specified as a function of boundary layer thickness and distance from the wall. Such a model can only be applied to boundary layers along solid surfaces and not to free shear flows, unless the empirical input is modified. Models of this kind are called "not complete" here. A model is denoted as a "complete model"[1] if the empirical statements are not related to specific boundary conditions or to global parameters, depending on the flow geometry.

I have divided the existing models into the two groups "not complete models" and "complete models", because there exists a great demand for methods which predict as wide a range of flow problems as possible with a single formulation, and since methods based on the transport equations are potentially capable of accomplishing these requirements if the set of equations forms a "complete model". In what follows, only such models for which numerical results are available, are listed. Completeness of the lists is not claimed.

5.1 "Not complete models"

"Not complete models" are compiled in Table 1. The number and kind of transport equations is given. In addition, it is indicated, whether the high Reynolds number approximation is used or whether Reynolds number effects on the turbulent motion are included.

[1] The notation is arbitrary and should not be taken literally,

Table 1. "Not complete models"

Authors	Year	Ref.	Number Transport Equations	Transport Equations for	Viscous Effects[1]	Applied to
Prandtl	1945	5	1	$\overline{q^2}/2$	0	plane channel flow plane
Glushko	1965	9	1	$\overline{q^2}/2$	1	boundary layer $p = \text{const}$
Beckwith/Bushnell	1968	10	1	$\overline{q^2}/2$	1	plane boundary layers $p \neq \text{const}$, laminar/turbulent transition
Bradshaw et al.	1967	11	1	$\overline{q^2}/2$	0	boundary layer, arbitrary $p(x)$
Nee/Kovasznay	1969	12	1	eddy viscosity	1	boundary layers, $p = \text{const}$
Launder/Ying	1971	13	1	$\overline{q^2}/2$	0	square duct
Donaldson et al.	1968	14	4	$\overline{u^2}, \overline{v^2}, \overline{w^2}, \overline{uv}$	1	boundary layer, $p = \text{const}$
Donaldson et al.	1971	15	4	$\overline{u^2}, \overline{v^2}, \overline{w^2}, \overline{uv}$	1	free jet, free shear layer
Donaldson et al.	1971	16	7	$\overline{u^2}, \overline{v^2}, \overline{w^2}, \overline{uv}, \overline{uT}, \overline{vT}, \overline{T^2}$	1	isolated vortex, atmospheric-shear layers

[1] 0 High Reynolds number approximation; 1 Reynolds number effects included.

5.1.1 Models with one transport equation. There are a number of models, which are based on a single transport equation.

Method of Prandtl

Several methods, which lean upon *Prandtl's* [5] idea, use the transport equation for the kinetic fluctuation energy, Eq. (5), in which relationships for the rate of energy dsisipation and turbulent diffusion are introduced. For two-dimensional mean flows this equation reads

$$\underbrace{U\frac{\partial(\overline{q^2}/2)}{\partial x}+V\frac{\partial(\overline{q^2}/2)}{\partial y}}_{\text{convection}}+\underbrace{\overline{uv}\frac{\partial U}{\partial y}}_{\text{production}}+\underbrace{c\frac{(\overline{q^2}/2)^{3/2}}{L}}_{\text{dissipation}}-\underbrace{\frac{\partial}{\partial y}\left\{k_q\sqrt{\frac{\overline{q^2}}{2}}\,L\frac{\partial(\overline{q^2}/2)}{\partial y}\right\}}_{\text{diffusion}}=0\,, \tag{13}$$

where c and k_q are empirical coefficients. U and V denote the components of mean velocity in direction of x and y coordinate, respectively. The Reynolds shear stress, $-\varrho\overline{uv}$, is approximated by the eddy viscosity formula:

$$-\overline{uv}=k\sqrt{\overline{q^2}/2}\,L\frac{\partial U}{\partial y}\,, \tag{14}$$

where k is another dimensionless coefficient. For the dissipation Eq. (8) is used, and the turbulent diffusion is approximated by a gradient exchange relation analog to Eq. (14). The application to the logarithmic law of the wall provides an important relation between the coefficients, c and k. *Wieghardt* has used these relations to treat the flow through a two-dimensional channel. *Glushko* [9] extended this model with respect to the Reynolds number effect and performed calculations for the two-dimensional boundary layer at constant pressure. *Beckwith* and *Bushnell* [10] repeated *Glushko*'s computations and extended them to boundary layers with arbitrary pressure distributions.

Method of Bradshaw

Bradshaw et al. [11] developed a calculation method for two-dimensional boundary layers. The authors replaced *Prandtl's* Eq. (14), by assuming a constant ratio of Reynolds shear stress to kinetic fluctuation energy

$$-\overline{uv}=\frac{\tau}{\varrho}=a_1q^2\,. \tag{15}$$ [1]

With this assumption, the transport equation for the kinetic fluctuation energy is converted into a transport equation for the shear stress:

$$\underbrace{U\frac{\partial}{\partial x}\left(\frac{\tau}{2a_1\varrho}\right)+V\frac{\partial}{\partial y}\left(\frac{\tau}{2a_1\varrho}\right)}_{\text{convection}}-\underbrace{\frac{\tau}{\varrho}\frac{\partial U}{\partial y}}_{\text{production}}+\underbrace{\frac{(\tau/\varrho)^{3/2}}{L}}_{\text{dissipation}}+\underbrace{\left(\frac{\tau_{\max}}{\varrho}\right)^{1/2}\frac{\partial}{\partial y}\left(\frac{\tau}{\varrho}G\right)}_{\text{diffusion}}=0\,. \tag{16}$$

The turbulent diffusion is approximated by

$$\overline{(q^2/2+p/\varrho)\,v}=\left(\frac{\tau_{\max}}{\varrho}\right)^{1/2}\frac{\tau}{\varrho}G\,, \tag{17}$$

where $\tau_{\max}$ is the maximum value of the shear stress and G a non-dimensional function, which is defined in Fig. 2. With this expression the equations have hyperbolic properties and a method of characteristics is developed for numerical integration.

The numerical results of *Prandtl*'s as well as *Bradshaw*'s method depend strongly on the choice of the empirical distributions of L. Figure 3 shows the distributions of L, used by *Glushko* (*Beckwith*/*Bushnell*) and *Bradshaw*. The assumptions of the two authors are, in fact, not very different. The method of *Bradshaw* has found many users and produces results in good agreement with real flows. The method is extended to temperature boundary layers, boundary

---

[1] The value of a_1 was put equal to 0.15.

layers in compressible flow, with suction and blowing, and on rough surfaces as well as to three-dimensional boundary layers. As an example, Fig. 4 shows the comparison of the predicted and experimental mean velocity profile of an equilibrium boundary layer in adverse pressure gradient.

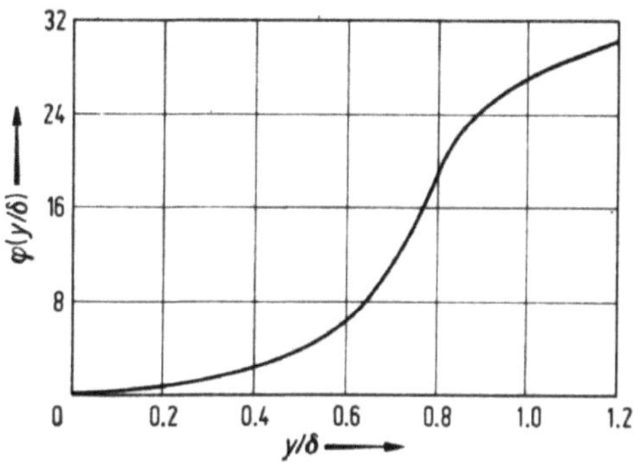

Fig. 2. *Bradshaw*'s diffusion function, G, of Eq. (17). $G = \left(\frac{\tau_{max}}{\varrho U_e^2}\right)^{1/2} \varphi(y/\delta)$. U_e = mean velocity at edge of boundary layer; δ = boundary layer thickness.

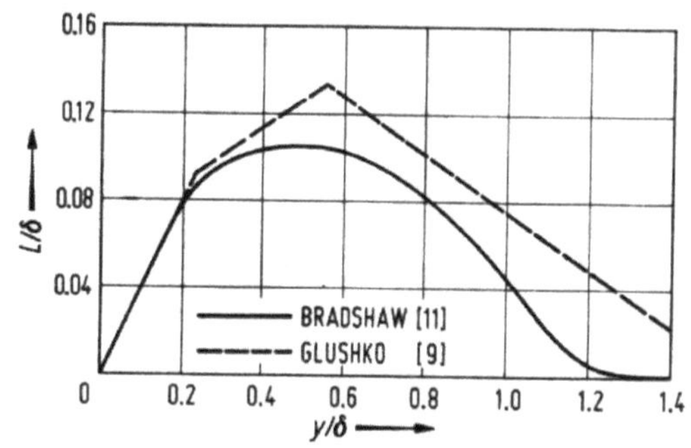

Fig. 3. Length scale distributions.

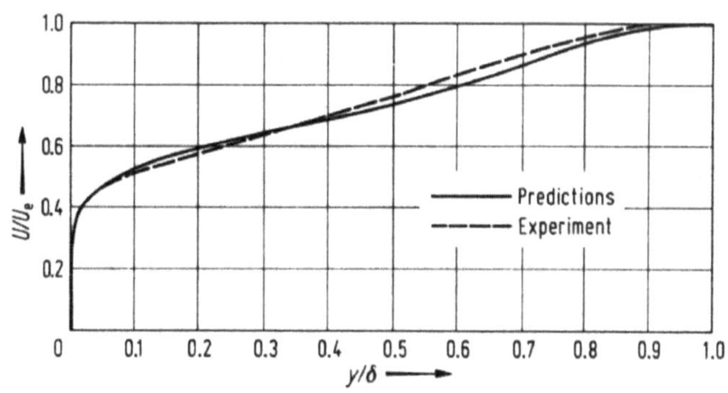

Fig. 4. Mean velocity profile of equilibrium boundary layer. $U_e \sim x^{-0.255}$ (*Bradshaw* et al. [11]).

Prandtl's and *Bradshaw*'s equations both reduce to the mixing length formula, if the transport terms are omitted. Since boundary layer calculation methods based on the mixing length formula are known to give reasonable results, it is not surprising that *Prandtl*'s and *Bradshaw*'s models yield good results too. The approximate inclusion of convective and diffusive transport effects, in any case, is an improvement, even if the determination of the transport is quantitatively not correct.

Transport equation for the eddy viscosity

A transport equation for the eddy viscosity which was applied to boundary layer calculations has been proposed by *Nee* and *Kovasznay* [12]. In principle, the transport equation for the eddy viscosity seems to be justified not more and not less than the *Prandtl* 1945 model, but it is a wholly empirical equation. As *Bradshaw* [1] has mentioned, attempts to provide a theoretical basis for the eddy viscosity transport equation lead to results which are at variance with the exact transport equations for the Reynolds stresses.

Secondary flows in ducts of square cross section

A very interesting work on the calculation of secondary flows in ducts of square cross section, which shows how powerful the Reynolds stress transport equations really are, has been provided by *Launder* and *Ying* [13].

In Table 1 the number 1 is inserted for the number of transport equations used (it is the transport equation for kinetic fluctuation energy) whereas, in fact three transport equations are used. But the two transport equations for the stresses normal to the axis of the duct reduce to the pressure strain term. It was found that the stress field in the cross sectional plane of the duct arises as a consequence of mean rates of strain in planes parallel to the longitudinal axis of the duct. The Reynolds shear stresses in planes parallel to the longitudinal axis are determined using *Prandtl's* eddy viscosity equation. The calculated flow properties are in remarkably good agreement with experimental data. This is shwon in Fig. 5, where the distributions of mean cross flow velocity, U_1, at different distances, x_1, from the centre line are plotted (h = half width of duet).

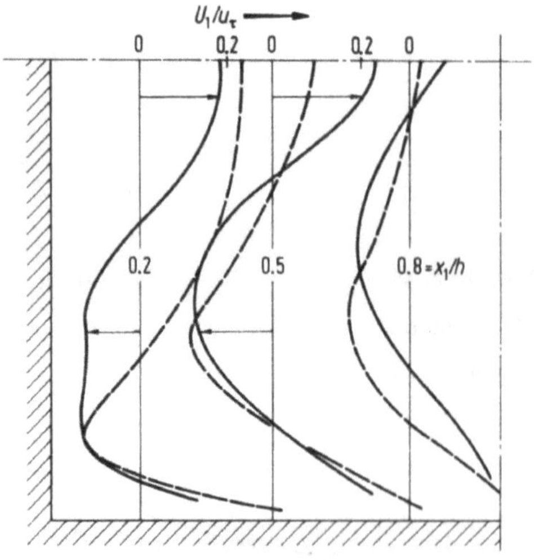

Fig. 5. Secondary velocity profiles in a duct of square cross section (*Launder/Ying* [13]). ——— Predictions Re = 225,000; – – – Experiment Re = 215,000; u_τ = average wall friction velocity; Re based on hydraulic diameter.

5.1.2 Models with many transport equations. *Donaldson* [14, 15, 16] has used a transport equation for each Reynolds stress. He neglects the contributions of the mean velocity gradients to the pressure strain terms and introduces simple relations for the length scales. This model has been applied to calculate the turbulent boundary layer at constant pressure flow, axisymmetric free jet and free shear layer. In addition, chemically reacting flows, atmospheric shear layer, and the decay of an isolated vortex were calculated by simultaneously integrating up to 7 transport equations.

5.2 "Complete models"

Results produced with the "not complete models" are superior to those of the mean field methods, but the disadvantage that special assumptions have to be made for each type of flow has not been overcome. The specific field of application for the transport equations are "com-

Table 2. "Complete models"

Authors	Year	Ref.	Number Transport Equations	Transport Equations for	Viscous Effects[1]	Applied to
Rodi/Spalding	1970	22	2	$\overline{q^2}/2, \overline{q^2 L}/2$	0	free jet, mixing layer
Ng/Spalding	1972	23	2	$\overline{q^2}/2, \overline{q^2 L}/2$	0	flat plate, pipe, two-dimensional channel, wall jet
Rotta	1971	25	3	$\overline{q^2}/2, \overline{uv}, \overline{q^2 L}/2$	0	wake, pipe, two-dimensional channel
Hanjalić/Launder	1972	21	3	$\overline{q^2}/2, \overline{uv}, \varepsilon$	0	diversity of two-dimensional flows
Daly/Harlow	1970	20	5	$\overline{u^2}, \overline{v^2}, \overline{w^2}, \overline{uv}, \varepsilon$	1	two-dimensional channel

[1] 0 High Reynolds number approximation; 1 Reynolds number effects included.

plete models", which apply to a wide class of flows. Formalisms of this kind are characterized by absence of any global parameters like boundary layer thickness or some similar quantity. All relations and coefficients appearing in the formalism are supposed to have universal validity (within a certain range of applicability). The methods compiled in Table 2 may be considered as being "complete" in this sense. It is seen that, in addition to the transport equations for the Reynolds stresses, there is a transport equation either for a length scale or for the rate of dissipation, ε. Therefore, the theoretical background of these equations shall be discussed next.

5.2.1 Length scale transport equation. The length scales appearing in the dissipation formula and other empirical relations are characteristic lengths of the energy containing eddies. Thus they are comparable to the integral length scales of two-point velocity correlation functions. This notion implies a variety of integral length scales depending on the choice of velocity components involved and the direction of integration. We shall restrict the consideration to flows for which the usual boundary layer simplifications are admissible and discuss a transport equation for the integral length scale defined by

$$\overline{q^2}L = \frac{3}{8} \int_{-\infty}^{\infty} R_{ii}(r_y)\, dr_y, \tag{18}$$

where

$$R_{ii}(r_y) = \overline{u_i(x, y, z)\, u_i(x, y + r_y, z)} \tag{19}$$

is the two point correlation function, Fig. 6, and

$$\overline{q^2} = \overline{u_i(x, y, z)\, u_i(x, y, z)} = R_{ii}(0). \tag{20}$$

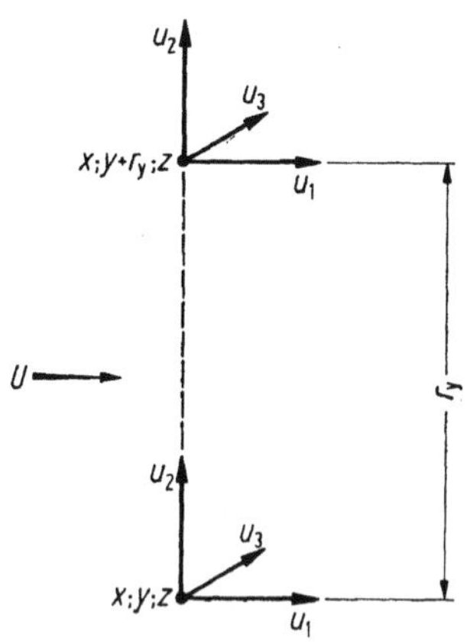

Fig. 6. Definition of two-point correlation function and integral length scale, Eq. (18).

The factor 3/8 in Eq. (18) causes L to be identical with the well known lateral integral length scale of an isotropic turbulence field. If solid walls are present, the range of integration is limited by the solid wall. For general flow cases, a spherically integrated length scale may perhaps be a more appropriate quantity.

The transport equation for the product $\overline{q^2}L/2$ is deduced from the time-dependent Navier-Stokes equations in a manner similar to that used to deduce the transport equation for the kinetic fluctuation energy [8]. The equation given here is a simplified form for two-dimensional mean flow at high Reynolds numbers:

$$\underbrace{\frac{1}{2}\frac{\partial}{\partial x}(U\overline{q^2}L) + \frac{1}{2}V\frac{\partial(\overline{q^2}L)}{\partial y}}_{\text{convection}} + \underbrace{\frac{3}{16}\left[\left(\frac{\partial U}{\partial y}\right)_y \int_{-\infty}^{\infty} R_{21}\, dr_y + \int_{-\infty}^{\infty}\left(\frac{\partial U}{\partial y}\right)_{y+r_y} R_{12}\, dr_y\right]}_{\text{production}}$$

$$\underbrace{-\frac{3}{16}\int_{-\infty}^{\infty}\frac{\partial}{\partial r_k}[R_{(ik)i} - R_{i(ik)}]\, dr_y}_{\text{destruction}} + \underbrace{\frac{3}{16}\frac{\partial}{\partial y}\left\{\int_{-\infty}^{\infty}\left[R_{(i2)i} + \frac{1}{\varrho}(\overline{pv} + \overline{vp})\right] dr_y\right\}}_{\text{turbulent diffusion}} = 0, \tag{21}$$

where the following correlation functions occur:

$$\begin{aligned} R_{12} &= \overline{u(y)\, v(y + r_y)}\,, \\ R_{21} &= \overline{u(y + r_y)\, v(y)}\,, \\ R_{(ik)j} &= \overline{u_i(y)\, u_k(y)\, u_j(y + r_y)}\,, \\ \overline{pv} &= \overline{p(y)\, v(y + r_y)}\,, \\ \overline{vp} &= \overline{p(y + r_y)\, v(y)}\,. \end{aligned} \tag{22}$$

It is seen that the product, $\overline{q^2L}/2$, and the fluctuation energy, $\overline{q^2}/2$, are subject to similar effects There is convection by mean flow, production, destruction and diffusive transport by turbulent interactions. The production term (generation by interaction of mean and turbulent motion) takes a very complicated form in contrast to the simple expression for the production term of the transport equation for the kinetic fluctuation energy, Eq. (5). Instead of the local value, the distribution of mean shear, $\partial U/\partial y$, of the whole plane, $x = \text{const}$, effects the production. With the aid of a Taylor expansion for $\partial U/\partial y$, the production term may be represented by a series, on the basis of which the approximation

$$\frac{3}{16}\left[\left(\frac{\partial U}{\partial y}\right)_y \int_{-\infty}^{\infty} R_{21}\, dr_y + \int_{-\infty}^{\infty} \left(\frac{\partial U}{\partial y}\right)_{y+r_y} R_{12}\, dr_y\right] = \overline{uv}\left[\zeta L \frac{\partial U}{\partial y} + \zeta_3 L^3 \frac{\partial^3 U}{\partial y^3}\right] \tag{23}$$

is suggested, where ζ and ζ_3 are two empirical constants.

The destruction term is physically a transport of kinetic energy in wave number space. In analogy to Eq. (8) the relationship

$$-\frac{3}{16}\int_{-\infty}^{\infty} \frac{\partial}{\partial r_k}\left[R_{(ik)i} - R_{i(ki)}\right] dr_y = c_L c \left(\frac{\overline{q^2}}{2}\right)^{3/2} \tag{24}$$

is used. An estimation of the coefficient, c_L, is gained from a comparison with the decay law for kinetic fluctuation energy in a uniform stream behind a grid.

5.2.2 Transport equation for the rate of turbulent energy dissipation. No length scale appears explicitly in the Reynolds stress transport equations. This is the main motivation for some authors to look at the transport equation for the rate of turbulent energy dissipation as an additional equation.

Davydov [17] and independently *Harlow* and *Nakayama* [18] derived the dissipation equation in exact form from the Navier-Stokes equations, using the simplified expression,

$$\varepsilon \approx \nu\overline{(\partial u_i/\partial x_k)\,(\partial u_i/\partial x_k)}\,, \tag{25}$$

for the dissipation. If one starts from the full expression, Eq. (7), the following somewhat different form is obtained:

$$\begin{aligned} &\underbrace{U_k \frac{\partial \varepsilon}{\partial x_k}}_{\text{convection}} + \underbrace{2\nu \frac{\partial U_i}{\partial x_k}\left[\overline{\frac{\partial u_k}{\partial x_j}\left(\frac{\partial u_i}{\partial x_j} + \frac{\partial u_j}{\partial x_i}\right)} + \overline{\frac{\partial u_j}{\partial x_i}\left(\frac{\partial u_j}{\partial x_k} + \frac{\partial u_k}{\partial x_j}\right)}\right] + 2\nu \frac{\partial^2 U_i}{\partial x_k\, \partial x_j}\overline{u_k\left(\frac{\partial u_i}{\partial x_j} + \frac{\partial u_i}{\partial x_j}\right)}}_{\text{generation by mean motion}} \\ &+ \underbrace{2\nu \overline{\frac{\partial u_k}{\partial x_j}\frac{\partial u_i}{\partial x_k}\left(\frac{\partial u_i}{\partial x_j} + \frac{\partial u_j}{\partial x_i}\right)}}_{\text{generation by vortex stretching}} + \underbrace{2\nu^2 \left[\overline{\frac{\partial^2 u_i}{\partial x_k\, \partial x_j}\frac{\partial}{\partial x_k}\left(\frac{\partial u_i}{\partial x_j} + \frac{\partial u_j}{\partial x_i}\right)}\right]}_{\text{dissipation of vorticity}} \\ &\underbrace{- \nu \frac{\partial}{\partial x_k}\left[\overline{u_k \frac{\partial u_i}{\partial x_j}\left(\frac{\partial u_i}{\partial x_j} + \frac{\partial u_j}{\partial x_i}\right)} + 4\,\overline{\frac{\partial p}{\partial x_j}\frac{\partial u_k}{\partial x_j}} - \frac{\partial \varepsilon}{\partial x_k}\right]}_{\text{diffusion}} = 0\,. \end{aligned} \tag{26}$$

The deviations of this equation from that given in the mentioned papers are immaterial with respect to the use as part of calculation procedures.

The equation has again all the terms typical for transport equations of turbulent flows. There is convective and diffusive transport, as well as generation and destruction by viscous forces (vorticity dissipation). The physical meaning of the terms has been repeatedly misinterpreted, as was shown by *Rodi* [19], and, consequently, strange closure approximations are offered in the literature. Most important are the generation and viscous destruction terms. There is a generation by the mean velocity gradients and a generation of vorticity by vortex stretching due to turbulent interactions. The latter effect is connected with the energy transfer between eddies of different sizes. In high Reynolds number flows the vorticity generation and the vorticity dissipation are the dominating terms. Both terms have a magnitude of order $\varepsilon^{3/2}/\nu^{1/2}$ and are almost in equilibrium with each other. The small difference between the two terms comes partly from the big eddies and partly from smaller eddies. The first member of the generation by mean velocity gradients reduces to

$$2\nu \frac{\partial U_i}{\partial x_k} \left[\overline{\frac{\partial u_k}{\partial x_j} \left(\frac{\partial u_i}{\partial x_j} + \frac{\partial u_j}{\partial x_i} \right)} + \overline{\frac{\partial u_j}{\partial x_i} \left(\frac{\partial u_j}{\partial x_k} + \frac{\partial u_k}{\partial x_j} \right)} \right] = \frac{8}{5} \frac{\partial U_i}{\partial x_k} \varepsilon\, \delta_{ik}, \tag{27}$$

if the concept of local isotropy is applied. Because there is the tendency of turbulence fields to approach the conditions of local isotropy with growing Reynolds number, it follows from Eq. (27) that in a simple shear flow the generation of vorticity by the mean velocity gradient tends to zero. The second member of the generation by mean velocity gradients also becomes very small at high Reynolds numbers. In the light of these facts, it appears that the relevance of the dissipation transport equation decreases as the Reynolds number increases. The appropriate closure of the dissipation transport equation relies much more on empiricism than does closure of the Reynolds stress transport equations, since terms which are governed by the small scale motion have to be related to quantities produced by the large scale motion. The closed dissipation transport equations which have been constructed by *Daly* and *Harlow* [20] and *Hanjalić* and *Launder* [21] can hardly be taken for faithful imitations of the exact dissipation equation (26).

So far, only the transport equation for the scalar quantity, ε, has been considered. But in the case of general flows at lower Reynolds numbers, where locally isotropic conditions do not prevail, six individual viscous destruction terms, $2\nu\overline{(\partial u_i/\partial x_k)\,(\partial u_j/\partial x_k)}$, appear for which a system of six coupled transport equations exists. The present author is not aware of any investigations of these equations, which may perhaps provide the theoretical framework necessary to incorporate the Reynolds number effects in the calculation method.

5.2.3 Models with two transport equations. In view of the success which has been achieved with "not complete models", based on one transport equation it appears to be a logical step to devise simple "complete models" by adding a second transport Equation [22, 23, 24]. The oldest two-transport-equation model is that proposed by *Kolmogorov* [7]. *Spalding* together with *Rodi* [22] and *Ng* [23], investigated a "complete model" which consists of *Prandtl*'s 1945 model, Eqs. (13), (14), supplemented by a length scale equation of the simplified form

$$\underbrace{\frac{1}{2} U \frac{\partial(\overline{q^2}L)}{\partial x} + \frac{1}{2} V \frac{\partial(\overline{q^2}L)}{\partial y}}_{\text{convection}} + \underbrace{\overline{uv}\zeta L \frac{\partial U}{\partial y}}_{\text{production}} + \underbrace{c_L c \left(\frac{\overline{q^2}}{2}\right)^{3/2}}_{\text{destruction}}$$

$$\underbrace{- \frac{\partial}{y^i} \frac{\partial}{\partial y} \left\{ y^j k_{qL} \sqrt{\overline{q^2}/2}\, L \left[L \frac{(\partial \overline{q^2}/2)}{\partial y} + \alpha_L \frac{\overline{q^2}}{\partial L} \frac{\partial L}{\partial y} \right] \right\}}_{\text{turbulent diffusion}} = 0, \tag{28}$$

which includes the four empirical coefficients, ζ, c_L, k_{qL}, α_L (c occurred already in the energy transport equation). This equation applies to plane ($j = 0$) and axisymmetric ($j = 1$) mean flows and contains the high Reynolds number approximation. This model implies that the structure of turbulence is described by two quantities, namely the kinetic fluctuation energy and the length scale L. The equations were solved first for various free jet flow cases. It turned out that no set of the empirical coefficients involved produced exact agreement with all the relevant experimental data. Thus a set of best-fit constants was determined, which gives fair agreement in different cases. Figure 7 shows the results for the plane jet. When the model was applied to boundary layers near walls, it was found that at least one of the coefficients must have a different value in the vicinity of a wall from that which prevails when walls are remote. Therefore, the coefficient c_L was allowed to vary in accordance with the relation

$$c_L = c_L' + c_w (L/y)^q, \tag{29}$$

where c_L', c_w, and q are truly constant. With this modification the predictions of the model for the flow cases of flat plate boundary layer, fully developed flow through a pipe and a two-dimensional channel, and of plane wall jet were found to compare favorably with the experimental data.

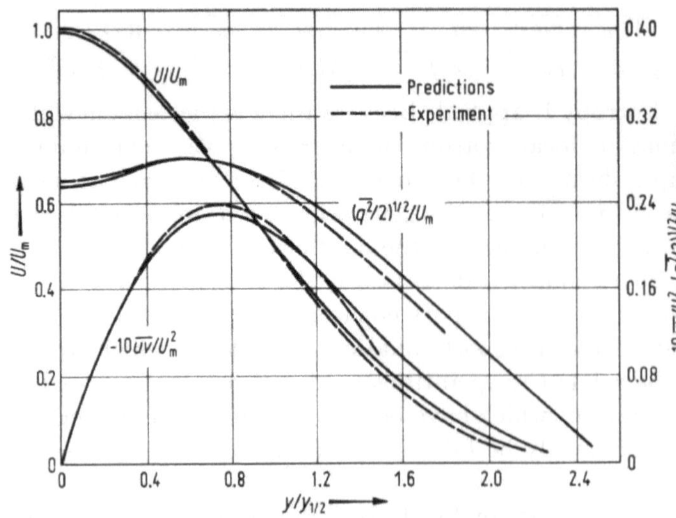

Fig. 7. Plane jet in a fluid at rest. Distributions of mean velocity, U, shear stress, $\overline{uv}$, and square root of kinetic fluctuation energy, $\sqrt{\overline{q^2}/2}$. Predictions by *Rodi/Spalding* [22]; $y_{1/2}$ = half width of jet (at $U = U_m/2$); U_m = maximum mean velocity.

There is, of course, no justification from theoretical arguments for the variation of c_L with y. It is not plausible why one coefficient should vary with distance from the wall while the others remain constant. Apart from this question, the merits of *Ng*'s, *Rodi*'s and *Spalding*'s work is that, for the first time, various turbulent flow cases could be predicted with the same empirical input.

5.2.4 Models with three transport equations. A "complete model", using three transport equations has been proposed by the author [25]. This model differs from the model just mentioned in that *Prandtl*'s eddy viscosity relation, Eq. (14), was replaced by a transport equation for the shear stress,

$$\underbrace{U\frac{\partial \overline{uv}}{\partial x} + V\frac{\partial \overline{uv}}{\partial y}}_{\text{convection}} + \underbrace{a_p \frac{\overline{q^2}}{2}\frac{\partial U}{\partial y}}_{\text{production}} + \underbrace{k_p c\sqrt{\overline{q^2}/2}\,\frac{\overline{uv}}{L}}_{\text{destruction}} - \underbrace{k_{q\tau}\frac{\partial}{\partial y}\left(\sqrt{\overline{q^2}/2}\,L\frac{\partial \overline{uv}}{\partial y}\right)}_{\text{turbulent diffusion}} = 0, \tag{30}$$

which again is a high Reynolds number approximation; a_p, k_p, and $k_{q\tau}$ are additional coefficients[1]. The production term includes the contribution of the pressure-strain term due to interaction with mean flow. The destruction is the pressure-strain term due to turbulent interaction, Eq. (10). In addition, a more complicated production term in accordance with Eq. (21) was used in the length scale equation. Calculations performed so far reveal the necessity to simulate the production term of the length scale equation properly. With the use of the two

[1] Note that Eq. (14) is the solution of Eq. (30), if convection and turbulent diffusion is neglected.

term approximation, Eq. (23), more refined finite difference schemes must be applied in order to avoid numerical instability.

With the "complete model" advanced by *Hanjalić* and *Launder* [21], the transport equation for the length scale is replaced by a transport equation for the rate of energy dissipation, while the transport equations for shear stress and kinetic fluctuation energy are basically the same as before (Eqs. (30) and (13) respectively). This model has been applied to the prediction of thin shear flow, including asymmetric channel flow, flow through a symmetric annulus, boundary layer at constant pressure, plane wall jet, plane jet in a fluid at rest, and plane mixing layer. Reasonable agreement with measured data has been achieved. Even such a feature as the non-coincidence of the positions of maximum velocity and zero shear stress in asymmetric flows are faithfully reproduced. To give an example, predicted and measured distributions of mean velocity and shear-stress in asymmetric channel flow are reproduced in Fig. 8. The asymmetry had been produced by fixing ribs of square cross-section to one wall aligned normally to the flow. These results provide another remarkable example of the fact that on the basis of transport equations calculation methods can be devised which can be used for a variety of flows.

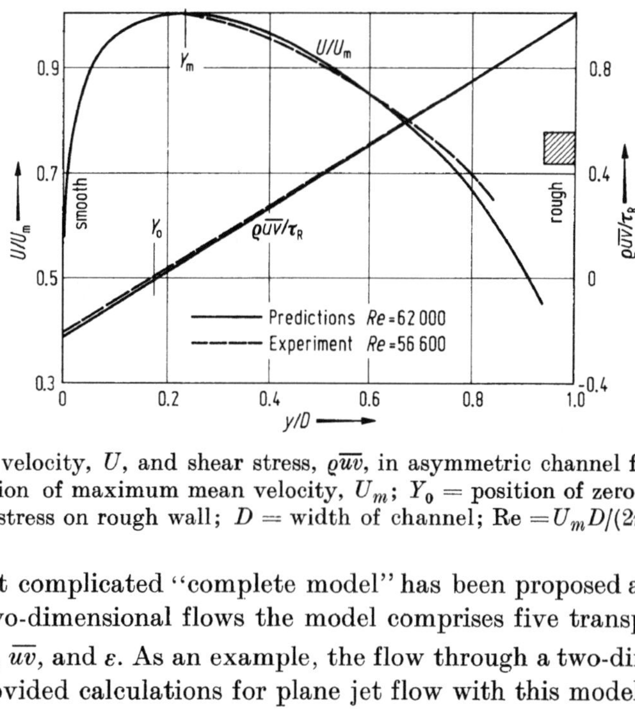

Fig. 8. Distribution of velocity, U, and shear stress, $\varrho\overline{uv}$, in asymmetric channel flow (*Hanjalić* and *Launder* [21]). Y_m = position of maximum mean velocity, U_m; Y_0 = position of zero shear stress; τ_R = shear stress on rough wall; D = width of channel; $\mathrm{Re} = U_m D/(2\nu)$.

Finally, the most complicated "complete model" has been proposed and tested by *Daly* and *Harlow* [20]. For two-dimensional flows the model comprises five transport equations for the quantities $\overline{u^2}$, $\overline{v^2}$, $\overline{w^2}$, $\overline{uv}$, and ε. As an example, the flow through a two-dimensional channel was calculated. *Rodi* provided calculations for plane jet flow with this model.

6. Concluding remarks

Within the space available for this lecture, it was not possible to give a complete account of the present state of the art. Many detailed questions had to be omitted, and the reader is referred to the original literature for further information. It is hoped that this survey, which mainly reflects a developer's point of view, will supplement the reviews of *W. C. Reynolds* [26] and *Bradshaw* [1].

The results achieved are encouraging and show that the transport equations for the Reynolds stresses may serve as a framework for generally applicable calculation methods. It must be stated, however, that only simple flow cases have been treated so far, for which models with two or three transport equations will suffice. Little work has become known with regard to complex flows, which require models with many transport equations .Further experimental work is needed before one can develop such refined methods. But much work remains to be done also in order to put the simpler models for two-dimensional layer-flows and axisymmetric flow on a firmer basis. Additional experiments to determine the terms of the Reynolds shear stress equation and the length scale equation would be helpful. Theoretical studies to improve the models should be undertaken. For example, any suitable proposals to incorporate the effect of intermittency of turbulence into the calculations are still missing. Furthermore, systematic investigations on how variations of empirical coefficients affect the predictions would be useful. Such investigations might start from given distributions of mean velocity. Preliminary results of such a study are described by the author [27]. Finally, the numerical procedures may be improved in order to overcome, the difficulties that are observed with higher order derivatives.

References

1. Bradshaw, P.: Aeron. J. 76 (1972) 403—418.
2. Buri A.: Eine Berechnungsgrundlage für die turbulente Grenzschicht bei beschleunigter und verzögerter Grundströmung. Diss. Eidgen. Tech. Hochschule Zürich 1931.
3. Gruschwitz, E.: Ing.-Arch. 2 (1931) 321—346.
4. Kline, S. J., Sovran, G., Morkovin, M. V., Cockrell, D. J. (eds.): 1968 AFOSR-IFP-Stanford Conference, I Stanford Univ., Cal., 1969.
5. Prandtl, L.: Nachr. Akad. Wiss. Göttingen, Math.-Phys. Kl. (1945) 6—19.
6. Rotta, J.: 1.Mitt.: Z. Phys. 129 (1951) 547—572; 2.Mitt.: Z. Phys. 131 (1951) 51—77.
7. Kolmogorov, A. N.: Izv. Akad. Nauk. SSSR, Ser. Fiz. VI (1942) 56—58.
8. Rotta, J. C.: Turbulente Strömungen. Stuttgart: Teubner 1972.
9. Glushko, G. S.: Izv. Akad. Nauk SSSR, Ser. Mek. No. 4 (1965) 13—23.
10. Beckwith, I.E., Bushnell, D. M.: NASA TN D-4815 (1968).
11. Bradshaw, P., Ferriss, D. H., Atwell, N. P.: J. Fluid Mech. 28 (1967) 593—616.
12. Nee, V. W., Kovasznay, L. S. G.: Phys. Fluids 12 (1969) 473—484.
13. Launder, B. E., Ying, W. M.: J. Fluid Mech. 54 (1972) 289—295.
14. Donaldson, C. duP., Rosenbaum, H.: NASA SP-216 (1969) 231—253.
15. Donaldson, C. duP.: AGARD-CP-93 (1972) B-1 to B-24.
16. Donaldson, C. duP.: AIAA Journ. 10 (1972) 4—12.
17. Davydov, B. I.: Soviet Physics-Doklady. Fluid Mech. 6 (1961) 10—12.
18. Harlow, F. H., Nakayama, P. I.: Phys. Fluids 10 (1967) 2323—2332.
19. Rodi, W.: Dept. Mech. Engng., Imperial College, London 1971. Report TM/TN/A/14.
20. Daly, B. J., Harlow, F. H.: Phys. Fluids 13 (1970) 2634—2649.
21. Hanjalić, K., Launder, B. E.: J. Fluid Mech. 52 (1972) 609—638.
22. Rodi, W., Spalding, D. B.: Wärme- und Stoffübertragung 3 (1970) 85—95.
23. Ng, K. H., Spalding, D. B.: Phys. Fluids 15 (1972) 20—30.
24. Saffman, P. G.: Proc. Roy. Soc. London A 317 (1970) 417—433.
25. Rotta, J. C.: AGARD-CP-93 (1972) A-1 to A-11.
26. Reynolds, W. C.: Dept. Mech. Engng, Stanford Univ., Cal., Rep. MD-27 (1970).
27. Rotta, J.C. DLR-FB 72-27, 1972, 347-362.

L. Sedov

Moscow, USSR

The increase of velocity of bodies moving in water and at its surface is one of the major technical tendencies which is of great importance for the development of hydrodynamics, marine science and technology, nowadays as well as it was in the past.

Recently, much attention has been paid to various hydrodynamic effects connected with high speed motion of solid bodies in water and, in particular, with effects of the dynamic interaction during fast body immersion into water.

Important problems arise when the impact and spreading of drops and jets on the surface of a body are investigated. The nature of these phenomena is closely connected with relative velocities and "solid" body properties.

The theory of gas bubble expansion and oscillation in water resulting from underwater explosions, electric discharges, laser beam focusing in liquids and so on is developed in numerous published papers.

Calculations show orders of magnitude of oscillation periods for large gas bubbles as well as for small ones. It was found that for small air bubbles in water (with radius up to 1 cm at atmospheric pressure) the duration of one oscillation is of the order of milliseconds. When the external water pressure decreases the oscillation periods of bubble radii increase and can attain values comparable with the time scale typical for macroscopic processes occurring in the case of motion of water with gas bubbles.

On the other hand, calculations showed some interesting properties of characteristic gas states in bubbles and in the liquid near the bubbles. For example, *H. G. Flynn* calculated the motion of argon bubbles in water when their radii increased many times due to perturbations and when the liquid pressure far from the bubble is constant and equal to atmospheric pressure. These calculations showed that if the initial equilibrium bubble radius was of the order of 10^{-4} cm then a bubble collapses in a time period of the order of microseconds, while peak pressures and temperatures within the bubble and in its vicinity are of the order of 10^4 atm and tens of thousands degrees Kelvin; these peak values arise when the radius is close to minimum and exist only for times of the order of 10^{-7} sec. Under these conditions it is necessary to take into account the water compressibility, "real" vapor or gas state equations in bubbles and heat conductivity effects in gas and water.

At the present time the dynamics of vapor bubble growth and oscillations in bubble chambers used for elementary particle track study has come into importance. In this connection, in particular, the behaviour of hydrogen gas bubbles in liquid hydrogen as well as vapor bubbles in water was investigated in the presence of an ultrasonic acoustic field; it has been shown by calculations that in these cases a non-linear resonance can arise for a number of frequencies.

The greatest difficulties arise when nonsymmetric bubble motions due to non-homogeneous pressure distribution and bubble instability are analyzed. Non-symmetry effects are strongly revealed at bubble collapse near a fluid boundary, at passing shock fronts and also in those cases in which the liquid's weight is essential.

The bubble surface is strongly deformed at bubble collapse near walls or because of a passing shock; as a result of this deformation a thin jet is formed at a bubble's surface and flows from there into the bubble. In case of a bubble collapse near a boundary wall this jet impinges upon the wall and can destroy it. The cavitation erosion is explained in many cases by the described effect.

Many papers are devoted to the theoretical and experimental analysis of the behaviour of a single bubble in water; they clarify a number of remarkable properties of this phenomenon, but up to date the main problems of dynamic growth, inception and collapse of gas and vapor bubbles still require additional clarification when the possibilities of gas diffusion from liquid into bubble, evaporation, condensation and unsteady heat transfer must be taken into account. The problems concerned with the motion of a liquid with a great number of gas or vapor bubbles belong to this class of problems, the liquid being considered a compressible continuum with special thermodynamic properties. Such motions have a number of practically important features resulting from the unusual compressibility nature of this medium and the unusual complicated mechanisms of macroscopic mechanical energy dissipation.

Some characteristic dynamical effects accompanied by cumulative jet formation at the merging of gas and liquid flows are very interesting. This can also occur at the edges of contact surfaces during solid body impacts accompanied by their melting; at the edges of wetted surfaces of bodies entering water; near the edges of wetted surfaces of water-gliding bodies; at the formation of a water fountain following explosions near or on the water surface, and so on.

The not so well known pictures of the formation of a thin liquid jet polarized by an electric field are very curious. Such jets were obtained and described by *G. Taylor* and later they were obtained at Moscow University by *V. V. Gogosov* and *A. K. Kazennov*; in their experiments the liquid jet discontinuities due to electric surface forces extracting the jet were observed (see Fig. 1). In practical situations extraction of liquid with an electric field is widely used to obtain colour jets when painting different objects. But there is still no well developed theory of these phenomena.

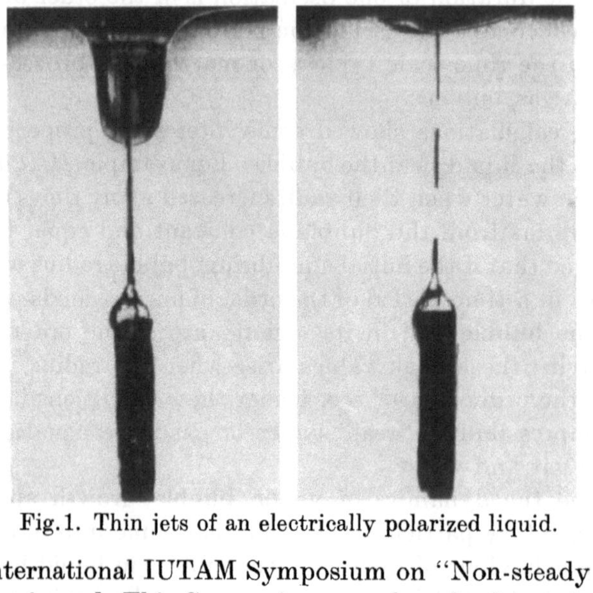

Fig. 1. Thin jets of an electrically polarized liquid.

In June 1971 the International IUTAM Symposium on "Non-steady flow of water at high speed" took place in Leningrad. This Symposium was largely devoted to some of the above listed phenomena. Modern investigations and achievements in this field were presented at the Symposium in many papers and, in particular, in excellent review papers by *G. Birkhoff* (USA), *V. E. Johnson* (USA), *J. E. Field* (UK), *H. G. Flynn* (USA) and in a joint paper by *G. V. Logvinovich* and *Yu. L. Yakimov* (USSR)[1]. On the one hand these materials make my

[1] All papers presented at the Symposium were published in the "Proceedings" of the Symposium (Moscow, 1973). The papers by Russian authors were published in English in a special rotaprint edition as an addition to the Symposium Proceedings. The review papers translated into Russian were published in the Collection of Translations on Mechanics "Mekhanika" (Moscow, 1972—1973).

task—to deliver a lecture devoted to the Leningrad Symposium—more complicated because it is difficult to say something new, but on the other hand this fact simplifies my task for it is not necessary to repeat many things which were recently discussed. In connection with this, I shall try, in my lecture, to emphasize and to develop in explicit form only some fundamental results directly or indirectly concerned with this Symposium.

The examples of hydrodynamic phenomena mentioned above are associated with investigations of steady and unsteady fluid flows for which the main features of the essential effects are determined by inertia properties of the fluid and the solid body in the flow. In many cases, when the relative flow velocities or accelerations are rather high, it is possible to consider water as an ideal incompressible fluid and this model allows to describe directly the more important real characteristic features of the phenomena or to obtain a base for further analysis and explanation of them. At the same time one needs to emphasize that in many cases heat conductivity, compressibility, gravity, viscosity and phase transformations can be of prime importance.

Let us consider the main problem of high speed body motion in water. The formulation and solution of problems of continuous perfect incompressible fluid flows near bodies is of fundamental value. The classical flow scheme is shown in Fig. 2. In this case, for constant-velocity steady flow about a body, D'Alembert's paradox is valid. Maximal pressures are obtained at critical points at the nose (point A) and at the afterbody (point B) where the ideal pressure is fully recovered.

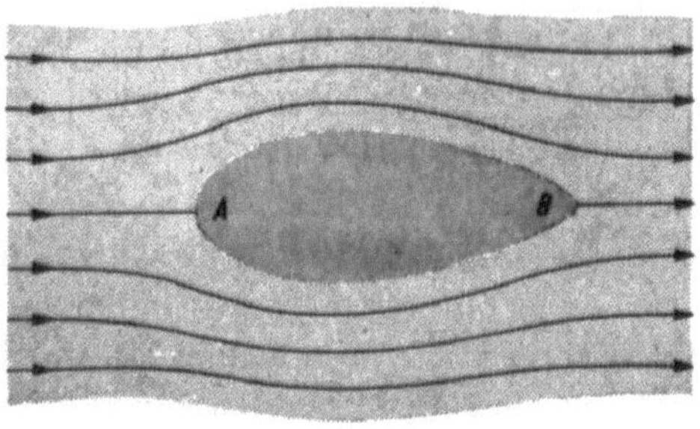

Fig. 2. Scheme of a continuous perfect fluid flow.

For streamlined oblong bodies of airship type this flow scheme allows to obtain correct pressure distributions at the nose and in the middle part of a body surface. But at the rear part, pressure recovery is not full because of viscous losses; this is why along with the main part of the drag (friction drag) the additional shape drag arises, and a vortex wake appears behind the body. In steady flows this vortex wake, slowly expanding, proceeds downstream to infinity.

It has been known for a long time that at rather high speed motion of bodies of a fixed volume realization of this scheme of smooth flow turns out to be disadvantageous because of the large friction drag.

At the dawn of aerodynamics, in the earliest theories devoted to plane problems of the origin of circulation and to steady-state flows past finite bodies and developed in the framework of ideal fluid conceptions, dynamic lift and "inductive drag" effects were explained by means of correct flow schemes with discontinuity interfaces for tangential velocities. Schemes of ideal fluid flows past bodies, constructed in such a way, with discontinuous velocity fields along the "vortex sheet", rather well simulated the really observed thin vortex sheets extending downstream into the liquid from a streamlined body's surface.

In this connection it is useful to direct the attention to the fact that in the framework of ideal fluid theory there is no unique solution of the problem of gas or liquid flows past bodies, especially for "badly streamlined" bodies when in the flow past these bodies discontinuities of the velocity field, stagnation regions, vacuum cavities, or cavities filled with gas or vapor are allowed. For a long time many authors treated flow past bodies according to Kirchhoff's flow scheme with a "stagnation fluid region" as a mathematical solution far from reality.

At present we have pictures of real flows past bodies penetrating the water surface at a great speed and pictures of high speed flows past bodies characterized by cavitation numbers

$$\sigma = \frac{2(p_\infty - p_c)}{\varrho v_\infty^2}$$

close to zero, and it became clear to everybody that it is just this scheme with gas or liquid vapor in a "stagnation region" (in a cavity) that corresponds very well to real situations. Here the following symbols are used: p_∞ is the pressure at infinity, p_c is the pressure in the cavity behind a body, ϱ is the liquid density, v_∞ is the free-stream velocity.

We shall describe below some other fluid flow schemes with discontinuities.

The realization of one or another global flow regime is, generally speaking, connected with a number of additional effects on fluid flows past bodies. It can be done, for example, by means of proper choice of a body's outline and, hence, by means of control of flow separation from a body's surface, by means of gas or liquid injection at proper flow locations, by means of fluid suction from the boundary layer, by means of using or taking into account the ejecting action of gas or liquid jets, by means of volume size and shape control of cavities near a streamlined body surface and by other means.

It is typical that earlier specialists in hydrodynamics tried to construct smooth streamlined contours, whereas nowadays, to realize high speed body motion in water, flows with jet separation are used which allow to reduce the wetted area of a body's surface resulting in a drastic reduction of skin friction drag which for streamlined bodies is close to the total drag and is proportional to the squared velocity and wetted area of a body's surface. Thus, in the case of a lifting body surface gliding upon water, flow separation and reduction of the wetted surface of a boat's bottom are obtained by means of steps.

In the case of high speed body motion in water one always has separation and cavitation flows. For moderately high velocities one can achieve similar effects by means of a cavitator with sharp edges providing flow separation, by means of liquid jets injected from within the body and by means of special gas blowing to drive away the outer relative water flow from the body's surface. Experiments show the possibility of artificial generation of gas cavities with inner pressure p_c by means of injection of the necessary quantity of gas into the flow.

On the base of theoretical analysis and of model and full scale experiments for "Big Volga" ships and for barges with plane bottoms *A. A. Butuzov* and other workers of the Leningrad Krylov Institute proved the possibility of hydrodynamic drag reduction for these ships by means of systems of thin stable air films at their bottoms. For very small air flow rates through such film cavities of "air lubrication" they managed to reduce the total barge drag by 20 to 28% for velocities of the order of 11—18 km/h.

As was theoretically shown long ago, already in the papers by *S. A. Chaplygin* in Moscow (1899) and *R. Thiry* in Paris (1921), it is possible to construct ideal fluid flows past bodies by introducing, instead of critical points, regions with a given constant pressure (generally larger than the pressure at infinity) bounded by streamlines. The boundary of such regions contains a cusp point where streamlines merge or branch. One can theoretically construct such flows with stagnation regions instead of the front critical point and also instead of the rear one (see Fig. 3).

In the thirties detailed numerical calculations of flows past a plane plate with a stagnation region at the rear part were carried out by me, in collaboration with *A. I. Chulanov*. Such flows behind bodies of revolution were obtained and studied experimentally by *T. G. Lang* and *H. V. L. Patrick* in the course of their investigation of water flow about a body of revolution with small air blowing into the cavity. In these cases the cavity boundary is concave with respect to the main flow direction. If the cavity is filled with gas, then the latter is moderately carried away by the water flow. In the case of short cavities their instability does not have enough time to develop. The sketch of such a cavity at the rear part of a streamlined body is shown in Fig. 4. Experiments of *T. G. Lang* and *H. L. V. Patrick* also provided data on the inner cavity pressure p_c as a function of the air blowing rate.

If $\sigma = -1$, i.e. $p_c = p_\infty + \varrho v_\infty^2/2$ then, in the theoretical solution, these regions degenerate into critical points. If $\sigma < -1$, i.e. $p_c > p_\infty + \varrho v_\infty^2/2$ then, in the considered region, a pressure rise to a value higher than the stagnation pressure is obtained. Such a level of pressure at the rear part of the body corresponds to the appearance of thrust. The corresponding increased pressure leads to the formation of a gas or liquid propulsive jet and can exist, for example, at the expense of a pump in hydrojet propulsion, at the expense of gas jet propulsion or in many cases at the expense of a propeller (see Fig. 5).

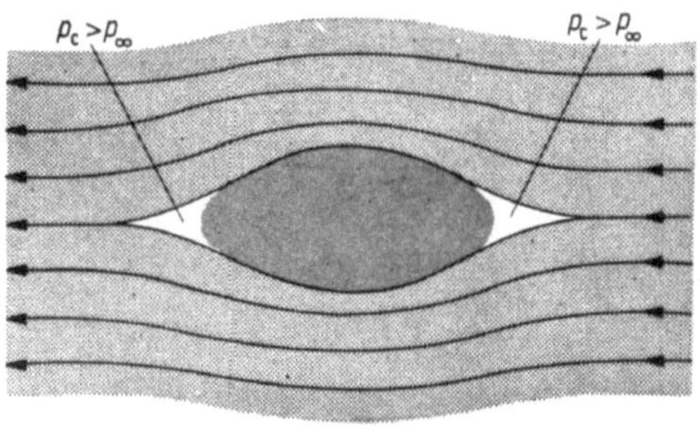

Fig. 3. Flow with regions of higher pressure ahead of the body or behind it ($\sigma < 0$).

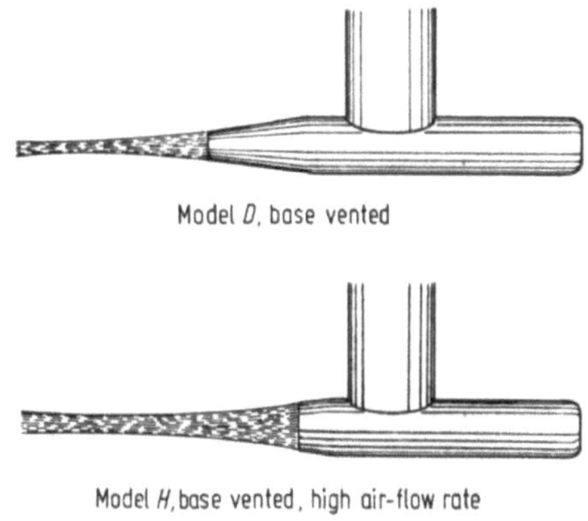

Fig. 4. Experimental data for cavities at p_c (sketched from photographs).

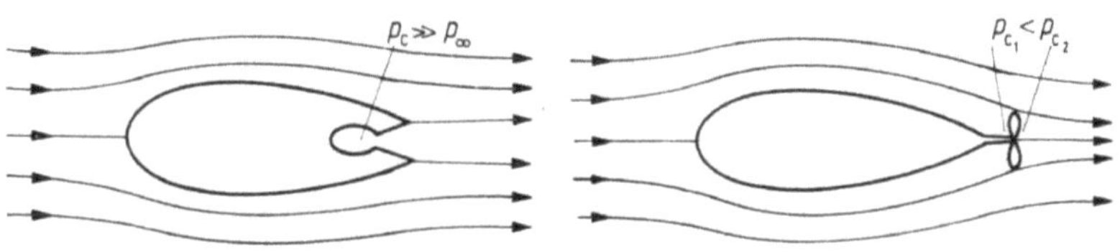

Fig. 5. The thrust to compensate drag is usually produced by a propulsive engine which provides pushing pressures at the rear part of the body.

The flow schemes with regions where $\sigma \leq 0$ were considered above. One can construct, according to D. P. Riabouchinsky's scheme (see Fig. 6), initial theoretical cavitational fluid flow past bodies at $\sigma > 0$. Along with a given body A the auxiliary body B is artificially introduced and between these two bodies a reduced pressure region develops. It is obvious that the total hydrodynamic force acting upon a system of two bodies $A + B$ in perfect fluid equals zero. But at small σ the cavity is long; therefore the presence of the body B weakly influences the flow near the body A and its drag. One can consider a flow scheme past two bodies A and B as a flow past a single body C presented in Fig. 7.

The presence of free boundaries corresponding to reduced pressure p_c and bounding gas cavities can provide a friction drag reduction.

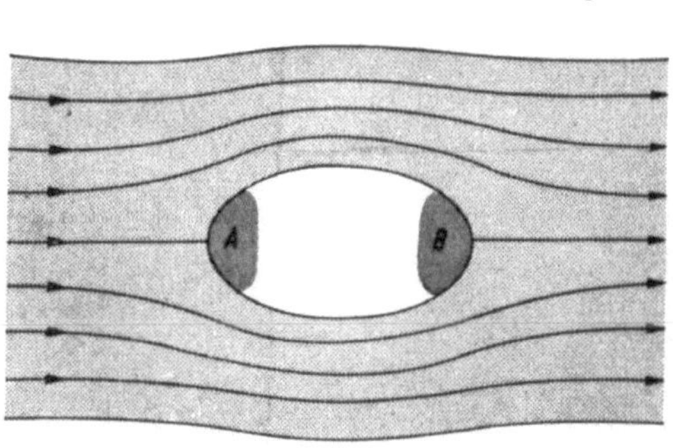

Fig. 6. Riabouchinsky flow scheme.

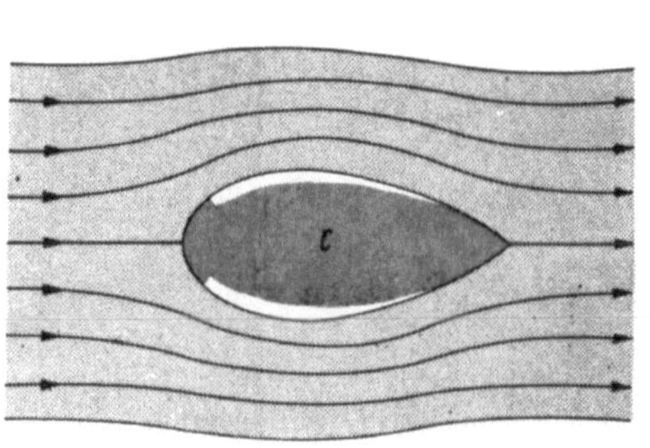

Fig. 7. Flow past body C with jets and cavity adjacent to the body surface.

An advantage of this flow scheme is connected with the solution of a practically difficult problem—to produce similar flow with rather high pressure recovery in the rear part of a body. It is obvious that this problem—to produce such a flow—becomes easier at high cavitation numbers, but real advantages at increasing values of σ are connected with a significant flow improvment, i.e., with relatively small losses at flow merging in the rear part of a body. Practically this problem is very complicated.

One can indicate a few other flow schemes with free surface formation at $\sigma > 0$. Among these flow schemes the scheme suggested by *D. A. Efros* (Moscow) is of special interest. It is a scheme with a jet flowing into a cavity (this scheme was suggested simultaneously by *D. Gilbarg* in USA); see Fig. 8. It is easy to get convinced that in this case drag in perfect fluid flow is determined by the formula

$$W = \varrho Q(v_\infty + v_c) = \varrho Q v_\infty(1 + \sqrt{1 + \sigma}),$$

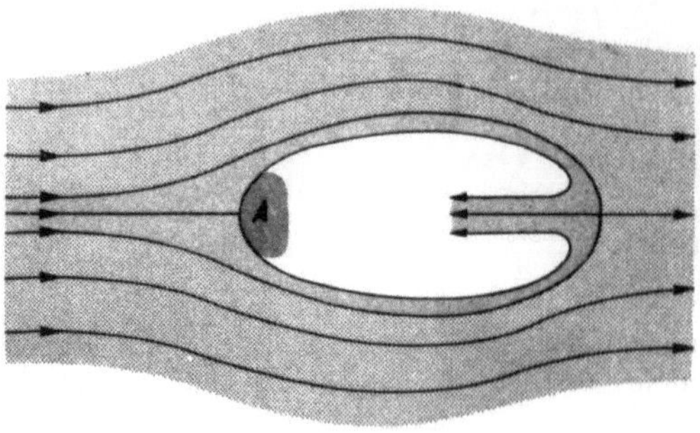

Fig. 8. Flow with re-entrant jet formation.

where $Q = Sv_c$ is the fluid volume rate in the recirculating jet, S is the jet cross section, $v_c = \sqrt{2(p_\infty - p_c)/\varrho + v_\infty^2} = v_\infty \sqrt{1 + \sigma}$ is the velocity at the cavity's surface, equal to the velocity of the recirculating jet. Experiments and theory show that in this case too, when $\sigma \to 0$, the drag of the body A is determined by the classical Kirchhoff scheme; thus, when $\sigma \to 0$, one obtains

$$W_{\text{Kirch}} = 2\varrho Q v_\infty$$

or, in the limit,

$$\varrho Q v_\infty = W_{\text{Kirch}}/2 \quad \text{and} \quad Q = W_{\text{Kirch}}/2\varrho v_\infty.$$

Starting from steady-state perfect fluid flow past body A, according to Efros' scheme, one can construct in an evident manner a new perfect fluid flow reversing only the directions of velocities of all fluid particles. Now one can consider Efros' jet to be a propulsion jet ejected by a body and moving alongside the oncoming fluid counterstream, which then can expand infinitely flowing downstream or merging at the contour of the body A (see schemes in Fig. 9).

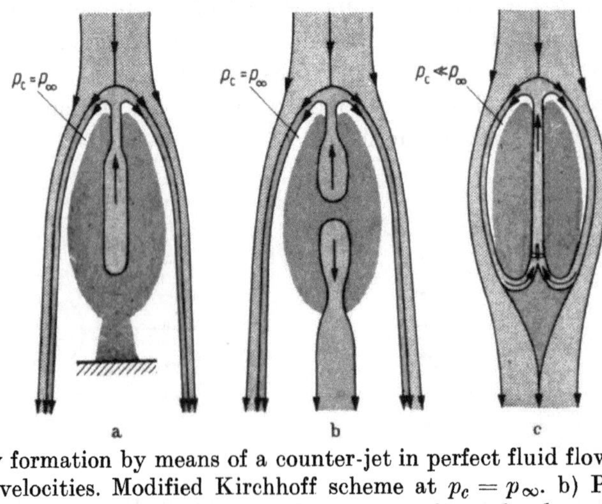

Fig. 9. Scheme of cavity formation by means of a counter-jet in perfect fluid flow. a) Limit case of two jets meeting with opposite velocities. Modified Kirchhoff scheme at $p_c = p_\infty$. b) Propulsion forces from jets directed forwards and backwards are balanced. If W corresponds to the drag of a given body with rigid cavitator then $\varrho Qv = W/r$. c) There is a fluid jet in front, circulating along a closed contour at $p_c < p_\infty$.

It is evident that for such fluid flow reversal all hydrodynamic equatons are satisfied; the pressure at all points preserves its value. Therefore the total force acting on the contour of the body A, which in the original flow is equal to the drag W, in the reversed flow is directed against the free stream and may be considered as a thrust. Thus, in reversed flow one can obtain thrust, e.g., a force totally directed forward and acting upon the body A, by means of ejection of a jet from the body which is directed forward and is interacting with the free stream.

Let us note that in the so constructed reversed flow at $\sigma \approx 0$ the momentum rate in the jet is equal only to a half of the corresponding Kirchhoff force, which in the present case is the obtained thrust.

Some modified flow schemes at $\sigma \geq 0$, based upon Efros' scheme of reversed motion, are presented in Fig. 9, where a jet of liquid ejected from the body opposite to the free stream is used instead of a rigid cavitator which is an important part of a streamlined solid body surface. In this case, for the above considered flow schemes, energy spent by the fully submerged moving body to overcome drag is replaced by energy spent to form and to support a counterjet. The mechanism of partition of a liquid by a counterflowing jet shown in Fig. 9a resembles the phenomenon of breaking through hard steel of a cumulative jet of liquid metal studied by *M. A. Lavrent'ev.*

Introducing nonsymmetric flows with liquid jets one can obtain transversal components of the hydrodynamic force acting upon a body. Inner flow and intensive jets inside the body can strongly influence its stabilization and control.

In the main problem of direct flow past body A at $\sigma > 0$ the mathematical solution according to this scheme implies that the liquid jet goes to the second sheet of the Riemann surface. In practice, if the jet can be formed, the considered flow can not practically exist because the cavity is filled by jet water and flow with a vorticity region behind the body results. However, many authors note that in the experiments at not so large values of σ the flow according to this scheme occurs with a clear tendency to form a recirculating jet.

In the experiments this recirculating jet is unsteady and very weak. The jet sprays and is removed downstream along the cavity's surface. Such fluid motion is connected with energy losses due to the viscous boundary layer at the cavity's surface with wave oscillations of this surface and with hydraulic losses reducing the counterpressure in the region of the unsteady shock cavity closing at the rear part of the cavity, where also the gas blown into the cavity to keep the necessary reduced pressure $p_c < p_\infty$ is removed.

Calculations show that at rather low σ the flows at the forebody and in the cavity near the body are—according to D. A. Efros' and D. P. Riabouchinsky's schemes (Fig. 6 and 7)—quite similar for the same body A; the same is true for the drag.

By means of body B (see Fig. 6), which can be placed into the flow and can be connected with body A, it is theoretically possible to recover the pressure at body B and to obtain, in such a way, the thrust at body B, compensating the drag W. In real flow with cavity formation the recirculating jet is practically not formed because of the above mentioned energy losses by which mechanical energy is transformed into heat; thus, these losses leading to drag are equal or higher than $\varrho Q v_c$ in any case.

In practice, at high values of the cavitation number σ, when energy losses are insufficient to supress a recirculating jet, Efros' scheme is not realized in steady flow. In the case of large values of $\sigma > 0$ one can hope to obtain thrust at the rear of the body to compensate partially the drag W, or at least not to obtain additional drag (as compared with forebody drag W) at the rear of the body. At present, the problem of detailed study of cavitation flows at $\sigma \gtreqless 0$, to clarify the mechanisms of losses in cavities and the possibilities of reduction of energy losses in fluid flow merging at the rear of the body, is put forward.

It seems attractive to obtain a good pressure recovery at comparatively high cavitation numbers $\sigma > 0$, which allows to obtain cavitation flows past bodies moving in water at greater depth.

The above mentioned schemes show that in flows with a velocity discontinuity and with special jet formation the problems of thrust and drag, due to close interaction of the streamlined body shape and the propulsive device system, form a unitary complex. As is well known, during the last decades the ship hull-propeller interaction was studied and taken into account. In the nearest futuie it would be necessary not only to take into account the interaction between separately designed ship elements, but to design a unitary mobile system of ship hull and propulsive device.

Currently, many experimental investigations of cavity properties behind bodies at $\sigma > 0$ are already available. The main results are concerned with the following data: The specific volume[1] gas rate, $Q/(d^2 v_\infty)$, the relative area of the cavity midsection, D^2/d^2, and its aspect ratio, l/D, are found, where d is a typical transversal body size, l is the cavity size and D is a typical transverse cavity midsection size. Experiments show that among the given basic parameters for a given cavitator shape the following ones are of major significance: cavitation number, σ, Froude number, $\mathrm{Fr} = v/\sqrt{gd}$, and relative submergence depth, h/d. In the papers published up to date the main attention was paid to the clarification of the mechanism of gas removal from the cavity and to the gravity influence on the cavity shape.

The main results of experiments with various disk diameters and cone angles, reported by *L. A. Epshtein* at the Leningrad Symposium, are as follows: A typical picture of a disk in water flow is shown in Fig. 10. The cavity shape in the vertical section is non-symmetric because

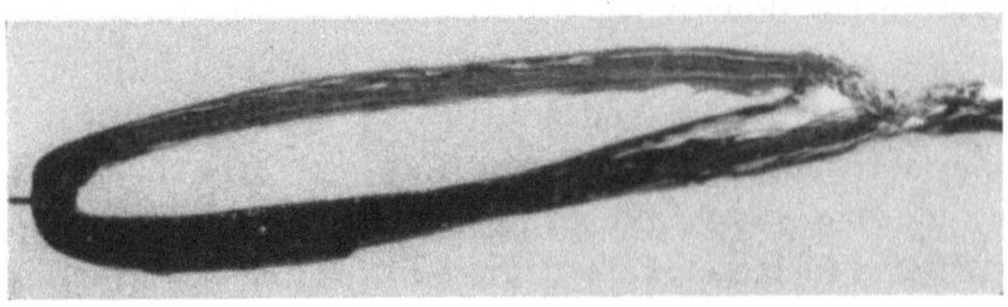

Fig. 10. Flow past a disk towed in a channel ($\sigma = 0.17$, $\mathrm{Fr} = 7.5$).

of the free boundary and gravity influence. At low values of σ and Fr two eddies are formed behind the cavity; through these eddies the main air-flow rate takes place. At these flow regimes, the lengthening of eddies filled with air is the main cause of the flow rate of air blown into the cavity. The scheme of the longitudinal and transverse cavity size transformations at low values of σ and Fr is presented in Fig. 11. The gravity influence on cavity shape and gas removal decreases when σ or Fr increases.

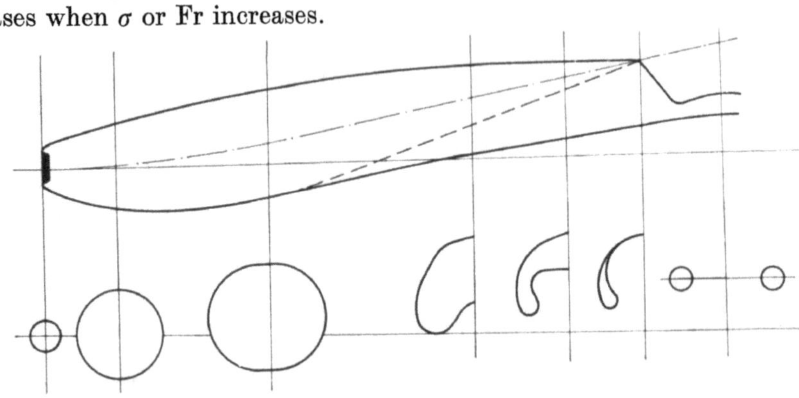

Fig. 11. Longitudinal cavity transformation with two eddy formations at low σ and Fr.

In Fig. 12 a typical diagram characterizing the experimentally obtained gas losses in cavities is shown. Experiments were carried out in a circular channel (of a radius large in comparison with model and cavity sizes) rotating about the horizontal axis together with water contained in it. A model was fixed at the channel's bottom. In such an experiment, centrifugal forces acting upon the water lead to an effective gravity increase.

The drag coefficient C_{x_0} corresponds to a given body drag according to Kirchhoff's flow, i.e., $C_x = C_{x_0}$ at $\sigma = 0$ and $\mathrm{Fr} = \infty$ (C_{x_0} was preliminary determined in other experiments).

[1] In the following, the gas volume Q is defined at water temperature and cavity pressure p_c.

It is shown in the experiments with cones that the position of the almost vertical branch of the curve giving the relation between the air-flow rate and σ in Fig. 12 depends on the Froude numbers and the relative submergence; however, it is obvious that its position essentially depends also on the geometric shape of the body contour. In Fig. 12 the vertical branch of the curve corresponds to the air flow rate due to the two hollow eddies into which the rear part of the cavity is transformed. For the horizontal part of the curve the gas removal occurs by separate portions and by bubbles formed after drop impact on the cavity. This mechanism depends also on the conditions of the merging of jet surfaces bounding the cavity. At these surfaces small oscillations and oscillations of the cavity as a whole are developed. Experimental data of great interest about these oscillations in the case of a plane problem of flow behind a wedge were reported by *J. M. Michel* at the Leningrad Symposium.

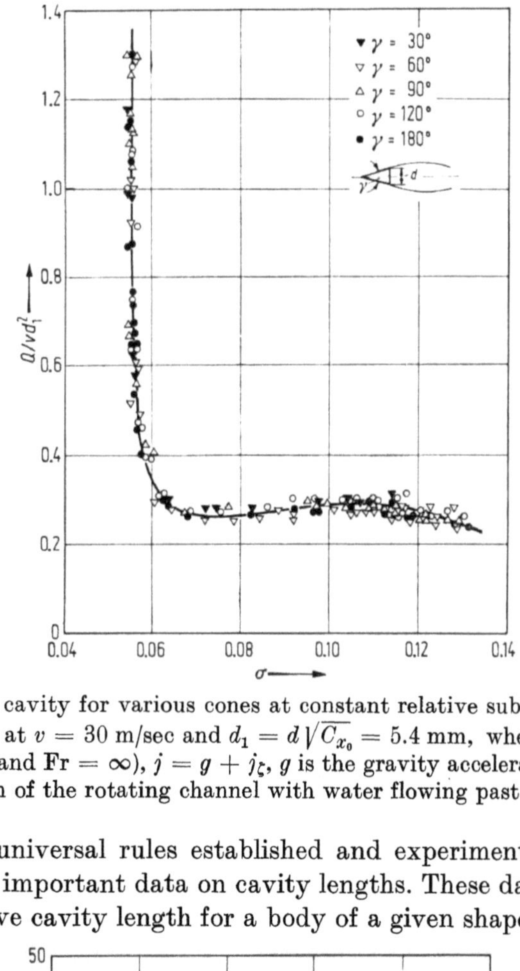

Fig. 12. Gas flow rate from a cavity for various cones at constant relative submergence $h/d_1 = 5.6$, Froude number $\mathrm{Fr} = v/\sqrt{g\,d_1} = 13.3$, at $v = 30$ m/sec and $d_1 = d\sqrt{C_{x_0}} = 5.4$ mm, where C_{x_0} is the drag coefficient of Kirchhoff's flow (at $\sigma = 0$ and $\mathrm{Fr} = \infty$), $j = g + j_\zeta$, g is the gravity acceleration and j_ζ is the centrifugal acceleration of the rotating channel with water flowing past the model.

Among a number of universal rules established and experimentally confirmed by *L. A. Epshtein* let us note some important data on cavity lengths. These data are represented at the diagram of Fig. 13. Relative cavity length for a body of a given shape depends on the velocity

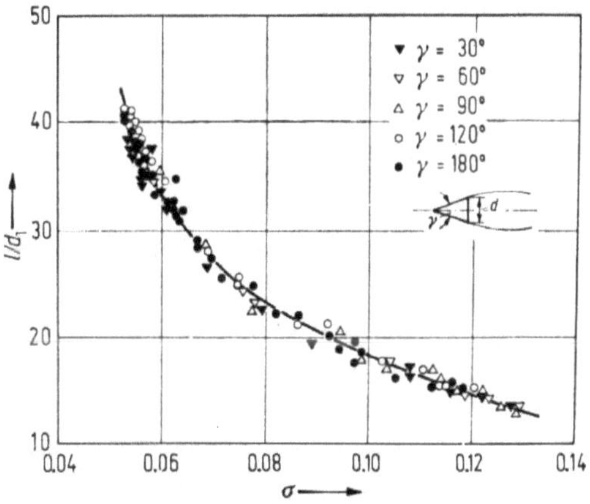

Fig. 13. Cavity aspect ratio behind the cones (at the same condition as in Fig. 12).

and pressure in the free stream and on the cavity pressure p_c. This dependence is especially obvious and stable for small test-body sizes and high Froude numbers. On the base of these relationships *V. P. Karlikov*, *V. P. Fedchenko* and *G. I. Sholomovich* developed a simple method to measure velocity distributions in water flows. Velocity distributions in the flow can be judged by the size of the well observed cavity which is formed behind a thin wire stretched across the flow. In Fig. 14 a picture of the cavity behind the wire crossing the flow and the large cavity in the main water flow is shown. Thus, a clear picture of the velocity distribution near a cavity is obtained (in the present case at $\sigma > 0$).

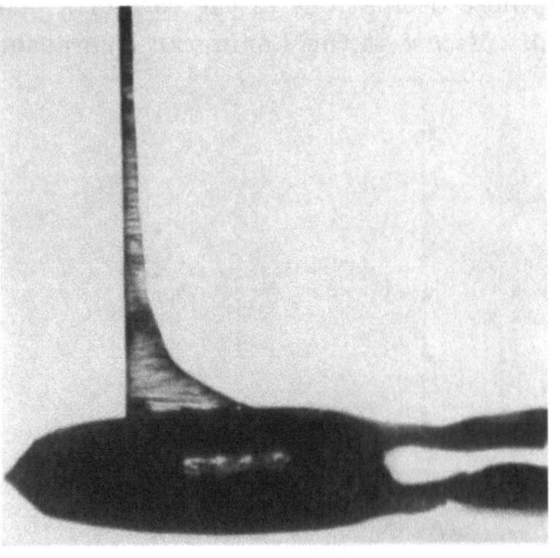

Fig. 14. Downstream cavity lengths behind a wire characterizing the flow velocity distribution.

In the published papers there are also results concerned with investigations and clarifications of many other typical effects, such as Reynolds number influence, the influence of surface tension and submergence depth on the behaviour of a cavity. The information on fundamental frequencies of cavity oscillations at some flow regimes and, connected with this, the change of flow rate necessary to keep up a fixed value of σ is obtained.

Experiments show that, generally, for bodies of revolution the influence of the nose cavitator shape is given by the parameter $d_1 = d\sqrt{C_{x_0}}$ only, where d is the circle diameter at which flow separation occurs, and C_{x_0} is the corresponding drag coefficient according to Kirchhoff's scheme at $\sigma = 0$ and $\mathrm{Fr} = \infty$.

However, the quality of the jets' surfaces and their smoothness are connected with the characteristics of the boundary layer flowing away from the body surface at the separation line and, partially, with the methods of supply of air into the cavity. In this connection the nose-part shape can, in a general case, influence the oscillations of the jet surface bounding a cavity and thus the jet merging at the cavity's rear part and hence the gas flow rate providing the necessary pressure p_c. In practical situations some methods were developed to reduce specific gas flow rate at cavity closing on the rear part of the body surfaces and also some methods to obtain lift. The joint paper presented at the Leningrad Symposium by *G. S. Migirenko*, *G. S. Kozyuk*, *L. I. Mal'tsev*, *V. I. Mikuta* and *B. G. Novikov* dealt with this subject.

Currently, besides experimental investigations systematical numerical computations are being carried out for plane and axisymmetric flows past bodies with flow separation and formation of cavities bounded by constant pressure surfaces. The joint papers by *A. V. Kuznetsov*, *O. M. Kiselev*, *L. M. Kotlyar* and *A. G. Terent'ev* and by *V. A. Barabanov*, *A. A. Butuzov*, *A. N. Ivanov* and *I. A. Titov*, as well as the paper by *P. R. Garabedian*, deal with such calculations. In the paper presented by *G. Birkhoff* one can find a detailed review and discussion of the advantages of different numerical methods for the calculation of incompressible perfect fluid flows past bodies with cavity formation.

G. V. Logvinovich suggested an effective approximate method to calculate cavity shape and water motion behind a body by means of the hypothesis of independent plane-parallel flows in the plane cross sections normal to the direction of the velocity of the main translational body motion. The best results are obtained using this method by means of numerical analysis of translational body motion in water at a constant speed or at a speed changing only its magnitude. Such an approximate method for calculation of unsteady water motion was applied to determine the cavity shape behind the body up to its closure and to determine the magnitude of the force acting on the body during its vertical water entry.

The numerical method corresponding to the theory of plane cross sections with some natural additional simplifications was successfully applied by *G. V. Logvinovich*, and after him by other authors, to clarify the nature of the hydrodynamic suction force compensating the drag of swimming fish. It turned out that the oscillating wing thrust due to the suction force at the wing leading edge analysed in the thirties was similar to the suction force for fish. This force arises at the back of a fish due to transversal motions which propagate in the shape of a wave along the fish's body. It should be pointed out here that the works by *G. V. Logvinovich* were preceded by those by *M. A. Lavrent'ev* and *M. M. Lavrent'ev* who analysed some possible sources of thrust developed by fish.

In conclusion, let us consider very briefly some peculiarities of water impact and fast solid body submergence.

From the thirties on, many experiments were dedicated to these phenomena, the general theory was developed and approximate numerical methods were designed. Attention was paid to the determination of impact forces acting on the body as functions of entry conditions, mainly of the body velocity relative to water; valuable data were obtained on the influence of the wetted surface shape and of the dynamic properties of impacting bodies on hydrodynamic forces.

Recently it was established that the presence of atmospheric air or some other gas over the free surface in the case of the impact of a body on water may be essential for the motion of thin jets ejected near the edges of the wetted body surface. In Fig. 15 a picture of such a jet is shown. One can also see in this picture that at the body immersion into water a cavity is formed behind the body. The development of the cavity shape and its closure is also essentially connected with the interaction between water and gas moving over the water's free surface.

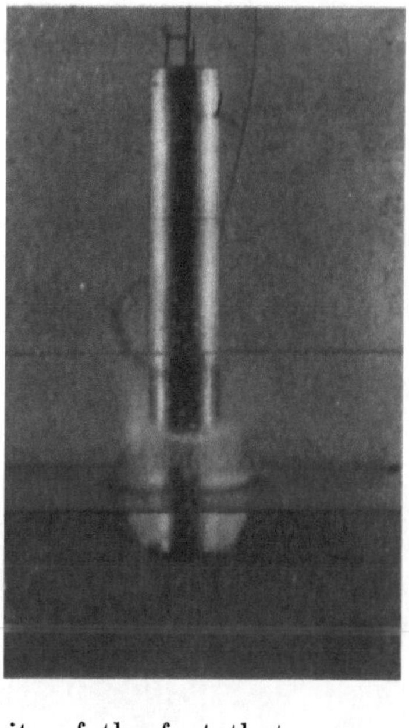

Fig. 15. Cylinder entering water at velocity 15 m/sec, air pressure 1 atm, cylinder diameter 53 mm.

It was experimentally established that in spite of the fact that a governing parameter, equal to the ratio of air and water densities, is small and has the order of $10^{-3}-10^{-4}$, the influence of this parameter on the behaviour of a sprayed jet is significant. It was experimentally established for various gases (air, helium and freon) that it was the gas density to water density ratio that is important in this phenomenon, and that the difference in gas pressures over the water surface is only of secondary importance. This fact was noted in publications of many authors but a mechanism of small-density effects has not been clarified up to date. To explain this phenomenon, *Yu. L. Yakimov* analyzed asymptotic laws of thin jet formation and development near an edge of a falling body. It was clarified theoretically that one can approximately consider this phenomenon to be self-similar. Aerodynamic forces, roughly proportional to gas density, influence a fluid particle moving in the upper part of the jet and depend on the velocity of the falling body and on the fluid velocity in the jet which is essentially larger than the body's immersion velocity.

In Figs. 16 and 17 the spark photographs of a jet formed near the edge of a falling prismatic flat-based cylinder are given (the motion is approximately plane-parallel); these pictures were obtained by *Yu. L. Yakimov* in the Research Institute of Mechanics, Moscow University. The comparison of such pictures for different moments of time shows that the fluid jet motion can be considered to be a self-similar one.

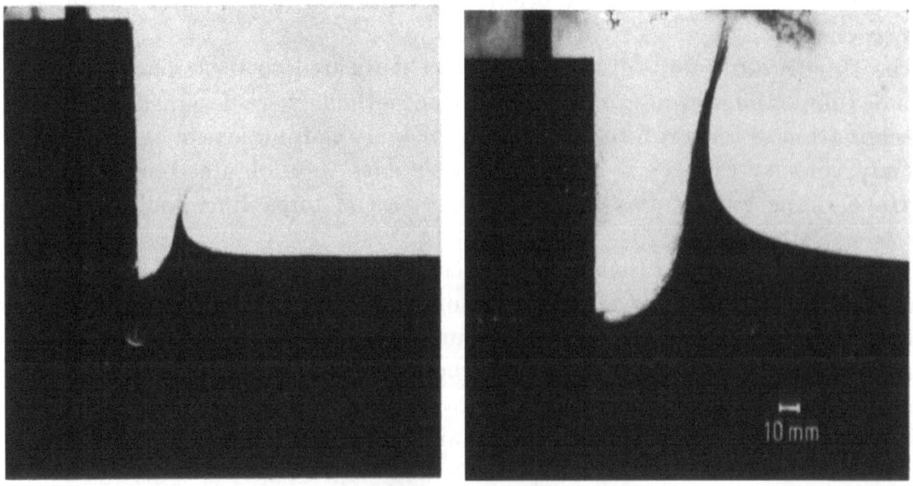

Fig. 16. Jet similitude at the impact of a sharp-edged body on water (width 140 mm, velocity 4.5 m/sec, air pressure 0.05 atm).

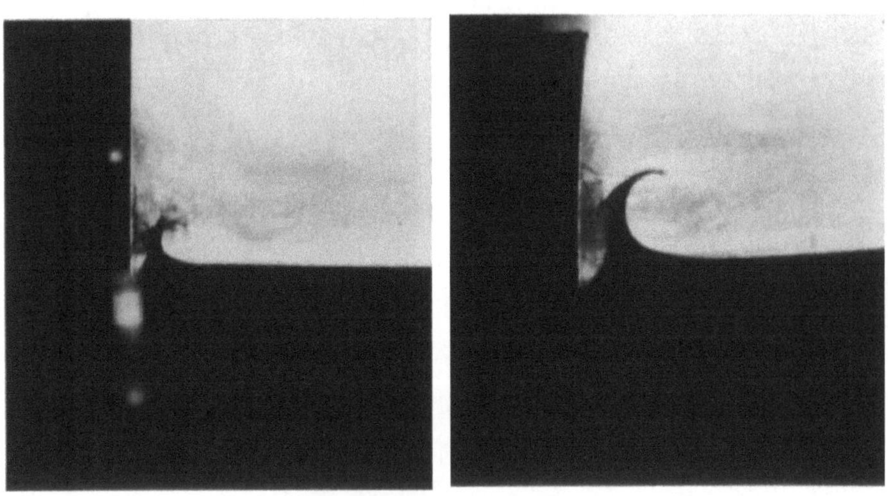

Fig. 17. Jets at the impact of a sharp-edged body on water (width 140 mm, velocity 7.5 m/sec, air pressure 1 atm).

References

Ackeret, J.: Techn. Mech. und Thermodynamik 1 (1930) 1–22, 63–72.
Akulitchev, V. A., et al.: Proc. 7th Intern. Congr. Acoustics 1971, Vol. 2, Budapest 1971, 481–484.
Barabanov, V. A., et al.[1]
Besant, W. H.: Hydrostatics and hydrodynamics. Cambridge, 1859.
Birkhoff, G.: Hydrodynamics. 2nd ed. Princeton University Press 1960.
Birkhoff, G., Zarantonello, E. N.: Jets, wakes and cavities. Academic Press 1957.
Birkhoff, G.[1]
Butuzov, A. A.: Mekhanika zhidkosti i gaza 2 (1966) 167–170.
Chaplygin, S. A.: Trudy otdeleniya fizicheskikh nauk Obshchestva lyubitelei estestvoznaniya 10 (1899) 35–40 (Complete works, Vol. 1. Leningrad 1933, 254–263).
Cole, R. H.: Underwater explosions. Princeton University Press 1948.
Efros, D. A.: Doklady, Acad. Sci. SSSR 51 (1946) 263–266.
Epshtein, L. A., Blyumin, V. I.: Trudy, Central Aero-Hydrodynamic Institute 950 (1965) 3–40.
Epshtein, L. A.[1]
Field, J. E.[1]
Flynn, H. G.: Physics of acoustic cavitation in liquids. Physical acoustics, Vol. 1B. Academic Press 1964, 57–172.
Flynn, H. G.[1]
Garabedian, P. R.[1]
Gilbarg, D.: Jets and cavities. Encycl. Phys., Springer 1960, 311–443.
Heuckroth, L. E., Glass, I. I.: Phys. Fluids 11 (1968) 2095–2107.
Hickling, R., Plesset, M. S.: Phys. Fluids 7 (1964) 7–14.
Johnson, V. E.[1]
Karlikov, V. P., Fedchenko, V. P., Sholomovich, G. I.: Author's certificate, SSSR, No. 224929.
Kogarko, B. S.[1]
Kornfeld, M., Suvorov, L.: J. Appl. Phys. 15 (1944) 495–506.
Kuznetsov, A. V., Kiselev, O. M., Kotlyar, L. M., Terent'ev, A. G.[1]
Lang, T. G., Patrick, H. V. L.: ASME Paper 64-WA/UNT-5 (for meeting 29. XI–4. XII. 1964).
Lauterborn, W.[1]
Lavrent'ev, M. A., Keldysh, M. V., Markushevich, A. I., Sedov, L. I., Lotov, A. B.: Trudy, Central Aero-Hydrodynamic Institute, 1935, No. 152.
Lavrent'ev, M. A.: Uspekhi matematicheskikh nauk 12 (1957) 41–56.
Lavrent'ev, M. A., Lavrent'ev, M. M.: Zhurnal prikladnoi mekhaniki i tekhnicheskoi fiziki 4 (1962) 3–9.
Logvinovich, G. V.: Uchenye zapiski, Central Aero-Hydrodynamic Institute 1 (1970) 11–17.
Logvinovich, G. V., Epshtein, L. A., Gurevich, M. I., Stepanov, G. Yu.: Papers in "Mechanics in the USSR during 50 years", Vol. 2, Moscow 1970.
Logvinovich, G. V., Yakimov, Yu. L.[1]
Michel, J. M.[1]
Migirenko, G. S. et al.[1]
Plesset, M. S., Hsieh, D. Y.: Phys. Fluids 7 (1964) 1099–1108.
Plesset, M. S.: Phil. Trans. Roy. Soc. London A 260 (1966) 241–244.
Plesset, M. S.: Naval Hydrodynamics, 1970, Publ. 515, Ch. XII.
Plesset, M. S., Chapman, R. B.: J. Fluid Mech. 47 (1971) 283–290.
Rayleigh, Lord: Phil. Mag. 6/34 (1917) 94–98.
Riabouchinsky, D. P.: Proc. 3rd Intern. Congr. Appl. Mech. Stockholm 1930, Vol. 1, 1931, 137–158.
Sedov, L. I.: Trudy, Central Aero-Hydrodynamic Institute, 1934, No. 187.
—: ibidem, 1936, No. 252.
—: ibidem, 1937, No. 325.
—: Trudy (Proceedings), Conference on wave resistance. Central Aero-Hydrodynamic Institute, 1937, 7–30.
—: Two-dimensional problems in hydrodynamics and aerodynamics. 2nd ed., Moscow 1966. English translation of the 1st ed.: Wiley 1965.
Silberman, E., Song, C. S.: J. Ship Research 5 (1961) 13–33.
Solomon, L. P., Plesset, M. S.: Intern. Shipbuild. Progr. 14 (1967) 98–103.
Taylor, G.I.: Proc. Roy. Soc. London A 313 (1969) 453–475.
Thiry, R.: Ann. Ecole norm. (3) 38 (1921) 229–339.
Tkachev, L. G., Shestakov, V. D.: Akustichesky zhurnal 18 (1972) 433–440.
Trilling, L.: J. Appl. Phys. 23 (1962) 14–17.
Wagner, H.: ZAMM 12 (1932) 193–215.
van Wijngaarden, L.: J. Fluid Mech. 33 (1968) 465–474.
Wu, T. Y., Whitney, A. K., Brennen, C.[1]
Zhukovsky, N. E.: Matematichesky sbornik 15 (1890) 121–276 (Complete works, Vol. 3. Moscow-Leningrad, 1936, 195–341.)
Zwick, S. A., Plesset, M. S.: J. Math. and Phys. 33 (1955) 308–330.

[1] See: Non-steady flow of water at high speeds (Proceedings of the IUTAM Symposium held in Leningrad, June 22–26, 1971). Moscow, Nauka 1972.

On two ways of constructing the theory of elastic shells

I. N. Vekua

Tbilisi, USSR

1.

If the theory of shells is constructed on the basis of the hypothesis that it is sufficient to define the resulting forces and couples, it is desirable to derive a system of partial differential equations which would provide the fulfilment of five independent boundary conditions; the system of equations should be correctly solvable when the boundary values of normal, tangential and transversal forces, and the bending and twisting moments are arbitrarily given. We shall call the theory that satisfies these requirements the consistent theory of shells.

Investigations devoted to the construction of the shell theory have a long history; they date back to the works of *L. Euler*, *J. Bernoulli*, *J. L. Lagrange*, *A. L. Cauchy* and *S. D. Poisson*. Later *G. R. Kirchhoff*, *A. Clebsch*, *H. Aron* and *A. E. H. Love* worked out the basis of the modern general theory of shells. But investigations in this connection are still being carried out. This is due to the existence of different versions of the shell theory. In particular, according to the commonly adopted approach in modern scientific-technical works, and based on the Kirchhoff-Love assumptions, it is impossible to prescribe independently the boundary values of the transversal force and twisting moment. Some of the scientists who do not share these assumptions have made successful attempts to construct a theory of shells which would be compatible with the five independent boundary conditions (see [4, 5, 6]).

In the present paper we shall discuss two different methods of constructing a consistent theory of elastic shells which lead to elliptic systems of 10th order. These equations are compatible with the boundary conditions on face surfaces on which the stresses $\overset{(+)}{\boldsymbol{P}}$ and $\overset{(-)}{\boldsymbol{P}}$ are prescribed. Both of these systems of equations have the same structure but are not identical. The differences between them are not so essential; they can be reduced by the choice of the values of some special numerical parameter, which may be included in one of the systems of equations.

Mention should be made of the differences and similarities between this theory and the classical Kirchhoff-Love theory. The former leads to the elliptic system of 8th order and therefore ensures the fulfilment of only four independently given boundary conditions. This fact carries in itself an essential point of divergence. In other respects our constructions bear a great resemblance to the classical theory. A number of corollaries, resulting from new equations, are in fairly good agreement with the classical results. For instance, for the bending (normal deflection) of the plate, loaded by a constant normal force, we obtain the classical equation of *S. Germain-Lagrange*.

Thus, the constructions given below, being in complete agreement with the main conditions of the mechanical problem, make certain refinements to the classical theory, thus overcoming some of its drawbacks. But still many classical results remain valid, which are widely recognized in scientific-technical works, and have the expanded range of practical applications.

The classical theory is often preferred for its comparatively simple mathematical apparatus. The importance of this fact cannot be neglected. An extremely complicated mathematical apparatus may almost fully deprive the theory of its practical significance. But comparative mathematical simplicity is inherent also in our constructions. For example, in the case of a plate we have equations of rather a simple structure which may be integrated in an explicit form.

2.

2.1. A very simple and natural way of constructing the general theory of elastic shells is the method, based on expansions of the sought fields of displacements and stresses into orthogonal Fourier series by Legendre polynomials (see [1, 2]). Below we shall make use of a special case of this general method for constructing one of the versions of the consistent moment theory of elastic shells. In this section we shall give a schematic but sufficiently full survey of the foundations of this general theory.

Below, we shall systematically use notations and symbols employed in tensor analysis (see, for instance, [7]). In particular, we shall use the well-known summation rule but agree that Latin and Greek indices run the values 1, 2, 3 and 1,2, respectively.

Using the vector notations, one can write the equations of equilibrium of continuous media and elasticity relations (Hooke's generalized law) in the form (see [2])

$$\frac{1}{\sqrt{g}}\,\partial_i\left(\sqrt{g}\,\boldsymbol{P}^i\right) + \boldsymbol{\Phi} = 0 \qquad \left(\partial_i = \frac{\partial}{\partial x^i}\right), \tag{1a}$$

$$\boldsymbol{P}^i = \lambda(\boldsymbol{R}^k\,\partial_k\boldsymbol{U})\,\boldsymbol{R}^i + \mu(\boldsymbol{R}^i\,\partial_k\boldsymbol{U})\,\boldsymbol{R}^k + \mu(\boldsymbol{R}^i\boldsymbol{R}^k)\,\partial_k\boldsymbol{U}. \tag{1b}$$

Here λ and μ are Lamé's constants,

$$\lambda = \frac{E\sigma}{(1+\sigma)\,(1-2\sigma)}, \qquad \mu = \frac{E}{2(1+\sigma)},$$

where E is Young's modulus, σ is Poisson's ratio. $\boldsymbol{R}$ is the radius-vector of the point, $\boldsymbol{R}^i$ is the contravariant basis of the coordinate system, g is the discriminant of the metric quadratic form of the space,

$$ds^2 = g_{ij}\,dx^i\,dx^j, \qquad g_{ij} = \boldsymbol{R}_i\boldsymbol{R}_j, \tag{2}$$

where $\boldsymbol{R}_i = \partial_i\boldsymbol{R} = g_{ij}\boldsymbol{R}^j$ is the covariant basis of the system of coordinates. Finally, $\boldsymbol{\Phi}$ is the volume force, $\boldsymbol{U}$ is the displacement vector, $\boldsymbol{P}^i$ are contravariant components of the stress field. If $\boldsymbol{l}$ is the unit vector, applied at the point (x^1, x^2, x^3), then the stress in this point, acting on the surface element with the normal $\boldsymbol{l}$, is expressed as

$$\boldsymbol{P}^{(l)}(x^1, x^2, x^3) = l_i\boldsymbol{P}^i(x^1, x^2, x^3), \qquad l_i = \boldsymbol{l}\,\boldsymbol{R}_i. \tag{3}$$

Relations (1b) may also be written as

$$\boldsymbol{P}^i = \mathbb{C}^{ij}\,\partial_j\boldsymbol{U}, \tag{1c}$$

where $\mathbb{C}^{ij}$ are the dyadic operators of the form

$$\mathbb{C}^{ij} = \lambda\boldsymbol{R}^i \mid \boldsymbol{R}^j + \mu\boldsymbol{R}^j \mid \boldsymbol{R}^i + \mu g^{ij}\mathbb{E}, \qquad g^{ij} = \boldsymbol{R}^i\boldsymbol{R}^j. \tag{4a}$$

Here $\boldsymbol{R}^i \mid \boldsymbol{R}^j$ are the base contravariant dyads of the coordinate system under consideration, and $\mathbb{E}$ is the operator of identical transformation. If ϕ is a function, and $\boldsymbol{U}$ is a vector, which, in general, may depend on the choice of the system of coordinates, we have the formulas

$$\boldsymbol{U}(\boldsymbol{R}^i \mid \boldsymbol{R}^j) = U^i\boldsymbol{R}^j, \qquad (\boldsymbol{R}^i \mid \boldsymbol{R}^j)\,\boldsymbol{U} = U^j\boldsymbol{R}^i \quad (U^i = \boldsymbol{U}\boldsymbol{R}), \tag{4b}$$

$$\phi(\boldsymbol{R}^i \mid \boldsymbol{R}^j) = (\boldsymbol{R}^i \mid \boldsymbol{R}^j)\,\phi = (\phi\boldsymbol{R}^i) \mid \boldsymbol{R}^j = \boldsymbol{R}^i \mid (\phi\boldsymbol{R}^j), \tag{4c}$$

$$\phi\mathbb{E} = \mathbb{E}\phi = \phi, \qquad \boldsymbol{U}\mathbb{E} = \mathbb{E}\boldsymbol{U} = \boldsymbol{U}. \tag{4d}$$

For $\mathbb{E}$ one can use the formula of dyadic expansion

$$\mathbb{E} = \boldsymbol{R}^p \mid \boldsymbol{R}_p. \tag{4e}$$

The operators $\mathbb{C}^{ij}$ may also be represented as

$$\mathbb{C}^{ij} = E^{ijpq}\boldsymbol{R}_p \mid \boldsymbol{R}_q, \tag{4f}$$

where

$$E^{ijpq} = \lambda g^{ip} g^{jq} + \mu g^{jp} g^{iq} + \mu g^{ij} g^{pq}; \tag{4g}$$

E^{ijpq} is obviously the contravariant tensor of 4th rank.

Formulas (1c) may be generalized to the case of an arbitrary homogeneous anisotropic elastic medium. Then the tensor E^{ijpq} will have a more complicated structure. Due to this fact our further arguments remain valid for any non-homogeneous and anisotropic elastic shell. But, for the sake of simplicity we shall consider the case of the homogeneous isotropic elastic shell. That is why elasticity relations will be used in the form (1b).

2.2. Consider inside the shell some smooth surface $\mathcal{S}$, which will be conditionally called the middle surface. Generally speaking, the shell is not situated symmetrically with respect to $\mathcal{S}$ (see Fig. 1). Considering the system of coordinates, normally connected with the surface $\mathcal{S}$, we have formulas (see, for instance, [7])

$$\boldsymbol{R}_i = (a_{ij} - x^3 b_{ij})\, \boldsymbol{r}^j = (a_i^j - x^3) b_i^j\, \boldsymbol{r}_j, \tag{5a}$$

$$\boldsymbol{R}^i = \Theta^{-1}[(1 - 2Hx^3)\, a_j^i - x^3 b_j^i]\, \boldsymbol{r}^j = \Theta^{-1}[(1 - 2Hx^3)\, a^{ij} - x^3 b^{ij}]\, \boldsymbol{r}_j, \tag{5b}$$

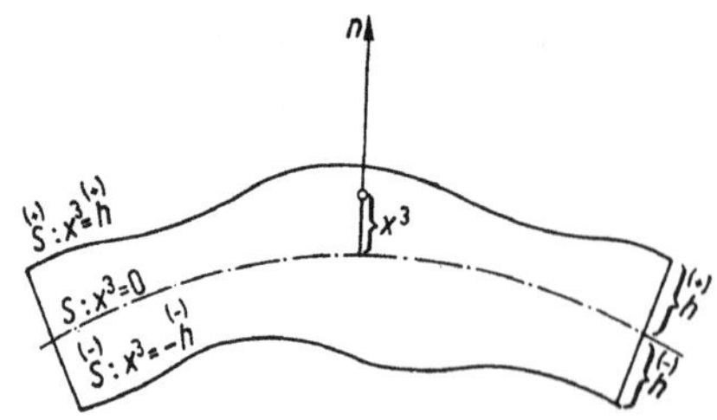

Fig. 1. Denotations.

where $\boldsymbol{r}$ is the radius-vector of the point of the surface $\mathcal{S}$, $\boldsymbol{r}_3 = \boldsymbol{r}^3 = \boldsymbol{n}$ is the unit vector of the normal to $\mathcal{S}$, $\boldsymbol{r}_\alpha = \partial_\alpha \boldsymbol{r}$ and $\boldsymbol{r}^\alpha = a^{\alpha\beta} \boldsymbol{r}_\beta$ are the covariant and contravariant bases of coordinates (Gaussian parameters) x^1, x^2 of the point of the surface $\mathcal{S}$,

$$a_{ij} = \boldsymbol{r}_i \boldsymbol{r}_j,\ a_j^i = \boldsymbol{r}^i \boldsymbol{r}_j = \delta_j^i, \qquad a^{ij} = \boldsymbol{r}^i \boldsymbol{r}^j, \tag{6a}$$

$$b_{ij} = -\boldsymbol{r}_i\, \partial_j \boldsymbol{n}, \qquad b_j^i = -\boldsymbol{r}^i\, \partial_j \boldsymbol{n}, \qquad b^{ij} = -\boldsymbol{r}^i\, \partial^j \boldsymbol{n} \quad (\partial^i = a^{ij}\, \partial_j), \tag{6b}$$

$$\Theta = (1 - k_1 x^3)(1 - k_2 x^3) = 1 - 2Hx^3 + K(x^3)^2, \tag{6c}$$

$$H = \frac{1}{2} b_\alpha^\alpha = \frac{1}{2}(k_1 + k_2),\ K = b_1^1 b_2^2 - b_1^2 b_2^1 = k_1 k_2. \tag{6d}$$

Here k_1 and k_2 are the main curvatures, H and K represent the middle and main (Gaussian) curvatures of the surface $\mathcal{S}$. It is easily seen that

$$a_{ij}, a^{ij} = \begin{cases} a_{\alpha\beta}, a^{\alpha\beta}, & \text{if } \ i = \alpha, j = \beta, \\ \delta_{ij}, \delta^{ij}, & \text{if } \ i = 3 \ \text{ or } \ j = 3, \end{cases} \tag{7a}$$

$$b_{ij}, b_j^i, b^{ij} = \begin{cases} b_{\alpha\beta}, b_\beta^\alpha, b^{\alpha\beta}, & \text{if } \ i = \alpha, j = \beta, \\ 0, 0, 0, & \text{if } \ i = 3 \ \text{ or } \ j = 3, \end{cases} \tag{7b}$$

where $a_{\alpha\beta}$ and $b_{\alpha\beta}$ are the coefficients of the first and second main quadratic forms of the surface $\mathcal{S}$,

$$\mathrm{I} = ds^2 = a_{\alpha\beta}\, dx^\alpha\, dx^\beta, \quad a_{\alpha\beta} = \boldsymbol{r}_\alpha \boldsymbol{r}_\beta, \tag{8a}$$

$$\mathrm{II} = k_s\, ds^2 = b_{\alpha\beta}\, dx^\alpha dx^\beta, \quad b_{\alpha\beta} = -\boldsymbol{r}_\alpha\, \partial_\beta \boldsymbol{n}, \tag{8b}$$

where k_s is the normal curvature of the surface $\mathcal{S}$ in the direction of the tangential element (dx^1, dx^2).

Let $x^3 = \overset{(+)}{h}(x^1, x^2)$ and $x^3 = -\overset{(-)}{h}(x^1, x^2)$ be the equations of the face surfaces of shells (see Fig. 1), where x^1 and x^2 are the Gaussian parameters of the surface $\mathcal{S}$. Introduce the notations

$$h = \left(\overset{(+)}{h} + \overset{(-)}{h}\right)/2, \quad \bar{h} = \left(\overset{(+)}{h} - \overset{(-)}{h}\right)/2 \tag{9a}$$

$2h$ is evidently the thickness of the shell. Further we shall consider $\overset{(+)}{h}(x^1, x^2)$ and $\overset{(-)}{h}(x^1, x^2)$ to be piecewise smooth functions of the variables x^1, x^2. In the domain Ω_S, occupied by the shell, the following conditions are satisfied;

$$\Omega_S: -\overset{(-)}{h}(x^1, x^2) \leq x^3 \leq \overset{(+)}{h}(x^1, x^2), \quad (x^1, x^2) \in S. \tag{9b}$$

Considering the case of thin or shallow shells, one can assume that

$$1 \pm |k_\alpha| h \cong 1 \quad (\alpha = 1, 2). \tag{9c}$$

According to this, let us assume

$$\boldsymbol{R}^i \cong \boldsymbol{r}^i, \qquad \boldsymbol{R}_i \cong \boldsymbol{r}_i \tag{9d}$$

and take Eqs. (1) in the form

$$\frac{1}{\sqrt{a}}\, \partial_i \left(\sqrt{a}\, \boldsymbol{P}^i\right) + \boldsymbol{\Phi} = 0, \tag{10a}$$

$$\boldsymbol{P}^i = \overset{(0)}{\mathbb{C}}{}^{ij}\, \partial_j \boldsymbol{U}, \tag{10b}$$

where a is the discriminant of the metric quadratic form (8a),

$$\overset{(0)}{\mathbb{C}}{}^{ij} = \lambda \boldsymbol{r}^i \mid \boldsymbol{r}^j + \mu \boldsymbol{r}^j \mid \boldsymbol{r}^i + \mu a^{ij} \mathrm{E}, \quad a^{ij} = \boldsymbol{r}^i \boldsymbol{r}^j. \tag{10c}$$

Formula (10b) may also be written as

$$\boldsymbol{P}^i = \lambda(\boldsymbol{r}^k\, \partial_k \boldsymbol{U})\, \boldsymbol{r}^i + \mu(\boldsymbol{r}^i\, \partial_k \boldsymbol{U})\, \boldsymbol{r}^k + \mu\, \partial^i \boldsymbol{U}. \tag{10d}$$

It should be noted that when $\mathscr{S}$ is the plane (e.g., for the plate), the Eqs. (10) are exactly fulfilled.

2.3. Let $P_k(x)$ be the Legendre polynomial of degree k. Multiplying both parts of Eq. (10a) by $\left(k + \frac{1}{2}\right) h^{-1} P_k(h^{-1}(x^3 - \bar{h}))$ and then integrating them with respect to x^3 within $-\overset{(-)}{h}$ and $+\overset{(+)}{h}$, we shall obtain (see [2])

$$\frac{1}{\sqrt{a}}\, D_i \left(\sqrt{a}\, \overset{(k)}{\boldsymbol{P}}{}^i\right) + \overset{(k)}{\boldsymbol{\Phi}} = 0, \quad \overset{(k)}{\boldsymbol{p}}{}^i = \overset{(0)}{\mathbb{C}}{}^{ij}\, D_j \overset{(k)}{\boldsymbol{u}} \tag{11a}$$

$$(k = 0, 1, \ldots),$$

where the following notations are used:

$$\overset{(k)}{\boldsymbol{p}}{}^i = \left(k + \frac{1}{2}\right) \frac{1}{h} \int_{-\overset{(-)}{h}}^{\overset{(+)}{h}} \boldsymbol{P}^i(x^1, x^2, x^3)\, P_k\left(\frac{x^3 - \bar{h}}{h}\right) dx^3, \tag{12a}$$

$$\overset{(k)}{\boldsymbol{u}} = \left(k + \frac{1}{2}\right) \frac{1}{h} \int_{-\overset{(-)}{h}}^{\overset{(+)}{h}} \boldsymbol{U}(x^1, x^2, x^3)\, P_k\left(\frac{x^3 - \bar{h}}{h}\right) dx^3, \tag{12b}$$

$$\overset{(k)}{\boldsymbol{\Phi}} = \left(k + \frac{1}{2}\right) \frac{1}{h} \int_{-\overset{(-)}{h}}^{\overset{(+)}{h}} \boldsymbol{\Phi}(x^1, x^2, x^3)\, P_k\left(\frac{x^3 - \bar{h}}{h}\right) dx^3, \tag{12c}$$

$$D_j \overset{(k)}{\boldsymbol{u}} = \begin{cases} \partial_\alpha \overset{(k)}{\boldsymbol{u}} - \dfrac{1}{h}\, \partial_\alpha \bar{h}\, \overset{(k)}{\boldsymbol{u}}{}' - \dfrac{1}{h}\, \partial_\alpha h \overset{(k)}{\boldsymbol{u}}{}'', & \text{if } j = \alpha, \\ \dfrac{1}{h} \overset{(k)}{\boldsymbol{u}}{}', & \text{if } j = 3. \end{cases} \tag{13a}$$

Here

$$\overset{(k)}{\boldsymbol{u}}{}' = (2k+1)\left(\overset{(k+1)}{\boldsymbol{u}} + \overset{(k+3)}{\boldsymbol{u}} + \cdots\right), \tag{13b}$$

$$\overset{(k)}{\boldsymbol{u}}{}'' = k\overset{(k)}{\boldsymbol{u}} + (2k+1)\left(\overset{(k+2)}{\boldsymbol{u}} + \overset{(k+4)}{\boldsymbol{u}} + \cdots\right). \tag{13c}$$

If into (13) we insert $\overset{(k)}{\boldsymbol{u}}$ instead of $\overset{(k)}{\boldsymbol{p}}{}^i$ we shall get expressions for $D_j\overset{(k)}{\boldsymbol{p}}{}^i$.

Consider the operators

$$\underline{D}_j\overset{(k)}{\boldsymbol{u}} = \begin{cases} \partial_\alpha \overset{(k)}{\boldsymbol{u}} - \dfrac{1}{h}\,\partial_\alpha \bar{h}\,\underline{\overset{(k)}{\boldsymbol{u}}} - \dfrac{1}{h}\,\partial_\alpha h\,\underline{\underline{\overset{(k)}{\boldsymbol{u}}}}, & \text{if } j = \alpha, \\ -\dfrac{1}{h}\,\underline{\overset{(k)}{\boldsymbol{u}}} & \text{if } j = 3, \end{cases} \tag{14a}$$

where

$$\underline{\overset{(k)}{\boldsymbol{u}}} = (2k+1)\left(\overset{(k-1)}{\boldsymbol{u}} + \overset{(k-3)}{\boldsymbol{u}} + \cdots\right) \quad \left(\overset{(k)}{\boldsymbol{u}} = 0, \quad \text{if} \quad k < 0\right), \tag{14b}$$

$$\underline{\underline{\overset{(k)}{\boldsymbol{u}}}} = -(k+1)\,\overset{(k)}{\boldsymbol{u}} + (2k+1)\left(\overset{(k-2)}{\boldsymbol{u}} + \overset{(k-4)}{\boldsymbol{u}} + \cdots\right). \tag{14c}$$

Then the system of Eqs. (11) may be written as

$$\frac{1}{\sqrt{a}}\,\underline{D}_i\left(\sqrt{a}\,\overset{(k)}{\boldsymbol{p}}{}^i\right) + \overset{(k)}{\boldsymbol{F}} = 0, \quad \overset{(k)}{\boldsymbol{p}}{}^i = \overset{(0)}{\mathrm{C}}{}^{ij}\underline{D}_j\overset{(k)}{\boldsymbol{u}} \tag{11b}$$

$$(k = 0, 1, \ldots),$$

where

$$\overset{(k)}{\boldsymbol{F}} = \overset{(k)}{\boldsymbol{\Phi}} + \frac{2k+1}{2h}\left\{\left(1 + \partial_\alpha \overset{(+)}{h}\,\partial^\alpha \overset{(+)}{h}\right)^{1/2}\overset{(+)}{\boldsymbol{P}} - (-1)^k\left(1 + \partial_\alpha \overset{(-)}{h}\,\partial^\alpha \overset{(-)}{h}\right)^{1/2}\overset{(-)}{\boldsymbol{P}}\right\}. \tag{14d}$$

Here $\overset{(+)}{\boldsymbol{P}}$ and $\overset{(-)}{\boldsymbol{P}}$ are the given stresses, applied to the face surfaces $x^3 = \overset{(+)}{h}$ and $x^3 = -\overset{(-)}{h}$, respectively. Therefore $\overset{(k)}{\boldsymbol{F}}$ are the given vector fields on $\mathscr{S}$. In deriving (14d) we have used the approximate equalities (9d).

2.4. We can write the second relations (11b) in an extended form as

$$\overset{(k)}{\boldsymbol{p}}{}^i = \lambda\left(\boldsymbol{r}^j D_j\overset{(k)}{\boldsymbol{u}}\right)\boldsymbol{r}^i + \mu\left(\boldsymbol{r}^i D_j\overset{(k)}{\boldsymbol{u}}\right)\boldsymbol{r}^j + \mu(\boldsymbol{r}^i\boldsymbol{r}^j)\,D_j\overset{(k)}{\boldsymbol{u}}. \tag{15a}$$

According to this, for the components of the vectors $\overset{(k)}{\boldsymbol{p}}{}^i$ we have the expressions

$$\overset{(k)}{p}{}^{ij} = \lambda\overset{(k)}{\Theta}a^{ij} + 2\mu\overset{(k)}{e}{}^{ij}, \tag{15b}$$

where

$$\begin{aligned} \overset{(k)}{e}{}^{\alpha\beta} = \overset{(k)}{e}{}^{\beta\alpha} &= \frac{1}{2}\left(\nabla^\alpha \overset{(k)}{u}{}^\beta + \nabla^\beta \overset{(k)}{u}{}^\alpha\right) - b^{\alpha\beta}\overset{(k)}{u}_3 \\ &\quad - \frac{1}{h}\left(\nabla^\alpha \bar{h}\,\overset{(k)}{u}{}'^\beta + \nabla^\beta \bar{h}\,\overset{(k)}{u}{}'^\alpha\right) - \frac{1}{h}\left(\nabla^\alpha h\,\overset{(k)}{u}{}''^\beta + \nabla^\beta h\,\overset{(k)}{u}{}''^\alpha\right), \end{aligned} \tag{15c}$$

$$\overset{(k)}{e}{}^{\alpha 3} = \overset{(k)}{e}{}^{3\alpha} = \frac{1}{2}\left(\nabla^\alpha \overset{(k)}{u}_3 + b^\alpha_\beta \overset{(k)}{u}{}^\beta - \frac{1}{h}\nabla^\alpha \bar{h}\,\overset{(k)}{u}{}_3' - \frac{1}{h}\nabla^\alpha h\,\overset{(k)}{u}{}_3'' + \frac{1}{h}\overset{(k)}{u}{}'^\alpha\right), \tag{15d}$$

$$\overset{(k)}{e}{}^{33} = \frac{1}{h}\overset{(k)}{u}{}_3', \tag{15e}$$

$$\overset{(k)}{\Theta} = \overset{(k)}{e}{}^i_i = \nabla_\alpha \overset{(k)}{u}{}^\alpha - 2H\overset{(k)}{u}_3 - \frac{1}{h}\nabla_\alpha \bar{h}\,\overset{(k)}{u}{}'^\alpha - \frac{1}{h}\nabla_\alpha h\,\overset{(k)}{u}{}''^\alpha + \frac{1}{h}\overset{(k)}{u}{}_3'. \tag{15f}$$

Here ∇_α and ∇^α are the symbols of covariant and contravariant derivatives on $\mathscr{S}$, $\nabla^\alpha = a^{\alpha\beta}\nabla_\beta$; $\overset{(k)}{u}{}^i$, $\overset{(k)}{u}{}'^i$, $\overset{(k)}{u}{}''^i$ are the components of the vectors $\overset{(k)}{\boldsymbol{u}}$, $\overset{(k)}{\boldsymbol{u}}{}'$, $\overset{(k)}{\boldsymbol{u}}{}''$, $\overset{(k)}{u}_3 = \overset{(k)}{u}{}^3$, $\overset{(k)}{u}{}_3' = \overset{(k)}{u}{}'^3$, $\overset{(k)}{u}{}_3'' = \overset{(k)}{u}{}''^3$ being the scalars, and $\overset{(k)}{u}{}^\alpha$, $\overset{(k)}{u}{}'^\alpha$, $\overset{(k)}{u}{}''^\alpha$ are the contravariant vectors on $\mathscr{S}$; $\overset{(k)}{u}_\alpha = a_{\alpha\beta}\overset{(k)}{u}{}^\beta$, $\overset{(k)}{u}{}_\alpha' = a_{\alpha\beta}\overset{(k)}{u}{}'^\beta$, $\overset{(k)}{u}{}_\alpha'' = a_{\alpha\beta}\overset{(k)}{u}{}''^\beta$ represent the covariant vectors on $\mathscr{S}$.

Thus, we have an infinite system of Eqs. (11) with an infinite number of unknown vector fields $\overset{(k)}{\boldsymbol{u}}$, depending on the point (x^1, x^2) of the surface $\mathscr{S}$.

The advantage of the system (11) is that it contains only two independent variables. But the decrease of the number of independent variables by one is realized by increasing the number of equations to infinity. Quite naturally this causes obvious practical inconveniences. That is why we have to make a further step — reduce the problem to a system of equations of finite order with two independent variables. Such a reduction can be effected in many different ways. Any version of the theory of shells is constructed on the basis of considering the finite set of Eqs. (11) which constitute only the necessary condition for shell equilibrium. For this reason, none of the versions of the shell theory may claim to be perfect. The common shortcoming for all these versions is that a great number of other necessary conditions of shell equilibrium are not considered at all. Below we suggest a very simple way of reducing the problem to a system of equations of finite order.

2.5. Consider the set of the first $N + 1$ Eqs. (11b) and (11c) and assume that $\overset{(k)}{\boldsymbol{u}} = 0$ if $k > N$, where N is some non-positive integer. Then we shall have

$$D_j\overset{(k)}{\boldsymbol{u}} = 0, \quad \overset{(k)}{\boldsymbol{p}}{}^i = 0, \quad \text{if} \quad k > N.$$

Hence, we have obtained the elliptic system of equations of order $6N + 6$ for the $3N + 3$ components $\overset{(k)}{u}{}^i$ of the vectors $\overset{(k)}{\boldsymbol{u}}$ $(k = 0, 1, \ldots, N)$ (see [1, 2]). We shall denote it for brevity by (E_N). This system can also be derived on the basis of the variational principle of the minimum of the energy functional [10].

The energy functional of elastic deformation has the form

$$I(U) = \frac{1}{2} \iiint_{\Omega_s} \boldsymbol{P}^i(U)\, \partial_i \boldsymbol{U}\, d\Omega - \iiint_{\Omega_s} \boldsymbol{\Phi}\, U d\Omega, \tag{16a}$$

where

$$\boldsymbol{P}^i(\boldsymbol{U}) = \mathbb{C}^{ij}\, \partial_j \boldsymbol{U}. \tag{16b}$$

If we approximate the displacement vector by the expressions of the following form:

$$\boldsymbol{U}_N = \sum_{k=0}^{N} \overset{(k)}{\boldsymbol{u}}(x^1, x^2)\, P_k\left(\frac{x^3 - \bar{h}}{h}\right), \tag{16c}$$

then the corresponding approximate expressions for the energy functional will have the form

$$I_N = I(\boldsymbol{U}_N) = \sum_{k=0}^{N} \frac{2}{2k+1} \left\{\frac{1}{2} \iint_S \overset{(k)}{\boldsymbol{p}}{}^i(\boldsymbol{U}_N)\, D_i\overset{(k)}{\boldsymbol{u}}\, dS - \iint_S \overset{(k)}{\boldsymbol{\Phi}}\overset{(k)}{\boldsymbol{u}}\, dS\right\}. \tag{16d}$$

If we set $\overset{(0)}{\boldsymbol{P}}{}^i \simeq \mathbb{C}^{ij}\, \partial_j \boldsymbol{U}$, we shall have

$$I_N = \sum_{k=0}^{N} \frac{2}{2k+1} \left\{\frac{1}{2} \iint_S \overset{(0)}{\mathbb{C}}{}^{ij} D_j\overset{(k)}{\boldsymbol{u}}\, D_i\overset{(k)}{\boldsymbol{u}}\, dS - \iint_S \overset{(k)}{\boldsymbol{\Phi}}\overset{(k)}{\boldsymbol{u}}\, dS\right\}. \tag{16e}$$

Let $\boldsymbol{U} \in C^{m+2}(\overline{\Omega}_S)$. Then it can be proved (see [10]) that

$$I(\boldsymbol{U}) - I(\boldsymbol{U}_N) = 0(N^{-m-1/2}). \tag{16f}$$

If we consider now the problem for the minimum of the functional I_N in the class of sets of $N + 1$ vectors $\overset{(0)}{\boldsymbol{u}}, \overset{(1)}{\boldsymbol{u}}, \ldots, \overset{(N)}{\boldsymbol{u}}$, dependig on (x^1, x^2) we shall obtain the system of Equations (E_N).

Thus, Eqs. (E_N) represent the system of Euler-Lagrange equations of the variational problem on the minimum of the energy functional of elastic strain for the displacement fields of type (16e).

The system of Eqs. (E_N) has a number of important advantages. Firstly, to deduce this system we have used none of the physical and kinematic hypotheses. This deduction is based on a simple idea, using only the Weierstrass theorem on the approximation of continuous functions by polynomials. Secondly, this system is compatible with rather a wide class of natural boundary conditions. For corresponding boundary value problems the theorems of existence and uniqueness of solutions are proved. In the case $N = 1$ the fulfilment of all the five physical boundary conditions of the classical moment theory of shells is fully guaranteed (see [2]).

2.6. Having solved some boundary value problem for the system (E_N), we shall find the components $\overset{(k)}{u}_i$ of the vectors $\overset{(k)}{\boldsymbol{u}}$ $(k = 0, 1, \ldots, N)$. The corresponding fields of displacements and stresses will be expressed as

$$\boldsymbol{U}_N(x^1, x^2, x^3) = \sum_{k=0}^{N} \overset{(k)}{\boldsymbol{u}}(x^1, x^2)\, P_k\left(\frac{x^3 - \bar{h}}{h}\right), \tag{17a}$$

$$\boldsymbol{P}^i_N = \overset{(0)}{\mathbb{C}}{}^{ij}\, \partial_j \boldsymbol{U}_N = \sum_{k=0}^{N} \overset{(k)}{\boldsymbol{p}}{}^j(x^1, x^2)\, P_k\left(\frac{x^3 - \bar{h}}{h}\right), \tag{17b}$$

where $\overset{(k)}{\boldsymbol{p}}{}^i$ are expressed by (11b) by means of the vectors $\overset{(k)}{\boldsymbol{u}}$, $k \leq N$.

Expressions (17) represent polynomials of degree N for x^3. Therefore we shall call them approximations of order N. These formulas are not, in general, compatible with the boundary conditions on the face surfaces $x^3 = \overset{(+)}{h}$ and $x^3 = -\overset{(-)}{h}$. But the addition of summands to the right-hand sides of these formulas can ensure the compatibility with the boundary conditions on the face surfaces; these summands may be taken in such a way that the values of the Fourier coefficients of order $\leq N$ of displacement fields and stresses will not be changed, and besides, in their absolute values, these summands will be arbitrarily small inside the shell. All the necessary conditions will be satisfied if for the displacement and stress fields we take expressions of the form

$$\boldsymbol{U}(x^1, x^2, x^3) = \boldsymbol{U}_N(x^1, x^2, x^3), + \boldsymbol{U}_0(x^1, x^2, x^3), \tag{18a}$$

$$\boldsymbol{P}^i(x^1, x^2, x^3) = \boldsymbol{P}^i_N + \overset{(0)}{\mathbb{C}}{}^{ij}\, \partial_j \boldsymbol{U}_0, \tag{18b}$$

where

$$\boldsymbol{U}_0 = \boldsymbol{A}_m\left(P_{m+2}\left(\frac{x^3}{h}\right) - P_m\left(\frac{x^3}{h}\right)\right) + \boldsymbol{A}_{m+1}\left(P_{m+3}\left(\frac{x^3}{h}\right) - P_{m+1}\left(\frac{x^3}{h}\right)\right), \tag{18c}$$

m is any fixed integer $> N$,

$$\begin{aligned} \boldsymbol{A}_k = \frac{h}{2(2k+3)} \Bigg\{ &\frac{1}{\lambda + 2\mu}\left[\left(\overset{(+)}{P^3} - \overset{(+)}{P^{33}_N}\right) - (-1)^k \left(\overset{(-)}{P^3} - \overset{(-)}{P^{33}_N}\right)\right] \boldsymbol{n} \\ &+ \frac{1}{\mu}\left[\left(\overset{(+)}{P^\alpha} - \overset{(+)}{P^{3\alpha}_N}\right) - (-1)^k \left(\overset{(-)}{P^\alpha} - \overset{(-)}{P^{3\alpha}_N}\right)\right] \boldsymbol{r}_\alpha \Bigg\} \\ &(k = m,\ m+1). \end{aligned} \tag{18d}$$

Here $\overset{(+)}{P^i}$, $\overset{(-)}{P^i}$, $\overset{(+)}{P^{3i}_N}$, $\overset{(-)}{P^{3i}_N}$ are the components of the vectors $\overset{(+)}{\boldsymbol{P}}$, $\overset{(-)}{\boldsymbol{P}}$, $\overset{(+)}{\boldsymbol{P}^3_N}$, $\overset{(-)}{\boldsymbol{P}^3_N}$, respectively.

For the purpose of simplifying formulas it has been suggested that $\bar{h} = 0$, $h = \text{const}$. Taking into account the formula

$$P'_{k+2}(x) - P'_k(x) = (2k+1)\, P_{k+1}(x),$$

it is easy to find out that the Fourier coefficients $\overset{(k)}{U_0} = \overset{(k)}{\boldsymbol{p}}{}^i(\boldsymbol{U}_0) = 0$, if $k \leq N$. Besides, using the inequalities (see e.g. [8])

$$\sqrt[4]{1 - x^2}\, |P_k(x)| < 4\sqrt{\frac{2}{\pi}}\, \frac{1}{\sqrt{k}}, \quad -1 \leq x \leq 1,$$

it is easily proved that for any fixed $\delta > 0$ the estimates

$$\boldsymbol{U}_0(x^1, x^2, x^3) = 0(m^{-3/2}), \quad -h + \delta \leq x^3 \leq h - \delta, \tag{18e}$$

$$\boldsymbol{P}^i(\boldsymbol{U}_0) = 0(m^{-1/2}), \quad -h + \delta \leq x^3 \leq h - \delta, \tag{18f}$$

hold.

Thus, if we take m sufficiently large, for any fixed N, the additional summands $\boldsymbol{U}_0$ and $\boldsymbol{P}^i(\boldsymbol{U}_0)$ in (16) may be chosen so that they are arbitrarily small in their absolute value inside the shell.

2.7. In the case $N = 1$ we have an elliptic system of 12th order. The corresponding version of the theory of shells is much related to the classical Kirchhoff-Love theory. For instance, considering our equations for the plate, loaded along the face surface $x^3 = h$ by the constant normal force q, for the bending (normal deflection) we shall get the equation (see [1, 2])

$$\Delta\Delta w = \frac{q}{D'}, \tag{19}$$

where

$$D' = \frac{(1-\sigma)^2}{1-2\sigma} D, \qquad D = \frac{2Eh^3}{3(1-\sigma^2)}. \tag{20}$$

Here D is the cylindrical stiffness which occurs in the classical equation of *S. Germain-Lagrange*. Our theory gives somewhat larger values for the cylindrical stiffness (for example, if $\sigma = 0{,}3$, then $D' = 1.225D$). Therefore, the bending (normal deflection) of the plate decreases correspondingly. Such discrepancies are evidently quite admissible within the framework of classical approximations. But, by a slight modification of the above method of constructing, one can deduce other systems of equations, which are in better agreement with the classical theory.

2.8. In concluding this section, we should note that for integration of the elliptic system (E_N) one can apply the following iteration method: If we write the systems (E_N) with respect to the isometric Gaussian parameters and use the iteration, the solution of the boundary value problem may be reduced to the solution of the sequence of boundary value problems for the system of equations of equilibrium of a plate. For each step of iteration it is necessary to solve the plane problem $N+1$ times and the same number of Poisson equations with regard to completely defined (homogeneous) boundary conditions. In the transition from the nth step to the $(n+1)$th step, only the right-hand sides of the equations are changed, which are expressed by the solutions of the boundary value problems of the previous nth step.

This allows, in principle, to compile standard programs for solving the boundary value problems by means of the computers.

3.

3.1. Consider the set of the first two equations of the system (2.11a) assuming that $\overset{(k)}{\boldsymbol{u}} = 0$ when $k > 2$. Then, writing this system with respect to the components of vectors contained in it, we shall have (we assume, for simplicity, that $\bar{h} = 0$, $h = \text{const}$)

$$\nabla_\alpha \overset{(0)}{p}{}^{\alpha\beta} - b_\alpha^\beta \overset{(0)}{p}{}^{\alpha 3} + \overset{(0)}{F}{}^\beta = 0 \quad (\beta = 1, 2), \tag{1a}$$

$$\nabla_\alpha \overset{(0)}{p}{}^{\alpha 3} + b_{\alpha\beta} \overset{(0)}{p}{}^{\alpha\beta} + \overset{(0)}{F}{}^3 = 0, \tag{1b}$$

$$\nabla_\alpha \overset{(1)}{p}{}^{\alpha\beta} - b_\alpha^\beta \overset{(1)}{p}{}^{\alpha 3} - \frac{3}{h} \overset{(0)}{p}{}^{3\beta} + \overset{(1)}{F}{}^\beta = 0 \quad (\beta = 1, 2), \tag{1c}$$

$$\nabla_\alpha \overset{(1)}{p}{}^{\alpha 3} + b_{\alpha\beta} \overset{(1)}{p}{}^{\alpha\beta} - \frac{3}{h} \overset{(0)}{p}{}^{33} + \overset{(1)}{F}{}^3 = 0, \tag{1d}$$

where

$$\overset{(0)}{p}{}^{\alpha\beta} = \lambda \nabla^\gamma \overset{(0)}{u}_\gamma a^{\alpha\beta} + \mu\left(\nabla^\alpha \overset{(0)}{u}{}^\beta + \nabla^\beta \overset{(0)}{u}{}^\alpha\right) + \frac{\lambda}{h} \overset{(1)}{u}_3 a^{\alpha\beta} - 2(\lambda H a^{\alpha\beta} + \mu b^{\alpha\beta}) \overset{(0)}{u}_3, \tag{2a}$$

$$\overset{(0)}{p}{}^{\alpha 3} = \overset{(0)}{p}{}^{3\alpha} = \mu\left(\nabla^\alpha \overset{(0)}{u}_3 + \frac{1}{h} \overset{(1)}{u}{}^\alpha\right) + \mu b_\beta^\alpha \overset{(0)}{u}{}^\beta, \tag{2b}$$

$$\overset{(0)}{p}{}^{33} = \lambda \nabla^\gamma \overset{(0)}{u}_\gamma + (\lambda + 2\mu) \frac{1}{h} \overset{(1)}{u}_3 - 2\lambda H \overset{(0)}{u}_3, \tag{2c}$$

$$\overset{(1)}{p}{}^{\alpha\beta} = \lambda \nabla^\gamma \overset{(1)}{u}_\gamma a^{\alpha\beta} + \mu\left(\nabla^\alpha \overset{(1)}{u}{}^\beta + \nabla^\beta \overset{(1)}{u}{}^\alpha\right) + \frac{3\lambda}{h} \overset{(2)}{u}_3 a^{\alpha\beta} - 2(\lambda H a^{\alpha\beta} + \mu b^{\alpha\beta}) \overset{(1)}{u}_3, \tag{2d}$$

$$\overset{(1)}{p}{}^{\alpha 3} = \overset{(1)}{p}{}^{3\alpha} = \mu\left(\nabla^\alpha \overset{(1)}{u}_3 + \frac{3}{h} \overset{(2)}{u}{}^\alpha\right) + \mu b_\beta^\alpha \overset{(1)}{u}{}^\beta, \tag{2e}$$

$$\overset{(1)}{p}{}^{33} = \lambda \nabla^\alpha \overset{(1)}{u}_\alpha + 3(\lambda + 2\mu) \frac{1}{h} \overset{(2)}{u}_3 - 2\lambda H \overset{(1)}{u}_3. \tag{2f}$$

In deducing the above relations, Gauss and Weingarten derivative formulas (see e.g. [7]) have been used

$$\partial_\alpha \partial_\beta \boldsymbol{r} = \Gamma^\gamma_{\alpha\beta} \boldsymbol{r}_\gamma + b_{\alpha\beta} \boldsymbol{n}, \qquad \partial_\alpha \boldsymbol{n} = - b^\beta_\alpha \boldsymbol{r}_\beta, \tag{3}$$

where $\Gamma^\gamma_{\alpha\beta}$ are Christoffel's symbols of 2nd order of the surface $\mathcal{S}$; use has been made, too, of the differentiation formula for the surface vector

$$\partial_\alpha \boldsymbol{u} = (\nabla_\alpha u_\beta - b_{\alpha\beta} u_3)\, \boldsymbol{r}^\beta + (\nabla_\alpha u_3 + b^\beta_\alpha u_\beta)\, \boldsymbol{n}. \tag{4}$$

In the six equations of system (1) nine unknown functions $\overset{(0)}{u_i}, \overset{(1)}{u_i}, \overset{(2)}{u_j}$, the components of the vectors $\overset{(0)}{\boldsymbol{u}}, \overset{(1)}{\boldsymbol{u}}, \overset{(2)}{\boldsymbol{u}}$, are contained as well as their partial derivatives of 1st and 2nd order. Obviously, this system does not allow to determine all the nine unknown functions. That is why it is necessary to add three more relations to it which would allow to obtain the definite system of equations. For this purposs, let us turn to the boundary conditions on the face surfaces of the shell. Apparently, they are satisfied if the components of the stress tensor have the form

$$P^{\alpha\beta} = \overset{(0)}{p^{\alpha\beta}} + \overset{(1)}{p^{\alpha\beta}} P_1\left(\frac{x^3}{h}\right), \tag{5a}$$

$$P^{i3} = \left(1 - P_2\left(\frac{x^3}{h}\right)\right) \overset{(0)}{p^{i3}} + \frac{1}{2}\left(\overset{(+)}{P^i} - \overset{(-)}{P^i}\right) P_1\left(\frac{x^3}{h}\right) + \frac{1}{2}\left(\overset{(+)}{P^i} + \overset{(-)}{P^i}\right) P_2\left(\frac{x^3}{h}\right), \tag{5b}$$

where $\overset{(+)}{P^i}$ and $\overset{(-)}{P^i}$ are the components of the given vectors $\overset{(+)}{\boldsymbol{P}}$ and $\overset{(-)}{\boldsymbol{P}}$, which represent stresses on the face surfaces $x^3 = \pm h$. From (5b) it follows that

$$\overset{(1)}{p^{i3}} = \frac{1}{2}\left(\overset{(+)}{P^i} - \overset{(-)}{P^i}\right). \tag{6}$$

Substituting into the left-hand sides expressions (2e) and (2d) we shall get equations from which it is seen that

$$\overset{(2)}{u^\alpha} = -\frac{h}{3} \nabla^\alpha \overset{(1)}{u_3} - \frac{h}{3} b^\alpha_\beta \overset{(1)}{u^\beta} + \frac{h}{2\mu}\left(\overset{(+)}{P^\alpha} - \overset{(-)}{P^\alpha}\right), \tag{7a}$$

$$\overset{(2)}{u_3} = -\frac{\lambda h}{3(\lambda + 2\mu)} \nabla^\gamma \overset{(1)}{u_\gamma} + \frac{2\lambda h}{3(\lambda + 2\mu)} \overset{(1)}{u_3} + \frac{h}{6(\lambda + 2\mu)}\left(\overset{(+)}{P^3} - \overset{(-)}{P^3}\right). \tag{7b}$$

In this way, three unknown functions $\overset{(2)}{u_i}$ are expressed by $\overset{(1)}{u^i}$ in an explicit form. By substituting expressions (7) into relations (2) and inserting the latter into Eqs. (1), we shall obtain a system of six equations in which six unknown functions $\overset{(0)}{u^i}, \overset{(1)}{u^i}$ are contained.

Now it is easily seen that, if we take into account formulas (2f) and (6), the sixth Eq. (1) will have the form

$$\begin{aligned} &(\lambda + 2\mu)\left[1 + \frac{2h^2}{9(\lambda + 2\mu)} (2\lambda' H^2 + \mu(k_1^2 + k_2^2))\right] \frac{\overset{(1)}{u_3}}{h} \\ &= -\lambda\left(\nabla^\alpha \overset{(0)}{u_\alpha} - 2H \overset{(0)}{u_3}\right) + \frac{2h}{3}\left(\lambda' H \nabla^\alpha \overset{(1)}{u_\alpha} + \mu b_{\alpha\beta} \nabla^\alpha \overset{(1)}{u^\beta}\right) \\ &\quad + \frac{h}{6} \nabla_\alpha \left(\overset{(+)}{P^\alpha} - \overset{(-)}{P^\alpha}\right) + \frac{\lambda h H}{3(\lambda + 2\mu)}\left(\overset{(+)}{P^3} - \overset{(-)}{P^3}\right) + \frac{h}{3} \overset{(1)}{F^3}, \end{aligned} \tag{8a}$$

where

$$\lambda' = \frac{2\lambda\mu}{\lambda + 2\mu} = \frac{E\sigma}{1 - \sigma^2}. \tag{8b}$$

Due to this the function $h^{-1} \overset{(1)}{u_3}$ is uniquely determined. It expresses the elongations of transverse fibres of the shell.

By substituting the value $\overset{(1)}{u_3}$, obtained from (8a), into relations (2) and inserting these expressions into Eqs. (1a)—(1c), we shall have an elliptic system of equations of 10th order for five unknown functions $\overset{(0)}{u_1}, \overset{(0)}{u_2}, \overset{(0)}{u_3}, \overset{(1)}{u_1}, \overset{(1)}{u_2}$. For brevity, denote this system by (E).

The system of Eqs. (E) allows to define all the components of the vector $\overset{(0)}{\boldsymbol{u}}$ and tangential components of the vector $\overset{(1)}{\boldsymbol{u}}$. The importance of the system (E) lies in the fact that it ensures the fulfilment of all the five physical boundary conditions of the classical moment theory of shells.

3.2. The stress and moment acting on the transversal area with unit normal $\boldsymbol{l}$ can be expressed in the following form

$$\boldsymbol{T}_{(l)} = \int_{-h}^{h} \boldsymbol{P}_{(l)} dx^3, \qquad \boldsymbol{M}_{(l)} = \boldsymbol{n} \times \int_{-h}^{h} \boldsymbol{P}_{(l)} x^3 \, dx^3. \tag{9a}$$

According to (2.3.) we will have

$$\boldsymbol{T}_{(l)} = \boldsymbol{T}^{\alpha} l_{\alpha}, \qquad \boldsymbol{M}_{(l)} = \boldsymbol{M}^{\alpha} l_{\alpha}, \tag{9b}$$

where

$$\boldsymbol{T}^{\alpha} = 2h\overset{(0)}{\boldsymbol{p}}{}^{\alpha}, \qquad \boldsymbol{M}^{\alpha} = \frac{2h^2}{3}\left(\boldsymbol{n} \times \overset{(1)}{\boldsymbol{p}}{}^{\alpha}\right). \tag{9c}$$

For the components of the vectors $\boldsymbol{T}^{\alpha}$ and $\boldsymbol{M}^{\alpha}$, according to (9c), we have the formulas

$$T^{\alpha\beta} = 2h\overset{(0)}{p}{}^{\alpha\beta}, \qquad T^{\alpha 3} = 2h\overset{(0)}{p}{}^{\alpha 3}, \qquad M^{\alpha\beta} = \frac{2h^2}{3} c^{\gamma\beta} \overset{(1)}{p}{}^{\alpha}_{\gamma}, \tag{9d}$$

where $c^{\alpha\beta}$ is the contravariant discriminant tensor of the surface $\mathscr{S}$,

$$c^{11} = c^{22} = 0, \qquad c^{12} = -c^{21} = \frac{1}{\sqrt{a}}. \tag{10}$$

Taking into account that $\nabla_{\gamma} c^{\alpha\beta} = 0$, in virtue of Eqs. (6) and formulas (9d), we can write the Eqs. (1a)—(1c) (the sixth Eq. (1) has already been used) as follows

$$\nabla_{\alpha} T^{\alpha\beta} - b^{\beta}_{\alpha} T^{\alpha\beta} + \overset{(0)}{F}{}'^{\beta} = 0 \quad (\beta = 1, 2), \tag{11a}$$

$$\nabla_{\alpha} T^{\alpha 3} + b_{\alpha\beta} T^{\alpha\beta} + \overset{(0)}{F}{}'^{3} = 0, \tag{11b}$$

$$\nabla_{\alpha} M^{\alpha\beta} - c^{\alpha\beta} T_{\alpha 3} + \overset{(1)}{F}{}'^{\beta} = 0 \quad (\beta = 1, 2), \tag{11c}$$

where

$$\overset{(0)}{F}{}'^{i} = 2h\overset{(0)}{F}{}^{i} = \overset{(+)}{P}{}^{i} - \overset{(-)}{P}{}^{i} + \int_{-h}^{h} \Phi^{i} dx^3, \tag{12a}$$

$$\overset{(1)}{F}{}'^{i} = \frac{2h^2}{3}\overset{(1)}{F}{}^{i} = h\overset{(+)}{P}{}^{i} + h\overset{(-)}{P}{}^{i} + \int_{-h}^{h} \Phi^{i} x^3 dx^3. \tag{12}$$

Using (9d), (2a), (2b), (2d) and inserting there the expressions for $\overset{(2)}{u}_j$ and $\overset{(1)}{u}_3$, from (7) and (8), we shall get formulas expressing $T^{\alpha\beta}$, $T^{\alpha 3}$, $M^{\alpha\beta}$ in an explicit form by five unknown functions $\overset{(0)}{u}_1$, $\overset{(0)}{u}_2$, $\overset{(0)}{u}_3$, $\overset{(1)}{u}_1$, $\overset{(1)}{u}_2$ and stresses given on the face surfaces.

3.3. For thin and flat shells these expressions may be considerably simplified if we neglect the summands which are small in absolute value and whose presence can have no essential effect. According to the assumption (2.9), for thin and shallow shells the expressions of the kind $(1 \pm c_1 k_1 h)\, A$ and $(1 \pm c_2 k_2 h)\, A$ can be equated to A, where c_1 and c_2 are the constants which do not depend on h. From this assumption it naturally follows that in previous equations one can neglect the summands of order $(hk_1)^2$, $(hk_2)^2$, $h^2 k_1 k_2$. By realizing the corresponding simplifications in (7b) and (8a) we shall have

$$\begin{aligned} \frac{1}{h}\overset{(1)}{u}{}^{3} = &-\frac{\lambda}{\lambda + 2\mu}\left(\nabla^{\alpha}\overset{(0)}{u}_{\alpha} - 2H\overset{(0)}{u}_{3}\right) + \frac{2h}{3(\lambda + 2\mu)}\left(\lambda' H \nabla^{\alpha}\overset{(1)}{u}_{\alpha} + \mu b_{\alpha\beta}\nabla^{\alpha}\overset{(1)}{u}{}^{\beta}\right) \\ &+ \frac{1}{2(\lambda + 2\mu)}\left(\overset{(+)}{P}{}^{3} + \overset{(-)}{P}{}^{3}\right) + \frac{h}{6(\lambda + 2\mu)}\nabla_{\alpha}\left(\overset{(+)}{P}{}^{\alpha} - \overset{(-)}{P}{}^{\alpha}\right) + \frac{1}{2h}\int_{-h}^{h}\Phi_3\, x^3\, dx^3, \end{aligned} \tag{13}$$

$$\begin{aligned} \frac{3}{h}\overset{(2)}{u}_{3} = &-\frac{\lambda}{\lambda + 2\mu}\nabla^{\gamma}\overset{(1)}{u}_{\gamma} - \frac{2\lambda^2 hH}{(\lambda + 2\mu)}\nabla_{\alpha}\overset{(1)}{u}{}^{\alpha} \\ &+ \frac{\lambda}{2(\lambda + 2\mu)}\left(\overset{(+)}{P}{}^{3} - \overset{(-)}{P}{}^{3}\right) + \frac{\lambda h^2 H}{3(\lambda + 2\mu)}\nabla_{\alpha}\left(\overset{(+)}{P}{}^{\alpha} - \overset{(-)}{P}{}^{\alpha}\right) + \frac{\lambda H}{\lambda + 2\mu}\int_{-h}^{h}\Phi_3 x^3\, dx^3. \end{aligned} \tag{14}$$

Substituting these expressions into (2a) and (2d) and making similar simplifications, we will get the formulas

$$\overset{(0)}{p}{}^{\alpha\beta} = \lambda'\nabla^\gamma \overset{(0)}{u}_\gamma a^{\alpha\beta} + \mu\left(\nabla^\alpha \overset{(0)}{u}{}^\beta + \nabla^\beta \overset{(0)}{u}{}^\alpha\right) - 2(\lambda' H a^{\alpha\beta} + \mu b^{\alpha\beta}) \overset{(0)}{u}_3$$
$$+ \frac{2\lambda h}{3(\lambda + 2\mu)} \left(\lambda' H \nabla^\gamma \overset{(1)}{u}_\gamma + \mu b_{\gamma\nu} \nabla^\gamma \overset{(1)}{u}{}^\nu\right) a^{\alpha\beta} \tag{15}$$
$$+ \left[\frac{1}{2(\lambda + 2\mu)} \left(\overset{(+)}{P}{}^3 + \overset{(-)}{P}{}^3\right) + \frac{\lambda h}{6(\lambda + 2\mu)} \nabla_\gamma \left(\overset{(+)}{P}{}^\gamma - \overset{(-)}{P}{}^\gamma\right) + \frac{\lambda}{2h} \int_{-h}^{h} \Phi_3 x^3\, dx\right] a^{\alpha\beta},$$

$$\overset{(1)}{p}{}^{\alpha\beta} = \lambda'\nabla^\gamma \overset{(1)}{u}_\gamma a^{\alpha\beta} + \mu\left(\nabla^\alpha \overset{(1)}{u}{}^\beta + \nabla^\beta \overset{(1)}{u}{}^\alpha\right) + \frac{\lambda}{2(\lambda + 2\mu)} \left(\overset{(+)}{P}{}^3 - \overset{(-)}{P}{}^3\right) a^{\alpha\beta}$$
$$+ \frac{\lambda' h}{\mu} (2\lambda' H a^{\alpha\beta} + \mu b^{\alpha\beta}) \nabla^\gamma \overset{(0)}{u}_\gamma \tag{16}$$
$$- \frac{\mu h^2}{3(\lambda + 2\mu)} b^{\alpha\beta} \nabla_\gamma \left(\overset{(+)}{P}{}^\gamma - \overset{(-)}{P}{}^\gamma\right) + (\lambda' H a^{\alpha\beta} - \mu b^{\alpha\beta}) \int_{-h}^{h} \Phi_3 x^3\, dx^3.$$

Since the tangential components of the stress tensor are expressed by the formulas

$$P^{\alpha\beta} = \overset{(0)}{p}{}^{\alpha\beta} + \frac{x^3}{h} \overset{(1)}{p}{}^{\alpha\beta}, \tag{17}$$

it is easily verified that in Eqs. (16) there is no sense in retaining the summands which have as multipliers the main curvatures k_1 and k_2 of the surface $\mathscr{S}$. In the right-hand sides of Eqs. (17) we shall obtain summands of the form $(1 + c_\alpha k_\alpha h)\, A$, which will be equated to A as it has been agreed above. If we neglect the above mentioned summands, (16) will take the form

$$\overset{(1)}{p}{}^{\alpha\beta} = \lambda'\nabla^\gamma \overset{(1)}{u}_\gamma a^{\alpha\beta} + \mu\left(\nabla^\alpha \overset{(1)}{u}{}^\beta + \nabla^\beta \overset{(1)}{u}{}^\alpha\right) + \frac{\lambda}{2(\lambda + 2\mu)} \left(\overset{(+)}{P}{}^3 - \overset{(-)}{P}{}^3\right) a^{\alpha\beta}. \tag{18}$$

By means of (17) one can easily find out that the summands in the right-hand side of (15) which depend on $\overset{(1)}{u}_\alpha$ have no essential effect on values of tangential components of the stress tensor in that part of the shell where $0.08h \leq x^3 \leq h$. They have influence only on rather a narrow layer near the middle surface. But even there it will not be essential for shallow shells. For this reason, one can neglect the above-mentioned summands in (15).

Suppose the volume force is equal to zero, $\boldsymbol{\Phi} = 0$, and the surface loads have the form

$$\overset{(+)}{P}{}^3 = -\overset{(+)}{P}, \qquad \overset{(-)}{P}{}^3 = \overset{(-)}{P}, \qquad \overset{(+)}{P}{}^\alpha = \overset{(+)}{P}{}^\alpha = 0, \tag{19}$$

where $\overset{(+)}{P}$ and $\overset{(-)}{P}$ are given scalars. In other words, we consider the shell to be loaded only by normal surface forces. Under these conditions, formulas (15) and (18) may be taken in the form

$$\overset{(0)}{p}{}^{\alpha\beta} = \lambda'\nabla^\gamma \overset{(0)}{u}_\gamma a^{\alpha\beta} + \mu\left(\nabla^\alpha \overset{(0)}{u}{}^\beta + \nabla^\beta \overset{(0)}{u}{}^\alpha\right) - 2\left(\lambda' H a^{\alpha\beta} + \mu b^{\alpha\beta}\right) \overset{(0)}{u}{}^3 - \frac{\lambda}{2(\lambda + 2\mu)} \left(\overset{(+)}{P} - \overset{(-)}{P}\right) a^{\alpha\beta}, \tag{20a}$$

$$\overset{(1)}{p}{}^{\alpha\beta} = \lambda'\nabla^\gamma \overset{(1)}{u}_\gamma a^{\alpha\beta} + \mu\left(\nabla^\alpha \overset{(1)}{u}{}^\beta + \nabla^\beta \overset{(1)}{u}{}^\alpha\right) - \frac{\lambda}{2(\lambda + 2\mu)} \left(\overset{(+)}{P} + \overset{(-)}{P}\right) a^{\alpha\beta} \tag{20b}$$

Now it is easily seen that formulas (13) and (14) can be simplified too and written in the form

$$\overset{(1)}{u}_3 = -\frac{\lambda h}{\lambda + 2\mu} \left(\nabla^\alpha \overset{(0)}{u}_\alpha - 2H \overset{(0)}{u}_3\right) - \frac{h}{2(\lambda + 2\mu)} \left(\overset{(+)}{P} - \overset{(-)}{P}\right), \tag{21a}$$

$$\overset{(2)}{u}_3 = -\frac{\lambda h}{(3\lambda + 2\mu)} \nabla^\alpha \overset{(1)}{u}_\alpha - \frac{h}{6(\lambda + 2\mu)} \left(\overset{(+)}{P} - \overset{(-)}{P}\right). \tag{21b}$$

If we introduce the notations

$$u_i = \overset{(0)}{u}_i, \quad v_i = \overset{(1)}{u}_i / h \tag{22}$$

and insert into (9) the expressions (2), (18) and (20), we shall have

$$T^{\alpha\beta} = 2h[\lambda'\nabla^\alpha u_\gamma\, a^{\alpha\beta} + \mu(\nabla^\alpha u^\beta + \nabla^\beta u^\alpha) - 4h(\lambda' H a^{\alpha\beta} + \mu b^{\alpha\beta})\, u_3 - \frac{\lambda h}{\lambda + 2\mu}\Big(\overset{(+)}{P} - \overset{(-)}{P}\Big) a^{\alpha\beta}, \tag{23a}$$

$$T^{\alpha 3} = 2\mu h(\nabla^\alpha u_3 + b^\alpha_\beta u^\beta + v^\alpha), \tag{23b}$$

$$M^{\alpha\beta} = \frac{2h^3}{3}[\lambda'\nabla^\gamma\, v_\gamma c^{\alpha\beta} + \mu(\nabla^\alpha c^{\gamma\beta} v_\gamma + c^{\gamma\beta}\nabla_\gamma v^\alpha)] - \frac{\lambda h^3}{3(\lambda + 2\mu)}\Big(\overset{(+)}{P} - \overset{(-)}{P}\Big) c^{\alpha\beta}. \tag{23c}$$

If we introduce expressions (23) into Eqs. (11) and neglect the summands of order k_1^2, k_2^2, $k_1 k_2$ as well as the expressions with the multipliers $\partial_\alpha k_1$, $\partial_\alpha k_2$, the system of equations will have the form

$$\mu\nabla_\alpha\nabla^\alpha u_\beta + (\lambda' + \mu)\,\nabla_\beta\nabla^\alpha u_\alpha - \mu b^\alpha_\beta(\nabla_\alpha u_3 + v_\alpha) - 2(\lambda' H a^\alpha_\beta + \mu b^\alpha_\beta)\,\nabla_\alpha u_3$$
$$- \frac{\lambda}{2(\lambda + 2\mu)}\,\nabla^\beta\Big(\overset{(+)}{P} - \overset{(-)}{P}\Big) = 0\,, \tag{24a}$$

$$\mu\nabla_\alpha\nabla^\alpha u_3 + \mu\nabla_\alpha\, v^\alpha + 2\lambda' H\nabla_\alpha\, u^\alpha + 3\mu b_{\alpha\beta}\nabla^\alpha\, u^\beta - \frac{1}{2h}\Big(\overset{(+)}{P} - \overset{(-)}{P}\Big) = 0\,, \tag{24b}$$

$$\mu\nabla_\alpha\nabla^\alpha\, v_\beta + (\lambda' + \mu)\,\nabla_\beta\nabla^\alpha v_\alpha - \frac{3\mu}{h^2}(\nabla_\beta u_3 + b^\alpha_\beta u^\beta + v_\beta) - \frac{\lambda}{2(\lambda + 2\mu)h}\,\nabla_\beta\Big(\overset{(+)}{P} + \overset{(+)}{P}\Big) = 0\,. \tag{24c}$$

These equations will be considered as the system of equations of equilibrium of flat shells. For brevity, denote this system by (E_0).

4.

4.1. In the case of the plate, $b_{\alpha\beta} = 0$, the system of equations (E), as well as the simplified system (E_0) will take the form

$$\mu\nabla_\alpha\nabla^\alpha\, u_\beta + (\lambda' + \mu)\nabla_\beta\nabla^\alpha\, u_\alpha - \frac{\lambda}{2(\lambda + 2\mu)}\,\nabla_\beta\Big(\overset{(+)}{P} - \overset{(-)}{P}\Big) = 0\,, \tag{1}$$

$$\nabla_\alpha\nabla^\alpha u_3 + \nabla^\alpha v_\alpha - \frac{1}{2\mu h}\Big(\overset{(+)}{P} + \overset{(-)}{P}\Big) = 0\,, \tag{2a}$$

$$\mu\nabla_\alpha\nabla^\alpha\, v_\beta + (\lambda' + \mu)\nabla_\beta\nabla^\alpha\, v_\alpha - \frac{3\mu}{h^2}(\nabla_\beta u_3 + v_\beta) - \frac{\lambda}{2(\lambda + 2\mu)\, h}\,\nabla_\beta\Big(\overset{(+)}{P} + \overset{(+)}{P}\Big) = 0\,. \tag{2b}$$

Thus, in the case of a plate of constant thickness the system of Eqs. (E_0) is splitted into two independent systems of Eqs. (1) and (2). As it will be seen later, their general solution may be expressed in an explicit form.

Similar systems of equations are obtained in a somewhat different way by a number of other authors (see e.g. [5, 6]). In *E. Reissner*'s work in the left-hand side of (2b) instead of the summand $3\mu h^{-2}(\nabla_\beta u_3 + v_\beta)$ we have $2.5\mu h^{-2}(\nabla_\beta u_3 + v_\beta)$.

4.2. Consider first the system (2). It is easy to derive the following equations

$$\Delta\phi = \frac{1}{D}\Big(q + \frac{\sigma h^2}{3(1 - \sigma)}\Delta q\Big), \quad \phi = \nabla_\alpha v^\alpha, \quad q = \overset{(+)}{P} + \overset{(-)}{P}, \tag{3}$$

$$\Delta\Delta u_3 = -\frac{1}{D}\Big(q - \frac{(2 - \sigma)\, h^2}{3(1 - \sigma)}\Delta q\Big), \tag{4}$$

$$\Delta\omega - 3h^{-2}\omega = 0, \quad \omega = c^{\alpha\beta}\nabla_\alpha v_\beta, \tag{5}$$

where $\Delta = \nabla_\alpha\nabla^\alpha = \nabla^2$ is the Laplace operator. By means of Eq. (4), from (3) we have

$$\phi = \nabla_\alpha v^\alpha = \Delta\Omega, \tag{6}$$

where

$$\Omega = -u_3 - \frac{2h^2}{3(1 - \sigma)}\Delta u_3 + \frac{2(2 - \sigma)\, h^4}{3(1 - \sigma)}\,\frac{q}{D}. \tag{7}$$

Evidently, Ω satisfies the equation

$$\Delta\Delta\Omega = \frac{1}{D}\Big(q + \frac{\sigma h^2}{3(1 - \sigma)}\Delta q\Big). \tag{8}$$

Now it is easily proved that the general solution of the system of Eqs. (2) is expressed by formulas

$$v_\alpha = \nabla_\alpha \Omega + c_{\alpha\beta} \nabla^\beta \omega, \tag{9a}$$

$$u_3 = -\Omega + \frac{2h^2}{3(1-\sigma)} \Delta\Omega + \frac{2h^4(3\sigma^2 - 9\sigma + 4)}{9(1-\sigma)^2} \frac{q}{D}, \tag{9b}$$

where ω and Ω are the arbitrary solutions of Eqs. (5) and (8).

In the version of the theory of shells considered above, the components of the displacement vector are expressed by the formulas

$$U_i(x^1, x^2, x^3) = \overset{(0)}{u}_i(x^1, x^2) + \overset{(1)}{u}_i(x^1, x^2) P_1\left(\frac{x^3}{h}\right) + \overset{(2)}{u}_i(x^1, x^2) P_2\left(\frac{x^3}{h}\right). \tag{10}$$

Therefore, the bending (normal deflection) of the plate is expressed as

$$w = w(x^1, x^2) = -U_3(x^1, x^2, 0) = -\overset{(0)}{u}_3 + \overset{(2)}{u}_3/2. \tag{11a}$$

By substituting here expressions (3.21) we obtain

$$w = -u_3 - \frac{\lambda h^2}{6(\lambda + 2\mu)} \nabla_\alpha v^\alpha - \frac{h}{12(\lambda + 2\mu)} q. \tag{11b}$$

Applying to both parts of this equality the operator $\Delta\Delta$, with the help of (3) and (4) we shall have

$$\Delta\Delta w = \frac{1}{D}\left(q - \frac{(4-\sigma) h^2}{3(1-\sigma)} \Delta q - \frac{h^4}{18} \Delta\Delta q\right). \tag{12}$$

For $q = \text{const}$ we have the classical equation of *S. Germain-Lagrange* (see e.g. [4]). Thus, the version of the moment theory of shells, considered above, is in full agreement with the well-known classical results.

Now it is not difficult to derive the relations

$$w = \Omega - \frac{(4+\sigma) h^2}{6(1-\sigma)} \Delta\Omega - \frac{(4+\sigma)(33 - 35\sigma + 12\sigma^2) h^4}{36(1-\sigma)^2} \frac{q}{D}, \tag{13a}$$

$$\Omega = w + \frac{(4+\sigma) h^2}{6(1-\sigma)} \Delta w + \frac{(50 - 20\sigma + \sigma^2) h^4}{36(1-\sigma)} \frac{q}{D} + \frac{(4+\sigma) h^6}{108(1-\sigma)} \frac{\Delta q}{D}. \tag{13b}$$

4.3. Let us turn now to system (1). It represents the system of equations of generalized plane stress state of the elastic body. Let $\overset{(-)}{P} = 0$ and $\overset{(+)}{P} = q$. Then the general solution of system (1) is expressed by the formula

$$u_\alpha = \underline{u}_\alpha - \frac{2h^4}{3} \nabla_\alpha \left(\Delta\Omega_0 - \frac{\sigma h}{3(1-\sigma)} \frac{q}{D}\right), \tag{14}$$

where $\underline{u}_\alpha$ represent an arbitrary solution of the corresponding homogeneous system of equations, and Ω_0 is some particular solution of Eq. (8).

With respect to Cartesian coordinates x and y we have the formula (see [9])

$$\underline{u}_1 + i\underline{u}_2 = \varkappa' f(z) - z\overline{f'(z)} - \overline{f_0(z)}, \tag{15}$$

where f and f_0 are arbitrary analytic functions of $z = x + iy$

$$\varkappa' = \frac{\lambda' + 3\mu}{\lambda' + \mu} = \frac{3+\sigma}{1-\sigma}. \tag{16}$$

The preceding formulas make it clear that the general solution of the systems of Eqs. (1) and (2) is expressed by means of five arbitrary analytic functions of $z = x + iy$. This reveals the fact that by a suitable choice of these functions one can ensure the fulfilment of five boundary conditions which have been given independently.

In this way, the moment theory of shells, constructed above, eliminates the principal shortcoming of the classical theory, makes certain refinements to it, but at the same time it is in agreement with a number of classical results which are widely applied in practice.

5.

We proceed now to the consideration of the second way of deriving the equations of the moment theory of shells, which is based on successive differentiation of the equilibrium equations of a continuous medium and the elasticity relations. In order to derive general equations there is no necessity to restrict the class of shells. The necessary restrictions may be introduced later with the aim of simplifying the equations already obtained.

5.1. As before, we shall use the system of coordinates which is normally connected with the middle surface of the shell. Then, taking into account that

$$\sqrt{g} = \Theta \sqrt{a}, \quad \Theta = (1 - k_1 x^3)(1 - k_2 x^3), \tag{1}$$

the equations of the theory of elasticity (II.1) can be written in the form

$$\frac{1}{\sqrt{a}} \partial_\alpha(\sqrt{a}\, \underline{Q}^\alpha) + \partial_3 \underline{Q}^3 + \underline{\Phi} = 0, \tag{2a}$$

$$\underline{Q}^i = \lambda(\underline{R}^k \partial_k U)\, \underline{R}^i + \mu(\underline{R}^i \partial_k U)\, \underline{R}^k + \mu(\underline{R}^i \underline{R}^k)\, \partial_k U, \tag{2b}$$

where

$$\underline{Q}^i = \Theta P^i, \quad \underline{R}^i = \sqrt{\Theta} R^i, \quad \underline{\Phi} = \Theta \Phi. \tag{2c}$$

Formula (2b) may be written as

$$\underline{Q}^i = \underline{\mathbb{C}}^{ij} \partial_j U, \quad \underline{\mathbb{C}}^{ij} = \Theta \mathbb{C}^{ij}. \tag{3a}$$

Taking into account the equalities

$$(R^i)_{x^3=0} = r^i, \quad (\partial_3 R^i)_{x^3=0} = b^i_j r^j, \tag{3b}$$

we shall easily find out that

$$(\underline{R}^i)_{x^3=0} = r^i, \quad (\partial_3 \underline{R}^i)_{x^3=0} = \underline{b}^i_j r^j, \tag{3c}$$

where

$$\underline{b}^i_j = b^i_j - H a^i_j = b^p_q \left(a^i_p a^q_j - \frac{1}{2} a^p_q a^i_j\right). \tag{3d}$$

Since (see p. 323)

$$\mathbb{C}^{ij} = \lambda R^i \mid R^j + \mu R^j \mid R^i + \mu g^{ij} \mathrm{E}, \quad g^{ij} = R^i R^j, \tag{4a}$$

we have

$$(\underline{\mathbb{C}}^{ij})_{x^3=0} = \overset{(0)}{\mathbb{C}^{ij}} = \lambda r^i \mid r^j + \mu r^j \mid r^i + \mu a^{ij} \mathrm{E}, \tag{4b}$$

$$(\partial_3 \underline{\mathbb{C}}^{ij})_{x^3=0} = \overset{(1)}{\underline{\mathbb{C}}^{ij}} = \{\lambda(\underline{b}^i_p a^j_q + a^i_p \underline{b}^j_q) + \mu(\underline{b}^j_p a^i_q + a^j_p \underline{b}^i_q) + \mu \underline{b}^{ij} a_{pq}\} \times r^p \mid r^q. \tag{4c}$$

This formula can also be written as

$$(\partial_3 \underline{\mathbb{C}}^{ij})_{x^3=0} = \overset{(1)}{\underline{\mathbb{C}}^{ij}} = \underline{b}_{mn} \overset{(1)}{\underline{\mathbb{C}}^{ijmn}}, \tag{4d}$$

where

$$\overset{(1)}{\underline{\mathbb{C}}^{ijmn}} = \{\lambda(a^{im} a^n_p a^j_q + a^i_p a^{jm} a^n_q) + \mu(a^{jm} a^n_p a^i_q + a^j_p a^{im} a^n_q) + 2\mu a^{im} a^{jn} a_{pq}\}\, r^p \mid r^q \tag{4e}$$

or

$$\overset{(1)}{\underline{\mathbb{C}}^{ijmn}} = E^{ijmnpq}\, r_p \mid r_q, \tag{4f}$$

where E^{ijmnpq} is the contravariant tensor of the 6th rank,

$$E^{ijmnpq} = \lambda(a^{im} a^{np} a^{jq} + a^{ip} a^{jm} a^{nq}) + \mu(a^{jm} a^{np} a^{iq} + a^{jp} a^{im} a^{nq}) + 2\mu a^{im} a^{jn} a^{pq}. \tag{4g}$$

5.2. Below, for simplicity, we shall confine ourselves to considering the case of thin shells. Then it would be sufficient to find the values of the displacement fields and stresses and their partial derivatives of 1st order with respect to x^3 on the surface $\mathscr{S}$, i.e., when $x^3 = 0$. This means that the approximations will have the form

$$U(x^1, x^2, x^3) = (U)_{x^3=0} + x^3(\partial_3 U)_{x^3=0}, \tag{5a}$$

$$P^i(x^1, x^2, x^3) = (P^i)_{x^3=0} + x^3(\partial_3 P^i)_{x^3=0}. \tag{5b}$$

These formulas show that our constructions may ensure only accuracy of order h^2.

Differentiating Eqs. (2) with respect to x^3, and then putting $x^3 = 0$ in (2) and in the equations obtained by differentiation, we will have

$$\frac{1}{\sqrt{a}}\,\partial_\alpha\Big(\sqrt{a}\,\overset{(k)}{\boldsymbol{q}}{}^{\alpha}\Big) + \overset{(k+1)}{q}{}^{3} + \overset{(k)}{\underline{\boldsymbol{\Phi}}} = 0, \tag{6a}$$

$$\overset{(k)}{\boldsymbol{q}}{}^{i} = \overset{(0)}{\mathbb{C}}{}^{ij}\,\partial_j\overset{(k)}{\boldsymbol{v}} + \overset{(1)}{\underline{\mathbb{C}}}{}^{ij}\,\partial_j\overset{(k-1)}{\boldsymbol{v}} = \overset{(0)}{\mathbb{C}}{}^{ij}\,\partial_j\overset{(k)}{\boldsymbol{v}} + \underline{b}_{mn}\overset{(1)}{\underline{\mathbb{C}}}{}^{ijmn}\,\partial_j\overset{(k-1)}{\boldsymbol{v}}, \tag{6b}$$

$$\Big(k = 0,\,1;\quad \overset{(-1)}{\boldsymbol{v}} = 0\Big),$$

where we have used the notations

$$\overset{(k)}{\boldsymbol{q}}{}^{i} = (\partial_3^k\boldsymbol{Q}^i)_{x^3=0},\qquad \overset{(k)}{\boldsymbol{v}} = (\partial_3^k\boldsymbol{U})_{x^3=0},\qquad \overset{(k)}{\underline{\boldsymbol{\Phi}}} = (\partial_3^k\underline{\boldsymbol{\Phi}})_{x^3=0}, \tag{6c}$$

$$\partial_j\overset{(k)}{\boldsymbol{v}} = \begin{cases} \partial_\alpha\overset{(k)}{\boldsymbol{v}}, & \text{if}\quad j = \alpha,\\ \overset{(k+1)}{\boldsymbol{v}}, & \text{if}\quad j = 3. \end{cases} \tag{6d}$$

Relations (6b) may also be written as

$$\overset{(k)}{\boldsymbol{q}}{}^{i} = \lambda\Big(\boldsymbol{r}^j\,\partial_j\overset{(k)}{\boldsymbol{v}}\Big)\,\boldsymbol{r}^i + \mu\Big(\boldsymbol{r}^i\,\partial_j\overset{(k)}{\boldsymbol{v}}\Big)\,\boldsymbol{r}^j + \mu(\boldsymbol{r}^i\boldsymbol{r}^j)\,\partial_j\overset{(k)}{\boldsymbol{v}} + \underline{b}_{mn}\overset{(k)}{\boldsymbol{q}}{}^{imn}, \tag{6e}$$

where

$$\overset{(k)}{\boldsymbol{q}}{}^{imn} = \lambda\Big(\boldsymbol{r}^m\,\partial^n\overset{(0)}{\boldsymbol{v}}\Big)\,\boldsymbol{r}^i + \lambda a^{im}\Big(\boldsymbol{r}^j\,\partial_j\overset{(0)}{\boldsymbol{v}}\Big)\,\boldsymbol{r}^n + 2\mu a^{im}\,\partial^n\overset{(0)}{\boldsymbol{v}} + \mu a^{im}\Big(\boldsymbol{r}^n\,\partial_j\overset{(0)}{\boldsymbol{v}}\Big)\,\boldsymbol{r}^j + \mu\Big(\boldsymbol{r}^i\,\partial^m\overset{(0)}{\boldsymbol{v}}\Big)\,\boldsymbol{r}^n \tag{6f}$$

$$\Big(\partial^i = a^{ij}\,\partial_j,\quad \overset{(0)}{\boldsymbol{q}}{}^{imn} = 0\Big).$$

It is easily seen that Eqs. (6a) and (6b) represent six equations in which nine components of three vector fields, depending on the point (x^1, x^2) belonging to the surface $\mathscr{S}$, are contained, as well as their partial derivatives of 1st and 2nd order. In order to reduce the number of unknown functions to six, we shall turn to boundary conditions on face surfaces $x^3 = \pm h$ (we assume that $\bar{h} = 0$), i.e., the shell is situated symmetrically with respect to the surface $\mathscr{S}$.

According to the Taylor formula, let us represent $\boldsymbol{Q}^3(x^1, x^2, x^3)$ approximately in the form

$$\boldsymbol{Q}^3 = \overset{(0)}{\boldsymbol{q}}{}^{3} + x^3\overset{(1)}{\boldsymbol{q}}{}^{3} + \frac{(x^3)^2}{2}\overset{(2)}{\boldsymbol{q}}{}^{3}. \tag{7}$$

Substituting here $x = \pm\varepsilon h$, where ε is the numerical parameter $0 < \varepsilon \leq 1$ not yet defined, we shall have

$$\overset{(1)}{\boldsymbol{q}}{}^{3} = \frac{1}{2\varepsilon h}\overset{(+)}{\boldsymbol{Q}}_{\varepsilon},\qquad \overset{(2)}{\boldsymbol{q}}{}^{3} = -2(\varepsilon h)^{-2}\overset{(0)}{\boldsymbol{q}}{}^{3} + (\varepsilon h)^{-2}\overset{(-)}{\boldsymbol{Q}}_{\varepsilon}, \tag{8a}$$

where

$$\overset{(+)}{\boldsymbol{Q}}_{\varepsilon} = \boldsymbol{Q}^3(x^1, x^2, \varepsilon h) - \boldsymbol{Q}^3(x^1, x^2, -\varepsilon h),\qquad \overset{(-)}{\boldsymbol{Q}}_{\varepsilon} = \boldsymbol{Q}^3(x^1, x^2, \varepsilon h) + \boldsymbol{Q}^3(x^1, x^2, -\varepsilon h). \tag{8b}$$

If we assume that $\varepsilon = 1$, which may be allowed in the case of small values of h (thin shells), then, as it is evident from (8b), $\overset{(+)}{\boldsymbol{Q}}_{\varepsilon}$ and $\overset{(-)}{\boldsymbol{Q}}_{\varepsilon}$ will take the prescribed values

$$\overset{(+)}{\boldsymbol{Q}}_{1} = \overset{(+)}{\Theta}\overset{(+)}{\boldsymbol{P}} - \overset{(-)}{\Theta}\overset{(-)}{\boldsymbol{P}},\qquad \overset{(-)}{\boldsymbol{Q}}_{1} = \overset{(+)}{\Theta}\overset{(+)}{\boldsymbol{P}} + \overset{(-)}{\Theta}\overset{(-)}{\boldsymbol{P}}, \tag{9}$$

where $\overset{(+)}{\boldsymbol{P}}$ and $\overset{(-)}{\boldsymbol{P}}$ are the given stresses on the surfaces

$$x^3 = \pm\varepsilon h,\qquad \overset{(+)}{\Theta} = (1 - k_1h)\,(1 - k_2h),\qquad \overset{(-)}{\Theta} = (1 + k_1h)\,(1 + k_2h).$$

Considering the values of the parameter ε, close to 1, it will be assumed that $\overset{(+)}{\boldsymbol{Q}}_{\varepsilon}$ and $\overset{(-)}{\boldsymbol{Q}}_{\varepsilon}$ are defined by formulas (9). Such an assumption is practically quite admissible; it makes no essential distortion of the state of stressed equilibrium of the shell. Inserting expressions (8a) into (7) we shall have

$$\boldsymbol{Q}^3 = \left(1 - \left(\frac{x^3}{\varepsilon h}\right)^2\right)\overset{(0)}{\boldsymbol{q}}{}^{3} + \frac{x^3}{2\varepsilon h}\overset{(+)}{\boldsymbol{Q}}_{\varepsilon} + \frac{(x^3)^2}{2(\varepsilon h)^2}\overset{(-)}{\boldsymbol{Q}}_{\varepsilon}. \tag{10}$$

This expression satisfies the boundary conditions on face surfaces, if we assume that $\varepsilon = 1$. Therefore, for the values of ε which are less but close to 1, the boundary conditions will be satisfied approximately, with sufficient accuracy.

In further relations we shall retain the arbitrary parameter ε. The presence of this parameter will allow to reduce some discrepancies which are revealed in comparing the results, obtained below, with classical equations, as well as with the corollaries of the above constructions.

When $i = 3$, formula (6e) yields

$$\overset{(1)}{\boldsymbol{q}}{}^3 = \left(\left(\lambda \boldsymbol{r}^\alpha \partial_\alpha \overset{(1)}{\boldsymbol{v}}\right) + (\lambda + 2\mu)\, \overset{(2)}{v}{}^3\right) \boldsymbol{n} + \mu \left(\left(\boldsymbol{n}\, \partial^\alpha \overset{(1)}{\boldsymbol{v}}\right) + \overset{(2)}{v}{}^\alpha\right) \boldsymbol{r}_\alpha + \underline{b}_{mn} \overset{(1)}{\boldsymbol{q}}{}^{3mn}. \tag{11}$$

As it is seen from (6f) in the expressions for $\overset{(1)}{\boldsymbol{q}}{}^{3mn}$ there are no components of the vector $\overset{(2)}{\boldsymbol{v}}$. From (11), taking into account (8a), we shall have

$$\overset{(2)}{v}{}^3 = -\frac{\lambda}{\lambda + 2\mu}\left(\boldsymbol{r}^\alpha\, \partial_\alpha \overset{(1)}{\boldsymbol{v}}\right) - \frac{1}{\lambda + 2\mu} \underline{b}_{mn} \overset{(1)}{q}{}^{33mn} + \frac{1}{2(\lambda + 2\mu)\, \varepsilon h} \overset{(+)}{Q}{}^3_\varepsilon, \tag{12a}$$

$$\overset{(2)}{v}{}^\alpha = -\left(\boldsymbol{n}\, \partial^\alpha \overset{(1)}{\boldsymbol{v}}\right) - \frac{1}{\mu} \underline{b}_{mn} \overset{(1)}{q}{}^{3\alpha mn} + \frac{1}{2\mu\varepsilon h} \overset{(+)}{Q}{}^\alpha_\varepsilon, \tag{12b}$$

where $\overset{(1)}{q}{}^{3jmn}$ are the components of the vectors $\overset{(1)}{q}{}^{mn}$; their expressions will be given below.

Thus, the vector $\overset{(2)}{\boldsymbol{v}}$, and therefore $\overset{(0)}{\boldsymbol{q}}{}^i$, $\overset{(1)}{\boldsymbol{q}}{}^i$, are always expressed by the vector fields $\overset{(0)}{\boldsymbol{v}}$ and $\overset{(1)}{\boldsymbol{v}}$, if $\overset{(+)}{\boldsymbol{Q}}_\varepsilon$ is assumed to be expressed by formula (9).

If we represent now $\overset{(k)}{\boldsymbol{q}}{}^i$ as

$$\overset{(k)}{\boldsymbol{q}}_i = \overset{(k)}{q}{}^{i\alpha} \boldsymbol{r}_\alpha + \overset{(k)}{q}{}^{i3} \boldsymbol{n} \tag{13}$$

and use the derivative formulas of Gauss and Weingarten (3.3), we can write Eqs.(6), taking into account (8), in the following form

$$\nabla_\alpha \overset{(0)}{q}{}^{\alpha\beta} - b^\beta_\alpha \overset{(0)}{q}{}^{\alpha 3} + \overset{(0)}{\underline{\Phi}}{}'^\beta = 0 \qquad (\beta = 1, 2), \tag{14a}$$

$$\nabla_\alpha \overset{(0)}{q}{}^{\alpha 3} + b_{\alpha\beta} \overset{(0)}{q}{}^{\alpha\beta} + \overset{(0)}{\underline{\Phi}}{}'^3 = 0, \tag{14b}$$

$$\nabla_\alpha \overset{(1)}{q}{}^{\alpha\beta} - b^\beta_\alpha \overset{(1)}{q}{}^{\alpha 3} - 2(\varepsilon h)^{-2} \overset{(0)}{q}{}^{3\beta} + \overset{(1)}{\underline{\Phi}}{}'^\beta = 0 \qquad (\beta = 1, 2), \tag{14c}$$

$$\nabla_\alpha \overset{(1)}{q}{}^{\alpha 3} + b_{\alpha\beta} \overset{(1)}{q}{}^{\alpha\beta} - 2(\varepsilon h)^{-2} \overset{(0)}{q}{}^{33} + \overset{(1)}{\underline{\Phi}}{}'^3 = 0, \tag{14d}$$

where

$$\overset{(0)}{\underline{\Phi}}{}'^k = \overset{(0)}{\underline{\Phi}}{}^k + \frac{1}{2}(\varepsilon h)^{-1} \overset{(+)}{Q}{}^k_\varepsilon, \qquad \overset{(1)}{\underline{\Phi}}{}'^k = \overset{(1)}{\underline{\Phi}}{}^k + (\varepsilon h)^{-2} \overset{(-)}{Q}{}^k_\varepsilon. \tag{14e}$$

Now, taking into account formulas (12), we shall have

$$\overset{(0)}{q}{}^{\alpha\beta} = \lambda\left(\nabla^\gamma \overset{(0)}{v}_\gamma + \overset{(1)}{v}_3\right) a^{\alpha\beta} + \mu\left(\nabla^\alpha \overset{(0)}{v}{}^\beta + \nabla^\beta \overset{(0)}{v}{}^\alpha\right) - 2(\lambda H a^{\alpha\beta} + \mu b^{\alpha\beta})\, \overset{(0)}{v}_3, \tag{15a}$$

$$\overset{(0)}{q}{}^{\alpha 3} = \overset{(0)}{q}{}^{3\alpha} = \mu\left(\nabla^\alpha \overset{(0)}{v}_3 + b^\alpha_\beta \overset{(0)}{v}{}^\beta + \overset{(1)}{v}{}^\alpha\right), \tag{15b}$$

$$\overset{(0)}{q}{}^{33} = \lambda\left(\nabla^\gamma \overset{(0)}{v}_\gamma - 2H \overset{(0)}{v}_3\right) + (\lambda + 2\mu)\, \overset{(1)}{v}_3, \tag{15c}$$

$$\begin{aligned}\overset{(1)}{q}{}^{\alpha\beta} = {} & \lambda'\, \nabla^\gamma \overset{(1)}{v}_\gamma a^{\alpha\beta} + \mu\left(\nabla^\alpha \overset{(1)}{v}{}^\beta + \nabla^\beta \overset{(1)}{v}{}^\alpha\right) - 2(\lambda' H a^{\alpha\beta} + \mu b^{\alpha\beta})\, \overset{(1)}{v}_3 + \\ & + \underline{b}_{mn}\left(-\frac{\lambda}{\lambda + 2\mu} \overset{(1)}{q}{}^{33mn} a^{\alpha\beta} + \overset{(1)}{q}{}^{\alpha\beta mn}\right) + \frac{\lambda a^{\alpha\beta}}{2(\lambda + 2\mu)\, \varepsilon h} \overset{(+)}{Q}{}^3_\varepsilon,\end{aligned} \tag{15d}$$

$$\overset{(1)}{q}{}^{\alpha 3} = \underline{b}_{mn}\left(\overset{(1)}{q}{}^{\alpha 3mn} - \overset{(1)}{q}{}^{3\alpha mn}\right) + \frac{1}{2\varepsilon h} \overset{(+)}{Q}{}^\alpha_\varepsilon, \tag{15e}$$

where

$$\overset{(1)}{q}{}^{ijmn} = \lambda\left(\boldsymbol{r}^m\, \partial^n \overset{(0)}{\boldsymbol{v}}\right) a^{ij} + \lambda a^{im} a^{jn}\left(\boldsymbol{r}^k\, \partial_k \overset{(0)}{\boldsymbol{v}}\right) + 2\mu a^{im}\left(\boldsymbol{r}^j\, \partial^n \overset{(0)}{\boldsymbol{v}}\right) + \mu a^{im}\left(\boldsymbol{r}^n\, \partial^j \overset{(0)}{\boldsymbol{v}}\right) + \mu a^{jn}\left(\boldsymbol{r}^i\, \partial^m \overset{(0)}{\boldsymbol{v}}\right). \tag{16}$$

Now it is easily seen that Eq. (14) is algebraic with respect to $\overset{(1)}{v}_3$. Defining $\overset{(1)}{v}_3$ from this equation and substituting the obtained expression into Eqs. (14a)—(14c) we shall have an elliptic system of equations of 10th order for five unknown functions $\overset{(0)}{v}_i$, $\overset{(1)}{v}_\alpha$, which is, for brevity, denoted by (E'). This system of equations bears much resemblance to the system (E), which has been established in Section 3, but they are not identical. This will be illustrated below in considering the case of a plate.

6.

Consider, for simplicity, a plate loaded by normal surface forces of type (3.19), being assumed that the volume force is equal to zero. Then formulas (5.15) will take the form

$$\overset{(0)}{q}{}^{\alpha\beta} = \lambda\left(\nabla^\gamma \overset{(0)}{v}_\gamma + \overset{(1)}{v}_3\right) a^{\alpha\beta} + \mu\left(\nabla^\alpha \overset{(0)}{v}{}^\beta + \nabla^\beta \overset{(0)}{v}{}^\alpha\right), \tag{1a}$$

$$\overset{(0)}{q}{}^{\alpha 3} = \overset{(0)}{q}{}^{3\alpha} = \mu\left(\nabla^\alpha \overset{(0)}{v}_3 + \overset{(1)}{v}{}^\alpha\right), \tag{1b}$$

$$\overset{(0)}{q}{}^{33} = \lambda\, \nabla^\gamma \overset{(0)}{v}_\gamma + (\lambda + 2\mu)\, \overset{(1)}{v}_3, \tag{1c}$$

$$\overset{(1)}{q}{}^{\alpha\beta} = \lambda'\, \nabla^\gamma \overset{(1)}{v}_\gamma a^{\alpha\beta} + \mu(\nabla^\alpha \overset{(1)}{v}{}^\beta + \nabla^\beta \overset{(1)}{v}{}^\alpha) + \frac{\lambda a^{\alpha\beta}}{2(\lambda + 2\mu)\, \varepsilon h} \overset{(+)}{Q}{}^3_\varepsilon, \tag{1d}$$

$$\overset{(1)}{q}{}^{\alpha 3} = 0. \tag{1e}$$

Considering now Eq. (5.14d), by virtue of (1c) and (5.14e) we shall have

$$\overset{(1)}{v}_3 = -\frac{\lambda}{\lambda + 2\mu} \nabla^\alpha \overset{(0)}{v}_\alpha + \frac{1}{2(\lambda + 2\mu)} \overset{(-)}{Q}{}^3_\varepsilon. \tag{2}$$

Introducing this expression into (1a) we get

$$\overset{(0)}{q}{}^{\alpha\beta} = \lambda'\, \nabla^\gamma \overset{(0)}{v}_\gamma a^{\alpha\beta} + \mu\left(\nabla^\alpha \overset{(0)}{v}{}^\beta + \nabla^\beta \overset{(0)}{v}{}^\alpha\right) + \frac{\lambda}{2(\lambda + 2\mu)} \overset{(-)}{Q}{}^3_\varepsilon. \tag{3}$$

By substituting expressions (1b), (1d) and (3) into Eqs. (5.14) we shall have the system of equations (being regarded that $b_{\alpha\beta} = 0$)

$$\mu\, \nabla_\alpha \nabla^\alpha \overset{(0)}{v}_\beta + (\lambda' + \mu)\, \nabla_\beta \nabla^\alpha \overset{(0)}{v}_\alpha - \frac{\lambda}{2(\lambda + 2\mu)} \nabla_\beta \left(\overset{(+)}{P} - \overset{(-)}{P}\right) = 0, \tag{4a}$$

$$\mu\, \nabla_\alpha \nabla^\alpha \overset{(0)}{v}_3 + \mu \nabla_\alpha \overset{(1)}{v}{}^\alpha - \frac{1}{2\varepsilon h} q = 0, \tag{4b}$$

$$\mu\, \nabla_\alpha \nabla^\alpha \overset{(1)}{v}_\beta + (\lambda' + \mu)\, \nabla_\beta \nabla^\alpha \overset{(1)}{v}_\alpha - \frac{2\mu}{(\varepsilon h)^2} \left(\nabla_\beta \overset{(0)}{v}_3 + \overset{(1)}{v}_\beta\right) - \frac{\lambda}{2(\lambda + 2\mu)\, \varepsilon h} \nabla_\beta q = 0. \tag{4c}$$

Here we have adopted the notations

$$\overset{(+)}{Q}{}^3_\varepsilon = -\left(\overset{(+)}{P} + \overset{(-)}{P}\right) = q, \quad \overset{(-)}{Q}{}^3_\varepsilon = -\left(\overset{(+)}{P} - \overset{(-)}{P}\right). \tag{5}$$

Comparing the system of Eqs. (4) with Eqs. (4.1) and (4.2) we shall see that with the exception of the three summands in Eqs. (4b) and (4c), which depend on ε, all the other summands have the same form. As it will be seen further, owing to this difference, for the bending of the plate we shall have equations with different right-hand sides.

Consider for simplicity the case $q = \text{const}$. Then for the bending $w' = -\overset{(0)}{v}_3$ we have the equation

$$\Delta\Delta w' = \frac{2}{3\varepsilon^3} \frac{q}{D}. \tag{6}$$

If the parameter ε is taken to be equal to

$$\varepsilon = (2/3)^{1/3} \cong 0.874, \tag{7}$$

we shall come to the full agreement with the classical equation of the bending of the plate.

Thus, the discrepancies between the systems of equations (E) and (E') of the moment theory of shells can be considerably reduced by means of the choice of the parameter ε.

We must recall that when $\varepsilon < 1$, the boundary conditions on the face surfaces are satisfied approximately. In fact, we consider the stresses applied to the face surfaces to be transferred without change to the parallel surface $x^3 = \pm\varepsilon h$ along the normal to $\mathscr{S}$. The order of error of such approximations is easily estimated. In the case of the constant normal surface load

$$\overset{(+)}{\boldsymbol{P}} = -\overset{(+)}{q}\boldsymbol{n}, \quad \overset{(-)}{\boldsymbol{P}} = \overset{(-)}{q}\boldsymbol{n} \quad \left(\overset{(+)}{q} = \text{const}, \overset{(-)}{q} = \text{const}\right) \tag{8}$$

it follows from the equilibrium equations in the stresses that

$$(\partial_3 P^{33})_{x^3=h} = 0, \quad (\partial_3 P^{33})_{x^3=-h} = 0. \tag{9}$$

Applying Taylor's formula, we shall have the equalities

$$P^{33}(x^1, x^2, \varepsilon h) - \overset{(+)}{q} = 0((\varepsilon - 1)^2 h^2), \quad P^{33}(x^1, x^2, -\varepsilon h) - \overset{(-)}{q} = 0((\varepsilon - 1)^2 h^2). \tag{10}$$

These equalities are fully in agreement with formulas (5.5). Therefore, for any choice of ε, $0 < \varepsilon \leq 1$, the error does not go beyond the framework of the admissible accuracy of approximations of the considered version of shell theory.

For the circular plate of radius a with fixed edge, the bendings are expressed by the formula

$$w' = \frac{q}{96\varepsilon^3 D}(a^2 - r^2)^2, \quad r^2 = x^2 + y^2. \tag{11}$$

Hence for the maximal bending we have the expression

$$w_0' = \frac{qa^4}{96\varepsilon^3 D}. \tag{12}$$

If one determines experimentally the values of w_0', the limits of variation of values of ε for a certain class of real shells can be established with more accuracy.

References

1. Vekua, I. N.: Trudy, Tbilisi Mathematical Institute 21 (1955) 191–259.
2. —: ibidem 30 (1965) 1–104.
3. —: Annotatsii dokladov (Abstracts), Seminar of the Institute of Applied Mathematics, Tbilisi State University, 1971, No. 5.
4. Timoshenko, S. P., Woinowsky-Krieger, S.: Theory of plates and shells. 2nd ed. New York: McGraw-Hill 1959.
5. Reissner, E.: J. Appl. Mech. 12 (1945) A68–A100.
6. Green, A. E., Zerna, W.: Theoretical elasticity. Oxford: Clarendon Press 1954.
7. Vekua, I. N.: Fundamentals of the tensor analysis. Tbilisi State University 1967.
8. Sansone, G.: Orthogonal functions. New York: Interscience Publishers 1959.
9. Muskhelishvili, N. I.: Some basic problems of the mathematical theory of elasticity. 5th ed. Moscow, 1966. English translation of the 4th ed., Groningen: Noordhoff 1963.
10. Vekua, I. N.: Variational principles of constructing the theory of shells. Tbilisi State University 1970.

The problem of non-uniqueness and stability in the non-linear mechanics of continuous media

I. I. Vorovich

Rostov-on-Don, USSR

1.

The exceptional interest for the problems of the non-linear mechanics of continuous media is dictated at present by a great number of engineering applications. We can cite as examples stability and the post-critical behaviour of thin-walled constructions, flows of an ideal and viscous fluid and the problem of turbulence, great deformations of continua under the influence of geometric and physical non-linearity, interaction of elastic and fluid media with the beginning of auto-oscillations (flutter), etc.

Today the non-linear mechanics of continuous media develops in several directions; we will mention some of them:

(i) The general physico-mechanical theory of deformation of continuous media,

(ii) the statement of boundary-value problems and their mathematical analysis,

(iii) the elaboration of numerical methods.

In this paper we dwell mainly on the second of the above-mentioned directions, which recently has seriously developed. A certain success attained here is due, to a large extent, to the important achievements we have in abstract mathematics—mainly in non-linear functional analysis.

Though continuous media differ appreciably in their physical essence and consequently in their mathematical description, there exists a great quantity of distinctive features common for many processes and phenomena in the non-linear mechanics of continuous media. All this allows us to speak about the formulation of a certain "non-linear" system of views. The foundation of this system was laid by *Leonhard Euler*. He was the first to formulate the definite idea of linearization, which appeared to be a determinant factor for non-linear mechanics in general. He was the first to discover the most important properties of linearized problems: spectrum, its discreteness, connection between properties of linear and non-linear problems.

The next stage in the formation of the non-linear views is represented by the works of *A. M. Lyapunov* and *Henry Poincaré*. In their works the local methods were developed essentially, and (what is especially important) they founded the non-local methods which gave us the possibility to survey mathematically the whole phenomenon in general without any suppositions about the smallness of determining parameters.

In many respects the development of the stability problem in the mechanics of continuous media was predetermined by the investigations of *Th. von Kármán*, *W. T. Koiter* in elasticity theory and *N. E. Zhukovsky*, *O. Reynolds*, *G. I. Taylor*, *A. I. Nekrasov*, *M. A. Lavrent'ev* in hydrodynamics. It may be also noted that such universal methods as the Bubnov-Galerkin and Krylov-Bogolyubov methods were of great importance for the development of the non-linear mechanics.

At present the mechanics of continuous media disposes of a great number of examples where the loss of stability occurs every time in a specific way. The loss of stability is possible at the transition from one equilibrium state to another, from a state of equilibrium to a periodic one, from one periodic state to another, and so on.

Guided by the available facts we shall try to formulate the problem in a sufficiently general form which should cover a broad class of phenomena. Let us consider a continuous medium, the state of which is described by elements a of some functional space that can almost always be regarded as a Banach space. The motion of the medium is described by a certain function $\boldsymbol{a}(t)$ satisfying an equation of the form

$$\ddot{\boldsymbol{a}} = A(\boldsymbol{a}, \dot{\boldsymbol{a}}, t, \boldsymbol{p}), \tag{1}$$

where A is an operator that characterizes the inner interactions in the medium and is determined by its physico-mechanical properties, boundary conditions and external factors $\boldsymbol{p}$, with respect to which the stability is investigated.

These can be force parameters, given velocities, and so on. External influences may depend on time. In particular, we shall limit ourselves to periodic external influences. Let, further, $\{\boldsymbol{a}(t)\}$ be a set of solutions (1) for arbitrary initial data, and let us consider its structure when $|t|$ is sufficiently large. For a properly posed problem of continuous media with damping it may be asserted that there exists a finite number of maximum regimes (principal states) of motion $\{\boldsymbol{a}^*(t)\}$ such that $\{\boldsymbol{a}(t)\} \to \{\boldsymbol{a}^*(t)\}$. We shall not define here exactly the meaning of these maximum transitions. The set of principal states contains all the equilibrium states of the system, auto-oscillations and forced periodic motions. Its qualitative and quantitative structure, of course, is defined by the value of the parameter $\boldsymbol{p}$. When it changes, the set of the principal states changes as well. To solve completely the stability problem, it is necessary:

(a) to decompose the n-dimensional space $P(p_1, \ldots, p_n)$ of determining parameters into such domains Y_i where the set of the principal states has a definite structure and changes continuously,

(b) to specify the degree of practicability of every principal state in the experiment or in real conditions of exploitation of a mechanical object.

Henceforth we shall call the operation (a) a Y-decompositi on (the name is taken from the theory of automatic control). The investigation of the problems (a) and (b) to a certain extent is characteristic to all stability theories. For instance, when proving asymptotic stability of solutions of the boundary-value problems describing certain phenomena, we confirm a greater degree of its reality. When proving the inverse, we prove its unreality. Of course, the numerical aspect of the problem, which has been elaborated unsufficiently, is of great importance here.

We shall explain the essence of the problem by the example of the Euler problem for a hinge-supported non-compressible rod compressed by the longitudinal force p_1. Here we have one parameter of loading and the space P is a straight line $-\infty < p_1 < \infty$. Eq. (1) in this case (the energy of longitudinal motions is not taken into consideration) becomes

$$\varrho v_{tt} + 2h v_{sst} + B v_{ssss} = B\left\{\frac{v_{ss}}{\sqrt{1-v_s^2}} - v_s\left(\frac{v_{ss}v_s}{\sqrt{1-v_s^2}}\right)_s \frac{1}{\sqrt{1-v_s^2}} + \frac{v_{ss}}{\sqrt{1-v_s^2}}\right\}_s. \tag{2}$$

The displacement of the rod, $v(s, t)$, when t is fixed, can be considered as an element of the Hilbert space with the norm

$$\|v\|^2 = \int_0^l v_{ss}^2\, ds. \tag{3}$$

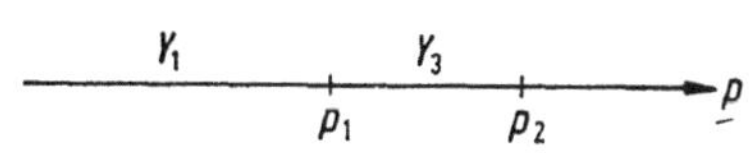

Fig. 1. Y-decomposition in the Eulerian case.

Since in the absence of friction the system is conservative, the set of the principal states contains here only the points of equilibrium. The Y-decomposition is given by the sequence of the eigenvalues p_i of the Eulerian linear boundary—value problem. Obviously, a half-line $-\infty < p < p_1$ (Fig. 1) is a Y_1, for here there is only one form of the equilibrium of the rod (a momentless one). The interval $p_1 < p < p_2$ is a Y_3, for here we have three forms of the equilibrium of the rod, and so on. These results are quite strict and therefore the problem (a) of the Y-decomposition is solved here completely. This is one of few examples when it can be done. As to the problem (b), the criteria of practicability of an equilibrium form in an experiment can be obtained through the investigation of the behaviour of the rod at small displacements. However, the quantitative criteria of the degree of stability or instability of equilibrium forms can be obtained in some cases if we take, as a guide, the level of potential energy of the system. The connection between these two approaches will be examined below.

In view of such a statement of the stability problem, a series of mathematical problems arises in each case, and though they are of a specific nature for every medium, some points are common for all of them:

(a) statement and solvability of boundary-value problems, general properties of solutions,

(b) evaluation of the number of solutions,

(c) choice of the solution having the greatest chance to be realized in an experiment,

(d) behaviour of solutions under the change of parameters defining the problem (degeneration of boundary-value problems, behaviour of solutions at $t \to \infty$),

(e) approximation methods.

In our paper we consider, for certain media, some particular applications of the stated stability problem.

We shall dwell mainly on the item (b) whose mathematical content consists in the evaluation of the number of solutions of corresponding boundary-value problems.

2.

One of the principal domains of mechanics, where the stability problem has a cardinal significance, is the theory of plates and shells. A series of peculiarities in the behaviour of thin-walled elements that makes the stability problem especially difficult, can be noted.

(i) Owing to the specific character of the form (thickness is very small in comparison with other dimensions) the thin-walled elements are disposed to great displacements, comparable with their thickness and even greater.

(ii) Under the given conditions of loading and support there are often several forms of equilibrium and motion (and even several stable ones), especially for interaction with fluid and gas flows.

The equilibrium of the shell under special assumptions is described by the Marguerre-Vlasov theory [3, 90] based on the following relations:

$$\varepsilon_1 = \frac{u_{1\alpha_1}}{A_1} + \frac{A_{1\alpha_2}}{A_2} u_2 + \frac{w}{R_1} + \frac{w_{\alpha_1}^2}{2A_1^2} \quad (1 \rightleftarrows 2),$$

$$\omega = \frac{A_1}{A_2}\left(\frac{u_1}{A_1}\right)_{\alpha_2} + \frac{A_2}{A_1}\left(\frac{u_2}{A_2}\right)_{\alpha_1} - \frac{2w}{R_{12}} + \frac{w_{\alpha_1} w_{\alpha_2}}{A_1 A_2},$$

$$\varkappa_1 = -\frac{1}{A_1}\left(\frac{w_{\alpha_1}}{A_1}\right)_{\alpha_1} - \frac{A_{1\alpha_2} w_{\alpha_2}}{A_2^2 A_1} \quad (1 \rightleftarrows 2),$$

$$\tau = -\frac{w_{\alpha_1\alpha_2}}{A_1 A_2} + \frac{A_{2\alpha_1} w_{\alpha_2}}{A_1 A_2^2} + \frac{A_{1\alpha_2} w_{\alpha_1}}{A_1^2 A_2},$$

$$T_1 = E_1^*(\varepsilon_1 + \sigma\varepsilon_2) \quad (1 \rightleftarrows 2), \qquad S = \frac{1}{2} E_1^*(1-\sigma)\,\omega,$$

$$\begin{aligned} A_2 \left\{\frac{1}{A_1 A_2}\left[(A_2 u_1)_{\alpha_1} + (A_1 u_2)_{\alpha_2}\right]\right\}_{\alpha_1} &+ \frac{1-\sigma}{2} A_1 \left\{\frac{1}{A_1 A_2}\left[(A_1 u_1)_{\alpha_2} - (A_2 u_2)_{\alpha_1}\right]\right\}_{\alpha_2} \\ &+ \frac{1-\sigma}{R_1 R_2} A_1 A_2 u_1 = -\left(\frac{A_2 w}{R_1} + \frac{A_2 w_{\alpha_1}^2}{2A_1^2}\right)_{\alpha_1} - \sigma\left(\frac{A_2 w}{R_2} + \frac{w_{\alpha_2}^2}{2A_2}\right)_{\alpha_1} \\ &- \left[\frac{A_1(1-\sigma)}{2}\left(-\frac{2w}{R_{12}} + \frac{w_{\alpha_1} w_{\alpha_2}}{A_1 A_2}\right)\right]_{\alpha_2} - \frac{A_{1\alpha_2}(1-\sigma)}{2}\left(-\frac{2w}{R_{12}} + \frac{w_{\alpha_1} w_{\alpha_2}}{A_1 A_2}\right) \\ &+ A_{2\alpha_1}\left[\frac{w}{R_2} + \frac{w_{\alpha_2}^2}{2A_2^2} + \sigma\left(\frac{w}{R_1} + \frac{w_{\alpha_1}^2}{2A_1^2}\right)\right] - \frac{A_1 A_2}{E^*} X_1 \quad (1 \rightleftarrows 2), \end{aligned} \tag{4}$$

$$A_1A_2\left\{\nabla^4 w + \frac{1-\sigma}{A_1A_2}\left[\left(\frac{A_2 w_{\alpha_1}}{A_1R_1R_2}\right)_{\alpha_1} + \left(\frac{A_1 w_{\alpha_2}}{A_2R_1R_2}\right)_{\alpha_2}\right]\right\}$$
$$= -\frac{A_1A_2}{E^{**}}\left[T_1\left(\frac{1}{R_1}+\varkappa_1\right) + T_2\left(\frac{1}{R_2}+\varkappa_2\right) - 2S\left(\frac{1}{R_{12}}-\tau\right) - Z + X_1\frac{w_{\alpha_1}}{A_1} + X_2\frac{w_{\alpha_2}}{A_2}\right]$$

Introducing the force function Φ we obtain the following system:

$$A_1A_2\nabla^4 w = -\frac{A_1A_2}{E^*}\left\{\left[T_1^0 + \frac{1}{A_2}\left(\frac{\Phi_{\alpha_2}}{A_2}\right)_{\alpha_2} + \frac{A_{2\alpha_1}\Phi_{\alpha_1}}{A_1^2A_2}\right]\left(\frac{1}{R_1}+\varkappa_1\right)\right.$$
$$+\left[T_2^0 + \frac{1}{A_1}\left(\frac{\Phi_{\alpha_1}}{A_1}\right)_{\alpha_1} + \frac{A_{1\alpha_2}\Phi_{\alpha_2}}{A_2^2A_1}\right]\left(\frac{1}{R_2}+\varkappa_2\right)$$
$$\left.-2\left(S^0 - \frac{\Phi_{\alpha_1\alpha_2}}{A_1A_2} + \frac{A_{1\alpha_2}}{A_1^2A_2}\Phi_{\alpha_1} + \frac{A_{2\alpha_1}}{A_1A_2^2}\Phi_{\alpha_2}\right)\left(\frac{1}{R_{12}}-\tau\right) + Z + X_1\frac{w_{\alpha_1}}{A_1} + X_2\frac{w_{\alpha_2}}{A_2}\right\}, \qquad (5)$$
$$A_1A_2\nabla^4\Phi = 2Eh\left[-\left(\frac{\varkappa_1}{R_1}+\frac{\varkappa_2}{R_2}+\frac{2\tau}{R_{12}}\right)+\tau^2-\varkappa_1\varkappa_2\right].$$

For the systems (4) and (5) some appropriate boundary conditions have to be formulated; we shall not write them down here.

According to the general conception stated above, we must evaluate the number of solutions of the boundary-value problem (4), (5); this is a rather difficult task because of their essential non-linearity.

Of course, the first and principal question that is to be answered, is the ascertainment of the existence conditions of at least one solution. This problem found a sufficiently complete solution in the author's works [4—8, 10, 12, 13, 19]. We shall describe in a few words the method used. The problem was reduced to an operational equation with a completely continuous operator of the form $w = Aw$. Then a certain topological invariant—rotation of the vector field $w - Aw$ on the spheres of a large radius was calculated. It was established that the rotation is $+1$; hence, by virtue of the Leray-Schauder principle in Krasnosel'sky's form [29], it followed that there existed at least one solution under very broad suppositions as to the form of the shell, the loading and the contour limiting the shell. It should be noted also that the solvability of the main boundary-value problems is established in this way without any smallness restrictions on the charcateristic data of the problem (external loading, the form of the shell and its contour) and with smoothness conditions which are minimal from the mechanical point of view. Such general solvability conditions prove a good structure of the Vlasov-Marguerre theory.

It should be noted, that simultaneously with the author *F. N. Morozov* carried out the investigations of solvability of boundary-value problems for the Kármán equations (the case of a plate) [39, 40, 42]. Solvability problems under some conditions, with due regard for the temperature, were solved by *W. Piechocki* [94, 95]. A series of results in this field belongs to *P. Fife* [70] and *S. S. Antman* [65].

Later, the solvability theorems for plates were retold many times. It must be said that one of the main elements of the demonstration of the existence of the solution is an a priori evaluation. In the case of plates it is much simpler than in the case of shells. That is why this problem served as a model for the treatment of many theorems in the theory of non-linear operational equations, for instance in the case of monotone operators. Recently, the solvability theorem was extended by *I. I. Vorovich* and *L. P. Lebedev* [19] to the most general case of anisotropic shells of variable thickness with general mixed boundary conditions.

To specify the features of solvability of essentially non-linear problems we shall consider the case in which the system is subjected to the influence of external loading $\boldsymbol{F}(\boldsymbol{a})$ depending on the displacement vector, rotation angles and so on. Let $\boldsymbol{F}(0)$ be non-autoequilibrated; then the system in the course of deformation must find itself the configuration in which not only internal, but also external forces become balanced. The existence theorem of the soluton is obtained here upon the condition that sufficiently great external forces act. Consequently, we have to state in terms of Y-decompositions that the segment $|p_1| \leq \mu^*$ (where μ^* is a certain number) belongs to Y_0. This theorem shows that in essentially non-linear problems we can come across a great peculiarity of the statement of the boundary-value problems, when they become solvable under sufficiently great loadings.

The essence of such restrictions can be understood from the following elementary example. Let us consider the equilibrium of a Kirchhoff rod, on the contour of which two forces act which remain perpendicular to the rod under the deformation (Fig. 2). It is easy to see that the solution of the problem exists if the force satisfies the inequality

$$p > 11.37\, EJ/l^2, \tag{6}$$

and the form of equilibrium of the rod is given in Fig. 2.

The same example shows that the number of equilibrium forms of an essentially non-linear elastic system depends on loading and increases with the increase of loading. More exactly, the number of forms depends on a certain dimensionless parameter, directly proportional to the force factor and inversely proportional to the parameter characterizing the thinness of the wall.

Thus, the problem of solvability of the main boundary-value problems has been investigated with sufficient completeness. We shall note also that in most cases it is possible to justify a series of approximation methods of solving these problems as well.

We pass now to the evaluation of the number of solutions, i.e., to the Y-decomposition. Here the Eulerian linearization principle is of great importance; hence we shall first consider its efficiency. In this connection a series of important works of *R. Hill* and his collaborators devoted to the linearization principle [78—80] should be mentioned. Here the problem will be considered from another point of view. Suppose that we have an n-parametric loading of a shell and that a certain point in $P(p_1, \ldots, p_n)$ corresponds to each state of loading. Let us suppose also that the shell in all points of P admits a momentless (membrane) state of stresses. Let us linearize the problem near this momentless state of stresses, i.e. let us use the Eulerian method and consider the appropriate Eulerian linear boundary-value problem for its eigenvalues. Under some conditions it can be proved that in space P we shall have a denumerable set of surfaces σ_k, each of them dividing P into two parts, Σ_{1k} and Σ_{2k}, the former containing the origin of coordinates (Fig. 3). The surfaces are star-shaped in their structure and besides

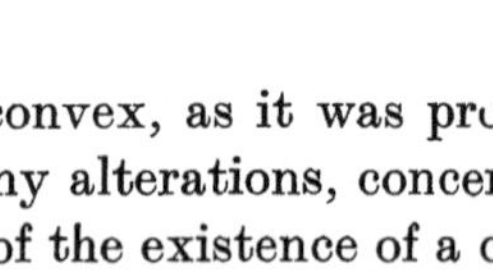

Fig. 2. Bending of a Kirchhoff rod by forces perpendicular to its axis.

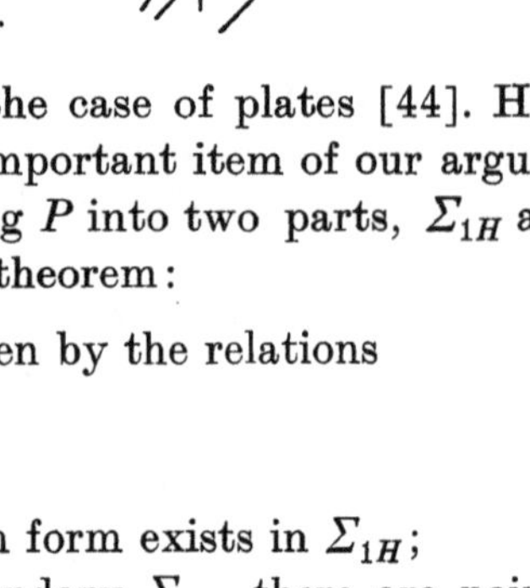

Fig. 3. Y-decomposition in a general case.

they are convex, as it was proved by *P. F. Papkovich* in the case of plates [44]. His proof, without any alterations, concerns shells as well. The most important item of our arguments is the proof of the existence of a certain surface σ_H also dividing P into two parts, Σ_{1H} and Σ_{2H}; its role in Y-decompositions is illustrated by the following theorem:

Theorem I. The mutual location of domains Σ is given by the relations

$$\Sigma_{1H} \subseteq \Sigma_{11} \subset \Sigma_{12} \subset \cdots. \tag{7}$$

Moreover,

(i) $\Sigma_{1H} \subseteq Y_1$, i.e. the unique momentless equilibrium form exists in Σ_{1H};

(ii) in an arbitrary small neighbourhood of the boundary Σ_{2H} there are points of P belonging to $Y_2 + Y_3 + \cdots$, i.e. points where both the momentless (membrane) and moment (bending) states of stresses exist;

(iii) $\Sigma_{12} \subseteq Y_2 + Y_3 + \cdots$, i.e. in any point of Σ_{12} the shell has both the moment equilibrium form and the momentless equilibrium form;

(iv) in each domain Σ_{2k} $(k > 1)$ we can point out such a neighbourhood of its boundary σ_k, which belongs to $Y_3 + Y_4 + \cdots$; at the same time there exists a solution whose norm vanishes at approaching the boundary (in other words, each point of the boundary is a branching point).

In the case of plates and shells subject to one-parametric loading, a part of the statements of the Theorem I was proved by *M. S. Berger* [66—68]. In his work it is also proved for plates that if the eigenvalue is a double one, then at least two solutions branch out of it additionally.

Theorem I, besides the facts related to the Y-decomposition, shows that the Eulerian principle of linearization near a momentless state can be applied to the solution of a stability problem if $\sigma_H \equiv \sigma_1$. In the opposite case, the linearization will not determine the domain of existence of the unique momentless state of stresses. Therefore, from the point of view of the general theory, the establishment of criteria of coincidence and non-coincidence of σ_H and σ_1 is of great importance. We shall cite some sufficient conditions. For this purpose we shall note that the relations (4), (5) lead to the existence of the potential J of the system "shell + exterior forces". This functional can be considered as expressed through the transversal displacement w and having the following structure:

$$J(w) = J_2(w) + J_3(w) + J_4(w), \tag{8}$$

where $J_k(w)$ is a homogeneous relative to the w-functional of order k. Let, further, $\phi(P)$ be an eigenelement of the boundary-value problem linearized in the Eulerian sense. Naturally, it will depend on point P of surface σ_1.

Theorem II. Let condition

$$J_3(\phi) \neq 0 \tag{9}$$

be satisfied in every point of surface σ_1. Then there exists a surface σ_B dividing P into the parts Σ_{1B} and Σ_{2B} so that

(i) $\Sigma_{1H} \subseteq \Sigma_{1B} \subset \Sigma_{11}$ and Σ_{1B} does not coincide with Σ_{11};

(ii) the domain Σ_{11}/Σ_{1B} belongs to $Y_2 + Y_3 + \cdots$, i.e. the shell has here at least one moment equilibrium form, with a lower energy level;

(iii) a certain neighbourhood of the boundary Σ_{12} belongs to $Y_3 + Y_4 + \cdots$, i.e. the shell has here at least three equilibrium forms, two of them being moment forms and one having an energy level lower than the momentless form.

Theorem II gives, in a strict generalized form, a criterion of the impossibility to use the Eulerian linearization method near the momentless states of stresses since here $\sigma_H \neq \sigma_1$. It should be noted that to verify the condition (9), only the Eulerian boundary-value problem must be solved; therefore, this condition in a great number of stability problems for cylindrical, conical or spherical shells can be verified exactly or with the necessary degree of exactness.

In many works, beginning with *Th. von Kármán, H. S. Tsien, D. Yu. Panov, V. I. Fedos'ev, Kh. M. Mushtari, W. T. Koiter, A. S. Vol'mir, F. Niordson, G. A. Thurston, J. W. Hutchinson, N. J. Hoff*, this fact was verified on the base of one or another approximate method of solving the problem.

Theorem III. For the plates $\sigma_H \equiv \sigma_1$, hence Σ_{11} coincides with the domain Y_1, while Σ_{12} coincides with the domain $Y_2 + Y_3 + \cdots$.

The Theorem III completely justifies the linearization in the problems of stability of plates under multiparametric loading.

We shall illustrate the above theorem for the case in which there is one parameter of loading p and the space of determining parameters is a straight line $-\infty < p < \infty$ [10, 12, 13, 15, 17]. Then the surfaces σ_1, σ_H, σ_B are determined by three points $p_H \leq p_B \leq p_1$ (Fig. 4). Hence the half-line $-\infty < p < p_H$ belongs to Y_1, and there exists a unique momentless form of the shell equilibrium. On the segment $p_H \leq p \leq p_B$ there exist points, however close they may be to p_H, which belong to $Y_2 + Y_3 + \cdots$ and for which there exist at least two equilibrium forms of the shell. It is very probable that this whole section belongs to $Y_2 + Y_3 + \ldots$

The half-line $p_B \leq p < \infty$ belongs completely to $Y_2 + Y_3 + \cdots$, in other words: here exists simultaneously with a momentless equilibrium form of the shell at least one form having an energy level lower than a moment form and in this sense a more stable one. At last, an interval $p_k < p < p_k + \Delta p_k$ belongs to $Y_3 + Y_4 + \cdots$ and there are at least three equilibrium forms of the shell on it. With a great degree of probability we can assert that $\Delta p_1 = \infty$ and hence, for all $p > p_1$, there exist at least three equlibrium forms, but this is not proved. It is proved only that the half-line $p > p_1$ entirely belongs to the domain $Y_2 + Y_3 + \cdots$, and along with a momentless equilibrium form there exists a moment form, which is more stable and has a lower energy level. We shall refer to the result of *J. H. Wolkowisky* [102], which seems to be very interesting. He showed that for a round plate, compressed by a longitudinal pressure p, the domain $p > p_n \subset Y_{2n+1}$ under some boundary conditions.

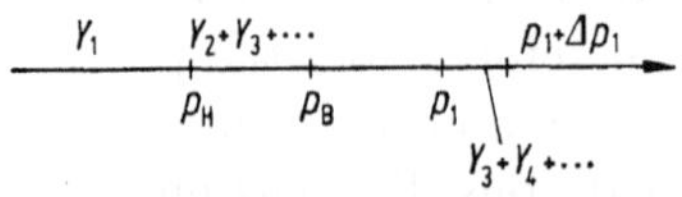

Fig. 4. General pattern of the Y-decomposition for shells in the one-parametric case.

Since the considerations listed here justify the impossibility to resolve the equilibrium problem by the Eulerian method, the question may be raised about the magnitude of the error arising if we still try to resolve the problem in this way. In some cases it can be estimated. For instance, in a one-dimensional case we have

$$p_1 - p_H > \max \frac{J_3^2(\phi)}{4J_4(\phi)\, p_1}. \tag{10}$$

Here, the maximum is taken over the whole variety of its eigenelements corresponding to p_1.

Now we pass to Y-decompositions for the case of moment states of stresses of a shell. Here the results are even more scanty and it is more difficult to obtain them. We shall dwell on the problem of the rigidity of shells which in our opinion is of great theoretical and practical interest.

It is known that under the natural conditions of support the plates have a unique equilibrium form if the loading is small. In our terms it means that a certain neighbourhood of zero of the space of determinant parameters is in the domain Y_1. For shells, in the general case, this does not take place. There exist convex shells of a sufficiently high elevation for which the neighbourhood of the zero of the space of determining parameters belongs already to class Y_2. Hence, when there is no loading, shells may have equilibrium forms distinct from the trivial one. Such shells will be called below non-rigid shells. If the zero of space P belongs to Y_1, shells are called rigid shells. Such shells without loading have, therefore, a unique non-stressed equilibrium form.

The concept of shell rigidity has a geometric origin. A geometric rigidity means uniqueness of the solution of the boundary-value problem of bending. We may ascribe some mechanical properties to surfaces in geometry, too, but they should be limit properties. Namely, the potential energy of deformation of such shells proves to be equal to zero. In shell theory we have the same statement of the problem, but the potential energy accumulates here in the system according to a quite definite law depending on the surface deformation. We shall give two general criteria of rigidity of a thin-walled construction. Let the following conditions on the contour be satisfied for a shell:

$$u_1|_\Gamma = u_2|_\Gamma = 0 \ (*), \quad \text{or} \quad T_n|_\Gamma = S_n|_\Gamma = 0. \ (**)$$

The transversal displacements can satisfy a rather broad class of homogeneous boundary conditions.

Theorem IV. The shell is rigid, if the following condition is satisfied:

$$4h^2\lambda_1 > M \max_{\Omega} (2H^2 - K), \; M = \begin{cases} 3(1-\sigma^2) & \text{for } (*), \\ 3(1-\sigma) & \text{for } (**). \end{cases} \tag{11}$$

In the formula (11) λ_1 is the least eigenvalue of the boundary-value problem for the equation

$$\nabla^4 w + \frac{1-\sigma}{A_1 A_2}\left[\left(\frac{A_2}{A_1} w_{\alpha_1} K\right)_{\alpha_1} + \left(\frac{A_1}{A_2} w_{\alpha_2} K\right)_{\alpha_2}\right] = \lambda w \tag{12}$$

under appropriate boundary conditions (H and K are the mean and Gaussian curvatures, $2h$ is the shell thickness). In (11) the maximum is taken over the whole domain of shell parametrization.

We shall cite some consequences:

Let a given shell satisfy the rigidity condition (11). Then the shell will remain rigid under the imposition of constraints limiting the transversal displacements of the shell.

At the same time, the computations with the help of an electronic computer show that the imposition of longitudinal constraints can effect non-rigidity of the shell. Let us give some examples. Let us consider an entire cylindrical shell of the length L and radius R. At $\sigma = 1/3$ the rigidity condition takes the following form:

$$\begin{aligned} Rh/L^2 &> 0.072 \quad (w|_\Gamma = w_{\alpha_1\alpha_1}|_\Gamma = u_1|_\Gamma = u_2|_\Gamma = 0), \\ Rh/L^2 &> 0.059 \quad (w|_\Gamma = w_{\alpha_1\alpha_1}|_\Gamma = T_n|_\Gamma = S_n|_\Gamma = 0), \\ Rh/L^2 &> 0.031 \quad (w|_\Gamma = w_{\alpha_1}|_\Gamma = u_1|_\Gamma = u_2|_\Gamma = 0), \\ Rh/L^2 &> 0.025 \quad (w|_\Gamma = w_{\alpha_1}|_\Gamma = T_n|_\Gamma = S_n|_\Gamma = 0). \end{aligned} \tag{13}$$

The table clearly illustrates the above-mentioned considerations concerning the influence of longitudinal and transversal constraints on rigidity. It shows as well that at sufficient lengths of the shell all rigidity conditions are violated, and it is reasonable to suppose that cylindrical shells of a great length are not rigid. This is confirmed by the results of calculations with the help of an electronic computer.

As the next example we shall consider the rigidity conditions of a closed cylindrical shell fastened by plane bottoms of the same material and the same thickness (Fig. 5):

$$\frac{Rh}{L^2} > 0.975 \left[\frac{1 + \frac{1394}{153}\frac{L}{R} + \frac{512}{6795}\left(\frac{L}{R}\right)^2}{1 + \frac{22}{9}\frac{L}{R} + \frac{32}{27}\left(\frac{L}{R}\right)^2} \right]^{\frac{1}{2}}. \tag{14}$$

This condition is violated when the thickness of the shell or its radius is too small, or its length is too large. It is reasonable to suppose that in all these cases the shell would not be rigid. It is interesting to note that in the light of the proved facts the complexity of the influence of additional fastenings on the shell rigidity is revealed. Their installation may result even in the loss of rigidity.

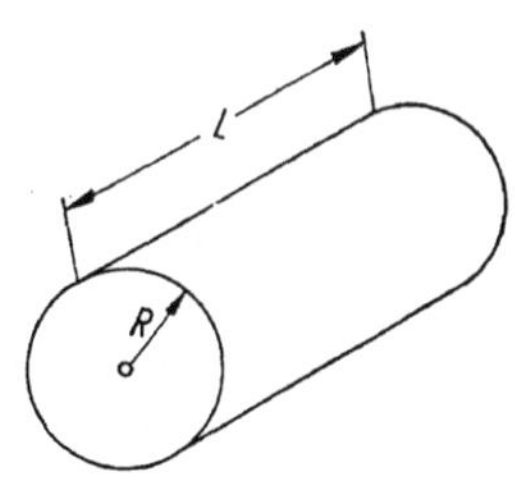

Fig. 5. Closed cylindrical shell fastened by plane bottoms.

The criteria of non-rigidity are of great interest. They are not numerous. The more interesting strict results were obtained by the asymptotic method. The first investigations in this field are due to *K. O. Friedrichs* and *J. J. Stoker* [74—76]. Afterwards the method was developed by *W. Z. Chien* [69], *P. Fife* [70], *V. I. Yudovich* and *L. S. Srubshchik* [55], *L. S. Srubshchik* [51—54]. In the works of the latter the asymptotic method was essentially generalized and justified. He also used the method to obtain the criteria of non-rigidity.

It is known that the equations of the non-linear theory of shells contain a natural small parameter $\varepsilon = (h/(L\gamma))^{1/2}$ ($\gamma^2 = 12(1 - \sigma^2)$) connected with the higher derivatives.

The essence of the asymptotic method consists in using representations of solutions of non-linear boundary-value problems of the shell theory for $\varepsilon \to 0$. It is found [55] that depending on the relation of ε to the dimensionless parameter of loading, $p/(E\gamma)$, a certain kind of expansion holds true. But under smooth loadings an expansion given by the first iterative process is always valid within the domain occupied by the plane of the shell. In the neighbourhood of the boundary and of the sets of loading discontinuity it is necessary to construct solutions of the boundary-layer type. Unlike the linear case, the solutions must be matched here at each stage of the expansion. Considerable difficulties arise in justifying the method as well. *L. S. Srubshchik* [51, 52] constructed an asymptotic representation of solutions for a hinge-supported sperical dome with fixed edge, i.e., when tangential displacements vanish on the boundary. It follows from his expansions quite strictly that without loading such a shell, with a sufficiently small thickness, has a non-trivial form of equilibrium, close to the symmetrically reflected one. The same effect can be achieved at the given thickness as well, but for shells of a sufficiently large elevation.

This proves the non-uniqueness of solutions in the non-linear theory of shells and the non-rigidity of a class of shells. It should be noted that virtually the cited paper states the negativeness of the lower critical number of shells. In another paper [53] this method was essentially developed; first, by *L. S. Srubshchik* who passed from spherical shells to strongly convex shells of arbitrary form. Moreover, a much more difficult problem of rigidity of shells without tangential forces on the contour was investigated there. We have noted already that in this case the tendency of shells to non-rigidity is not expressed so obviously, and it necessitates a more subtle investigation of the problem. It was found by *L. S. Srubshchik* that, if a structural condition is satisfied as to the form of the middle surface, a strongly convex hinge-supported shell stops being rigid in the absence of tangential forces on the boundary at a sufficiently small thickness. Moreover, its lower critical number proves to be negative. In the case of a shell of revolution the structural condition is abolished. Thus, it is established that all strictly convex shells of revolution with the above-mentioned kind of fastening are non-rigid when their thickness is sufficiently small.

The asymptotic method can be used successfully for the direct construction of Y-decompositions as well, including the most difficult case of moment states of stresses. Let us consider [54] a strictly convex shell of revolution (subjected to the action of axisymmetric loading p_1 which for the sake of simplicity is supposed to be constant. It has been proved that at sufficiently small ε there exist two such values p_{1B}, p_{1H} (upper and lower critical loadings) that the intervals

$$p_{1B} - \Delta < p_1 < p_{1B}, \qquad p_{1H} < p_1 < p_{1H} + \Delta \tag{15}$$

at some Δ belong to $Y_2 + Y_3 + \cdots$, i.e. at these values of p_1 the equilibrium forms of the shell are non-unique. Analogous results for non-axisymmetric strictly convex shells have been found later. It should be noted that the ideas of the geometric theory of shells, developed by *A. V. Pogorelov* [45], are of essential significance for these considerations.

The analysis of Y-decompositions justifies the following hypothesis: the domains Y_k with an odd index k always have a space dimension equal to the dimension of the space of loading parameters, n, while the domains with an even index are always degenerated and their dimension is less than n. The fact seems to be of importance that, as we have mentioned, the investigation of boundary-value problems of the non-linear shell theory reduces to an operator equation with a completely continuous operator or, in other words, to finding a fixed point of a vector field. The rotation of this field on the spheres of a large radius appears to be equal to $+1$.

Investigations giving estimates of the number of equilibrium forms of shells in one or another class are of essential interest. Thus, *N. F. Morozov* [41] proved several uniqueness theorems for axisymmetric equilibrium forms of plates under axisymmetric loading and support. Such theorems are proved by him for the case when the plate has no longitudinal displacements on the contour, or if the longitudinal forces are non-compressing. These theorems can be generalized. Namely, the uniqueness of the equilibrium forms of plates is valid in the classes where the main tangential forces are positive. The proof is based on the fact that here the topologic index of each possible solution is $+1$ and the sum of all indices must be equal to $+1$ as well.

N. F. Morozov has proved [42] a few theorems on the generation of non-symmetric forms under axisymmetric loadings. Thus, it has been found that under the action of a sufficiently large transversal loading of a particular type, non-symmetric equilibrium forms appear in the cases of a rigidly fastened plate, a plate without tangential forces on the contour, and so on.

A general theorem guaranteeing, under certain conditions, the appearance of solutions of boundary-value problems of the non-linear shell theory in more broad classes can be proved. These classes differ by the groups of transformations, relative to which the prescribed data of the problem are invariant.

We have mentioned above that, in the non-linear shell theory, the increase of the number of solutions of the boundary-value problem with the increase of loading was noticed in many cases. To illustrate it, let us consider the problem of the stability of a momentless state of stresses for a closed shell of arbitrary shape and let us try to observe the change of the number of equilibrium forms for such a shell when its thickness decreases while the form of the middle surface and external loading remain invariable. Evidently, the problem is determined by the following parameters: external loading p_1, elasticity constants of the shell E and μ, thickness h, characteristic curvature radii R_1 and R_2; they can be assembled into dimensionless combinations: p_1/E, h/R_1, R_2/R_1, μ. Using methods of the similarity theory and some heuristic considerations, we can propose as a hypothesis the following asymptotic estimate for the number N of equilibrium forms, when $h \to 0$:

$$N \sim \left(\frac{p_1}{E}\right)^{\frac{1}{2}} \frac{R_1}{h} \Phi\left(\frac{R_2}{R_1}, \mu\right). \tag{16}$$

This estimate was verified for a great number of problems, but we have not yet any strict proof of it. An essential fact follows from (16): when $h \to 0$ and all other data remain unchanged, the number of forms increases inversely proportional to h. It can be shown also that the solutions are condensed in the energetical sense too. It explains a great part of the shell stability problem as well as numerous phenomena arising during experimentation with rather thin shells. Since the equilibrium forms are numerous here and poorly separated in the energetical sense, it is difficult to make pure experiments and obtain stable results. The same circumstances impede the use of the majority of numerical methods, as the probability of a jump to the next form during the calculation is great; this becomes apparent in the instability of the calculations. At last, it should be noted that we have scarcely touched important conclusions which can be made in the Y-decomposition problem on the basis of the method of a small parameter, as a detailed review in this field was compiled by *W. T. Koiter* and *J. W. Hutchinson* [82].

3.

In this section we shall consider some aspects of stability in hydrodynamics, or more precisely in the theory of viscous flow. This problem is chosen because of its significance for the theory of turbulence as well as for many other practical issues. The up-to-date state of this problem was described in the books of *C. C. Lin* [89], *A. S. Monin* and *A. M. Yaglom* [38]. We shall concentrate our attention on the flow behaviour after the loss of stability, the problem that has been hardly investigated but seems to have a fundamental significance for turbulence. It would be reasonable to cite here the words of *J. Kampé de Fériet* whos said that "our actual lack of existence and uniqueness theorems for the Navier-Stokes equations is a stumbling stone on our way towards a statistical fluid mechanics [83]". Besides, it should be true to say that the insufficiently investigated problem of Y-decompositions in hydrodynamics is one more "stumbling stone". The absence of a variational intrepretation of the problem, comparable to the one that we have in the theory of elasticity in the form of theorems of the minimum of potential energy, seems to bring about some considerable difficulties as well.

Construction of Y-decompositions is reduced to the evaluation of the number of solutions and their structure for the boundary-value problems of the Navier-Stokes equations:

$$\boldsymbol{v}_t + \boldsymbol{v}(\nabla\, \boldsymbol{v}) = -\frac{1}{\varrho} \operatorname{grad} p + \nu \Delta \boldsymbol{v} + \boldsymbol{F}. \tag{17}$$

The force parameters as well as geometrical ones and most often the kinematic parameter (the Reynolds number) can appear as determining parameters here. When the heat convection or any other phenomena are taken into consideration, some other parameters should be certainly added.

The first problem we come across is that of the solvability of the main boundary-value problems for the Navier-Stokes equations. This problem was first investigated by *L. Lichtenstein* [88] and *F. K. G. Odqvist* [91]. *J. Leray* [84, 85] reached the next stage in the development of the problem. It should be said that this problem, as it had happened many times in the history of sciences, was a starting point to the creation of general methods of investigations of non-linearized equations with a completely continuous operator, which are widely used now (the Leray-Schauder principle). These methods were improved while trying to solve this very problem. Recently, a detailed investigation of the stationary flows of viscous fluids has been carried out by *I. I. Vorovich* and *V. I. Yudovich* [9, 11], *K. K. Golovkin* [23], *S. G. Krein* [30], *O. A. Ladyzhenskaya* [32], *V. A. Solonnikov* [49], *R. Finn* [71], *E. Hopf* [81], *G. Prodi* [96] and others. Solvability theorems have been stated and many properties of solutions were studied. The problem of uniqueness has been solved in the small. It is quite evident now that the problem cannot be solved in the large [24, 56]. The solvability theorem for a non-stationary case was given by *E. Hopf* in the most general form, though the uniqueness of the solution in this case has not yet been proved. The analogous situation is observed in the non-linear theory of shell oscillations [8].

An important step in the non-stationary case was made by *O. A. Ladyzhenskaya* [31] who proved the existence and uniqueness in the same class, but only for the case of two space variables. The case of the solution depending on all three generalized coordinates is not actually clear up till now. It is possible that the physical scheme posed by the Navier-Stokes boundary-value problem does not ensure the solvability in natural classes for arbitrary initial data and all moments of time.

As it is known [11, 32], the investigation of stationary flows of viscous fluids can be reduced to the study of an operational equation with a completely continuous multilinear operator.

Y-decompositions are constructed and the number of solutions is evaluated mainly by perturbation theory methods; therefore it is necessary to analyse the spectrum of linearized problems, in particular to establish the odd-foldness of eigenvalues [29]. In some cases the evaluation of the number of solutions is based on the comparison of the index of the solution with a small norm with the rotation of a vector field on spheres of a large radius.

As the first example, let us consider the Kolmogorov flow [37], i.e., the stationary plane flow described by the equation

$$\Delta\psi = \psi_{x_2}(\Delta\psi)_{x_1} - \psi_{x_1}(\Delta\psi)_{x_2} - \frac{\gamma L^3}{8\pi^3\nu^2}\cos x_2 . \tag{18}$$

This flow is realized in the plane x_1, x_2 under the action of the body force $F_1 = -\gamma \sin x_2$ directed along the axis x_1 ($F_2 = 0$). In Eq. (18) L is a characteristic linear dimension, ν is the kinematic viscosity. This equation has always the solution

$$\psi_0 = -\frac{\gamma L^3}{8\pi^3\nu^2}\cos x_2 . \tag{19}$$

Our purpose is to study the whole set of its solutions. In such a general formulation the problem has not yet been analysed. *V. I. Yudovich*, using results of *L. D. Meshalkin* and *Y. G. Sinay* concerning the linearized problem [37], has investigated the probability of the generation of new secondary flows, periodical in x_1 (with period T). In this case the problem has two determining parameters: $\lambda = \gamma L^3/(8\pi^3\nu^2)$ and T/L. In Fig. 6 the T-plane with the data on Y-decompositions following from [56] is given. The existence of lines of the form $\Phi_k = f(L/k)$ $(k = 1, \ldots, \infty)$, dividing the quadrant $0 < \lambda$, $T < \infty$ into domains Σ_0, Σ_1, ... (Fig. 6), is ascertained; ($f(x)$ is a monotone function). Domain Σ_0 is limited by the axes of coordinates and the line Φ_1. Domain Σ_k is confined between Φ_k and Φ_{k+1}. The part of the domain Σ_0 $(0 \leq T \leq L)$ undoubtedly belongs to Y_1 and here we have a unique periodic state of flow, as established by *A. N. Kolmogorov*. All the remaining part of Σ_0 seems to belong to Y_1 too, but this has not been proved.

The domains $\Sigma_1, \Sigma_3, \ldots$ belong to $Y_2 + Y_3 + \cdots$, and small neighbourhoods of the boundaries of domains $\Sigma_2, \Sigma_4, \ldots$, adjoining correspondingly $\Phi_1, \Phi_3, \ldots$, belong to $Y_2 + Y_3 + \cdots$ as well.

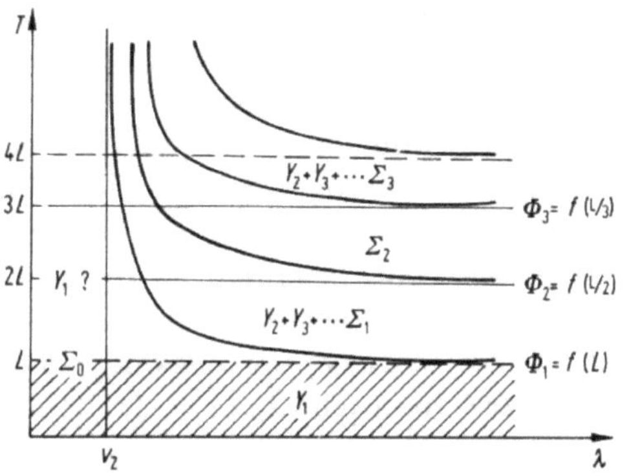

Fig. 6. Y-decomposition for the Kolmogorov flow.

Let us consider the results for the flows of the Couette type, i.e. the plane flows between two rotating cylinders under the action of a vortical body force $\gamma F(r)$ [60]. In this case the boundary-value problem for the Navier-Stokes equations always has the solution $v_0(r)$ determined by the equations

$$\left(\frac{d^2}{dr^2} + \frac{1}{r}\frac{d}{dr} - \frac{1}{r^2}\right)v_0 = -\gamma F(r), \quad v(r_i) = \omega_i r_i \ (i = 1, 2) \tag{20}$$

We shall be interested in the conditions when parallel to the solutions of (20) there appear flows periodical in x_3 (with period L). The problem is characterized by the following dimensionless parameters, which should be considered as determining parameters:

$$r_2/r_1, \ \omega_1/\omega_2, \ \lambda = \gamma r_1/(\nu\omega_1), \ L/r_1.$$

We shall give below the results of the investigation of Y-decompositions [60] relative to the parameter λ, other parameters being fixed. As it was shown in [60], if conditions

$$v_0(r) > 0, \frac{dv_0}{dr} + \frac{v_0}{r} < 0 \quad (r_1 < r < r_2) \tag{21}$$

are satisfied, a denumerable set of eigenvalues λ_k determining intervals Σ_0 $(0 < \lambda < \lambda_1)$, Σ_k $(\lambda_k < \lambda < \lambda_{k+1})$ corresponds to each value of L/r_1. A neighbourhood of the left hand end of Σ_0 (and may be the whole interval Σ_0) is contained in Y_1, a neighbourhood of the lefthand end of each interval Σ_k is contained in $Y_2 + Y_3 + \cdots$ $(k \geq 1)$. At last, intervals Σ_k are contained in $Y_2 + Y_3 + \cdots$ for odd k.

These results were obtained by extensive use of the properties of integral equations with oscillatory kernels [20]. For instance, it is strictly proved that for the proper Couette flows with $\omega_2 r_2^2 > \omega_1 r_1^2$ only a unique plane-parallel flow (20) exists. As it is known, for this case *J. L. Synge* [100] showed that the linearized equation of small oscillations has roots in the left half-plane. If $\omega_2 r_2^2 < \omega_1 r_1^2$, we have the above-mentioned picture of Y-decompositions.

The problem of non-uniqueness and stability is the main problem for the phenomena of convection. The basic equations are of the form

$$\nu \Delta \boldsymbol{v} = \boldsymbol{v}(\nabla \boldsymbol{v}) + \nabla p + \beta T \boldsymbol{g}, \quad \chi \Delta T - \boldsymbol{v}\nabla T = c v_3, \tag{22}$$

here c is the temperature gradient, χ is the heat-conduction coefficient, β is a coefficient characterizing thermal expansion of the fluid. The boundary conditions that we do not give here should be added to Eq. (22). The problem is reduced to the evaluation of the number of solutions of (22) according to the temperature gradient c or to its dimensionless analogue that is a natural determining parameter.

The convection theory starts from the works of *Lord Rayleigh*, *A. Pellew* and *R. Southwell*. The mathematical analysis of the problem can be found in the works of *W. Velte*, *P. H. Rabinowitz*, *H. Görtler*, *K. Kirchgässner*, *P. Sorger* and *V. I. Yudovich* [59, 61, 77, 98, 101]. In Yudovich's work the problem is investigated most completely. We shall mention here some results. Let the conditions $v = T = 0$ be satisfied on the boundary of the domain. Also let C_0 (the first eigen-value of the linearized problem) be a prime number. In this case the following statement is true: the entire half-line $C \leq C_0 \subset Y_1$, and a half-interval $C_0 < C \leq C_1 \subset Y_3 + Y_4 + \ldots$ The first eigenvalue of the convection problem should be supposed to be always a prime number. This is strictly proved for high rectangular and circular cylinders [61].

The case of so-called cellular (hexagonal) convection in a liquid layer [59] is of great interest. It is the case of convection with formation of cells of quite determined symmetry (Fig. 7). Fluid motions of hexagonal structure being sought from the outset, Y-decompositions of the same class may draw attention. It is known that the linearized problem has here a denumerable discrete spectrum C_k ($k = 0, 1, \ldots, \infty$). We have the following results: the half-line $C \leq C_0 \subset Y_1$, and a number C_k^* can be indicated for each C_k ($k = 1, \ldots, \infty$) so that the half-intervals $C_k < C \leq C_k^* \subset Y_3 + Y_4 + \cdots$.

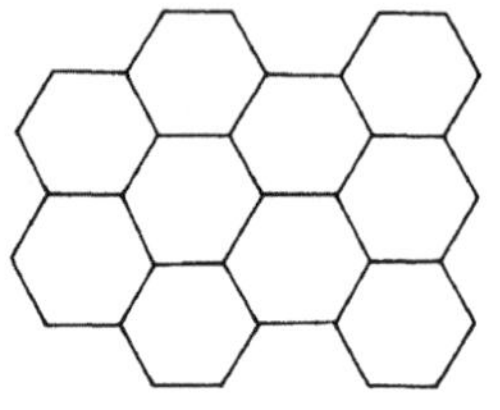

Fig. 7. Hexagonal convection.

All above-mentioned cases of non-uniqueness are characterized by the stationary state of newly generated flows. The appearance of non-stationary flows was studied much less. To date, only few classes have been investigated thoroughly. So, when a translational flow along the axis x_1 with velocity u is superimposed on the Kolmogorov flow, a time-periodic flow appears at the loss of stability [56]. The whole picture of Y-decompositions given in Fig. 6 is quite true, the only difference is that instead of the secondary flow the secondary time-periodic regime appears, and the ordinate axis should carry the values $\tau = T/u$ (period of generated auto-oscillations; Fig. 6) [56].

Another example gives a translational flow with velocity u_3 along the axis of cylinders superimposed on the Couette flow [60]. It is proved that, condition (21) being satisfied, a non-stationary periodic flow is generated for sufficiently small u_3 and for parameters of the rotational flow near to those of the loss of stability of the Couette flow. Some interesting phenomena at the loss of stability are also connected with the generation of regular non-stationary flows of the wavetype, but we cannot dwell on them here.

Interesting researches of *S. M. Zen'kovskaya* connected with the generation of periodic convection flows under the action of a temperature field should also be noted here [28].

4.

Now we pass to the analysis of problem (b) in our interpretation of stability. As it was mentioned above, analysis of the stability of each principal state leads to important qualitative conclusions concerning the problem (b). Therefore, the development of general methods of the stability analysis in continuum mechanics and creation of the qualitative theory of differential equations in functional spaces become quite necessary. Recently, *V. I. Yudovich* [57, 63, 64] transferred the whole theory of the first Lyapunov method to a broad class of operational equations including the boundary-value problems for a viscous fluid. As a result, we have got a general basis of the linearization method both in the autonomous and in the periodical case. This theory explains many important qualitative regularities of the phenomena of loss of stability in continuum mechanics.

At the same time it is important to develop methods giving quantitative estimates of the extent of feasibility of principal states in experiments. The fact is that, as it has been shown by the cited examples, in continuum mechanics we can have several stable equilibrium forms, several stable periodic states etc. When they exist simultaneously the problem (b) acquires all its significance. The experimental realization of a definite equilibrium form of a shell under the non-uniqueness conditions, of a definite cellular form of convection, and so on, depends on the same problem. It is clear that the problem can be completely solved only on the base of the information contained in the boundary-value problem itself. It is necessary to take into account all random factors influencing our system. It should be noted that until recently a statistical statement of the problem was characteristic in general only for the turbulence theory [38]. Recently, statistical methods began to be used very often in the mechanics of deformable bodies. Here the theory of media with randomly changing parameters, which is closely related to the turbulence theory [33, 34], acquires especial significance. The use of statistical methods in the theory of destruction storage, in the theory of random dislocation fields, and so on, refers to the same range of ideas. A somewhat different aspect arises in connection with the problem (b). Analysing boundary-value problems of continuum mechanics, we usually deal with a functional space H which is most closely associated with the problem and in which the study of the problem attains a natural form. The most typical spaces in this sense are energetical ones. Taking into consideration the random character of the phenomenon makes it necessary to consider distributions in these functional spaces. The theory of such distributions is still developed insufficiently and meets some difficulties [21, 22]. Anyhow, this theory has not yet any practical applications. Hence we shall use here a different method based on the following idea. In a number of problems of mechanics (e.g., in the non-linear shell theory) the theory of direct methods (*Ritz, Bubnov-Galerkin*) is developed quite well. With the help of these methods the solution of problems with an infinite number of degrees of freedom can be reduced to the consecutive solution of systems of algebraic equations, or ordinary differential equations in the case of oscillations. Under natural conditions, from the point of view of mechanics, these methods are justified and it is proved that finite-dimensional problems, solvable by direct methods, approximate well the corresponding infinite-dimensional problems. All this suggests that the same methods could be useful in constructing a statistical theory when random factors are taken into account. In this case any possible characteristic of an infinite-dimensional process should be obtained as a limit of corresponding characteristics of finite-dimensional problems. Thus, we connect the limit transition with the approximate solution of the problem itself. This can be illustrated by a scheme covering a broad class of problems of elasticity theory. Let the motion of a medium be described by an equation of the following form:

$$\boldsymbol{w}_{tt} + K\boldsymbol{w}_t + A_1\boldsymbol{w} + A_2\boldsymbol{w} = \boldsymbol{F}(P, t), \tag{23}$$

where $\boldsymbol{w}$ is the vector of displacements of points of the continuous medium occupying the domain Ω. Let us dwell on Eq. (23) in detail. Here A_1 is a positively definite, symmetric, in general unbounded operator given on the functions determined in Ω and independent of t. Using A_1 we can form a Hilbert space H_0 with a scalar product

$$(\boldsymbol{w}_1 \cdot \boldsymbol{w}_2)_{H_0} = \int_\Omega A_1\, \boldsymbol{w}_1 \cdot \boldsymbol{w}_2\, d\Omega\,. \tag{24}$$

Now let $\boldsymbol{w}(t)$ be a set of elements of H_0, depending on t and having a derivative $\boldsymbol{w}_t \in L_2$ in Ω. Let us introduce a Hilbert space H with a scalar product

$$(\boldsymbol{w}_1 \cdot \boldsymbol{w}_2)_H = \int_0^T \left[\int_\Omega \boldsymbol{w}_{1t} \cdot \boldsymbol{w}_{2t}\, d\Omega + (\boldsymbol{w}_1 \cdot \boldsymbol{w}_2)_{H_0} \right] dt\,. \tag{25}$$

The operator A_1 symbolizes the main linear part of elasticity forces developing in the medium. E.g., in the case of oscillations of a plate it is the biharmonic operator. Operator K allows for friction forces. We shall consider it to be linear and in some sense negative. The operator A_2 should be considered acting from H_0 in L_2 quite continuously. A_2 symbolizes non-linear forces of elasticity. Let us assume as well that the exterior forces $\boldsymbol{F}(P, t)$ satisfy the condition

$$\int_0^T \int_\Omega \boldsymbol{F}^2(P, t)\, dP\, dt < \infty\,. \tag{26}$$

The assumed scheme reflects the most typical situation which could be met, e.g., in the non-linear theory of shells. According to this scheme a generalized statement of the Cauchy problem for Eq. (23) is given and the solvability of the problem is proved [8]. In this scheme it is natural to pass to a probabilistic interpretation of the phenomenon. Firstly we have to define the causes of scattering of data in experiments. Primarily it is dispersion of mechanical and other characteristics of the medium, boundary conditions, etc. In this connection we suggest that the operators K and A_2 depend on a certain number of random parameters $\xi_1, \xi_2, \ldots, \xi_N$ considered for the sake of generality to be functional. Let ξ_k be expanded into a series over a basis ϕ_{kn}

$$\xi_k = \sum_{n=1}^{\infty} \alpha_{kn}\phi_{kn} \; (k = 1, \ldots, N). \tag{27}$$

We shall not go into details connected with the character of convergence in (27). It is supposed that the probability properties of ξ_k are specified by a law of distribution $\psi(\alpha_{kn})$ of any set of coefficients α_{kn}. According to most typical situations in continuum mechanics, let us assume that the exterior forces $\boldsymbol{F}(P, t)$ have the following structure:

$$\boldsymbol{F}(P, t) = \boldsymbol{F}_1(P, t) + \boldsymbol{F}_2(P, t) + \boldsymbol{F}_3(P, t) \tag{28}$$

where $\boldsymbol{F}_1(P, t)$ is the mathematical expectation of $\boldsymbol{F}(t)$, $\boldsymbol{F}_2(P, t)$ is a continuous process and $\boldsymbol{F}_3(P, t)$ is a white time-noise Further, it should be stiputaled that $\boldsymbol{F}_1(P, t)$ and each realization $\boldsymbol{F}_2(P, t)$ satisfy condition (26). Besides, let the following expansion be valid:

$$\boldsymbol{F}_2(P, t) = \sum_{k_1=1}^{\infty} \sum_{k_2=1}^{M} f_{k_1k_2} \chi_{k_1}(P) \sin \frac{2\pi k_2 t}{T} \tag{29}$$

where $\chi_{k_1}(P)$ is a basis in H_0. E.g., sometimes eigenelements of the operator A_1 can be taken as such a basis, and we shall do it henceforth. In the formula (29) a finite number of time harmonics is taken only for the simplicity of the scheme. According to (29) the description of the probability properties of $\boldsymbol{F}_2(P, t)$ gives the law of distributions $\theta(f_{k_1k_2})$ of any set of coefficients $f_{k_1k_2}$.

Using the Bubnov-Galerkin method, we should solve Eq.(23) assuming that

$$\boldsymbol{w}_m = \sum_{l=1}^{m} q_{ml}(t) \chi_l(P). \tag{30}$$

The equations for q_{ml} have the following form:

$$\ddot{q}_{mp} + \sum_{l=1}^{m} \dot{q}_{ml} \int_\Omega K\chi_l \cdot \chi_p \, d\Omega + \int_\Omega A_2 \boldsymbol{w}_m \cdot \chi_p \, d\Omega = \int_\Omega \boldsymbol{F}_1 \cdot \chi_p \, d\Omega + \sum_{k_2=1}^{M} f_{pk_2} \sin \frac{2\pi k_2 t}{T}$$
$$+ \int_\Omega \boldsymbol{F}_3 \cdot \chi_p \, d\Omega. \tag{31}$$

Evidently, random functions $\lambda_p = \int_\Omega \boldsymbol{F}_3 \cdot \chi_p \, d\Omega$ form white noise. Its correlation matrix $\lambda_{s_1s_2}$ $(s_1, s_2 = 1, \ldots, \infty)$ can be computed if the correlation matrix of the process $\boldsymbol{F}_3(P, t)$ is given.

If we assume further that the parameters α_{kn} and $f_{k_1k_2}$ have taken fixed values, then the conditional distribution law f_m for $q_m, \dot{q}_m$ can be found with the help of the Kolmogorov-Focker-Planck equation for the system (31) (cf. the fundamental work [1]):

$$\frac{\partial f_m}{\partial t} + \sum_{p=1}^{m} \frac{\partial f_m}{\partial q_{mp}} \dot{q}_{mp} - \sum_{p=1}^{m} \frac{\partial f_m}{\partial \dot{q}_{mp}} \Big[\int_\Omega (A_2 \boldsymbol{w}_m - \boldsymbol{F}_1) \chi_p \, d\Omega$$
$$+ \gamma_p q_{mp} - \sum_{k_2=1}^{m} f_{pk_2} \sin \frac{2\pi k_2 t}{T} + \sum_{l=1}^{m} \dot{q}_{ml} \int_\Omega K\chi_l \cdot \chi_p \, d\Omega \Big] - f \sum_{p=1}^{m} K\chi_p \cdot \chi_p \, d\Omega = \frac{1}{2} \sum_{i,j=1}^{m} \lambda_{ij} \frac{\partial^2 f_m}{\partial q_i \, \partial q_j}. \tag{32}$$

When composing Eq. (32) all functional parameters ξ_k are assumed to be segments of the series (27); for simplicity, we assume that each segment contains m terms.

From (32) we shall find the conditional distribution law $f_m(t, q, \dot{q}, \alpha_{kn}, f_{pk_2})$. It is natural to assume that $\alpha_{kn}, f_{pk_2}, \boldsymbol{F}_3(P, t)$ are statistically independent; hence the unconditional distribution law $\Phi_m(q, \dot{q})$ is given by

$$\Phi_m(q, \dot{q}) = \int_{-\infty}^{\infty} \cdots \int f_m(t, q, \dot{q}, \alpha_{kn}, f_{pk_2}) \, \psi(\alpha_{kn}) \, \theta(f_{pk_2}) \, d\alpha_{kn} \, df_{pk_2}. \qquad (33)$$

With the help of $\Phi_m(t, q, \dot{q})$ we can compute the probability of an event characterizing our system. In particular, we can study the problem of location of the system in the neighbourhood of one of the principal states. Therewith it is possible to obtain a comparative evaluation of the degree of feasibility in experiments for both equilibrium and periodic states, forced and auto-oscillational. As a very special case we obtain the criteria of evaluation of feasibility of equilibrium states by the level of potential energy. This approach was already used, e.g., in the theory of shells [14], where it appeared to be effective. So, here there is at least an outlook for the solution of problem (b).

It should be noted that the boundary-value problem for f_m is a linear one, though we studied a general non-linear case of the motion of a medium. The distribution of stationary points of the system can be obtained from (32) if we neglect the terms with time derivatives and derivatives with respect to $\dot{q}_{mp}$. The extrema of the function f_m correspond in this case to the stationary points. Thus, the solution of the non-linear boundary-value problem reduces here to the solution of a linear problem with subsequent determination of extreme points of the obtained function. It is not difficult to extend this solution to the cases when $F_3(P, t)$ has a more complicated random nature, being an integro-differential operator of white noise.

Naturally, we come across a problem of limit transition at $m \to \infty$. In some very special cases it can be retraced. These special cases suggest that using limit transitions at $m \to \infty$ we can compute the probability of location in any domain, natural from the point of view of mechanics, of spaces H_0, H, if λ_{ij} satisfy certain conditions. In particular, the faster λ_{ij} decrease for $i, j \to \infty$, the more the limit distribution concentrates on increasingly smooth functions from H. To all appearance we deal here with a peculiar probability analogue of known theorems stating that the smoother the initial data of the boundary-value problems for elliptic systems are, the smoother are their solutions.

It is not easy to forecast the further development of the problems treated above. But there can be no doubt that the non-local topological approaches, originated in the works of *Henri Poincaré* and *A. M. Lyapunov*, still may be of great importance for the elucidation of many complicated problems in the mechanics of continuous media. In particular, methods developed by *L. A. Lyusternik, L. G. Shnirel'man* [35, 36], *M. Morse, R. Bott* [91] seem to be greatly hopeful.

References

1. Andronov, A. A., Vitt, A. A., Pontryagin, L. S: Zhurnal eksperimental'noi i teoreticheskoi fiziki 3 (1933) 165–180.
2. Bolotin, V. V.: Trudy (Proceedings), 4th All-Union Conference on the Theory of Shells and Plates. Erevan 1964, 15–63.
3. Vlasov, V. Z.: The general theory of shell and its applications in engineering. Moscow-Leningrad, 1949.
4. Vorovich, I. I.: Izvestiya, Ser. Math., Acad. Sci. USSR 19 (1955) 173–183.
5. —: Prikladnaya matematika i mekhanika 20 (1956) 449–474.
6. —: Trudy (Proceedings), 3rd All-Union Mathematical Congress, Vol. 1. Moscow, 1956, 201–202.
7. —: Doklady, Acad. Sci. USSR 117 (1957) 203–206.
8. —: Izvestiya, Ser. Math., Acad. Sci. USSR 21 (1957) 747–784.
9. —: Doklady, Acad. Sci. USSR 122 (1958) 37–40.
10. —, Yudovich, V. I.: ibidem 124 (1959) 542–545.
11. —, —: Matematichesky sbornik 53 (1961) 393–428.
12. —: Trudy (Proceedings), 4th All-Union Mathematical Congress Vol. 2. Moscow 1964, 541–545.
13. —: Annotatsii dokladov (Abstracts), 2nd All-Union Congress on Theoretical and Applied Mechanics. Moscow 1964, 56–57.
14. —: Trudy (Proceedings), 4th All-Union Conference on the Theory of Shells and Plates. Erevan 1964, 64–94.
15. —: Trudy, Institute of Applied Mathematics, Tbilisi State University 2 (1969) 49–55.
16. —, Srubshchik, L. S.: Trudy (Proceedings), 7th All-Union Conference on the Theory of Shells and Plates. Moscow 1970, 156–159.
17. —: Problems on Continuum Mechanics. Mos-

cow 1961, 111–118 (English ed. SIAM, 1961, 762–771).
18. —: Problemy mekhaniki tverdogo deformirovannogo tela. Leningrad 1970, 113–119.
19. —, Lebedev, L. P. In press.
20. Gantmakher, F. R., Krein, M. G.: Oscillatory matrices, and kernels, and small vibrations of mechanical systems, 2nd ed., Moscow-Leningrad 1950.
21. Gel'fand, I. M., Yaglom, A. M.: Uspekhi matematicheskikh nauk 11 (1956) 77–114.
22. Gel'fand, I. M., Minlos, R. A., Yaglom, A. M.: Trudy (Proceedings), 3rd All-Union Mathematical Congress, Vol. 3. Moscow 1958, 521–531.
23. Golovkin, K. K.: Trudy, Mathematical Institute, Acad. Sci. USSR 73 (1964) 118–138.
24. —: Zhurnal vychislitel'noi matematiki i matematicheskoi fiziki 4 (1964) 773–775.
25. Goncharenko, V. M.: Ukrainsky matematichesky zhurnal 14 (1962) 198–202.
26. —: Izvestiya, Acad. Sci. USSR Mekhanika i mashinostroenie 1 (1962) 159–160.
27. Daletsky, Yu. V.: Matematichesky analiz, 1966. Moscow 1967, 83–123.
28. Zen'kovskaya, S. M.: Matematichesky analiz i ego prilozheniya, Vol. 2. Rostov University, 1970, 132–142.
29. Krasnosel'sky, M. A.: Topological methods in the theory of non-linear integral equations. Moscow 1956.
30. Krein, S. G.: Doklady, Acad. Sci. USSR 93 (1953) 969–972.
31. Ladyzhenskaya, O. A.: ibidem 123 (1958) 427–429.
32. —: Mathematical problems of the dynamics of a viscous incompressible fluid. 2nd ed., Moscow 1970.
33. Lomakin, V. A.: Doklady, Acad. Sci. USSR 155 (1964) 1274–1277.
34. —: Prikladnaya matematika i mekhanika 29 (1965) 888–893.
35. Lyusternik, L. A.: Uspekhi matematicheskikh nauk 1 (1946) 30–56.
36. —: Trudy, Mathematical Institute, Acad. Sci. USSR 19 (1947) 1–96.
37. Meshalkin, L. D., Sinai, Ya. G.: Prikladnaya matematika i mekhanika 25 (1961) 1140–1143.
38. Monin, A. S., Yaglom, A. M.: Statistical hydrodynamics, Vol.1–2, Moscow 1965–1967.
39. Morozov, N. F.: Doklady, Acad. Sci. USSR 114 (1957) 968–971.
40. —: Vestnik, Leningrad University 19 (1958) 100–124.
41. —: Doklady, Acad. Sci. USSR 123 (1958) 417–419.
42. —: Izvestiya vuzov, Matematika 6 (1960) 170–173.
43. —: Doklady, Acad. Sci. USSR 147 (1962) 1318–1319.
44. Papkovich, P. F.: Structural mechanics of a ship, pt 2. Leningrad 1941.
45. Pogorelov, A. V.: Geometric methods in the non-linear theory of elastic shells. Moscow 1967.
46. Postnikov, M. M.: Variational theory of geodesics, Moscow 1965.
47. Sobolevsky, P. E.: Doklady, Acad. Sci. USSR 128 (1959) 45–48.
48. —: ibidem 155 (1964) 50–53.
49. Solonnikov, V. A.: Trudy, Mathematical Institute, Acad. Sci. USSR 73 (1964) 221–291.
50. Sorokin, V. S.: Prikladnaya matematika i mekhanika 18 (1954) 197–204.
51. Srubshchik, L. S.: ibidem 31 (1967) 723–729.
52. —: ibidem 32 (1968) 435–444.
53. —: Izvestiya, Ser. Math., Acad. Sci. USSR 36 (1972) 890–909.
54. —: Prikladnaya matematika i mekhanika 36 (1972) 705–716.
55. Yudovich, V. I., Srubshchik, L. S.: ibidem 26 (1962) 913–922.
56. —: ibidem 29 (1965) 453–467.
57. —: Doklady, Acad. Sci. USSR 161 (1965) 1037–1040.
58. —: ibidem 169 (1966) 306–309.
59. —: Prikladnaya matematika i mekhanika 30 (1966) 688–698.
60. —: ibidem 30 (1966) 1000–1005.
61. —: ibidem 31 (1967) 101–111.
62. —: Matematichesky sbornik 74 (1967) 565–579.
63. —: Doklady, Acad. Sci. USSR 195 (1970) 292–295.
64. —: ibidem 195 (1970) 574–576.
65. Antman, S. S.: Arch. Rational Mech. and Analysis 40 (1971) 329–372.
66. Berger, M. S.: Comm. Pure and Appl. Math. 20 (1967) 687–719.
67. Berger, M. S., Fife, P. C.: ibidem 21 (1968) 227–241.
68. Berger, M. S.: Indiana Univ. Math. J. 2 (1971) 591–602.
69. Chien, W. Z.: Sci. Rept. Nat. Tzing Hua Univ., Ser. A 5 (1948) 71–94.
70. Fife, P.: Comm. Pure and Appl. Math. 14 (1961) 81–112.
71. Finn, R.: Acta Math. 105 (1961) 197–244.
72. —: Proc. Symp. Pure Math. Amer. Math. Soc. 4 (1961) 143–148.
73. —: Arch. Rational Mech. and Analysis 19 (1965) 363–406.
74. Friedrichs, K. O., Stoker, J. J.: Proc. U.S. Nat. Acad. Sci. 25 (1939) 535–540.
75. —, —: Amer. J. Math. 63 (1941) 839–888.
76. —, —: J. Appl. Mech. 9 (1942) A7–A 14.
77. Görtler, H., Kirchgässner, K., Sorger, P.: Problems of hydrodynamics and continuum mechanis, Moscow 1969, 133–149 (SIAM ed. 353–374).
78. Hill, R.: J. Mech. and Phys. Solids 5 (1957) 229–241.
79. Hill, R., Sewell, M. J.: ibidem 8 (1960) 105–118.
80. Hill, R.: ibidem 9 (1961) 114–130.
81. Hopf, E.: Math. Nachr. 4 (1950/1951) 213–231.
82. Hutchinson, J. W., Koiter, W. T.: Appl. Mech. Rev. 23 (970) 1353–1366.

83. Kampé de Fériet, J.: Proc. Symp. Appl. Math. Amer. Math. Soc. 13 (1962) 165—198.
84. Leray, J.: J. math. pure et appl. 12 (1933) 1—82.
85. —: ibidem 13 (1934) 331—418.
86. —: Acta Math. 63 (1934) 193—248.
87. Leray, J., Schauder, J.: Ann. Sci. École Norm. Sup. 51 (1934) 45—78.
88. Lichtenstein, L.: Math. Z. 28 (1928) 387—415.
89. Lin, C. C.: The theory of hydrodynamic stability. London: Cambridge University Press 1955.
90. Marguerre, K.: Proc. 5th Intern. Congr. Appl. Mech. 1938, 1939, 93—101.
91. Morse, M.: The calculus of variations in the large. New York, Amer. Math. Soc. 1934.
92. Odqvist, F. K. G.: Math. Z. 32 (1930) 329—375.
93. —: Proc. 12th Intern. Congr. Appl. Mech. Stanford 1968, 1969, 77—99.
94. Piechocki, W.: Arch. mech. stosow. 21 (1969) 81—102.
95. —: ibidem 21 (1969) 813—825.
96. Prodi, G.: Ann. mat. pura ed appl. 48 (1959) 173—182.
97. —: Rend. Sem. Mat. Univ. Padova 30 (1960) 1—15.
98. Rabinowitz, P. H.: Arch. Ratiolna Mech. and Analysis 29 (1968) 32—57.
99. Serrin, J.: The initial vale problem for the Navier-Stokes equations. Non-linear problems. Univ. Wisconsin Press, 1963, 69—98.
100. Synge, J. L.: Proc. Roy. Soc. London, A167, (1938) 250—256.
101. Velte, W.: Arch. Rational Mech. and Analysis 16 (1964) 97—125.
102. Wolkowisky, J. H.: Comm. Pure and Appl. Math. 20 (1967) 549—560.

Contributed Papers
Presented at the 13th International Congress
of Theoretical and Applied Mechanics

G. N. Abramovich, V. I. Bazhanov, T. A. Girshovich, D. S. Kovner, S. Yu. Krasheninnikov, V. G. Lushchik, Moscow, USSR: Investigation of complex and non-self-similar turbulent flows.

J. D. Achenbach, Evanston, Ill., USA: On dynamic effects in brittle fracture.

E. Adams. See H. Spreuer.

L. K. Agbezuge, Kumasi, Ghana, H. Deresiewicz, New York, NY, USA: On the indentation of a consolidating half-space.

Mehmet Akyurt, Ankara, Turkey, Teoman Ariman, Notre Dame, Ind., USA: Constitutive relations for plant materials.

A. Ya. Aleksandrov, Novosibirsk, USSR: On the solution of fundamental three-dimensional problems of the theory of elasticity for solids of arbitrary shape.

D. Allwood. See H. Vaughan.

B. O. Almroth, F. A. Brogan, M. B. Marlowe, Palo Alto, Cal., USA: Stability analysis of cylinders with circular cutouts.

S. A. Ambartsumyan, G. E. Bagdasaryan, M. V. Belubekyan, Yerevan, USSR: On magnetoelasticity of thin solid.

S. I. Anisimov, A. Kh. Rakhmatulina, Moscow, USSR: Kinetics of evaporation of solids into vacuum.

R. A. Antonia, R. E. Luxton, Sydney, Australia: Some statistical properties of turbulence in smooth and rough wall boundary layers.

J. Arbocz, C. D. Babcock, Pasadena, Cal., USA: On the role of imperfections in shell buckling.

J. H. Argyris, M. Friedel, G. Malejannakis, Stuttgart, F. R. Germany: On the natural geometry of surfaces and the associated theory of thin shells.

T. Ariman. See M. Akyurt.

N. Kh. Arutyunyan, S. M. Mkhitaryan, Yerevan, USSR: Contact problems for solids with elastic stiffeners.

E. Arzoumanian. See R. Dumas.

R. S. Azad. See T. R. Heidrick.

C. D. Babcock. See J. Arbocz.

K. I. Babenko, V. F. D'yachenko, A. V. Zabrodin, G. P. Prokopov, Moscow, USSR: Investigation of some three-dimensional flows of fluids and gases.

V. A. Babeshko, Rostov-on-Don, USSR: A new effective method of solution of dynamic contact problems.

J. Bäcklund, Göteborg, Sweden: Limit analysis of reinforced concrete slabs by a finite-element method.

G. E. Bagdasaryan. See S. A. Ambartsumyan.

St. Bălan, V. Petcu, Bucharest, Roumania: Elastic-plastic design of redundant structures according to the concept of mechanical density.

S. Banerjee. See T. R. Heidrick.

P. G. D. Barkam, V. J. Medi, A. C. Soudack, Vancouver, Canada: The concept of reference loci applied to four-body dynamics.

N. N. Bautin, Gorky, USSR: Some field rotation methods for qualitative investigation of dynamic systems.

V. I. Bazhanov. See G. N. Abramovich.

J. A. Begley. See J. D. Landes.

V. V. Beletskii, Moscow, USSR: Resonances in rotations of celestial bodies and Cassini's laws.

S. M. Belotserkovskii, M. I. Nisht, Moscow, USSR: Analysis of nonstationary stalled flow arround a slender wing in a nonviscous incompressible fluid stream.

M. V. Belubekyan. See S. A. Ambartsumyan.

S. J. Bennett, K. L. DeVries, M. L. Williams, Salt Lake City, Utah, USA: Adhesive fracture mechanics.

J. P. Benthem, Delft, Netherlands: Three-dimensional state of stress at the tip of a quarter-infinite crack in a half-space.

L. T. Berezhnitskii. See V. V. Panasyuk.

D. Bershader. See W. J. Loubsky.

R. V. Birikh, G. Z. Gershuni, E. M. Zhukhovitskii, R. N. Rudakov, A. A. Yakimov, Perm, USSR: New investigations of the stability of stationary convective flows.

R. E. D. Bishop, R. K. Burcher, W. G. Price, London, England: The uses of functional analysis in ship dynamics.
R. F. Blackwelder, R. E. Kaplan, Los Angeles, Cal., USA: The intermittent structure of the wall region of a turbulent boundary layer.
V. V. Bolotin, Moscow, USSR: On the theory of viscoelasticity for time-hardening media.
A. Bosznay, Budapest, Hungary: Algebraic eigenvalue problems bracketing eigenfrequencies of structures consisting of rods of varying croos section.
M. Braun, Stuttgart, F. R. Germany: Acceleration waves in anisotropic hyperelastic materials.
R. Brepta, F. Valeš, Prague, Czechoslovakia: State of stress of twodimensional configurations due to non-steady-state shock loading.
R. S. Brodkey. See J. M. Wallace.
F. A. Brogan. See B. O. Almroth.
E. H. Brown, London, England: Postbuckling behaviour of elastic frames.
G. L. Brown, A. Roshko, Pasadena, Cal., USA: Structure of the turbulent mixing layer.
H. Buggisch, Darmstadt, F. R. Germany: The steady two-dimensional reflection of an oblique partly dispersed shock wave from a plane wall.
R. K. Burcher. See R. E. D. Bishop.
T. V. Burchuladze. See V. D. Kupradze.
E. Byskov, Copenhagen, Denmark: Cracked anisotropic finite elements.
Y. Y. Chan. See V. Ozarapoglu.
I. B. Chelpanov. See S. S. Rivkin.
C. F. Chen, New Brunswick, NJ, USA: Onset of cellular convection in a stratified fluid due to temperature gradients.
W. T. Chen, P. A. Engel, Endicott, NY, USA: Impact and contact stress analysis in multilayer elastic media.
Y. N. Chen. See J. Kempner.
G. P. Cherepanov, Moscow, USSR: On the theory of electrochemical stress corrosion.
G. G. Chernyi, S. M. Gilinskii, V. P. Korobeinikov, V. A. Levin, V. V. Markov, Moscow, USSR: Flows behind shock waves in combustible gaseous mixtures.
M. V. Chichinadze, E. Kh. Saringulyan, N. V. Gerasimov, N. V. Voroshilova, Moscow, USSR: On errors of corrected gyro compass during manoeuvring.
P. I. Chushkin. See V. P. Korobeinikov.
R. C. Cipolla. See F. A. McClintock.
J. Citavý, Prague, Czechoslovakia: Prediction of the separation bubble on a circular cylinder.
C. Ciuperca. See S. Sandulescu.
C. Clarion. See R. Pelissier.
J. P. Clay. See C. H. Gibson.
R. J. Clifton. See J. C. C. Hsu.
J. Coirier. See J.-L. Peube.
J. D. Cole, Los Angeles, Cal., USA, E. Murman, Mountain View, Cal., USA: Calculation of steady transonic flows.
M. Como, A. Grimaldi, Naples, Italy: A general approach to the post-buckling behaviour of framed structures.
G. Coupry, Chatillon, France: Prévision de la variation des modes consécutive à la modification d'une structure.
M. J. Crochet, Louvain, Belgium, P. M. Naghdi, Berkeley, Cal., USA: On a restricted nonisothermal theory of viscoelastic materials.
S. Damms, D. Küchenmann, Farnborough, Hampshire, England: On a vortex-sheet model for the mixing between two parallel streams.
I. M. Daniel. See R. E. Rowlands.
R. Dat, Chatillon, France: Application des méthodes linéarisées à l'étude du comportement aéroélastique des hélicoptères.
J. Délery, J. C. Le Balleur, M. Sirieix, Chatillon, France: Etude expérimentale de l'interaction choc-couche limite turbulente en écoulement supersonique.
H. Deresiewicz. See L. K. Agbezuge.
K. L. DeVries. See S. J. Bennett.
J. Dickinson. See V. Ozararpoglu.
R. C. DiPrima, Troy, NY, USA, J. T. Stuart, London, England: Stability and growth of Taylor vortices in the flow between eccentric rotating cylinders.
Dj. Djukić. See B. Vujanović.
B. I. Dolaptschiev, Sofia, Bulgaria: Bemerkungen über Gibbs-Appellsche Form der Bewegungsgleichungen nicht-holonomer mechanischer Systeme.
E. H. Dowell. See C. F. Kalmbach.
D. C. Drucker. See G. Maier.
R. Dumas, E. Arzoumanian, A. Favre, Marseille, France: Structure de la turbulence: corrélations spatio-temporelles doubles et triples.
M. K. Duszek, Warsaw, Poland: On the problem of minimum principles in finite plasticity.

V. F. D'yachenko. See K. I. Babenko.
H. Eckelmann. See J. M. Wallace.
P. A. Engel. See W. T. Chen.
Yu. K. Engelbrecht. See U. K. Nigul.
D. Euvrard, G. Tournemine, Rennes, France: Elaboration d'une méthode numérique pour la résolution de certains problèmes elliptiques-hyperboliques non-linéaires, avec application au calcul de l'écoulement sonique autour d'un profil d'aile donné.
A. Favre. See R. Dumas.
V. A. Filippov. See P. V. Kharlamov.
U. Fischer, Magdeburg, GDR: Dynamisches Verhalten linearer Schwingungssysteme mit Zufallserregung periodisch veränderlicher Dispersion.
W. Fiszdon, A. Wortman, Z. Walenta, Warsaw, Poland: An experimental and theoretical study of boundary-layer effects on travelling shock waves.
H. Försching, Göttingen, F. R. Germany: An analytical approach to the solution of the nonlinear aeroelastic galloping problem.
J. Förste, Berlin, GDR: Instationäre Strömung einer inkompressiblen mikropolaren Flüssigkeit.
M. Fortune. See A. Lagarde.
M. E. Fourney, Seattle, Wash. USA: Application of picosecond holography to photoelasticity.
J. C. Frauenthal Medford, Mass., USA: Initial postbuckling behaviour of optimally designed columns and plates.
C. S. Fray. See W. J. Shack.
M. Frémond, Paris, France: Solide reposant sur une structure stratifiée.
O. G. Fridlender. See M. N. Kogan.
M. Friedel. See J. H. Argyris.
N. A. Fufaev. See Yu. I. Neimark.
H. Fujita. See L. S. G. Kovasznay.
N. K. Galimov. See Kh. M. Mushtari.
L. A. Galin, Moscow, USSR: Effect of vibrational load on polymer structural elements.
V. S. Galkin. See M. N. Kogan.
M. Gaster, Teddington, Middlesex, England: The influence of taper on bluff body flows.
T. G. Gegelia. See V. D. Kupradze.
N. V. Gerasimov. See M. V. Chichinadze.
J.-P. Germain, Grenoble, France: Théorie des ondes de gravité en eau peu profonde.
E. A. Gershbein, N. N. Pilyugin, V. M. Ovsyannikov, O. N. Suslov, G. A. Tirskii, Moscow, USSR: Hypersonic flow around axisymmetric blunt bodies intensively destroyed by convective and radiative heating.
G. Z. Gershuni. See R. V. Birikh.
K. Gersten, Bochum, F. R. Germany: Second-order boundary-layer effects for large injection or suction.
C. H. Gibson, J. P. Clay, La Jolla, Cal., USA: Measurements of turbulent temperature in mercury, air and water.
E. Giencke, Berlin-West: A simple "mixed" method for plates and shells.
S. M. Gilinskii. See G. G. Chernyi.
T. A. Girshovich. See G. N. Abramovich.
V. V. Gogosov, V. A. Polyanskii, Moscow, USSR: Structure of shock waes in electrohydrodynamics.
J. R. Goldgraben, Brooklyn, USA, S. Weinbaum, New York, N.Y, USA: On the mixing of low Re biological jets with quiescent outer bathing solutions.
B. H. L. Gowda. See N. V. C. Swamy.
A. Grimaldi. See M. Como.
V. T. Grinchenko, A. F. Ulitko, Kiev, USSR: Dynamic problems for bounded elastic bodies with piezoeffect.
G. L. Grodzovskii, V. V. Skvortsov, Moscow, USSR: Plasma accelerators in gas dynamics.
P. Hagedorn, Karlsruhe, F. R. Germany: On the instability of equilibrium of partially damped holonomic mechanical systems.
M. Hamada, M. Mase, Suita, Japan: The influence of internal pressure on the free vibration of the general axisymmetric shell.
H. Hasimoto, Tokyo, Japan, H. Ono, Kyoto, Japan: Nonlinear modulation of gravity waves.
D. G. Heale. See D. V. Reddy.
T. B. Hedley, J. F. Keffer, Toronto, Canada: Investigation of the intermittent nature of a turbulent flow.
T. R. Heidrick, S. Banerjee, Pinawa, Canada, R. S. Azad, Winnipeg, Manitoba, Canada: "Burst" signatures in turbulent pipe flow.
K. Hennig, Berlin, GDR: Zur Berechnung deterministischer und stochastischer Flatterschwingungen von Hängedächern.
F. Higashino, Tokyo, Japan: Cylindrical blast waves in ideal dissociating gases.
J. O. Hinze, R. E. Sonnenberg, Delft, Netherlands: Memory effect in a turbulent boundary-layer flow due to a relatively strong axial variation of the mean-velocity gradient.
S. Hjalmars, Stockholm, Sweden: The origin of instability in air-lift pumps.
C. S. Hsu, Berkeley, Cal. USA: Methods of stability analysis for general periodic dynamical systems.

J. C. C. Hsu, R. J. Clifton, Providence, R.I., USA: Combined longitudinal and torsional plastic waves in a strongly rate-sensitive material.
N. C. Huang, Notre Dam, Ind. USA: Theories of elastic slender curved rods.
J. C. R. Hunt, Cambridge, England: A theory of turbulent flow round bluff bodies.
A. Ibbetson. See D. J. Tritton.
V. F. Illarionov. See L. M. Shkadov.
G. R. Inger, Blacksburg, Va. USA: A unified theory of small disturbances within compressible laminar and turbulent boundary layers.
G. Iooss, Chatillon, France: Existence et stabilité de la solution périodique secondaire intervenant dans les problèmes d'évolution du type Navier-Stokes.
V. A. Ioselevich, N. G. Vastskaya, Moscow, USSR: Contribution to the analysis of turbulent flows of weak solutions of dragreducing linear polymers.
T. P. Ivanov. See M. Kozarov.
E. C. James, Bethesda, Md. USA: Lifting-line theory for a wing in nonuniform motion as a singular perturbation problem.
F. Bruun Jensen, Lyngby, Denmark: Shock-excited pulsations of air bubbles in water.
J. A. Johnson, Norwich, England: Upwelling layers in a winddriven ocean.
O. Yu. Kalekin, A. N. Podgornyi, V. L. Rvachev, Kharkov, USSR: Methods of algebra of logic (R-functions) in boundary-value problems of mechanics for domains of complicated shape.
C. F. Kalmbach, Jr., E. H. Dowell, F. C. Moon, Princeton, N.Y., USA: The application of feedback control to the suppression of a dynamic instability of an elastic body.
A. Kalnins, D. P. Updike, Bethlehem, Pa., USA: Contact problem of two thin spherical shells.
J. Kampé de Fériet, Lille, France: Vecteurs aléatoires de la classe de Gram-Charlier et représentation d'un champ de vitesses turbulent au voisinage de l'équilibre statistique.
R. E. Kaplan. See R. F. Blackwelder.
J. F. Keffer. See T. B. Hedley.
B. S. Keller. See I. M. Yavorskaya.
J. B. Keller. See S. Kogelmann.
J. Kempner, Y. N. Chen, Brocklyn, N.Y., USA: Buckling and initial postbuckling of oval cilyndrical shells under combined axial compression and bending.
K. A. Kerimov. See Kh. A. Rakhmatulin.
P. V. Kharlamov, E. I. Kharlamova, D. M. Klimov, G. V. Mozalevskaya, V. A. Filippov, Donetsk, Moscow, USSR: Problems of motion of rigid-body systems; development of new classes of solutions.
E. I. Kahrlamova. See P. V. Kharlamov.
S. A. Khristianovich, Moscow, USSR: Deformation of a hardening plastic material.
D. M. Klimov. See P. V. Kharlamov.
J. K. Knowles, E. Sternberg, Pasadena, Cal., USA: An asymptotic finite-deformation analysis of the elastostatic stresses and displacements near the tip of a crack.
S. Kodama. See N. Yamaki.
T. Koga, Y. Noguchi, Tokyo, Japan: Small-strain deformations superimposed on finite deformations of rubbery materials.
M. N. Kogan, V. S. Galkin, N. K. Makashev, O. G. Fridlender, Moscow, USSR: Some new phenomena in continuum gas flows due to kinetic effects.
S. Kogelman, J. B. Keller, New York, N.Y., USA: Asymptotic theory of nonlinear wave propagation.
V. P. Korobeinikov, P. I. Chushkin, L. V. Shurshalov, Moscow, USSR: Two-dimensional unsteady problems of expansion of the compressed gas volume.
V. P. Korobeinikov. See G. G. Chernyi.
W. Kosiński, Warsaw, Poland: Thermo-mechanical coupling in materials with memory.
L. S. G. Kovasznay, H. Fujita, Baltimore, Md., USA: Unsteady response of an airfoil to wake cutting.
D. S. Kovner. See G. N. Abramovich.
M. Kozarov, A. Rachev, Ts. P. Ivanov, Sofia, Bulgaria: Propagation of elastic waves in circular cylinders subject to finite deformations.
S. Yu. Krasheninnikov. See G. N. Abramovich.
J. Kratochvíl. See L. Trávníček.
D. Küchemann. See S. Damms.
V. X. Kunukkasseril, R. Palaninathan, Madras, India: Impulsive loading of shallow spherical shells.
V. D. Kupradze, T. V. Burchuladze, T. G. Gegelia, Tbilisi, USSR: Basic initial-boundary-value problems of elasticity.
A. I. Kur'yanov, Yu. N. Zhelnin, Moscow, USSR: Analysis of optimum re-entry trajectories with constraints on the phase coordinates and limited state information.
A. Lagarde, M. Fortune, Poitiers, France: Sur certains problèmes de similitude en thermoélastostatique plane.
H. Lanchon, Paris, France: Torsion élastoplastique d'un arbre cylindrique de section multiconnexe.
J. D. Landes, J. A. Begley, Pittsburgh, Pa., USA: The J-integral approach to fracture.
M. Larcheveque. See J.-L. Peube.
T. J. Lardner. See W. J. Shack.

J. C. Le Balleur. See J. Delery.
F. A. Leckie, Leicester, England: On creep rupture of structures.
R. Legendre, Paris, France: Condition de Joukowsky en écoulement tridimensionnel.
M. Ya. Leonov, V. Ya. Molotnikov, B. A. Rychkov, Frunze, USSR: Development of the concept of slip in the theory of plasticity.
P. Le Roy, J.-M. Pierrard, Marseille, France: Lois de comportement de matérieux à mémoire; viscoélasticité non linéaire de corps obéissant à un principe de superposition.
H. J. Leutheusser. See F. Resch.
V. A. Levin. See G. G. Chernyi.
T. Liber. See R. E. Rowlands.
P. W. Likins. See D. L. Mingori.
G. M. Lilley, Southhampton, England: A new approach to the theory of jet noise.
F. F. Ling, Troy, NY., USA: A new model of articular cartilage in human joints.
T. R. Logan, Long Beach, Cal., USA: Asymptotic solutions for stress concentrations in shells with arbitrary boundaries.
G. V. Logvinovich, Moscow, USSR: Hydrodynamics of the swimming of fish.
W. J. Loubsky, S. G. Prakash, D. Bershader, Stanford, Cal., USA: Interaction of a shock-generated plasma with a transverse magnetic field.
K. V. Lukanin. See V. A. Zlatoustov.
K. A. Lur'e, Leningrad, USSR: Optimal design of nonuniform continuous bodies.
V. G. Lushchik. See G. N. Abramovich.
R. E. Luxton. See R. A. Antonia.
G. A. Lyubimov, Moscow, USSR: Some gasdynamic problems connected with the investigation of the cathode region of vacuum arc discharge.
F. A. McClintock, R. C. Cipolla, Cambridge, Mass., USA: Boundary conditions for fracture by hole growth near a crack tip.
R. E. McGorman. See L. A. Mysak.
G. Maier, Milan, Italy, D. C. Drucker, Urbana, Ill., USA: Effects of geometry change on the essential features of inelastic behaviour.
N. K. Makashev. See M. N. Kogan.
G. Malejannakis. See J. H. Argyris.
V. P. Malkov. See A. G. Ugodchikov.
J. Mandel, Paris, France: Equations constitutive set directeurs dans les milieux plastiques et viscoplastiques.
A. P. Markeev. See V. A. Zlatoustov.
V. V. Markov. See G. G. Chernyi.
M. B. Marlowe. See B. O. Almroth.
M. Mase. See M. Hamada.
L. Maunder. See D. Ormandy.
T. Maxworthy, Los Angeles, Cal., USA: The dynamics of buoyant vortex rings and thermals.
R. Mazet, Paris, France: Sur l'équation d'état d'un milieu continu hétéroplastique.
N. S. Mel'nikova. See I. M. Yavorskaya.
D. H. Michael. See G. I. Taylor.
J. Miklowitz, G. B. Sinclair, Pasadena, Cal., USA: The static and dynamic response of end loaded elastic strips.
R. W. Miksad, London, England: Experiments of nonlinear interactions in free shear layer transitions.
M. D. Millionshchikov, Moscow, USSR: On the model of turbulent flow.
D. L. Mingori, Los Angeles, Cal., G. T. Tseng, Princeton, NJ., P. W. Likins, Los Angeles, Cal., USA: Constant and variable amplitude limit cycles in dual-spin spacecraft.
Yu. A. Mitropol'skii, Yu. G. Ostapov, Kiev, USSR: On the investigation of a class of nonlinear dynamic systems with distributed parameters by the method of averaging.
S. M. Mkhitaryan. See N. Kh. Arutyunyan.
V. J. Modi. See P. G. D. Barkham.
W. Möhring, Göttingen, F. R. Germany: On energy, group velocity and small damping of sound waves in ducts with shear flow.
V. Ya. Molotnikov. See M. Ya. Leonov.
F. C. Moon. See C. F. Kalmbach, Jr.
V. M. Morozov. See V. V. Rumyantsev.
G. L. Morrison. See A. E. Perry.
M. P. Mortell, Bethlehem, Pa., USA: B. R. Seymour, New York, NY., USA: A nonlinear theory of resonance in a continuous system in the presence of damping.
V. I. Mossakovskii, Dnepropetrovsk, USSR: Contact problems with unknown and semi-unknown boundaries.
G. V. Mozalevskaya. See P. V. Kharlamov.
P. C. Müller, München, F. R. Germany: Stability theory of linear autonomous mechanical systems.
T. Mura. See J. A. Simmons.
M. Murakami, K. Oshima, Tokyo, Japan: An exact solution of the propagation of small disturbances in a radiating gray gas with isotropic scattering.

S. Murakami, Y. Ohashi, M. Suzuki, Nagoya, Japan: On the similarity conditions in the photo-rheological method of stress analysis.
E. Murman. See J. D. Cole.
Kh. M. Mushtari, N. K. Galimov, Kazan, USSR: On the construction of a precise theory of sandwich plates and shells.
L. A. Mysak, Vancouver, Canada, R. E. McGorman, St. John, Canada: Internal waves in a randomly stratified fluid.
P. M. Naghdi. See M. J. Crochet.
R. Narasimha, K. R. Screenivasan, Bangalore, India: Reversion from turbulent to laminar flow in highly accelerated boundary layers.
A. H. Nayfeh. See W. S. Saric.
K. W. Neale, Sherbrooke, Canada: J. Schroeder, Waterloo, Canada: A variational method for elastic-plastic buckling.
V. Ya. Neiland, Moscow, USSR: On the asymptotic theory of hypersonic flow with massive blow-in.
Yu. I. Neimark, N. A. Fufaev, Gorky, USSR: Analytical mechanics and dynamics of rolling and electromechanical systems.
K. Wiese Nielsen, Reno, Nevada, USA: Quantitative determination of varticity distribution and cancellation in a two-dimensional incompressible periodic wake.
U. K. Nigul, Yu. K. Engelbrecht, Tallinn, USSR: Nonlinear and thermal effects in the theory of transient elastic waves.
A. A. Nikol'skii, Moscow, USSR: Separated flows of an ideal fluid and a gas.
M. I. Nisht. See S. M. Belotserkovskii.
Y. Noguchi. See T. Koga.
L. Noordzij. See L. van Wijngaarden.
B. Novotný, Bratislava, Czechoslovakia: Geometrically nonlinear shell theory in the light of the method of asymptotic integration.
A. M. Obukhov, Moscow, USSR: On the problem of nonlinear interactions in fluid dynamics.
Y. Ohashi. See S. Murakami.
D. E. Okhotsimskii, A. K. Platonov, Moscow, USSR: Locomotion synthesis of a walker.
N. Olhoff, Copenhagen, Denmark: Optimal design of vibrating rectangular plates.
E. R. Arantes Oliveira, Lisbon, Portugal: The role of convergence in the theory of shells.
W. Olszak, Udine, Italy: P. Perzyna, Warsaw, Poland: Physical theory of viscoplasticity for small deformations.
H. Ono. See H. Hasimoto.
G. Ooms, Amsterdam, Netherlands: Fluid-mechanical studies on "core-annular" flow.
D. Ormandy, L. Maunder, Newcastle-upon-Tyne, England: Dynamics of the oscillogyro.
K. Oshima. See M. Murakami.
Yu. G. Ostapov. See Yu. A. Mitropol'skii.
V. M. Ovsyannikov. See E. A. Gershbein.
V. Ozarapoglu, Quebec, Canada, Y. Y. Chan, Ottawa, Canada, J. Dickinson, Cambridge, England: Measurements and theoretical predictions of velocity and shear stress in a highly nonequilibrium turbulent boundary layer.
R. Palaninathan. See V. X. Kunukkasseril.
A. C. Palmer, J. R. Rice, Cambridge, England: The growth of slip surfaces in the progressive failure of clay slopes.
V. V. Panaysuk, L. T. Berezhnitskii, Lvov, USSR: On the application of fracture mechanics to evaluating the strength if composite materials.
Yu. A. Panov. See A. I. Zubkov.
A. G. Parkinson, London, England: The balancing of flexible shafts supported on massive flexible foundations.
J. Pecheux. See J.-L. Peube.
T. J. Pedley, London, England: On the distribution of inspired gas in the lung.
R. Pelessier, C. Clarion, Marseille, France: Mouvement oscillatoire libre d'un fluide pesant et visqueux.
Z. Peradzyński, Warsaw, Poland: Nonlinear wave interaction problem illustrated on gas dynamics.
A. E. Perry, G. L. Morrison, Melbourne, Australia: The aeroelastic instability of hot-wire filaments.
P. Perzyna. See W. Olszak.
V. Petcu. See St. Bălan.
J.-L. Peube, J. Pecheux, J. Coirier, M. Larcheveque, Poitiers, France: Structure générale de l'écoulement entre deux disques fixes ou tournants.
J. M. Pierrard. See P. Le Roy.
N. N. Pilyugin. See E. A. Gershbein.
A. K. Platonov. See D. E. Okhotsimskii.
A. N. Podgornyi. See O. Yu. Kalekin.
V. A. Polyanskii. See V. V. Gogosov.
S. G. Prakash. See W. J. Loubsky.
W. G. Price. See R. E. D. Bishop.

G. P. Prokopov. See K. I. Babenko.

A. Rachev. See M. Kozarov.

Kh. A. Rakhmatulin, Moscow, USSR, K. A. Kerimov, Baku, USSR: On the propagation of focused "transverse" waves induced by the geometrical and physical nonlinearity.

A. Kh. Rakhmatulina. See S. I. Anisimov.

F. J. Resch, H. J. Leutheusser, Toronto, Canada: Turbulence characteristics of bubbly two-phase flow.

J. R. Rice. See A. C. Palmer.

J. A. Rickard, Melbourne, Australia: Planetary waves in rotating global fluid shells.

G. Rieder, Aachen, F. R. Germany: On Babuška's paradox of the simply supported Kirchhoff plate and the finite-element method.

S. S. Rivkin, A. A. Sveshnikov, I. B. Chelpanov, Leningrad, USSR: Use of stochastic methods in the theory of gyroscopic devices.

G. Romano: Postcritical behaviour of discrete conservative systems.

P. G. Rose. See R. E. Rowlands.

A. Roshko. See G. L. Brown.

N. Rott, Zürich, Switzerland: Heat balance and acoustic streaming in thermally driven gas oscillations.

R. E. Rowlands, T. Liber, I. M. Daniel, P. G. Rose, Chicago, Ill., USA: Holographic stress analysis of composite plates.

G. I. N. Rozvany, Clayton, Victoria, Australia: A new class of optimal structures.

V. N. Rubanovskii. See V. V. Rumyantsev.

R. N. Rudakov. See R. V. Birikh.

D. T. Rues, Aachen, F.R. Germany: Wechselwirkung instationärer Stoßwellen mit ebenen Hyperschallströmungen um stumpfe Körper.

V. V. Rumyantsev, V. M. Morozov, V. N. Rubanovskii, V. A. Samsonov, Moscow, USSR: On bifurcation and stability of steady motions of complex mechanical systems.

V. V. Rusanov, Moscow, USSR: Structure of the transonic region in the supersonic axisymmetric flow around blunt bodies.

V. L. Rvachev. See O. Yu. Kalekin.

B. A. Rychkov. See M. Ya. Leonov.

Yu. A. Sadov. See V. A. Zlatoustov.

H. Saito. See H. Sato.

V. A. Samsonov. See V. V. Rumyantsev.

S. Săndulescu, C. Ciupercă, Bucharest, Roumania: The incompressible flow around the wing with variable parameters due to lateral fluid jets.

W. S. Saric, Albuquerque, N. M., USA, A. H. Nayfeh, Blackburg, Va., USA: Stability of liquid films interacting with supersonic streams: theory and experiment.

E. Kh. Saringulyan. See M. V. Chichinadze.

A. V. S. S. S. R. Sarma. See L. S. Srinath.

G. S. R. Sarma, Freiburg i.Br., F. R. Germany: Barodiffusion in a binary mixture at a rotating disc.

V. A. Sarychev. See V. A. Zlatoustov.

H. Sato, H. Saito, Tokyo, Japan: Randomization of velocity fluctuations in the laminar-turbulent transition process of a two-dimensional wake.

D. A. Saville, Princeton, N.J., USA: Electrohydrodynamic stability of charged drops.

D. W. Schmidt, W. J. Wagner, Göttingen, F. R. Germany: Measurement of temperature fluctuations in turbulent wakes.

J. Schroeder. See K. W. Neale.

L. W. Schwartz, M. D. Van Dyke, Stanford, Cal., USA: Analytic continuation of Stoke's expansion for gravity waves.

G. Schweitzer, München, F. R. Germany: Stabilizing a high-speed rotor.

K. R. Screenivasan. See R. Narasimha.

Ya. I. Sekerzh-Zen'kovich, Moscow, USSR: On steady capillary-gravitational waves of finite amplitude on the surface of a liquid flow with a wavy bottom.

S. V. Serensen, R. M. Shneiderovich, Moscow, USSR: Criteria of the low-cycle fracture in a kinetic aspect.

M. J. Sewell, Reading, England: A yield surface corner lowers the buckling stress of an elastic/plastic compressed plate.

B. R. Seymour. See M. P. Mortell.

W. J. Shack, C. S. Fray, T. J. Lardner, Cambridge, Mass., USA: Observations on the hydrodynamics and swimming motions of mammalian spermatozoa.

L. M. Shkadov, V. F. Illarionov, Moscow USSR: General theory of optimal three-dimensional motion of a re-entry vehicle.

R. M. Shneiderovich. See S. V. Serensen.

Yu. I. Shokin. See N. N. Yanenko.

L. V. Shursholov. See V. P. Korobeinikov.

A. I. Shvets. See A. I. Zubkov.

J. A. Simmons, Washington, D.C. USA, T. Mura, Evanston, Ill., USA: Calculation of the dynamic Green's tensor for anisotropic elasticity.
G. B. Sinclair. See J. Miklowitz.
M. Sirieix. See J. Delery.
V. V. Skvortsov. See G. L. Grodzovskii.
J. H. B. Smith, Farnborough, Hampshire, England: Similar solutions for slender wings with leading-edge separation.
S. H. Smith, Toronto, Canada: On pressure-driven Stewartson layers.
J. J. Smolderen. See T. Ytrehus.
R. E. Sonnenberg. See J. O. Hinze.
R. Souchet, Poitiers, France: Contribution à l'étude de théories non classiques de l'élasticité linéaire plane.
A. C. Soudack. See P. G. D. Barkham.
D. A. Spence, Oxford, England: The Hertz contact problem with finite friction.
T. M. Spencer, Boulder, Col., USA: Attitude dynamics of a "nearly-spherical" dual-spin satellite and orbital results for OSO-7.
H. Spreuer, E. Adams, Karlsruhe, F. R. Germany: A relation between eigenvalue problems and the (in)stability of parabolic problems, e.g. in fluid mechanics.
V. M. Starzhinskii, Moscow, USSR: Applied methods in the theory of nonlinear oscillations.
G. A. Yu. Stepanov. See I. Zubkov.
E. Sternberg. See J. K. Knowles.
T. N. Stevenson, Manchester, England: Internal waves around a body moving in a curved path.
B. Storåkers, H. Strifors, Stockholm, Sweden: Uniqueness and stability at finite deformation of rigid-plastic thick-walled spherical shells under hydrostatic pressure.
A. M. Strauss. See B. Vujanović.
H. Strifors. See B. Storåkers.
V. V. Struminskii, Moscow, USSR: Influence of the diffusion velocity on the flows of gaseous mixtures.
J. T. Stuart. See R. C. DiPrima.
O. N. Suslov. See E. A. Gershbein.
M. Suzuki. See S. Mhrakami.
A. A. Sveshnikov. See S. S. Rivkin.
N. V. Chandrasekhara Swamy, B. H. Lakshamana Gowda, Madras, India: Three-dimensional wall jets.
V. V. Sychev, Moscow, USSR: On laminar separation.
K. Szuwalski. See M. Życzkowski.
G. I. Taylor, D. H. Michael, Cambridge, England: On making holes in a sheet of fluid.
R. H. Thaler, New York, N.Y., USA: Transient heat and mass transfer controlled condensation of a binary mixture.
G. A. Tirskii. See E. A. Gershbein.
G. Tournemine. See D. Euvrard.
L. Trávníček, J. Kratochvíl, Prague, Czechoslovakia: Finite deformation of elastic-plastic solids.
D. J. Tritton, Newcastle-upon-Tyne, England, A. Ibbetson, Reading, England: Experiments on the structure of turbulence in a rotating fluid.
G. T. Tseng. See D. L. Mingori.
V. Tvergaard, Lyngby, Denmark: Imperfection-sensitivity of a wide integrally stiffened panel under compression.
A. G. Ugodchikov, V. P. Malkov, Gorky, USSR: Problem of optimal material distribution in thin-walled structures.
A. F. Ulitko. See V. T. Grinchenko.
D. P. Updike. See A. Kalnins.
F. Vales. See R. Brepta.
M. D. Van Dyke. See L. W. Schwartz.
N. G. Vasetskaya. See V. A. Ioselevich.
H. Vaughan, D. Allwood, Vancouver, Canada: An axisymmetric contact problem for finite elastic bodies.
N. V. Voroshilova. See M. V. Chichinadze.
B. Vujanović, Dj. Djukić, A. M. Strauss, Novi Sad, Yugoslavia: A new variational principle for irreversible phenomena.
W. J. Wagner. See D. W. Schmidt.
Z. Walenta. See W. Fiszdon.
J. M. Wallace, H. Eckelmann, R. S. Brodkey, Göttingen, F. R. Germany: The structure of turbulence in the wall region.
F. Y. M. Wan, Cambridge, Mass., USA: A method for linear dynamical problems in continuum mechanics with random loads.
R. D. Weidmann, Los Angeles, Cal., USA: Nonlinear spin up and spin down of a contained fuid.
S. Weinbaum. See J. R. Goldgraben.
J. F. Wendt. See T. Ytrehus.

J. A. Whitehead, Jr., Woods Hole, Mass., USA: Observations of the dynamics of Rayleigh-Bénard convection.

L. van Wijngaarden, L. Noordzij, Enschede, Netherlands: Relaxation effects in shock waves in liquid-bubble mixtures.

P. Y. Willems, Louvain, Belgium: Stability analysis of dynamical systems with constraints and integrals of motion.

M. L. Williams. See S. J. Bennett.

M. B. Wilson, Washington, D.C., USA: A Havelock-Oseen theory for ship resistance.

J. Wittenburg, Hannover, F. R. Germany: The theory of the torque-free gyrostat.

A. Wortman. See W. Fiszdon.

A. A. Yakimov. See R. V. Birikh.

N. Yamaki, S. Kodama, Sendai, Japan: Buckling of circular cylindrical shells under compression considering the effect of prebuckling edge rotations.

N. N. Yanenko, Yu. I. Shokin, Novosibirsk, USSR: On application of invariant difference methods to problems of gas dynamics.

I. M. Yavorskaya, B. S. Keller, N. S. Mel'nikova, Moscow, USSR: On hydrodynamic motions in stratified convective zones of stars and large planets.

T. Ytrehus, Trondheim, Norway, J. J. Smolderen, J. F. Wendt, Rhode-Saint-Genèse, Belgium: Large-scale rarefied flows produced by Knudsen effusion from a perforated boundary.

A. V. Zabrodin. See K. I. Babenko.

S. Zahorski, Warsaw, Poland: Nonviscometric flows with proportional stretch history.

Yu. N. Zhelnin. See A. I. Kur'yanov.

E. M. Zhukhovitskii. See R. V. Birikh.

V. A. Zlatoustov, K. V. Lukanin, A. P. Markeev, Yu. A. Sadov, V. A. Sarychev, Moscow, USSR: Some problems of passive satellite stabilization.

H. Zorski, Warsaw, Poland: The continuum model of dynamical systems.

A. I. Zubkov, Yu. A. Panov, G. Yu. Stepanov, A. I. Shevets, Moscow, USSR: Investigation of separated turbulent supersonic flows.

M. Życzkowski, K. Szuwalski, Cracow, Poland: Inadmissible discontinuities in the theory of plasticity.

721/25/73

MIX
Papier aus verantwortungsvollen Quellen
Paper from responsible sources
FSC® C105338

If you have any concerns about our products,
you can contact us on
ProductSafety@springernature.com

In case Publisher is established outside the EU,
the EU authorized representative is:
Springer Nature Customer Service Center GmbH
Europaplatz 3, 69115 Heidelberg, Germany

Printed by Libri Plureos GmbH
in Hamburg, Germany